AF517402

Freshwater Aquaculture

Freshwater Aquaculture

3rd Revised and Enlarged Edition

Dr. Rajendra Kumar Rath

M.Sc., Ph.D., D.F.Sc.
College of Fisheries
Orissa University of Agriculture and Technology
Rangeilunda, Berhampur - 760 007
Orissa

New Delhi • Jodhpur

Published by:
Scientific Publishers

5 A, New Pali Road, P.O. Box 91
Jodhpur - 342 001 Rajasthan (India)
E-mail: info@scientificpub.com
Website: www.scientificpubonline.com

4806/24, Ansari Road,
Daryaganj,
New Delhi - 110 002 (India)

ISBN: 978-81-7233-694-3 (H/B)
eISBN: 978-93-8634-760-2 (E/B)

Printed in India

PREFACE TO THE THIRD EDITION

To fillup the gap due to advancement in the field of Aquaculture, it was felt to update the informations and to incorporate new findings and achievements in the book "Freshwater Aquaculture" by publishing 3rd edition. Due to the adoption of scientific fish farming for production of carp, catfish and prawn, the freshwater aquaculture sector at present contributes more than 2/3rd of total Inland fish production and as a result the country ranks 2nd in the world Aquaculture production. The recent development in technology addresses the issues like genetics, wetland management, aquaculture diversification, export oriented culture, maintenance of International Standard for competing the world trading pattern.

The subjects in the book are dealt with in such a manner that they could easily be understood by the students with an objective analysis of reviewed scientific informations collected from various sources. Hope this book would facilitate teaching of the subject in Colleges & Universities.

The author owes a sense of deep gratitude to a host of dynamic personalities in the international arena of aquaculture, Dr. V.R.P. Sinha, Dr. S.D. Tripathi, Dr. S. Ayyappan, DG Fisheries, ICAR, Dr. N. Sarangi Ex-Director, CIFA, Bhubaneswar, and Dr. A.E. Eknath, Director, CIFA, Bhubaneswar for their inspiration and guidance. The author expresses his sense of gratitude to Dr. I.C. Mohapatra, Ex-Vice-Chancellor, Dr. K. Pradhan, Ex-Vice-Chancellor, Dr. Bhagirathi Senapati, Ex-Vice-Chancellor and Dr. Debi Prasad Ray, Vice-Chancellor, O.U.A.T, Bhubaneswar for their help and suggestions in making this revised edition. I hope the readers will find the present revised edition much more useful.

The author pays a sense of deep homage to late Prof. G.N. Mitra, the legendary Fisheries Expert and Ex-Fisheries Development Advisor and Managing Director, Central Fisheries Corporation, who left the mortal world on 18th July 2006. This edition is dedicated to my mother late Smt. Indumati Devi who passed away on Sri Panchami i.e. 31.01.2009.

Rajendra Kumar Rath
College of fisheries (OUAT)
Rangeilunda, Berhampur

PREFACE TO THE FIRST EDITION

The importance of fish culture to the national economy and its contribution to the Gross National Product hardly needs emphasis. Aquaculture has already assumed the status of a fast expanding industry not only in India but also in several Asiatic countries over the recent past. The book presents a comprehensive body of information on the theory and practice of aquaculture industry with a direct bearing on the rural economy especially in developing countries. As such the emphasis in this book is consciously placed on every relevant data of general interest to the farmers, professional scientists as a whole. The contents of the book, styled in a new format, attempt to provide latest information on all aspects of freshwater aquaculture including artificial propagation, nutrition, health management and integration with agriculture and livestock components exclusively aiming at production and productivity. Some new frontiers in aquaculture research such as genetic improvement, transgenic gene transplantation, nucleic acid isolation, recombinant DNA and gene engineering have also been incorporated to satisfy the interest of students and researchers. The book explores innovative applications of biological and biotechnological methods to aquaculture management with an inter-disciplinary approach and each chapter is replete with an object analysis of reviewed scientific data from various sources such as universities, ICAR research institutes and scientific journals. The author hopes that the book would be of immense use to be practicising aquaculturists as well as the academics as a practical guide and also in stimulating further interest in various diverse fields related to aquaculture.

The author owes a sense of deep gratitude to a host of eminent personalities for their kind inspiration, help and guidance in various ways in the making of this book. The foremost among them who deserve a special mention include Prof. G.S. Ghosh, G.M. College, Sambalpur, Prof. P. Mohanty Hejmadi, Utkal University, Bhubaneswar, Dr. V.R.P. Sinha, Director, CIFE, Bombay, Dr. S.D. Tripathi, Director, CIFA, Kausalyaganga, Dr. A. Noble, CMFRI, Cochin, Prof. T.J. Varghese, Fisheries College, Mangalore and Dr. S. Barat, North Bengal University, W.B. My wife Smt. Rebati Rath deserves special thanks for her unusual patience and endurance during the preparation of this book.

The author would gratefully acknowledge suggestions from all quarters that may be useful for improvement of this book in future.

20th January, 1993 **R.K. Rath**

CONTENTS

1

GENERAL CONSIDERATIONS IN AQUACULTURE

1. INTRODUCTION

Increase in production is one of the basic points under different plan periods of National Policy. It has been remarked that Science must help us to speedily improve production, natural and human resources can be profitably developed and equally shared, to create more employment and reduce drudgery, to strengthen nation and reduce vulnerability. As per the scientific policy of Government of India, the need of our country is to utilise the resource potentials for enhancing the economic development of the country. This will result ultimately in increase of per capita income, per capita production, per capita consumption and better Socio-economic condition of the people. This increase in per capita production and consumption will reduce malnutrition and vulnerability to diseases among people. Further, the Millennium Development Goals (MDGS) adopted a target set of policies in 2000 for reducing hunger and malnutrition population to half by 2015. Although some progress has been made, hunger and malnutrition still remain the most devastating problems facing the developing world. Today we live in a world where poverty & hunger are still prevalent. Nutritional deficiency in one form or the other is affecting more than 2 billion people globally.

Globally, more than a billions live on less than 1 dollar a day and 840 million are classified as under nourished (Delgado *et al.*, 2003). India is home to 40% of world 's under weight children. Fifty percent of hungry are in small holding farming house holds and 20% among rural landless (Millennium Task Force on Hunger, 2004).

In this context fish and fisheries have been playing an important role in addressing nutritional and food security of poor in developing countries. In India, nearly 50% of the population being vegetarians, the average per capita consumption of fish is lowest among Asian countries. But consumption and demand in India is expected to increase with growing population, changing dietary habits and increasing income levels. At present the diet of an average Indian is very low in calories, the animal component in it is alarmingly deficient. Therefore, the main food required is animal protein. Fish is a source of animal protein. Over 2.6 billion people get at least 20% of their animal protein intake from fish. As against fish providing 13% of animal protein intake in industrialized countries, in Asia the intake is much higher, on an average at 30%. In some countries of Asia it is much higher, 51% in Bangladesh, 58% in Indonesia and 75% in Cambodia (Delgado *et al*, 2003).

Fish are rich sources of protein, essential fatty acids, vitamins and minerals. The fat and fatty acids in fish, particularly omega 3 fatty acids, are highly beneficial and difficult to obtain from other food sources. In many developing countries, fish in the cheapest source of animal protein to the poor and fish protein is of superior quality as it contains all the essential amino-acids for body building than plant proteins.

The fish production from Inland water whish are essentially culture fisheries and are less capital intensive have high growth rate at present when compared with 1950, where as the marine fisheries which are essentially capture and are capital intensive have relatively slow growth rate at present when compared with 1950.

Therefore, in the present context the culture fisheries are emphasised. Different aspect of culture fisheries are included under Aquaculture. Aquaculture includes all aspects of production of aquatic organisms in captivity comprising either some or all stages of their life cycle, their live foods and the resultant marketable products in the habit of fresh, brackish and sea water. Hickling (1962) has defined fish culture has the same objective of agriculture and stock breeding mainly to increase the production by all possible means than the natural wild level (Zero culture level) of production. Lagelar (1956) has defined fishes are the most numerous of the vertebrates constituting above 40% of the vertebrate phylum. Some have defined fish is most efficient among farm animals in converting feed in to nutritious food. Lone (1988) has described aquaculture as an underwater agriculture.

However aquaculture is the fastest growing food production sector accounting for about 50% of food fish production, with average annual growth of 8.8% during 1950-2004. FAO in 1998 predicted that world aquaculture production will reach between 35 to 40 million tons of finfish, crustaceans and molluscs by 2010. However, the global aquaculture (food fish) production reached 45.4 million tons by 2004 surpassing the predictions and it is further expected to increase. Of this production, 30.6 million tons was produced in China alone. India with 2.47 million tons of culture fish production stands second in world aquaculture production (FAO 2006). It is expected that by 2025, globally one out of every two fish consumed will be from aquaculture

2. HISTORY OF AQUACULTURE

Pisciculture devoted solely to fish which is as old as civilisation itself. The hunting of fish in the prehistoric period was prevalent during the stone age. Acceptance of fish culture as a lucrative commercial enterprise in the Indo-pacific region owes its inception to the Chinese. Even today many countries follow the Chinese fish culture practices or are inspired by them.

However, the pictorial engraving on an ancient Egyptian tomb shown tilapia being fished out of an artificial tank for about 2500 B.C. provides evidence that the people of Egypt were probably the first in the World to culture Fish (Marr *et al.,* 1966). But carp culture was wide spread in China in 2000 B.C. It had further developed in Chou Dynasty (1122-249 B.C.). The earliest clear record on Chinese literature is said to be the ``Classic of fish culture" believed to have been written by Fanli in 475 B.C. who made a summary of the experiences in structure of ponds, method of propagation, rearing and growth of fry of common carp. It was further developed in Han Dynasty (206 B.C. - 7 A.D.). But in Tang Dynasty (618-906 A.D.), there was a big change in fish culture in China and the people were prohibited for catching, selling and eating common carp because the common carp in Chinese is called as Li, which is same as that of the surname of the emperor. As a result the culture of common carp which had developed for 1000 years suddenly came to a stop. This brought a new change in fisheries and people went for catching other group of fishes from natural water bodies. This resulted consequently the shifting of monoculture of common carp to polyculture of several species. In Sung Dynasty (906-1120 A.D.), the collection and transportation of fish fry were popular around river basins. It was further developed. Rearing of fry to adult fish were recorded in Ming Dynasty (1368–1644). In Ching Dynasty (1644-1911), fry production season and biological basis of fry were found in the records. Commercial scale of fish farming and transport of carp fry in bamboo baskets is recorded in the book ``Wei-sin-chak-shik'' written in 1243 A.D. In ``A complete Book of Agriculture'' written in 1639 A.D., describes the collection of carp fry from rivers and the methods of rearing them in Pond. With the experience gained through generations and with infinite patience and endurance the fish culture is brought to a very high level of development. During 1619-1904 A.D., only four types of fishes were cultured in China. After this period, another three species of fishes were brought under culture, making total seven types of cultivable fishes in China.

These species are as follows :

1. Common carp — *Cyprinus carpio*
2. Grass carp — *Ctenopharyngodon idellus*
3. Silver carp — *Hypophthalmicthys molitrix*
4. Big head — *Aristichthys nobilis*
5. Mud carp — *Cirrhina molitorella*
6. Black carp/Snail carp — *Mylopharyngodon piseus*
7. Bream — *Parabromis pekinensis*

Those Chinese who migrated carring along with them the traditional knowledge of carp culture to Malayasia, Formosa, Indonesia and Thailand established the fish culture industry in those countries. They took carp seeds from China and motivated the local people to take up pisciculture.

The common carp has been exported to several countries all over the world and its culture has achieved a very high degree of perfection. The combination of species of carps for utilising all available natural food organisms in different niches of pond ecosystem appears to have been first developed in China. Although, fish culture were in existence in several countries in its impirical status, but it is only during the recent time that the plan of scientifically based enterprise has developed in fish culture systems. The utilisation of siluroid fishes for culture in bamboo enclosures in running water appears to be an innovative idea of cage culture in Cambodia.

Undoubtly the fish culture system must have been developed thousands of years ago in India. Khona, the daughter-in-law of Varahamihir recommended the cultivation of vegetables in the embankments of fish ponds. In Kautilya's Arthasasthra, written some where between 321 to 300 B.C. indicate regarding rendering fishes in a reservoir poisonous in times of war (Hora & Pillay, 1962). King Somesvara of Chanakya Dynasty in encyclopaedia Manasoltara, compiled in 1127 A.D. (Hora, 1951) described the method of fattening fish in the chapter as `matsya-vinod'. No further records are available till about the 19th century when collection and transportation of carp spawn from rivers in Bihar and West Bengal was described. Spawning of carps in wet and dry bundh and fish culture had mostly been restricted in the eastern part of the Indo-Pakistan subcontinent until about the end of 19th century. Even introduction of an exotic food fish *Osphronemus gourami* from Java was also, attempted as early as 1841 and 1865 (Raj, 1916) in India.

During the British rule some expertising Englishmen started to develop the sport fishes in some hill streams of India. Finding the suitable ecological conditions of these streams they successfully established the trouts in areas like Nilgiris. This was followed by similar successful efforts in other mountain areas such as Kashmir kullu valley and the high ranges of Travancore. With the formation of fisheries' department in some States the culture of both sports and food fishes received encouragements. As a result of food situation created by World War II (1939-1944), great attention was given to fish farming as means of producing cheap animal protein to feed the population. The first scientifically designed fish farm at Sunkesula in Madras presidencies (now in A.P.) was constructed in the year 1911 under the guidance of H.B. Wilson. The studies on artificial feeeds (Horneli; 1920), nesting and breeding habits of Large Cat- fishes (Horneli; 1922), rearing of *Etroplus suratensis* and *Chanos chanos* (Milk fish) in to fresh water ponds were undertaken (Hornell, 1922; Ranganathan; 1961).

Exhaustive surveys were taken for the development of fisheries in West Bengal, Punjab, Uttar Pradesh, Karnataka, Andhra Pradesh, Gujarat and gradually by respective state fisheries' departments.

Studies on bionomics, spawing, seed transportation and diseases in fish were taken up. Possibilities of pearl spot and murrel culture were emphasized for meeting the regionl demands.

The holding of Symposia on ``Utilisation of sewage for fish culture" and ``Factors influencing the spawning of carps" in 1944 and 1945 respectively organised by the National Institute of Science gave a new Stride for boosting up the aquaculture activities. This resulted in spread up of the traditional fish culture practice beyond the limit of Assam, Bengal, Bihar and Orissa especially to the then Bombay and Madras presidencies with massive transport of fish seed from Bengal. Gradually, the shifting from open seed transport to closed seed transport system under Oxygen packig was developed. Besides sports and food fishes, attention was also focused on the larvivours fishes for public health significance. The biological control of mosquitoes acting as vectors for malaria, filaria, dengue etc. two exotic larvivorous fishes, *Lebistes reticulatus* and *Gambusia affinis* were transplanted in India as early as 1909. A new species, *Horaichthys setnai,* was further added to the list of larvivorous fishes. For control of Guinea worm, *Megalops cyprinoides* and *Ambasis* spp.. which profusely feed on cyclopes were considered. Therefore, a fish farm was established in Madras city at Chetpur specifically for the culture and distribution of Larvicidal and cyclopscidal fishes. But it is not known how far these biological agents could control the vectors or the diseases.

3. DEVELOPMENT OF FISH CULTURE IN INDIA

The Advisory Board of the ICAR in 1944 commenced sponsoring of adhoc fisheries research scheme by State Governments and Universities on different aspects of fish cuture. But no much head way was made till about the time of India's Independence in 1947. A real boost in aquaculture activity, which largely centred on carp culture, was noted soon after independence. Various organisations in the different States of India or an independent Department of Fisheries initiated action plans for stocking fish seeds from Calcutta in various water bodies. The Government of India provided institutional support of replacing the impirical state of fish culture to scientific base of fish culture by establishing the CIFRI in March 17th, 1947 at Calcultta and its sub-station in 1949 at Cuttack Killa Fish Farm, which is now shifted to Kausalyagang, Dhauli. This institution deals exclusively with problems of pond culture. Research and extension in the field of fish culture are now given high priority in India and the success achieved is highly encouraging in the recent years.

Indian aquaculture in recent years has become a dynamically developing sector providing several new oppertunities. Over seven fold growth in last 25 years has made aquaculture as one of the most vibrating sector with an impressive annual growth rate of over 6%. Taking advantage of the strong traditional knowledge base, research and developmental backup, freshwater aquaculture has grown in area coverage, diversification of culture species and practices, besides intensification of farming system. Even with vast increase in the production over the years, it is able to provide about 8 kg/caput to the present populace (taking 56% as fish eaters) against the nutritional requirement of 11 kg. The projected domestic requirement of the country by 2020AD is estimated at 12 million tons more than 3/4th of which has to come from inland sector. Fish production range under different culture system as follow depicts about sustainability since it is environmentally nondegrading, technically appropriate, economically viable and socially acceptable. It confirms not only to the concept of environmental sustainability but also to the most recent perspectives of social sustainability and economic sustainability. It is because in India, aquaculture is practiced with utilization of low to moderate levels of input especially organic based feed and fertilizers, thus making the system largely environment friendly.

Fish production range under different culture systems

S.N.	Culture systems	Average produc-tion rates Tones/ ha/ year
1	Composite fish culture	4 – 6
2	Intensive pond culture with feeding & aeration	10 - 15
3	Clarias culture	3 – 5
4	Sewages fed fish culture	3 – 5
5	Integrated fish farming	3 – 5
6	Pen culture	4 – 5
7	Freshwater prawn culture	1 – 2
8	Shrimp farming	2 – 5
9	Running water fish culture	25 – 50 kg cubic meter of water

10	Cage culture	10 – 15
11	Biogas slurry based system	3 – 5
12	Aquatic weed based culture	3 – 4
13	Makhana and Air breathing fish culture	1.5 tons with 94kg Makhana

The development of aquaculture is due to operation of various National projects. During 1950, "grow more food compaign" was launched to develop fisheries institutional network and to ensure supply of quality fish seed to farmers through fish seed syndicate.

During 1960's "National Demonstration Project" was launched to generate awareness about fish culture technology through demonstration to farmers.

During 1970's to 80's various projects such as AICRP on fish culture, air breathing fish culture, creation of FFDA, operational research project, establishment of Krishi Vygian Kendra, Rural aquaculture project, lnland fisheries project and Lab to Land project were operated to popularize fish culture and seed production technology among the fish farmers of the country. During 1980's to 90's, National Agriculture Extension Project (NAEP) and, creation of BFDA for strengthening extension-farmer linkage and utilization of brackish water for fish and shrimp culture were taken up for aquaculture development. During 1990's to 2000's, other projects such as Women in aquaculture, Institute village linkage project, National agriculture technology mission were operated for involvement of women in aquaculture, assessment and reorientation of technology adoption in line with local condition, creation of technology information center, increased production and nutrition security were emphasized. During 2006 and 2008 Mega seed project and National agriculture integrated projects were launched for production of quality seed of carps, shrimps, prawns and technological advancement in farming practices. These projects laid sound foundation for the development of aquaculture in the country.

The recent tendency on intensification of farming practices is posing ecological concern. Further introduction and transfer of aquatic organisms carry the risk of introducing several exotic pests and disease agent. In addition, both intentional and unintentional introduction may have undesirable ecological and genetic effect in the aquatic system. CIFRI News Letter (Jan-June 2008) has reported that, apart from cultivable Chinese carps, other exotic fish species gaining dominance in the natural water bodies of West Bengal. These are: (1) *Oreochromis niloticus niloticus* (2) *Barbonymus gonionotus* (3) *Pangasianodon hypophthalmus* (4) *Clarias gariepinnus* (5) *Piaractus brachypomus* and (6) *Pterygoplichthys disjunctivus* (7) *Pterygoplicthys pardalis.* Pterygoplichthys belong to family Lorocariidae and fishes can be distinguished by their hard, rough, armoured skin, ventrally located sucker like mouth with prominent barbell and large sail type dorsal fin. The basic color pattern is numerous brown or dark grey spot on a lighter back ground. Presence of more than 10 fin rays distinguish sail fin cat fish from other closely related species. Such diversities of fish in wet lands of West Bengal is expressed by Shannon Index H which ranged from 1.3 to 3. Higher the value, lower is the diversity. In order to reduce the risk that could arise from such movement, implementation of code of practices is essential. The food and agriculture organization, Rome in 1995 had evolved the code of conduct for responsible fisheries which was signed by 168 countries, including India (FAO, 1995). This provides a useful frame work and set of guidelines for the promotion of sustainable aquaculture. It ensures that people involved in aquaculture are committed to its principle to implement them. The main objectives of the code concerning aquaculture are to:

1. Conserve genetic diversity.
2. Minimize the negative effect of farmed fish on wild populations.
3. Monitoring and assessing the environmental effects of aquaculture including application of feed, fertilizers, drugs and chemicals
4. Minimize the risk of spread of diseases while introducing and trans-boundary of aquatic animals.
5. Develop techniques for protection and restoration of endangered species.
6. Biodiversity of aquatic habitats and ecosystems conserved.
7. Ensure that excess fishing effort is avoided and exploitation of stock remains economically viable.
8. Interests of fisheries, including those engaged in subsistence, small scale and artisinal fisheries should be taken in to account.

9. Update aquaculture development strategies and legal measures which could ensure that aquaculture development is ecological sustainable.
10. Allow for the rational use of resources shared by aquaculture and other activities.
11. Establish data base and information networks in aquaculture for collection and sharing of information to facilitate cooperation on planning for aquaculture development at the National level.
12. Promote active participation of fish farmers in the development of responsible aquaculture management practices.
13. Ensure that the disposal of offal, sludge and diseased fish not to constitute any hazard to human health and environment are products and maintain product quality.
14. Ensure food safety of aquaculture

Thus, formulation of appropriate policies and imposition of code of conduct for responsible aquaculture are essential for sustaining the growth and development of aquaculture sector in years to come.

Creation of additional resources for aquaculture using water logged, saline alkaline and other under utilized lands need strategically be addressed for aquaculture development in the country. Keeping these issues in view, the National Fisheries Development Board (NFDB) was established on 10th July 2006. Suitable guidelines are developed for the benefit of the farming community for aquaculture development of the country. The activities developed include:

1. Intensive aquaculture in ponds and tanks
2. Fisheries development in reservoirs
3. Coastal aquaculture
4. Mari culture
5. Seaweed cultivation
6. Infrastructure for post harvest processing
7. Fish dressing centers and solar drying of fish
8. Domestic marketing
9. Technology up gradation
10. Human resource development programme in fisheries sector.

11. Deep sea fishing and tuna processing

NFDB shall provide a

1. Platform for public- private partnership in fisheries development
2. A mechanism for an end-end approach for ensuring proper production process and efficient marketing
3. Strategies for "win-win" situation for both the producers and consumers of fish.

There by making an overall development of aquaculture sector in the country.

4. PURPOSE OF AQUACULTURE

(i) To increase production for per capita consumption and per capita income by which National Income will be higher.

(ii) Ornamental purpose like culture of angel fish, black molly, red sword tail, Blue gourami, Kissing gourami etc.

(iii) Sports and Game purposes-like culture of trouts and mahseers.

(iv) Available natural water resource utilisation.

(v) Earning foreign exchange.

(vi) Upliftment of Socio-economic status of the people.

(vii) Create Employment opportunity.

(viii) Utilisation of by-products of fish like issinglass, pearlessence, fish liver oil, fish protein concentrate, fish glue etc.

(ix) Controlling parasites like mosquito larvae, dengue by Larvicidal fishes (*Labistes reticulatus, Gambusia affinis*).

(x) Utilisation of medicinal added value of fishery products.

5. IMPORTANCE OF AQUACULTURE

Geometric increase in population has resulted in shortage of foodstuff from the land resources and the Land : Man ratio is decreasing day-by-day. That too the fish crop of animal protein is one among the largest crops of foodstuff being produced presently. However, the cost of production in land based animal is more than fish culture. Russian Scientist S.V. Mikhailov, in his paper ``Comparative efficiency of production of some products of the Land and Sea" had indicated that the production of million calories would take 15-20 mandays by fishing and 56 mandays by beef farming. He further stated that an annual output of 100 kg live weight of beef required a capital investment of 2000-2500 roubles while for fish it was only 1500-1700 roubles. The Recurring production cost for same amount of beef had been estimated at 600 roubles, where as for fish it was 200 roubles. Fish thus can be supplied with fewer manhours and lesser capital investment.

6. ADVANTAGES OF FINFISH CULTURE

(i) Fish is a poikilothermic or cold blooded animal so need not spend any energy for temperature regulation. It does not spend energy to maintain its position in water column. The specific gravity of fish (1.02-1.06) is nearly same with that of water. But land animals are subjected to walking resistance and gravitational force.

(ii) There is security of catching fish from the cultured water, unlike the uncertainty of catching in natural vast waters. That too, culture of fishes is more rapidly expanding in interior land based water resources of the country away from the coastal belt.

(iii) Unlike major agricultural crops, fish do not consume water. Compared to any bird and mammals used for husbandry purposes, fish have the highest fecundity.

(iv) Edible tissues of Fish is appreciably greater than that in mutton, beef or poultry. The Lean meat percentage in fish (80.9%) is greater than chicken broiler (64.7%) and Beef (choice grade) 51%. The refuse loss from fish (13.7%) is lesser than Beef (15%), and chicken broiler (32%).

(v) Fish can convert food in to body tissues more efficiently than other farm animals. The reason for the superior feed conve-rsion efficiency of fish is that it assimilates diets with higher percentage of protein, apparently because of their lower dietary energy requirements. These requirements for metabolism are less in fish, which evolve in a protein rich energy deficient environment, than in warm blooded animals, because fish do not have to maintain constant body temperature. Fish excrete relatively little energy to maintain position in water.

(vi) Fish excrete nitrogenous wastes as Ammonia (NH_3) instead of urea or uric acid that is less energy expenditure in protein metabolism. Because of their low energy requirement, fish can channelise more energy towards anabolism, that is fish can synthesise more protein per calories of energy consumed than poultry or livestock. Thus the primary advantage of fish over land animal is low energy cost of protein gain and superior feed conversion efficiency.

7. CATEGORIES OF FARM TYPE

7.1 Based on characteristics of the Farm environment

1. Based on temperature of water

(i) Warm water (ii) Cold water

2. Water salinity

(i) Fresh water (ii) Brackish water

(iii) Marine water

Freshwater is defined when the salt content in it is less than 0.5 PPT. Brackish water is defined when the salt content in it is in the range of 10-30 PPT. Marine water is defined when the salt content in it is more than 30-35 PPT.

3. Water replacement

(i) Running (ii) Stagnant

4. Physiographical zone

(i) Inland fish farm (ii) Coastal fish farm

(iii) Marine fish farm

5. Land topography

(i) Upland (ii) Low land

(iii) Terraced

6. Land area coverage

(i) Extensive (ii) Intensive

7. Water source

(i) Rain fed farm (ii) Tide fed farm

(iii) Diverted water and sewage water

(iv) Seepage water (v) Ground water

(vi) Spring water

7.2 Characteristics of the physical structure

1. Nature of enclosures

(i) Pond type fish farm (ii) Cage fish farm

(iii) Pen fish farm (iv) Raceways

2. Kind of materials used for enclosure

(i) Plastic tank (ii) Concrete cemented tank

(iii) Earth excavated pond

7.3 Kind of cultured fish - By species or a group of species

(i) Cat-fish farm (ii) Carp farm

(iii) Tilapia farm (iv) Mullet farm

(v) Prawn farm

Besides this, based on number of species taken for culture in a farm it may be either monospecies of carnivorous fish culture or multispecies of carp culture. So far as sex of the fish is concerned, some fish like Tilapia is brought under monosex culture and carp are brought under dualsex culture. Based on type of organisms, culture can be of fish culture, Prawn culture, Oyster culture, Murrel culture, Sea weed culture, carp culture, milk fish culture etc.

7.4 Based on Developmental stages of the species

(i) Nursery farm

(ii) Carp breeding farm

(iii) Trout hatching farm

8. CATEGORIES OF FISH FARMING SYSTEM

8.1 Characteristics of fish stock population

1. Number of species

(i) Monospecies fish farming (Monoculture)

(ii) Multispecies fish farming (polyculture and mixed culture)

2. Size group of species

(i) Single size stocking

(ii) Multizise stocking

3. Harvesting procedure

(i) Single harvesting

(ii) Multiple harvesting

(iii) Rotational harvesting

8.2 Characteristics of feeds and feeding of fish

1. Source of food supply

(i) Based on natural food production system. It is similar with that of zero level (wild culture) if seed is not stocked or Ist level culture if fish is stocked.

(ii) Based on artificial feed supply system. It is either IInd level if seed is also stocked or IIIrd level culture, if both seed and fertilizer application is mde.

2. Kind of natural foods

(i) Based on benthic algal production system

(ii) Based on planktonic algal production system

(iii) Based on filamentous green algal production system.

3. Usage of artificial feed

(i) Based on balanced feed system

(ii) Based on supplementary feeding.

8.3 Management procedure of techniques

1. *For a species or a group of species*

(i) Asiatic carp farming. It may be Chinese system of carp culture, Indian system of IMC culture or European system of common carp culture.

(ii) Bluegill and bass farming

(iii) Eel farming

(iv) Trout farming

2. *Synergetic combination of other farming crops*

(i) Prawn and milk fish farming

(ii) Rice and fish farming

(iii) Rice and prawn farming

(iv) Integrated fish farming

(v) Fish-cum-agriculture farming

(vi) Fish-cum-Animal Husbandry farming

(vii) Fish-Agriculture-cum-Animal Husbandry

3. *Levels of management intensity of a farming system*

(i) *Traditional* - Simplest form with minimum inputs and management. No selection of species, fertilization and feeding.

(ii) *Extensive.* Any improvement over the traditional system where selected species in pre-determined numbers are stocked in the properly prepared fields. Use of fertilizers and supplementary feed is to a limited extend as a reliance is placed mostly on the natural production of food in the pond.

(iii) *Semi-intensive* – Here reliance on natural production of food is negligible. Stocking is done with selected post larvae reared in hatchery. Formulated compounded feed as per nutritional requirement is provided from external sources.

(iv) *Intensive* – Stocking is done with hatchery reared post larvae. Water quality is maintained by frequent changing or by providing water circulation together with constant aeration. Food requirement is met fully with formulated feed. This system is practiced in Raceways and artificial tanks.

(v) *Super intensive* – Carried out in cement cisterns with cent percent water exchange. Continuous water exchange is made through biological filter system, high stocking density more than intensive. Constant aeration or rotating agitators, encapsulated pelleted diet feeding and stocking manipulation.

8.4 Levels of input intensity

8.4.1.O – Level – Without management either for fish population or for fish food supply

8.4.2. I – Level – Seeds are stocked. These are supported by the fish food organisms produced only by natural soil productivity. Only one management of stocking seed is prevalent.

8.4.3.II – Level – Seeds are stocked and are supported by the fish food organisms produced by natural soil productivity and increased productivity by organic and incorganic manure application. Two management of stocking seed and manure application is prevalent.

8.4.4. III – Level – Stocked fish are supported by natural soil based productivity, increased productivity due to manure application and supplementary feeding.

However, in China, the levels of management are more to harvest higher per cent production from less per cent of culturable water area. The reasons suggested are :

1. Water quality
2. Seeds
3. Feeds and Fertilization
4. Polyculture
5. Density
6. Rotation
7. Prevention of diseases
8. Management

9. BASIC CONSIDERATION IN THE SELECTION OF SPECIES FOR CULTURE

The following criteria should be considered before selecting a fish for culture purpose. As there can not be an universal fish which will follow or can satisfy all the desirable qualities, but have maximum criteria being suggested can be taken for selection of fish species for culture purposes.

1. Rate of Growth – Fishes which grows to a larger size in shorter period are suitable for culture. Because of this the Indian major carps are taken under carp culture in India.

2. Short food chain – The energy from primary producer to successive consumers are in decreasing order. Therefore, the fishes having short food chain like planktonivorous are usually preferred. Silver carp and catla are phytoplanktonivorous and Zooplanktonivorous respectively. These are taken under carp culture in India. Fishes which live on detritus decaying matter through saprophytic food chain are also taken for culture.

3. Adaptation to climate – The fishes which will be adaptable to various climatic condition can be cultured in many areas than the fishes in particular area. Trout and salmon can perferably be raised in cold water of optimum temperature of 10–12°C, they can not be cultured in tropical waters except in high level of mountain, where as Indian major carps can preferably be raised in warm waters of optimum temperature of 26°–30°C. But common carp can be cultured throughout the world as it is adjusted to various climatic conditions.

4. Tolerance of the fluctuations of physico-chemical condition of water – The fishes which have wide range of physico-chemical tolerance can be cultured more than those have narrow range of tolerance. The fish adopting to the change of physico-chemical condition have great importance in intensive fish culture. Trout requires oxyges of 9 mg/litre of water but major carps 5-6 mg/litre, where as common carp requires 2-3 mg/litre for which it is cultured in many parts of world. Like wise for Brackishwater culture the main criteria for consideration is of Euryhaline nature of fish.

5. Acceptance of artificial feed – With passage of time and scientific advancement, the traditional culture practice has been changed to intensive culture practice. In intensive aquaculture the natural food availability in aquatic ecosystem is not sufficient. Therefore, the fish which like artificial feed should be preferred.

6. Resistance to common fish diseases and parasites – Although fish diseases are not uncommon in aquatic medium, still the fishes having resistance to common fish diseases and parasites are preferred.

7. Easy recruitment under controlled condition – To maintain fish culture as a continuous practice, the availability of assured quantities of healthy and pure seed from a dependable source is of great importance. To meet such need, the fish which can be breed easily under controlled conditions are to be preferred. The Indian Major carps (Rohu, Catla and Mrigala) and Exotic fishes like Silver carp, Grass carp and Common carp are cultured because of this quality also.

8. Amiability to live together – The fishes proposed to be cultured should live together without interferring each other. This fact is more important in polyculture or multispecies culture of fishes. Herbivorous fishes are not cultured along with carnivorous fishes in same pond system. As well as carnviorous fishes of different groups are not cultured together. Besides, this fact, the fishes having profuse breeding capacity like Tilapia are not taken for culture in waters suitable for carp culture.

9. Compatibility – The fish species which are put in combination for culture should not compete among themselves for space and food. Unlike agriculture, aquaculture is of three dimensional unit of Length, Breadth and Depth. The distance between Soil bed of pond to its upper limit of water is called depth. This can be conveniently divided in to upper column, middle column and lower column of water zone.

10. Conversion efficiency – The species of fish which will give more edible flesh per unit of food consumed is preferred than which gives less flesh per unit weight. Although conversion efficiency is an old conception, scientists have focussed on the protein efficiency ratio and meat produced in protein assimilation ratio in fish for their suitability in culture systems.

11. Consumer's preference – The culture practice of fish is undertaken in line with local condition and consumer's preference. Orissa and West Bengal people prefer freshwater carps than marine water fishes, where as South-people prefer live fishes over carps. American's prefer cat fishes than carps. Similarly milk fishes are highly liked by South East Asian Countries but not by Kenya.

Besides these basic considerations, other considerations like scale less carp, reduced vertebrate bone in different carps, color etc. are also preferred.

10. CULTIVABLE FRESHWATER FINFISHES

10.1 Cultivable freshwater fishes in India

Large varieties of suitable indigenous species of fishes are available in various parts of India for pond culture. Besides these, some exotic species of food fishes have also been introduced to the country. The culture value of these fishes varies considerably depending upon their relative growth, size and adaptability. The large majority of these fishes taken for culture belong to the group of carps. Besides these, few species of livefishes, perches and a few quick growing saltwater fishes are taken for culture in fresh and brackishwater ponds.

1. Carps

1. Catla (*Catla catla*) Hamilton.
2. Rohu (*Labeo rohita*) Hamilton.
3. Mrigal *(Cirrhinus mrigala)* Hamilton.

4. Calbasus (*Labeo calbasu*) Hamilton.
5. Fringed lipped carp (*Labeo fimbriatus*) Bloch.
6. White carp (*Cirrhina cirrhosa*) Bloch.
7. Pig mouthed carp (*Labeo kontius*) Jerdon.
8. Reba (*Cirrhina reba*) Hamilton.
9. Bata (*Labeo bata*) Hamilton.
10. Nagendram fish (*Osteochilus thomassi*) Day.
11. Sandkhol carp (*Thynichthys sandkhol)* Sykes.
12. The carnatic carp (*Puntius carnaticus*) Jerdon.
13. The chocolate mah seer (*Acrossocheilus hexagonolepis*) McClelland.
14. The kozhimeen (*Barbus dubius*)
15. Sarana (*Puntius sarana*) Hamilton.
16. Gonius (*Labeo gonius*) Hamilton.
17. Nandina (*Labeo nandina*) Hamilton.
18. Doctor fish (*Tinca tinca*) Linneaus.
19. Golden carp (*Carassius carassicus*) Linnaeus.
20. Common carp (*Cyprinus carpio)* Linnaeus.
21. Silver carp (*Hypophthalmichthys molitrix*) Valenciennes.
22. Grass carp (*Ctenopharyngodon idella*) Valenciennes.

2. *Live fishes*

1. The Gouramy – *Osphronemus gourami* (Lacepede)
2. Murrels –

   Channa striatus (Bloch)

   Channa marulius (Hamilton)

   Channa punctatus (Bloch)

   Channa gachua (Hamilton)
3. Climbing perch –

   Anabas testudineus (Bloch)
4. Singhi –

   Heteropneustes fossilis (Bloch)
5. Magur –

   Clarias batrachus (Linnaeus)

3. Catfishes

1. Chitala – *Notopterus chitala* (Hamilton)
2. Falli – *Notopterus notopterus* (Pallas)
3. The fresh water Shark – *Wallago attu* (Schneider)
4. *Mystus seenghala* (Sykes)
5. *Pangasius pangasius* (Hamilton)

4. Freshwater prawns

1. *Macrobrachium rosenbergii*
2. *Macrobrachium malcolmsonii*
3. Ganga river prawn- *M. gangaeticum* (syn. *M. birmanicum* Choprai)
4. *M. josephi* (reported from Veli lake and Kula thoor rivulets in Trivandrum, Kerala).

10.2 Major cultivable fishes in China

China is very rich in freshwater fish resources of which more than 50% are carps. It is recorded that in the Yangtze river, there are 300 species, 2/3 of which are carps. In Pearl river there are 260 species of which 150 species are carps. In Yellow river, there are 140 species of fish and in Heilong river there are 90 species of fishes.

Most species in China are warmwater fishes. Until now 20–30 fishes are cultured in the ponds. This focuses on four well known Silver carp, Grass carp, Big head and Black carp. Other good stocks are common carp, crucian carp, mud carp, Wuchang fish, Tilapia etc. Understanding the habits of fish, growth, development, propagation, feeding and ecological requirements has contributed significantly the farming techniques, production, genetic improvement and improvement of fish yield.

(i) Grass carp (*Ctenopharyngodon idellus*) – It is one of the major fisheries in China. This fish is cultured every where in the country. The production of Grass carp contributes about 20–30% of total fish production in China. The largest specimen so far found in China was about 35 kg.

(ii) **Black carp** (*Mylopharyngodon piceus*) – It is a carnivorous fish, feeds on snails and clams. Comparatively speaking

Black carp is not cultured as like that of Grass carp in China. The largest specimen found was about 70 kg.

(iii) Silver carp (*Hypophthalmichthys molitrix*) – The largest specimen found so far was 20 kg. The production is very high than that of Grass carp if the water is fertile.

(iv) Big head *(Aristichthys nobilis)* – Big heads are commonly cultured in South of China. Their feeding habits are like that of *Catla catla.*

(v) Common carp *(Cyprinus carpio)* – China has long history of culturing common carp which have wide distribution and strong adoptability. There are lot of morphological variations through artificial breeding and natural selection of this species, i.e., scale carp, mirror carp, Wuyuan red purse carp, Xing guo red carp etc. The largest specimen so far found was about 40 kg.

(vi) Crucian carp (*Carassius auratus*) : It is also called as Gold fish. It is very familiar in Asiatic countries. This carp is smaller in comparison to other carps. *Carassius auratus gibelio* is a sub-species of crucian carp distribution in northern part of China and the largest specimen found so far was 3 kg. Another sub-species is *Carassius auratus cuveri* introduced in China in 1976 from Japan. This grows faster than *Carassius auratus gibelio, Carassius auratus cuveri* is also called as white crucian carp in China as it is very silvery white.

(vii) Wuchang fish (*Megalobrama amblycephala*) – It is called as Chinese bream. It is a herbivours fish and it has been cultured in few places. The largest specimen so far found was 3 kg.

(viii) Mud carp (*Cirrhina molitorella*) – This species is cultured in extreme South part of China because it requires higher temperature for growth. It is a botom feeder and Chinese prefer to culture mud carp than that of common carp. Mud carp accounts 30% of local fresh water fishes in Guangdong and Guangxi provience of South China. This is not cultured in North and East China because of ecothermal limit of the fish.

Besides these, above species culture, other species cultured in ponds are : soft bream, white amur bream, finless eel (*Morpterus albus),* Loach (*Misgurnus anguillicaudtus*), Snakehead *(Ophiocephalus argus*), Mandarin fish *(Siniperca chautsi*), Catfish *(Clarias fuseus*), Small scale *(Plaiognathops microlepis*), *Acipenser sinensis, Mugil cephalus, M. soiug, Myxocyprinus asiaticus, Squaliobarus curriculus, Anguilla japanica, Megalobrama terminalis* etc. There are some species introduced from abroad such as : *Oreochromis mossambica* and *O. nilotica.* China has introduced Rohu and Catla from Bangaladesh with exchange of chinese carps for culture. These Rohu and Catla are now cultured and propagated in South part of China in an experimental basis. China has introduced freshwater prawn *Macrobrachium* in 1976 from Japan. It has also introduced *Clarias macroceh* from Egypt.

II. BIOLOGY OF CULTIVABLE FINFISHES

1. The Biology of Major Freshwater Cultivable Carps in India

The introduction focuses on six (3 Indian major and 3 Exotic) well known carps such as : Catla, Rohu, Mrigal, Silver carp, Grass carp and Common carp. Culture of medium carps such as *L. calbasu, L. gonius, L. bata, L. fimbriatus, L. dussmeri, Barbus carnaticus, Puntius sarana, P. jerdoni, Gonoproktopterus (Puntius) lurmuca, G. kolus* and *Cirrhinus cirrihosa* having regional importance are being considered important candidate species for introduction to carp culture system due to their consumer preference and high market price. Besides these medium carps , some success in breeding technology of *Ompok pabda* (butter fish) and mud eel (*Monopterus cuchia*) are achieved. Adequate trials are needed for standardizing the farming techniques for their culture in large scale. Understanding the habits of fish, growth pattern, development propagation, feeding and their ecological conditions will be of great practical significance to fisheries production in India. Besides these the specification of farming techniques including the bases of selection for propagation, hybridization, inbreeding depression or heterosis, gynogenesis or polyploidy, sex reversal etc. can further improve the fish yield.

Morphology

Catla (Fig. 1) – Rapid growing indigenous carp in India. Body deep, stout broad snout, conspicuous head, large upturned mouth, non-fringed lips, back greyish colour, silvery on sides, but tends to be rather darkish in weedy waters, fins dark, scales pink. Catla grows to over 1.5 meter length. The fish attains sexual maturity in the 2nd year and are ready to breed.

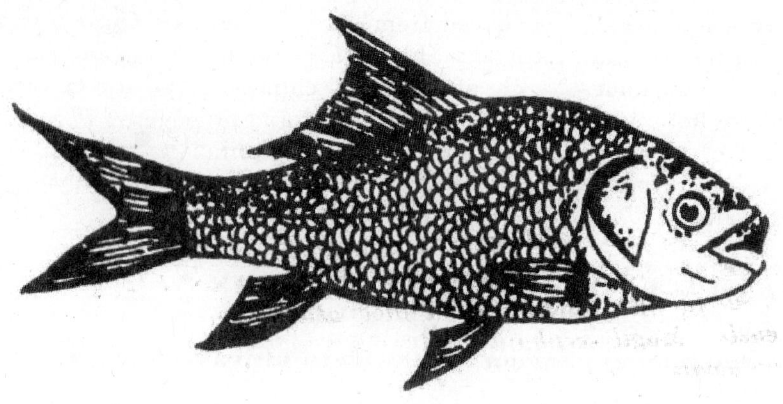

Fig. 1. *Catla catla* (Hamilton)

Rohu (Fig. 2) – Rohu is considered tasty among the Indian carps. It is distinguished by its relatively small or pointed head, almost terminal mouth with fringed cover lip. Body elongated with moderately convex abdominal. Back brownish grey. Dull reddish scales on the sides and pink reddish fins. The body is more linear than Catla. Sexual maturity is attained towards the end of the second year. Rohu grows over 91 cm. in length.

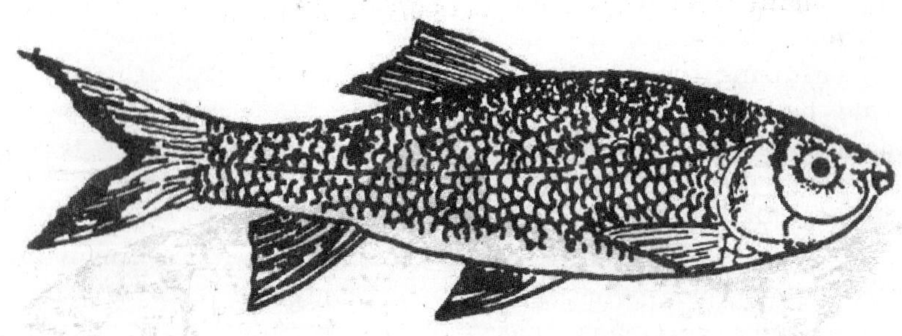

Fig. 2. *Labeo rohita* (Hamilton)

Mrigal (Fig. 3) – Mrigal is next in importance to Catla and Rohu for culture. Linear body, small head with blunt snout, subterminal mouth with thin nonfringed lips. Body silvery dark grey along back, fins orange tinged with black. Length over 66 cm.

Mrigal grows slower than Catla and Rohu. It attains sexual maturity in second year.

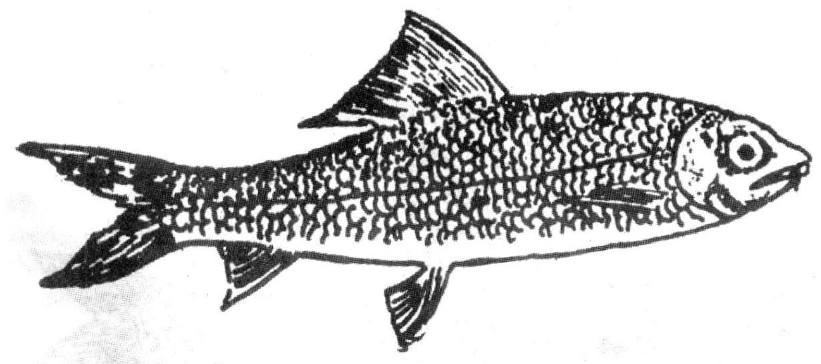

Fig. 3. *Cirrhina mrigala* (Hamilton)

Silver Carp (Fig. 4) - Silver carp belongs to Osteichthyes, Cypriniformes-cyprinidae. Body compressed, scales small, mouth sub-superior with lower jaw rather upturned. Eyes comparatively small, situated below horizontal axis of body. Gill rakers, dense, interlaced, connected and covered with a spongy sieve membrane. Abdominal keel extending from the base of pectoral fins to the anus.

Pharyngeal teeth one row in 4/4, surface of tooth flat with fine grooves. Intestinal length 6-10 times body length. Color of body in alive condition is silvery white, dorsally dark brown.

Fig. 4. *Hypophthalmicthys molitrix (*Valenciennes)

Grass Carp (FIg. 5) – Grass carp belongs to Osteichthtyes, cypriniformes, cyprinidae. Large sized, body almost cylindric with flat head and round abdomen. Scales big, mouth infront, the lower jaw shorter. Eyes small, Gill rakers short and sparse. Pharyngeal teeth comb like, in two rows left 2, 5/right 2, 4. Dorsal fin short. The back dark brown and abdomen white.

Fig. 5. *Ctenopharyngodon idella* (Valenciennes)

Common carp (Fig. 6) – There are lot of morphological variations through artificial breeding and natural selection of this species. Although 3 well known varieties of common carp like *Cyprinus carpio* var. *communis* (mirror carp), *Cyprinus carpio* var. *specularis* (Scale carp), *Cyprinus carpio* var. *nudas* (Leather carp), but a new genetical line of line carp has been developed in certain countries. Chinese claim of having Wu Yuan red purse carp and Xing Guo red carp which is being developed through artificial propagation and stabilisation of genic material in common carp. Body compressed, Dorsal projected in arch shape, Round abdomen, mouth slightly downwards with blunt snout and with two pairs of barbels on upper jaw, lower pair a little longer. Long dorsal fin, scales thick and big, color of the body in alive condition varying with different living conditions usually dark grey or yellowish brown dorsally, lateral golden yellow. Apex lining of caudal fin is slightly red.

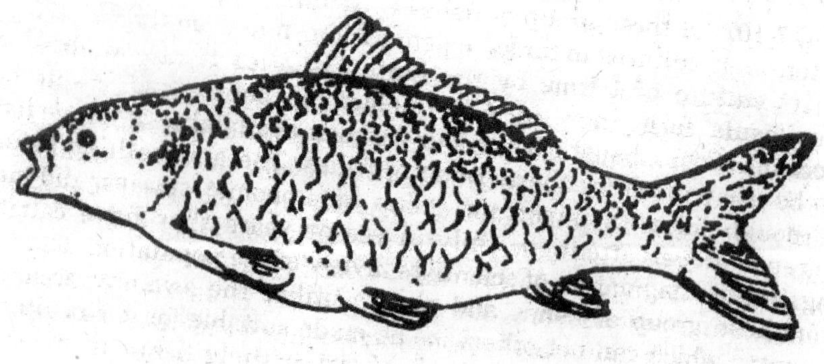

Fig. 6. *Cyprinus carpio* (Linnaeus)

2. Biology of important Airbreathing Fishes

Certain freshwater fishes, show unique environmental adaption for direct use of atmospheric oxygen and as such they are commonly known as airbreathing fishes. As these are marketed alive they are regarded as livefishes. Livefishes are preferred over Indian major carps in South part of India. That too, these fishes are known for their nutritive, invigorating and therapeutic qualities and are recommended by physicians as a diet during convalescence. The common airbreathing fishes are -

Family – Channidae, (*Channa marulius, Channa striatus),* Sacchobranchidae (*Heteropneustes fossilis*), Claridae, (*Clarias batrachus*), Notopteridae, (*Notopterus chitala*), (*N. notopterus*) Anabantidae, (*Anabas testudineus*) and goramy (*Osphranemus gourami*) etc.

The fishes which belong to family Channidae (*Ophiocephalidae*) are popularly known as Snake head murrels (Fig. 7-10). Of these group of fishes *C. striatus* and *C. marulius* are extensively cultured in tanks in penninsular India. On the early days of its culture first time by the State Fisheries Deptt. Madras at Sunkesula farm (now in A.P.) yielded no substantial result on account of cannibalistic behaviour and non-availability of forage fish to be use as food for these group of fishes. The airbreathing fishes although form an important group of economic species, did not receive proper attention. Only in recent years they have caught again the imagination of scientists inview of the population demand for these group of fishes and also to utilise the swampy areas of country which can not otherwise be made suitable for carp culture. Therefore, technology for culture of airbreathing fishes in cages of convenient sizes made up indigenously available split bamboos, kept partially immersed in the swampy water has been evolved. The fishes are given trash-fish and dead discarded silkworm pupae as the supplementary feed. The airbreathing fishes like murrel, magur and singhi have been successfully breed by hypophysation resulting new avenue for the availability of stock materials for culture which now no longer a problem.

Channa marulius (Fig. 7)

Body is cylindrical and compressed posteriorly. Head is depressed, cleft of mouth extends beyond the eye. Color of the back is greyish, abdomen is orange, a black ocellus encircled by whitish out line is present at upper part of the caudal fin and there are also black bands on lateral line, spots are present on posterior part of the body, dorsal fin, anal and caudal fin. The young ones have a brilliant orange coloured bands passing from the eye to the middle of head. Like other species of *channa,* this species is also very hardy and can stand extreme conditions of deteriorated water. It is also having great scope of culture in the derelict water. Its flesh is white in color, good flavour and is relished by people in many parts of country.

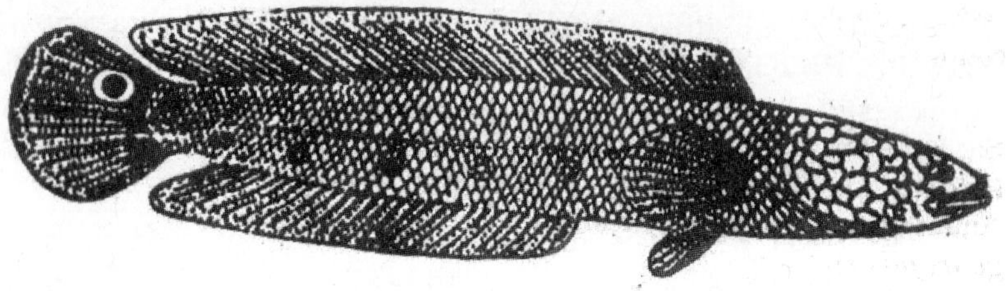

Fig. 7. *Channa marulius* (Hamilton)

Channa striatus (Fig. 8-10)

The lower jaw is longer. Cleft of mouth is oblique and hind edge of maxilla extends beyond the posterior margine of eye. There is an inner conical row of teeth on lower jaw and cardiform teeth on palate. Scales are cycloid. The lateral line is complete and curves down below 12 dorsal fin rays. Color of back is dusky or brown and dirty white on sides and beneath. Side possess stripes. The species is very hardy and can live for about 20 hrs. out of water with the help of accessory respiratory organs. It can tolerate extreme conditions of water and therefore has great potential of culture in derelict water.

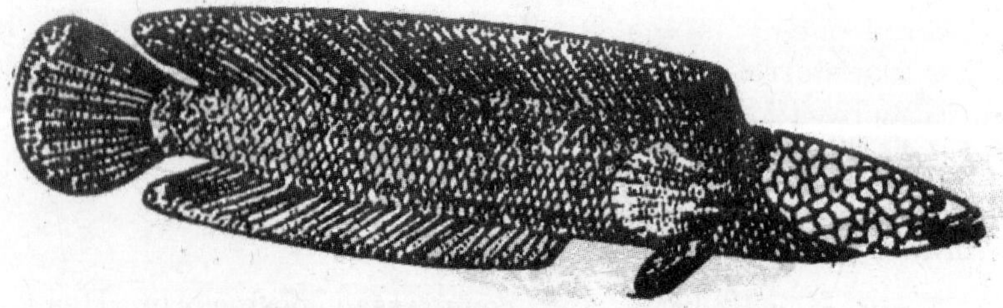

Fig. 8. *C. strisatus* (Bloch)

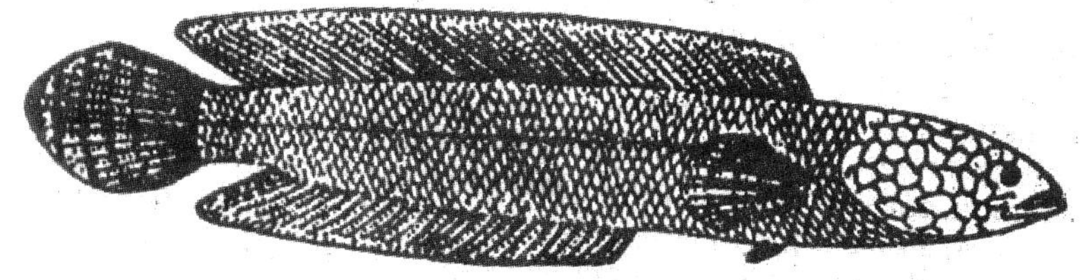

Fig. 9. *C. punctatus* (Bloch)

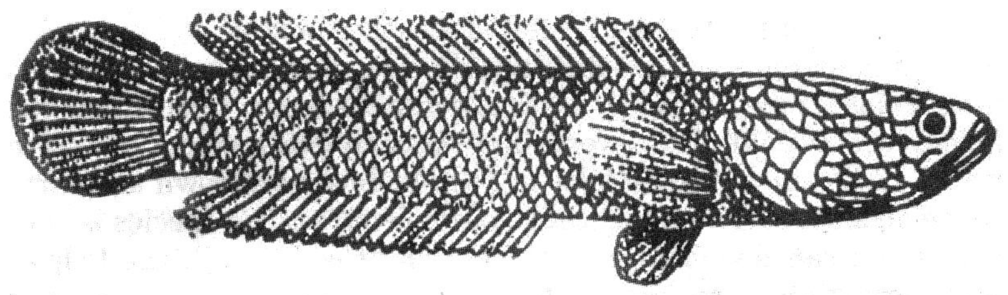

Fig. 10. *Channa gachua* (Hamilton)

Heteropneustes fossilis (Stining Cat fish) (Fig. 11)

The dorsal profile is almost straight but ventral is slightly convex. Jaws are of equal length. Lips are continuous and papillianated. There are 4 pairs of barbels, maxillary reaches to the middle of pectoral or even up to pelvic base. The dorsal fin is short and spineless. The pectoral spine is strong, osseous and pointed. The color is blackish above and lighter benath.

Fig. 11. *Heteropneustes fossilis* (Bloch)

Feather backs (Fig. 12 and 13)

These are represented by 2 species. *Notopterus chitala* (Fig. 12) and *Notopterus notopterus* (Fig. 13). The former grows to fairly a larger size, form a good fishery in West Bengal, Assam and

Bihar. It is highly esteemed as food fish especially when in fresh condition. The species especially *N. chitala* is carnivorous and grow well in swampy derelict water. The species has been cultured alongwith other airbreathing fishes in cage culture. In composite fish culture the species is introduced as a police fish.

Fig. 12. *Notopterus chitala* (Hamilton)

Fig. 13. *Notopterus notopterus* (Pallas)

Notopterus chitala

Dorsal profile is highly convex, while abdominal profile is almost straight. Dorsal fin is small in the caudal region and originates far behind the pelvic origin. The anal and caudal are united. Serrations along abdominal edge between pelvic fin and thorat are the taxonomical characters. Lateral line is well distinguished. Scales on body and head are of equal size approximately. Color is cupribrown above and silvery bands are present which join over the back. Fins are greyish, several black spots are present on the caudal region.

Clarias batrachus (Asian Cat fish) (Fig. 14)

It is commonly known as magur (Fig. 14). Magur attains sexual maturity when one year old. It breeds in innundated waters of rice fields, swamps ad marshy areas during monsoon and exhibit parental care. Sexual dimorphism is distinct as the genital papilla is pointed in male and swollen in female which gets vascularised during breeding season. The fish deposits the eggs in the holes made in the pond below the water surface. The size of the hole is 20 cm in diameter and 25 cm in width. Natural propagation and hypophysation technique in breeding *Clarias batrachus* is well documented in literature.

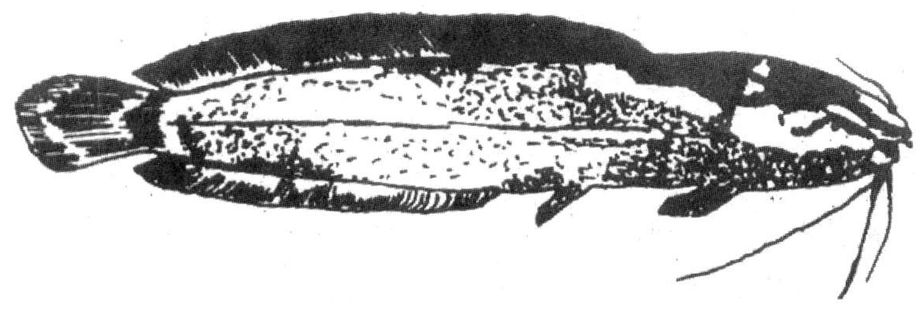

Fig. 14. *Clarias batrachus* (Linnaeus)

3. Biology of important Catfishes

U.S.A. has made considerable head way in culture of catfishes like channel catfish *Ictalurus punctatus* and white catfish *Ictalurus ictalurus.* However, the catfishes of great importance in India are *Mystus seenghala, M. oar, Wllago attu* and *Ompok bimiculatus.* Catfish constitutes an economically important group of species which contribute considerably to the fishery of national water. The species like *Wallago attu* (freshwater shark) and *Mystus seenghala* are highly carnivorus and piscivorus. Therefore, not desired to be taken in ponds having carp culture. Catfishes being less bony are preferred and are in great demand in many parts of the country.

Further, other suitable catfishes listed are –

1. Asian Cat Fish – *Calrias batrachus*
2. Stining Cat Fish – *Heteropneustes fossilis*
3. Thailand Cat Fish – *Calrias macrocephalus*
4. River or Silver Cat Fish – *Pangasius sutchi*

5. African Cat Fish (Thai magur) – *Clarias gariepinus* or *Clarias lazera*

6. European Cat Fish – *Silurus glanis*

However, in India, ``Thai Magur" is passing through a phase of controversial identity.

Mystus seenghala (Fig. 15)

Upper jaw longer and cleft of mouth shallow. The median longitudinl groove in head extends to the base of occipital process. Dorsal spine has grooves, weak and extends to snout. Pectoral spine is stronger than dorsal spine and serrated. Adipose dorsal fin present. Body brownish grey, silvery on sides and abdomen. A black spot on the hind end of adipose dorsal fin. It is very much liked in North-Western states of India.

Fig. 15. *Mystus seenghala* (Sykes)

Wallago attu (Fig. 16)

Snout is rounded and spatulate type. Mouth transverse and very large. The mouth cleft extends beyond eyes. Lower Jaw is more prominent. Teeth are large, pointed and depressible.

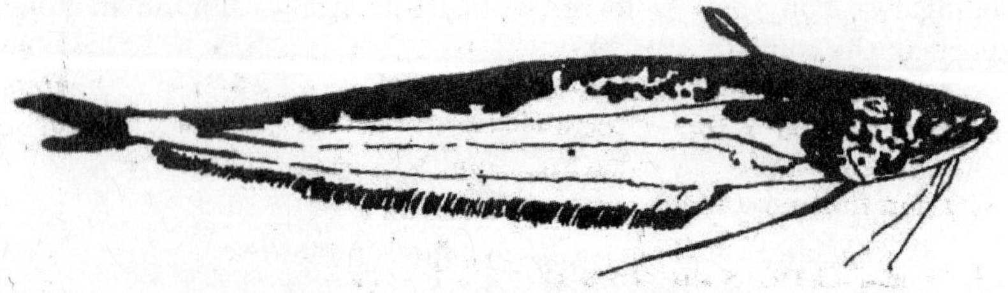

Fig. 16. *Wallago attu* (Schneider)

4. Note on important Larvicidal fishes

Larvivorous fishes can be used to control mosquito larva. Larvivorous are foremost natural anemies of mosquito larvae . So far three exotic and few indigenous genera of fishes have been mentioned in antimalaria literature. The three genera of exotic fishes include 1. Carassius 2. Gambusia and 3. Barbodos. Carassius has been brought from Siam and Gambusia from 1taly. It is better to make adequate attempt to determine the larvicidal value of native species before introducing exotic species.

The indigenous larvicidal fishes included for biological control are:

1. Chela (Ham)(family-Cyprinidae)
2. Laubuca (Bleeker)(family-cyprinidae)
3. Barilius(Ham.) (family-Cyprinidae)
4. Rasbora (Bleeker) (family-Cyprinidae)
5. Danio species
6. Esomus (Swaisnson) (family-Cyprinidae)
7. Puntius (Ham.) (family-Cyprinidae)
8. Cirrhina (Cuvier) (family-Cyprinidae)
9. Wallago (Bleeker)(family-Siluridae)
10. Panchax (Cuvier)
11. Aplocheilus (Mc celland) (family-Cyprinidae)
12. Aphanius (family-Cyprinidae)
13. Channa (family-Ophiocephalidae)
14. Anabas (family-Anabantidae)
15. Mugil (family-Mugilidae)
16. Ambassis (family-Ambassidae)
17. Therapan (family-Theraponlidae)
18. Badis (family-Ristolepidae)
19. Glossogobius (family-Gobidae)

No specific observations are available on the bionomics of these fishes for which the following are necessary for a useful larvicidal form.

1. Fishes must be small, hardy and can withstand shallow and weed infested waters.

2. Must be able to breed freely in confined water areas.
3. Able to withstand transportation and handling stress.
4. Difficult to catch because of their swift movement and either be surface feeder or carnivorous in habit.
5. Absolutely insignificant as food.

5. Note on exotic food fishes in India

Several exotic food fishes were introduced in India for culture, sport fishery and utilization of water bodies. A tabular list on exotic food fishes is presented as follows.

Sl. No.	Common name	species	year of introduction	source
1.	Common carp	*C. carpio*	1980	Srilanka
2.	Grass carp	*C. idella*	1939	Japan, Hong Kong
3.	Silver carp	*H. molitrix*	1959	Japan, Hong Kong
4.	Bighead carp	*A. nobilis*	1959	Bangladesh
5.	Golden carp	*C. carassius*	1968, 1970	England
6.	Doctor fish	*Tinca tinca*	1974	England
7.	Tawes	*P. javonicus*	1870	Indonesia
8.	Tilapia	*O. mossambica*	1972	Bangkok
9.	Nile tilapia	*O. niloticus*	1952	Thailand, Israel
10.	Red tilapia	*Hybrid*	1986	Phillipines, Israel
11.	African Cat fish	*C. gariepinus*	1978	Thailand
12.	Brown trout	*S. trutta ferio*	1863-1868	England, Japan
13.	Rainbow trout	*S. gairdneri*	1909-1974	Srilanka, England, Germany, Japan
14.	Brook Trout	*Salvelinus fontinalis*	1959	Canada
15.	Steel head	*S. g. trideus*	1867-1940	Europe
16.	Splake trout	*Salvelinus namayeush*	1968	Japan
17.	Sockeye Salmon	*O. nerka*	1968	Canada
18.	Gouramy	*Osphronemous gouramy*	1841, 1865	Java.

Some of the exotic fishes are also unauthorisely cultured in Indian tanks and ponds, the common ones are Clarias gariepineus (thai magur) and pacu (*Piaractus brachypomus*). Datta and Nandeesha (2006) made an observation of pacu (rupachanda) in the fish market of Tripura. The fish is being imported from Bangladesh and the live seeds of this is also sold in the market of Agartala. Pacu is also known as Red Pomfret and is mainly a vegetarian but will adapt itself to eat almost any thing. Pacu grows to more then 24" in open water tanks although pacu is more known as an ornamental fish world wide. It can be kept with other fishes too in aquarium where it can not grow bigger than 12 inches.

6. Food and Feeding Habits

Success in fish farming essentially lies on thorough knowledge about the food and feeding habits of the culturable varieties of fishes taken for culture. That too, food resources in ponds are varied, for which a judicious combination of species for rearing is essential.

Schaperclaus (1933) classfied the natural food of fishes in to three groups such as : (1) Main food, (2) Occasional food and (3) Emergency food. Main food is the natural food which the fish prefers under favourable conditions and in which it thrives best. The occasional food is that food which is consumed as and when available. But the emergency food is ingested when the preferred food items are not available and on which food the fish is just only to survive.

Nikolskii (1963) classified the natural food of fishes in to four categories on the basis of gut analysis such as : (1) Basic food, which comprises the main part of the gut content, (2) Secondary food, which is found frequently in gut content but less in quantity, (3) Incidental food, which rarely enters into gut and (4) Obligatory food which fish consumes in the absence of basic food. However, on the characteristics of diet the fishes are classified in to Herbivorous, Phytophagus, Detritophagus, Carnivorous, Predatory, Omvivorous, Coprophagus etc. Usually herbivorous are those which feed on green algal matter and such as Silver carp. The phytophagus are those which feed on relatively large green vegetables like weeds, grasses etc. The detritophagus are those which feed on detritus matter in the pond system. Carnivorus and predators are those which feed on invertebrate and fishes respectively. Omnivorous are those which feed on both plant and animal materials. The coprophagus are those which feed on dung applied as manure in pond system.

Based on extent of food variations consumed, fishes are classified as : Euryphagic of having wide range of food varieties, Stenophagic of having narrow range of food varieties, Monophagic of having single type of food. However, on the basis of trophic niche, the fishes have been grouped in to plankton or filter feeder (Catla, Silver carp), column feeder (Rohu, *L. bata*) and bottom feeder (*Cirrhina mrigala, C. reba, Labeo calbasu*). Usually herbivores and carnivores show definite peak period in feeding, while omnivores show little variation throughout the year.

6.1 Modifications due to different feeding habits

According to different feeding habits of fishes various modific-ations occur. For example Catla has a slightly upturned mouth showing its filter feeding habit. Similarly Silver carp also shows filter feeding habit. But the speciality of Silver carp is that is feeds preferably on phytoplankton where as Catla feeds on Zooplankton. Big head which is a cultivable species in China also feeds on Zooplankton. Rohu has a little sub-terminal mouth with thick fimbricated lips. The structure and position of mouth and lips are adopted for sucking the debris and for browsing on weeds. Mrigal has terminal mouth with thin lips. The carps which pick up or digout their foods from the mud at the pond bottom have the protrusible mouth as in *Barbus dubius.* These fish feed on worms, insects and snails at the bottom. In predators the mouth is generally large and provided with teeth which are not for chewing the fishes but preventing their escape. With terminal mouth the predators can easily snap at and catch relatively larger victims unlike herbivorous.

6.1.1. *Gills* – Gill although functions mainly as respiratory surface, still it contributes towards the function of major feeding organ. There is structural change in the gill rakers of Silver carp (phytoplankton feeder) and Bighead (Zooplankton feeder). In Silver carp, there are 1700 gill rakers per gill arch bone and in 1 mm length, there are about 12-13 gill rakers, where as in Bighead, there are 680 gill rakers per gill arch bone and in 1 mm length, there are 6-7 gill rakers. The distance between two gill rakers in

Silver carp is 34 micron and in Bighead it is about 84 micron. That too, there are minute bony bridges between gill rakers in Silver carp and the sieve membrane covering the bridges to form a fine net which is efficient in retaining phytoplankton. Therefore, the main food of Silver carp is phytoplankton. However, in Big head the gill rakers are short and sparse. That too, gill rakers are separated by large spaces and there is no minute bony bridges between the gill rakers. As a result, their main food are zooplankton.

In addition to the filtering organs (gill rakers), Silver carp and Big head have accessory organ, palatine fold. Palatine fold is located at the interior of the mouth cavity consisting of nine vertical ridges of mucus membrane, four on either side and one in between. The middle ridge is short and resembles an inverted letter-Y. Palatine folds act in coordination with gill rakers on filtering the food. The gill rakers of Rohu and mrigala are ill adapted for filtering water and retaining plankton. The gills of Catla are adopted for filter.

6.1.2. *Pharyngeal teeth* – Grass carp is a typical herbivorous species, consuming all sorts of aquatic and terrestrial grasses. The pharyngeal teeth are well developed, tough and strong. The pharyngeal teeth formula is 2,5/4, 2. The pharyngeal teeth are chopper shaped with saw toothed edges. Pharyngeal teeth at both sides are interlaced and are against the callous pad of the basioccipital, grinding food into pieces for digestion in intestine. Grass carp are voracious eaters, but can not digest cellulose. Common carp of about 50 mm in length are omnivorous, Common carp is inclined to be more carnivorous. The pharyngeal teeth of Common carp are relatively developed molar, shaped in 3 rows, with transverse grooves on the two inner row except the first tooth which is smooth. Common carp display a wide adaptation to foods. Their common natural foods are benthos and they take certain amount of detritus of higher aquatic plants and plant seeds.

6.1.3. *Nasal bone* – The nasal bone of common carp is well developed so that their premaxilla and mandible could be projected out like a tube to dig the mud for organic detritus.

6.1.4. *Intestine* – Intestine appears easily in early stages, in early stages i.e. under 15 mm in length the cultivable species of Silver carp, Grass carp, Common carp, Catla, Rohu and Mrigala are relatively more planktonivorous especially zooplankton and have straight tube like intestine. The highly coiled intestine is a characteristic of herbivorous adult fishes, seems to be formed latter on. But the adult predatory fishes have straight tube like intestine. The gastro somatic index (*length of intestine : length of body)* is higher in herbivorous fishes and it is less in predatory fishes.

7. Factors influencing feeding

7.1 *Temperature*

Rate of feeding, metabolism and growth are affected not only by the availbility of food but also directly by the water temperature. Variations in food intake is called as ingestive variation. Such ingestive variation is mainly influenced by the water temperature. The cultivable carps like Silver carp, Rohu, Grass carp and Mrigal shows less food intake when the water temperature falls below 15°C and below 8-10°C, the fish especially Silver carp and Grass carp almost stop feeding. However, these fishes have different adaptabilities to the variations of water temperature. The Indian major carps have a ecothermal limit of 17.5°C to 38°C where as Grass carp have tolerance range up to 40°C.

However, to summarize the ecothermal range, the survival temperature for Chinese carps is 0.5-38°C. The suitable temperature is 20–32°C, spawning temperature is 22-28°C and best growth is obtained when temperature ranges from 25-32°C. In Mud carp the best growth is obtained in temperature 30-32°C. But in Indian Major Carps the lower limit of thermal range is about 17.5 and higher limit is about 37.8°C.

7.2. *pH value* – When pH value of water drops below 5.5, metabolism lowers rapidly and the appetite of fish is affected. This ultimately affects the ingestive variation of fish. However, the cultivable carps adapt themselves to weak alkaline water of pH value 7-8.5.

7.3. *Oxygen (O_2)* – When dissolved oxygen falls below 2 mg/litre, carps show poor apetite. When the dissolved oxygen, falls below 1 mg/litre, fish will stop eating completely. However, fish can grow and develop normally when dissolved oxygen content is above 5.5-7 mg/litre.

However, growth rate is an important criteria in the evaluation of production efficiency. The carps like IMC and Exotic fishes taken under composite fish culture are selected because of their larger size and speedy growth. That's why these carps are taken as dominant culture species in India. However, in China, besides dominant cultured species of fish like, Silver carp, Grass carp, Black carp, Big head, Common carp, secondary species of crucian carp, wuchang fish, mud carp are taken under polyculture. Nevertheless, these secondary species are also important in the improvement of fish yield.

Growth rates are genetically controlled, as well as closely related to water temperature, water quality, four feed formula, feeding schedules, stocking density and management. As a rule growth rates are faster before the first sexual maturity. After that growth slows down or even stops after attaining considerable age.

8. Reproduction

The natural spawning grounds of major carps are vastly distributed in Ganga river system, Brahmahputra river system, Indus river system, West coast and East coast river systems. IMC usually like to habit in the lower and middle reaches of the river systems, or tributaries for fattening, where the water current is slow and the water is fertile with abundance of food. When the spawning season approaches especially during monsoon months, the spawning schools begin to move to middle and upper reaches for spawning. At that time the gonad, in a regat majority of brood fishes have reached to late maturing and mature stages. If the ecological conditions are suitable in the spawning grounds, they proceed to spawn. The spawning time in IMC are greatly influenced with the climatic conditions, but all spawn in monsoon when the water level of rivers rises and the temperature ranges from 25-30°C. There are different opinions about the optimum temperature for spawning of IMC, however, the optimum temperature for spawning, range from 22°C-28°C. Spawning starts from May to September in majority cases. The spawning grounds may be either of valley type or plain type. The characteristics of valley type spawning ground is that the width of river is narrow with high velocity of water current. The depth is more. But the characteristics of plain type spawning ground is that the width is more, the depth and current is relatively less.

The collection of spawn on a commercial scale is prevalent in Bihar, West Bengal and Uttar Pradesh. During past, Bihar alone was contributing over 50% of total spawn production. However, due to melting of snow rises the water level in the Kosi tributary of river Ganga during late May and early June makes the firat apperance of spawn, followed by the main Ganga. High yields of spawn were reported in the Eastern sector of the Ganga system in past years have almost defunct at present. Only Calcutta fish seed syndicate is collecting the spawn on commercial basis from Ganga river system.

In Brahmaputra system, the spawn collection were made mainly from South bank tributaries. South bank tributaries are comparatively deep and high current indicating like that of valley type spawning grounds where as the north bank tributaries are comparatively large with shallow channels of coarse sandy beds indicates like that of plain type spawning grounds. However, the percentage of major carp spawn in the collection is being poor and uneconomical as major carps appeared generally as stray specimens. In the lower reaches of Brahmaputra major carp spawn can be exploited commercially.

Only a small portion of Beas and Sutlej tributaries of Indus river system which lies in India provides scope for spawn collection purposes. There is no commercial fishery for major carps in Himachal Pradesh with the upper reaches of system harbours only coldwater fishes. The fish seed collection of major carps is done at Kanjli and Sultanpur lodi and Ludhiana in Indus system harbours in Punjab state.

In West coast river system, it is significant that the percentage of major carps is sufficiently high at the centres on the lower streches of Narmada in Gujarat than the middle stretches draining to Madhya Pradesh. The upper reaches are rocky unproductive and not suitable for spawn collection.

The deltaic stretches of the Godavari and Krishna rivers are sufficiently rich in carp fishery although the spawn of Indian major carps (IMC) are less. The Cauvery river harbours fairly good fishery of major carps on the middle and lower reaches. However, the IMC which were transplanted in Cauvery has established themselves. Spawns are collected by shooting nets, dip nets, basket traps, Cast nets and drag nets in shallow regions. However, the cheer fishing is used to catch seeds of carps which comprises a scare line and a composite net made by towing two cast nets.

Cheer fishing is seen during winter season particularly from November to January on river Tapti.

There are two common methods of spawning in natural habitat being reported by Chinese workers. Spawning on the surface is termed `floating spawning' and under water spawning is called `muffled spawning'. In floating spawning, the male chases the female, often bumping against the abdomen of the female with its head jumping out of water and then splashing into waves. Sometimes, both male and female would float on their backs with their pectoral fins vibrating violently. When the climax of spawning is achieved, female and male releases egg and milt respectively. But on muffled spawning chasing occurs underwater. Waves caused by chasing can be seen on the surface.

8.1 Living Habitat

Catla, Rohu and Mrigal remain in different layers of water due to their diversified feeding habits. Besides, other three Exotic carps like Silver carp, Grass carp and Common carp also remain compatible to IMC. Catla and Silver carp live in the upper layer, Rohu and Grass carp remain in the middle layer and Mrigal and Common carp remain on the bottom layer. However, the Big head carp being cultivated extensively in South China remain in surface layer, Mud carp and Black carp cultivated in China remain in bottom layer.

8.2 Water Temperature

The metabolic rate of fish is greatly affected by the temperature of water. When water temperature drops below 15°C, the appetite of cultivated carps reduces evidently and below 15°C, the fish almost stop feeding. But these fishes have different adaptabilities to the variations of water temperature. The ecothermal limit of IMC varies from 17.5°C to 38°C, where as Big head grow fast when average monthly temperature is 30-31°C and slowly when temperature drops below 10°C. Silver carp and Grass carp grow fast in higher temperature, but temperature above 39°C is fatal to both Grass carp and Silver carp. Mud carp can not withstand coldness and is being restricted extensively in South China. When water temperature drops to 5.5°C, they often die of cold.

8.3 Water quality

The dissolved oxygen and pH value are important factors with particular reference to the water quality for living habitat of cultivated carps. The congenial temperature of water (25-30°C) for best growth, the 6-7 mg oxygen/litre of water, 7.0-8.5 alkaline water is good for their food intake, metabolism and consequently the growth. When dissolved oxygen falls around 2 mg/litre of water, cultivated carps show poor appetite, when dissolved oxygen falls below 1 mg/litre, cultivated carps will stop feeding and below 0.5 mg/litre will cause fish to die. Similarly when pH value drops below 5 the appetite of fish is affected and the growth of fish also reduced.

8.4 Significance of propagation

1. Natural propagation is for auto stocking in large water bodies
2. Artificial propagation is for production of quality seed for stocking in culture ponds and also in large water bodies
3. To maintain genetic diversity and phenotypic variation in fish stocks

8.4.1 Site selection for riverine seed collection

Usually two distinct zones are identified in river course, such as erosion zone and shadow zone. Both these zones are unsuitable for seed collection due to high water current. Therefore, the site suitable for seed collection should be away from these zones where water current is mild and accessible.

8.4.2 Gears used for seed collection

Spawns are collected from river by a funnel shaped net called shooting net. Some times benchi jal is used for seed collection in North Eastern Bengal. However, the shooting net of Midnapur type is used by various State governments. The shooting net is operated in shallow margin flooded river with mouth facing the current, stretching both the arms of the net for large space area to allow maximum water volume to pass through. At the end part of the net (cod end), there is a ring stitched which is made up of split bamboo or cane. To this cod end, a receptacle is attached, called gamcha which helps for periodical scoop out of spawn collected. Jhingran (1982) have described different specifications of shooting net used in various states. Soon after spawns are collected, these were seived through a nonmetallic seiver to remove larger organisms and are transported.

In hatchery produced seeds, the spawns are collected in the spawn collection chamber by fixing a happa where water cushion is provided. These were measured, either stocked at farm site or transported under oxygen packing. In some cases the hatching pools are directly connected to the nursery tanks by suitable conduit for raising them to fry.

8.4.3 Behavior of spawn

Distribution and quality of spawn is greatly influenced by different hydrological conditions.

(a) **Flood Level:** Flood is the most significant character for spawn quality. Generally, the spawns of undesirable species is available in the first flood of the season and that of the major carps appear in subsequent flood. The spawn availability are seen in both receding and rising phase of flood, but abundant catches are encountered in the receding phases. Spawns are collected in both day and night, although suitable in night time.

(b) **Current and depth:** Distribution of spawn are significantly influenced by both current velocity and water depth in a riverine system. Mild velocity of water current is ideal where spawn distribution is not carried down stream. It is reported that when water current velocity is 0.086 Km/hr, spawns are stationary but facing the current, at 0.14-0.21Km/hr, these migrate upstream and some weak ones take shelter but when velocity of water current increases to 0.36- 0.42 Km/hr, these were carried down stream. However, many are in opinion that slow and gentle current velocity from 0.5-3.0 Km/hr is conducive to spawn catches. Usually spawns congregate at the water surface in shoal or in line in areas of very slow current near the bank of the river.

(c) **Current direction:** Spawns show restricted migration by the current velocity. A change in current direction sometimes drastically alters the suitability of spawn collection site either temporarily or permanently for the entire season. However, mild current parallel to river bank is suitable for good quantity of spawn catches.

(d) **Temperature, Turbidity, pH and Dissolved oxygen:** Indian major carp spawn has preferential distribution in water temperature range of 31-31.5°C when liberated in low water temperature. However, turbidity, pH and dissolved oxygen have no correlation with spawn availability.

(e) **Weather:** Stormy weather is unfavorable for spawn collection as it disrupts the current direction, wave pattern, net fixation and operation. Hence cloudy sky with gentle breeze, little drizzling coupled with other conditions cited above is ideal for spawn collection.

(f) **Spawn associates and plankton:** Abundance of Indian major carp spawn is neither correlated with spawn associates nor with plankton availability in riverine system. Although ***Frage-llaria oceania*** is related to the occurrence of shoals of Sardines in marine system.

2

AQUATIC ENVIRONMENT

1. INTRODUCTION

The physical features of the ``aquatic environment" will be the medium - that is, the material which immediately surrounds the organism and with which it has its all important exchange. Diversified media exists in any ecosystem. Some organisms live in the soil, some in ponds, some thrive in manure piles and some in blood streams of vertebrate animals. Certain nematodes live in vinegar and a fly larvae of genus psilpa grown in petroleum.

The medium in each of the above examples, are either a liquid or a gas and it is usually air or water. Although animals and plants inhabiting soil or mud may at first appear to be exceptions, a closer scrutinity shows that a film of air or water around each organism is actually the material in immediate contact with it. For example some small animals living in the wet sand of the sea shore shows that their essential exchange is with the water, percoloating between the sand grains and the medium for these animals is sea-water but not sand. Therefore, the term medium is thus used in a strict sense and is distinguished from the substratum or surface on or in which the organism lives.

Air and water exists as fundamental media which divides the world in to two major environments : Terrestrial and Aquatic. The media are not completely isolated from each other, however, some of the atmospheric gases are dissolved in all natural waters and some moisture is present almost everywhere in the atmospheres. These characteristic of intermixture of air and water play a role in sub-dividing the terrestrial environment into Arid and Humid climate and the aquatic environment into Stagnant and Running (areated) waters.

1.1 Properties of water

Air is composed of 79% Nitrogen, 21% Oxygen, 0.03% Carbon-dioxide and several other gases in much smaller quantities. These gases are not chemically combined, but exists as a simple physical mixture. Water by contrast, consists primarily of a single compound H_2O.

1.2 Unusual properties of water

The unusual qualities of water has been described by Henderson (1924) in his classic book ``The fitness of the environment". Water has a higher specific heat, latent heat of fusion and latent heat of evaporation than any common substance. these facts play a very important role in the heat regulation of organisms themselves and in the resistance of natural environments to temperature change. Another characteristic of water is its relatively high freezig point. Because of large amount of heat must be given-up before water can turn to ice and because of restricted stirring, Oceans and lakes freeze only at the surface. Even ponds rarely freez to the bottom. Therefore, the temperature of the medium can drop to 0°C in freshwater environments or to a few degrees lower in the ocean. But biological reactions of a great many plants and animals can go on still perfectly well at temperature down to the freezing point of water. Another unusual quality of water is its power as a solvent, no other common substance compares with water in this respect. Ionic substances in aquous medium can pass through into and out of body of an organism. Further more, the extent of ionisation of solutes in water is extremely high, providing the possibility of a great variety of radicals and of chemical combination. Water has the highest surface tension of any common substance except mercury. This high surface tension has many ecological influences, involving the movement of water into and through organisms as well as the rise of ground water in the soil.

The density representative of pure water is 1.000 g/cc at 4°C while that of pond water is 1.001 g/cc. The density representative of sea water (at 35 PPT salinity) is 1.028 g/cc and protoplasm is 1.028 g/cc. However, the density representative of air (at sea level) is 0.0013. This indicates that density of protoplasm is closely similar to that of sea water and only slightly greater than that of freshwater. But it is more than 850 times greater than that of air. These differences in densities influence the important differences in the pressure, inertia, viscosity and mobility of media.

1.3 Pressure

The differences in the density of the medium results in a great difference in the rate of change of pressure at increasing altitudes in the atmosphere and at increasing depth in the water. A rise of 300 meter in altitude from the earth's surface results in a reduction pressure of about 25 mm Hg. In contrast, for every increase of 10 m in depth in the water, pressure is increased by 760 mm Hg or 1 atmosphere. Therefore, tremedous pressure exists at the average depth of the ocean (3700 metre) is 370 atmospheric pressure. In the ocean deeps (10,860 m, marina trench) the pressure exists is 1086 atmospheric. It was believed that the tremendous pressure would annihilate all living beings, so that the greatest depths in the aquatic environment must be lifeless. But in 1951 the Danish, Galathea Expedition trawled aquatic organisms belonging to coelenterata (sea-anemone), Echinodermata (sea cucumber), molluscs, crustaceans from a depth of about 10,500 m off the Philippine islands. At this depth the pressure is 1050 atmospheric or about 1 ton on each square centimeter. But this terrific weight of water does not crush the organisms living at that depth because the pressure is the same inside their body as out-side. That too, the support and resistance to motion in air is less than that of water. Since water has nearly the same density as protoplasm, where as air is very much less dense, water furnishes much more buoyancy than air.

The coefficient of viscosity of water is 60 times that of air at the same temperature. That too because of the relatively great inertia of the water medium, some animals such as Jelly fish and scallops can propel themselves in one direction by pumping water in the opposite direction. The squid can dart backwards with remarkable rapidity by ejecting water from its siphon.

Water plays several different roles in the ecological relations of plants and animals. As a material entering the organisms water is important as a necessary and aboundant constituent of protoplasm. Water is essential also as one material taking part in the photosynthetic reaction, through which energy becomes available either directly or indirectly to all living beings.

1.4 Composition of water

A large amount of common ions like Na, K, Ca, Mg, Cl, SO_4, CO_3 etc. occurs in sea water than in typical freshwater. Hard freshwater contains more dissolved salts than soft freshwater. In sea water Cl and Na are the 1st and 2nd most abundant ions respectively. In typical hard freshwater, CO_3 is the most abundant with calcium second. In typical soft freshwater, the Ca and CO_3 are relatively less concentrated than Na and Cl. However, here the major concern is with the maintenance of the proper osmotic balance between the inside and the outside of the organism.

The osmotic pressure of a molar solution of a non-electrolyte is 22 atmosphere and 1700 cm of mercury. The freezing point of such a solution is depressed to -1.84°C and this fact provides a convenient method for measuring osmotic pressure.

The osmotic pressure of a teleost fish tends to remain within a relatively narrow range for both freshwater and marinewater. In freshwater the teleost is osmotically superior to its medium, where as it is osmotocially inferior in saltwater. In the sea a strong tendency exists for the fish to lose water from its tissues to the surroundings. A brief calculation will indicate the magnitude of the osmotic force with which such an animal must content inorder to maintain its water budget. If the osmotic pressure of the fish is represented by $\Delta = 0.7$ and that of the surrounding medium by $\Delta = 1.8$, the magnitude of the pressure corresponding to this difference of $\Delta = 1.1$.

This is $\frac{1.1}{1.84} \times 1700 = 1015 \ cm \ Hg$.

An osmotic pressure of 1015 cm Hg or 13 atmosphere is tending to extract water from the fish's tissues. Therefore, for the teleosts, other animals and plants with low internal osmotic pressure, seawater has the effect of being physiologically dry.

2. PRODUCTIVITY OF AQUATIC ECOSYSTEM AND ECOLOGICAL BIOENERGETICS

Population sustained capacity (carrying capacity) of an ecosystem is generally known as productivity. The productivity of a water body is characterised where by the living substance is manufactured through interaction of the constituents of the natural environment. One major factor concommitant to obtain optimum production levels from any given body of water is the maintenance of water quality. Therefore, the knowledge of water quality is extremely important to the aquaculturist. Nearly every problem that arises in aquaculture system is the result of degradation of water quality. Besides this the limno-chemical parameters such as climatic, edaphic and morphometric features affect greatly on the productivity of large impoundments. The climatic factors like rainfall, sunshine, wind velocity, air temperature and Edaphic factor-water and soil quality provide essential source of nutrients while morphometric features like depth, area, volume, water level fluctuations, regulate the supply of energy of and nutrients. In addition, the hydrological cycle (inflow and outflow) play important role in the productivity of impoundments. Circulation of water either by holomixis or neuromixis in relatively large impoundments is an important phenomenon that brings the chemical nutrients locked in the tropholytic zone up to nutrient rich trophogenic zone and facilitates fixation and utilisation of energy. Tropholytic layers are also characterised by strong O_2 decline and hence the distribution of O_2 is klinograde in deep impoundments, but are rich in energy budget indicating high productivity. On the other hand the uniform distribution of O_2 (orthograde) from surface to bottom in deep and large impoundments characterises the low reserve of bottom energy budget indicating low productivity. The decline of O_2 in tropholytic layers is accompanied by increased nitrogen levels and by the accumulation of CO_2. This enriched CO_2 and subsequent increases in H^+ $(CO_2 + H_2O)^2 = (2\ H^+ + CO_3)$ lowers the pH of the bottom layers. Hence, the bottom accumulation of CO_2 fall in pH, increase in HCO_3 conductivity and rise in bottom nutrient levels serve to reflect the large impoundment's high productivity.

2.1 Fundamental steps operated in productivity

The fundamental steps operated in the productivity are :

(a) reception of energy

(b) production of organic matter by producers

(c) consumption of this material by consumers and its further elaboration

(d) decomposition to inorganic compounds

(e) transformation into suitable forms for the nutrition of the producers.

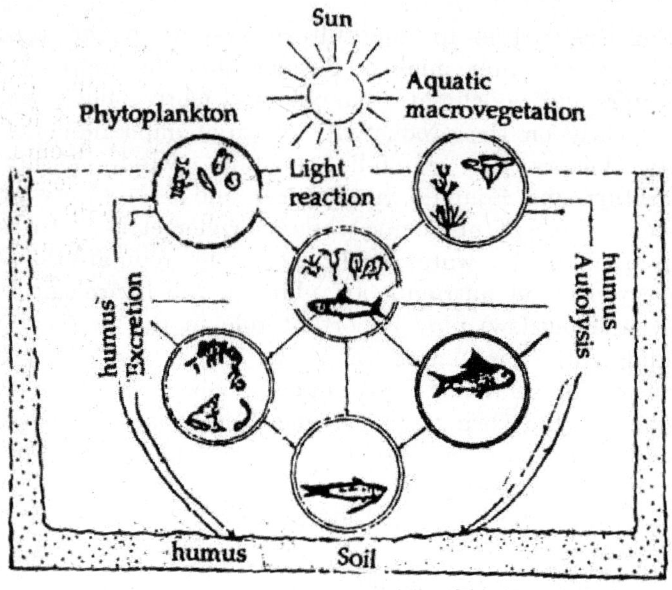

Fig. 17. Biotic Link (Food chain)

The energy generated within the sun is received on the earth's surface as electromagnetic radiation of different wave lengths ranging from 1A° to 1,35,000 A°. Some of these radiations are absorbed by plannets and reflect back the rest into the space. The atmosphere of our earth is so constituted that it permits radiations between 3700 A° to 8000 A° to reach to the earths surface. This radiation between 4000–8000 A° constitute the visible light or radiant energy. It is this energy that enters into the photosynthetic process of chlorophyll bearing plants. This is the first biotic link (Fig. 17) in the pond ecosystem that is the chlorophyll bearing organisms which can convert the solar energy into chemical energy through photosynthesis. On the basis of site of occurrence, plankton may be limnoplankton (lake plankton), rheoplankton/potamoplankton (running water plankton), heleopla-nkton (pond plankton), halioplankton (Saltwater plankton) and hypalmyro plankton (brackishwater plankton). These are therefore called as primary producers and the resultant is primary productivity. All subsequent production in any ecosystem depends on the primary production (Fig. 18). The rate at which photosynthesis occurs is dependent on the amount of chlorophyll available and the presence of light of the proper quality and quantity. The presence of plankton itself may affect light penetration. However, pond culture is practised in relatively shallow water and therefore, the light penetrates to the water to the extent of photic-zone or to a appreciable depth. Even in the trubid water of aquaculture ponds, light intensities at the bottom commonly exceed 1% of the incident, even if it is not so, light generally penetrates sufficiently to promote atleast some photosynthetic activity. However, the actual quantity of energy available for photosynthesis by green chlorophyll bearing plants varies from latitude to latitude in accordance with the path-length of solar beam through the atmosphere. This average amount of solar radiations reaching to earth's surface varies from 2.5×10^8 to 6.0×10^8 calories/m^2/day. As much as 95-99% of this energy is immediately lost from the plants and remaining 1 to 5% is used in photosynthesis, that is stored in the fom of organic matter.

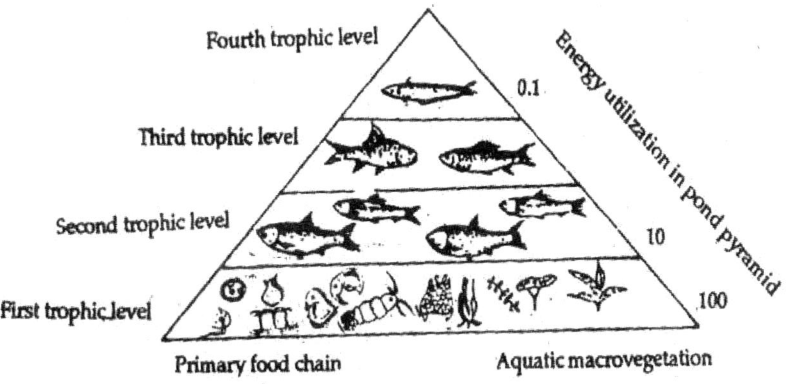

Fig. 18. Trophic zone in Aquatic system

The photosynthesis process can be divided into two parts, the light reaction and dark reaction. In the light reaction, hydrogen is withdrawn from water and passed through a series of hydrogen carriers to nicotinamide adenine dinucleotide phosphate (NADP), so that reduced nicotinamide adenine dinucleotide phosphate ($NADPH_2$) is formed. Associated with this oxygen transport there is a conversion of Adenosine diphosphate (ADP) and inorganic phosphate to Adenosine triphosphate (ATP). These chemical changes are associated with a considerable increase in free energy.

In the dark reaction, the $NADPH_2$ produced is used to reduce CO_2 to the level of carbohydrates. This too is associated with an increase in free-energy, the energy being supplied by the break down of ATP produced in the light reaction. Thus the overall photosynthesis process can be summarised as

$6CO_2 + 6H_2O \rightarrow C6H_{12}O_6 + 6\ O_2$

n CO_2 + n H donor solar energy $(CH_2O)n$ + n Oxidised donor.

The primary production can be estimated from the oxygen measurement by dark and light bottle techniques giving full day exposure. The results were integrated to obtain the values in gC/m^2/day which can be converted to calories/m^2/day. Using the following conversion standards, the productivity can be findout.

1 gm of Oxygen = 0.375 gm carbon.

1 gm of O_2 produced during photosynthesis = 3.68 K cal.

1 gm of carbon = 2 gm organic matter (Ryther 1956)

1 gm dry wt of organic matter = 4.5 K cal.

1 gm of carbon = 10 gm of wet wt of fish (Rodhe 1958)

1 gm of carbon is approximately equivalent to 10 K cal.

However, since the average decomposition of plant biomass differs some what from CH_2O due to the presence of protein, lipid and nucleic acids as well as carbohydrates, the O_2/CO_2 ratio is usually in the range of 1.1 to 1.2.

2.2 Photosynthetic efficiency

Photosynthetic efficiency (F) is nearly equal to the calories of energy (H) produced in the algal cells per unit space per day divided by the amount of visible solar energy (S) received at the water surface

$$F = \frac{H}{S}$$

Ganapati (1970) stated that `H' can be determined either from O_2 values or from the algal weight which may be found out by dividing the O_2 values by 1.63 (see Natarajan & Pathak, 1983). The general formula for the calculation of phytosynthetic efficiency is

$$F = \frac{\lambda\ n}{\lambda\ n - 1} \times 100$$

Where, λ n is the energy fixed by the primary producer and λ n -1 is the solar radiation received.

The energy fixed by producers can be obtained either from O_2 produced during photosynthesis (3.68 calories are required to produce 1 mg of O_2) or by converting O_2 to carbohydrates by multiplying with 0.937 (based on heat combustion of carbohydrate as glucose) and as 1 gm of carbohydrate is equivalent to 4.1 K cal. of energy, multiplying the carbohydrate values with 4.1 gives the energy fixed by producers as carbohydrate.

Light energy = energy by producers fixed + energy lost to the environmnet

$$\lambda n - 1 = \lambda n + H$$

λ n is the gross energy fixation or gross primary production. A part of this energy is used by plant themselves for their own metabolic activities (measured by respiration) and remaining is stored by them. The storage of energy by plants is termed as net energy fixation or net primary production and according to the law of thermodynamics.

Gross primary production = Net primary producton + energy spent on respiration

or

Gross energy fixation = Energy Assimilated

$$\lambda n = An + R$$

2.3 Other Source of Energy Input

2.3.1. Chemosynthesis - Some autotrophs rely on inorganic substrates for energy through oxidation. The inorganic substances that are usually oxidised by the autotrophs to obtain energy are : $H_2 S$, NH_3, HNO_2, N_2 and so on.

$2H_2 + O_2 = 2H_2 O = 136$ Cal.

$2S + 3O_2 + 2H_2O = 2 H_2SO_4 + 284$ Cal.

$2S + 2O_2 = SO_2 + 142$ Cal.

$2 NH_3 + 3O_2$ -- - - $2HNO_2 + 2H_2O + 158$ Cal.

$2HNO_2 + O_2$ - - - - $2HNO_3 + 42$ Cal.

However, the contribution of energy through this source may not be significant for heterotrophs, but it is important from the point of view of the organisms which get energy through this source.

2.3.2 Allochthonous Source - In carp ponds, application of fertilizers, manures and artificial feeds form an important source of energy. Hence to account for the total energy available in a carp culture pond, the energy other than that fixed by producers must also be accounted for. Thus,

Energy available at the lowest (plankton) trophic level of the Ecosystem	=	Chemical energy fixed by producers	+	Energy imported from allochthanous source

2.3.3 Energy release during decomposition process - Every water body has an indigenous bacterial flora that is maintained as a regular part of the biological complex of the system. The decomposing bacteria and other micro organisms obtain the energy for their metabolism from the oxidation of carbohydrates and other organic materials.

$$C_6H_{12}O_6 + 6O_2 \rightarrow 6CO_2 + 6H_2O + 674 \text{ Cal}$$

$$C + O_2 \rightarrow CO_2 + 94 \text{ Cal.}$$

2.4 Secondary Production

The further transformation of energy following primary production are known as secondary production. The intensity of energy flow decreases as it passes from the primary producers to consumers and so on (Fig. 18). The qualitative and quantitative composition of aquatic biota is not constant and definite which makes the situation more complicated to measure the exact amount of energy flow to successive trophic levels. If it is assumed that a portion of the energy requirement of the carps is met through the digestion of plant materials a substantial quantity of the production of the herbivores are left free. Thus a portion of energy produced which is not utilised by the consumers is utilised by the detritivores on the death of the producers. Therefore, there are two chains of energy flow in the aquatic system.

(1) Grazing food chain

(2) Saprophytic food/Detritus chain

Grazing food chain involves grazing of planktons by herbivores which are inturn taken by predators. The saprophytic food chain starts from dead organic matter by microbial organisms or detritivorous which are inturn taken by predators.

Usually the heterotrophs derive the energy from Oxidation of the organic material consumed by them. Usually the energy transformation in heterotrophs are represented as nutritional bioenergetic and follows as :

$$C = P + R + U + F$$

C = energy of food intake (Consumption)

P = energy for growth

(a) body growth = Pg

(b) reproductive growth = Pr which is passed out to environment

R = energy for respiration

F = energy of faeces

U = energy of Urine.

2.4.1 Instantaneous Energy Budget

However, if these values are converted in to common energy equivalent, the equation should be balanced in the following ways in different stages of life cycle of an organism.

(A) Egg stage

$$\frac{\Delta \ Energy \quad \inf \ low}{\Delta \ time} : 0 < \frac{\Delta \ Outflow}{\Delta \ time}$$

energy inflow is zero or less.

(B) Larval stage

$$\frac{\Delta \ \inf \ low}{\Delta \ time} < \frac{\Delta \ Outflow}{\Delta \ time}$$

energy inflow is less than outflow.

(C) Reproductive stage

$$\frac{\Delta \ \inf \ low}{\Delta \ time} > \frac{\Delta \ Outflow}{\Delta \ time}$$

energy inflow is more than outflow.

(D) Not growing adult male

$$\frac{\Delta \ \text{inf} \ low}{\Delta \ time} = \frac{\Delta \ Outflow}{\Delta \ time}$$

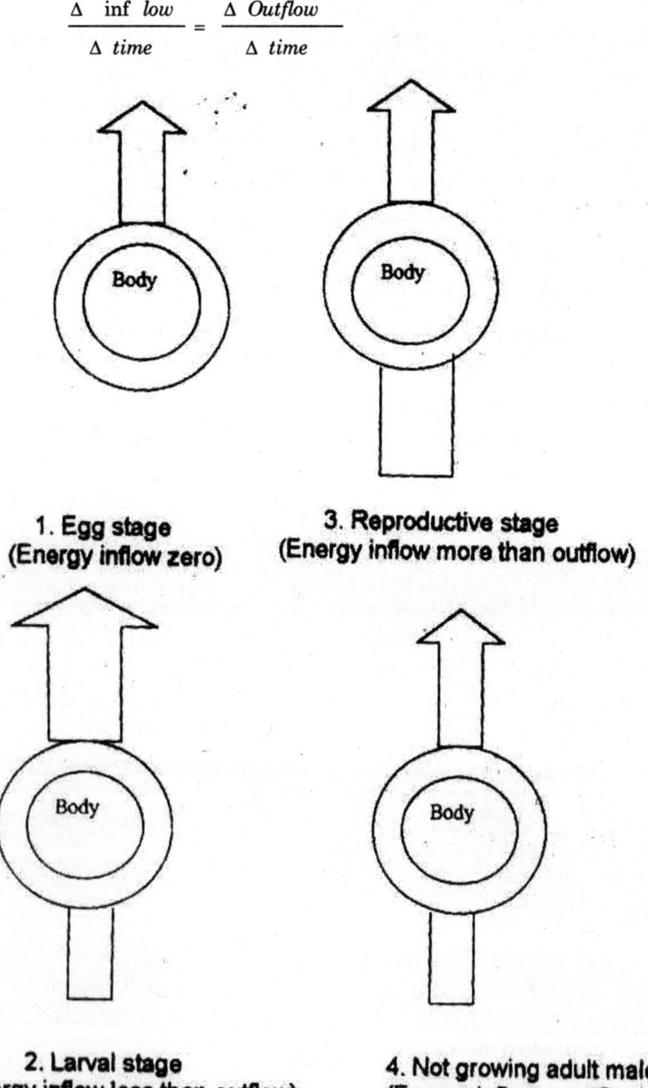

Energy flow diagram in Trilobium

The energy flow diagram in heterotrops as per the identified energy parameters can be represented as follow :

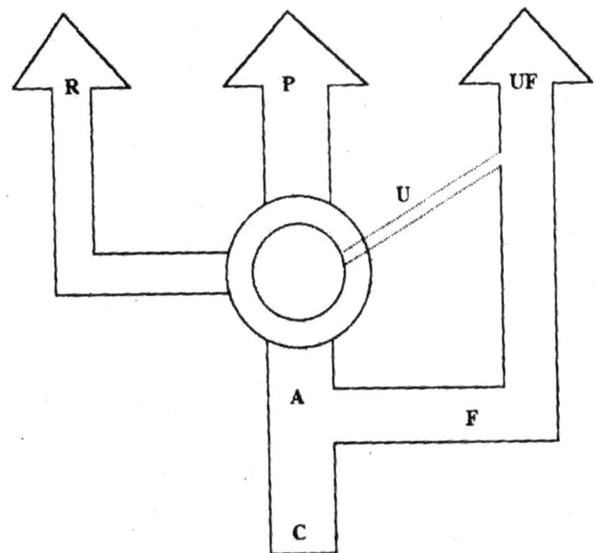

Energy flow diagram in Heterotrophs

Such an energy budget for a moment in time during the animals life cycle gives the instantaneous budget. But this will not be true. Since the unit time involved does not approximate to zero, at the time of measuring the true rate of consumption $\left(\frac{dc}{dt}\right)$.

Hence $\frac{dc}{dt}$ will not be equal to assimilation $\frac{dA}{dt} + \frac{dF}{dt}$ (defecation). Because the animal may feed at infrequent intervals or if the food is retained in the digestive tract for variable period of time. Thus, the useful measurement is the $\frac{\Delta C}{dt}$, $\frac{\Delta A}{dt}$ and $\frac{\Delta F}{dt}$. Howlong `t' should be, depend upon the feeding biology of the species. It my be either

t_C -- the period of feeding cycle

t_F -- the period of defaecation cycle.

t_R -- the period of respiratory cycle.

2.4.2 Energy budget Dynamics

The standing crop of aquatic ecosystem is represented by its population Biomass (B) or its numerical density (N). Assuming that the population is in its steady state (with constant age composition, growth rate, natality and mortality) for a defined period, then the equation of energy balance must be estimated for

(i) the survivors during the period of study

(ii) those died or were born during the period (commonly assumed that these lived for half of the period in question).

Therefore, the equation of energy balance can be obtained not only from the populations inflowing energy as consumption (C) and the outflowing (R, F and U) energy but also include the energy stored in the form of reproductive materials, body growth and the elimination of energy due to predation and decomposition during the defined period. Thus create a very complex situation even with steady state population which are probably rather rare in nature. But it is expected that the population biomass or numerical density may be increasing due to recruitment or decreasing due to predation. Still further complications are seen in certain species of fish because of coprophagus habit (eating on dungs) or excreta of grass carp and in some insects which even ingest their own fecal matter, what is termed as refaecation or because of cannibalistic habit of certain fish. So the energy balance in a species population with cannibalism and refaecation in a defined period of time can be represented as follows.

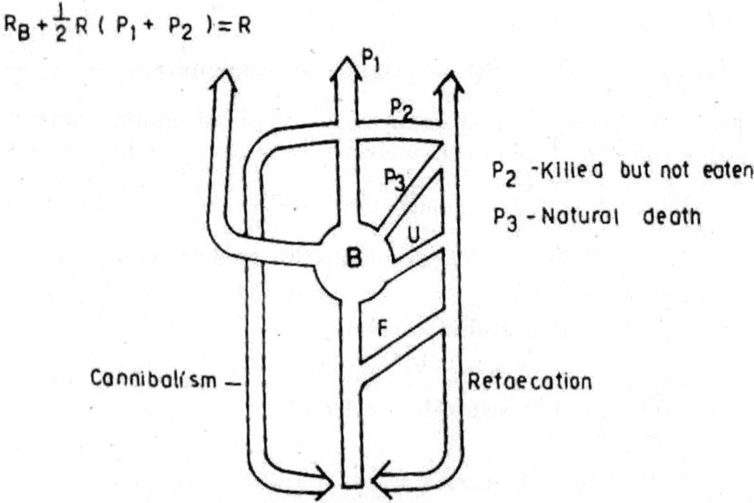

Energy balance in a population with cannibalism and refaecation

If cannibalism and refaecation is not seen, then the energy balance in a species population for a defined period of time can be represented as follows :

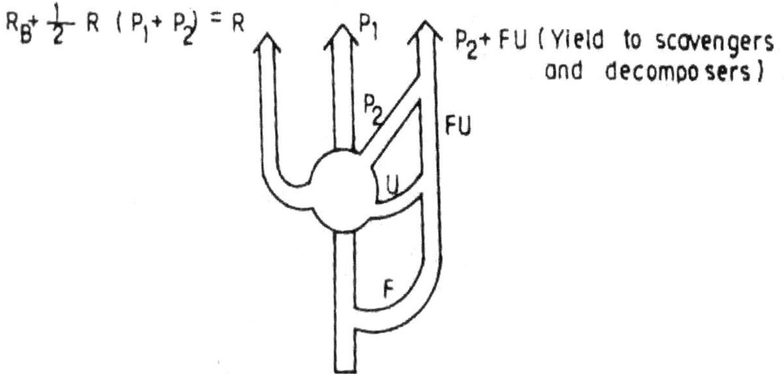

Energy balance in a species population

2.5. Cumulated Energy Budget

Therefore, to know the total energy budget dynamics, one must consider the cumulative energy budget. This was first proposed by Klekoswski *et al.,* 1967.

In order to have a clear idea of about cumulative energy budget, the following considerations were taken into account :

(i) Cumulated from the beginning of the life cycle (To)

(ii) to the end of life (Tn)

So this can led to get the common idea of

(i) cumulated food consumption from To to Tn

$$C_c = S_{To}^{Tn} \; C\,(T)\,dt \;\; \Omega \;\; \sum_{To}^{Tn} C$$

(ii) cumulated respiration from To to Tn

$$R_c = S_{To}^{Tn} \; R\,(T)\,dt \;\; \Omega \;\; \sum_{To}^{Tn} R$$

(iii) cumulated unassimilated food (Faeces from To to Tn)

$$F_c = S_{To}^{Tn} F(T)\, dt \;\Omega\; \sum_{To}^{Tn} F$$

(iv) cumulated production

$$P_c = S_{To}^{Tn} P(T)\, dt \;\Omega\; \sum_{To}^{Tn} P$$

This production is for both body growth and the reproductive products if any.

$$P_C = B_g + \sum_{To}^{Tn} P_r$$

Total production = (Body growth + reproductive grwoth)

Pc = Bg + Br from To to Tn.

(v) cumulated assimilation from To to Tn

= Ac = Pc + Rc

But the difficulties arise with the calculation of cumulative budget for one generation. For example, the fertilised egg belong to one generation but the cost of their production was paid for by the parents belonging to previous generation. Therefore, in cumulative budget, one must reflect the inclusion or exclusion of this egg production cost. However, to distinguish either inclusion or exclusion of egg production cost, the cumulative budget can be expressed either intrageneration cumulative energy budget where the budget exclude the cost of egg or young production or the intergeneration budget where the budget include the cost of egg or young production.

2.5.1 Efficiencies

In instantaneous energy budget, the parameters measured are -

$$\frac{\Delta C}{\Delta t'} \; \frac{\Delta P}{\Delta t'} \; \frac{\Delta R}{\Delta t}$$

So, the instantaneous efficiency can be defined as follows.

(i) Instantaneous coefficient of Assimilation efficiency $= \frac{A_i}{C}$

(ii) Instantaneous coefficient of consumed energy for growth

$$= \frac{P_i}{C_i} = K_1 i$$

(iii) Instantaneous coefficient of assimilated energy for growth

$$= \frac{P_i}{A_i} = K_2 i$$

In cumulative energy budget, the efficiency can be defined as follows,

(i) Cumulative coefficient of assimilation efficiency $= \frac{A_c}{C_c}$

(ii) Cumulative coefficient of utilization of an individual energy for growth which may be

$$\frac{P_C}{C_C} = \frac{W_g + W_r}{C_c}$$

(a) Body growth (Wg)

(b) reproductive growth (Wr)

(iii) Cumulative coefficient of utilisation of Assimilated energy for growth.

$$\frac{W_g + W_r}{A_c} = K_2 c$$

Cumulative energy budget provides the sums of the bioenergic characteristics of species which converts the biomass of food consumed throughout its life cycle in to biomass of body growth or biomass of reproductive products. It also measures the accumulated energy for growth and reproductive products. Hence, cumulative efficiencies are better indices than instantaneous efficiencies of species.

That too the cumulative energy budget provides information that how much energy is required for maintenance and production.

2.5.2 Biomass increment

The gross secondary production is understood as the sum of increments of all the individuals together with their reproductive product and other organic products present in population. This can be calculated for a series of intervals during the active existence of population or for a whole year. Some individuals may die during sampling interval. Therefore, their production can be taken in to consideration up to the moment of death. The production of one age group of the population for the sampling interval -

Ln-1 to Ln = Δt is expressed by the formula

$$P_{t\,n-1}^{\;-tn} = \left[N'tn \cdot \frac{\Delta W'}{\Delta t}\right] + \left[\frac{\Delta N'}{\Delta t} \cdot \frac{1}{2} \cdot \frac{\Delta W'}{\Delta t}\right] = 1$$

where $P_{t\,n-1}^{\;-tn}$ = production for the age group for the time interval Δ t

N'$_{tn}$ = number of individuals of the age group surviving at the moment t_n.

$\frac{\Delta N'}{\Delta t}$ = number of individuals of the age group that died during the interval Δ t.

$\frac{\Delta W'}{\Delta t}$ = average weight (or calorific) increment of the survivors of the age group during the interval Δ t.

If more than one age group is present as seen in rotational culture system during the sampling interval Δ t, their production can be calculated on similar manner and summed to give the population products for the concerned interval. Instead of weight increment and number decrements, the determined growth rate and mortality rate can be applied in the above formula.

Hence, the above formula can be written as -

$$P_{t\,n-1}^{\;-tn} = \left[N'tn \cdot \frac{\Delta W'}{\Delta t}\right] + \left[\frac{\Delta N}{\Delta t} \cdot \frac{1}{2} \cdot \frac{\Delta W}{\Delta t}\right] = II$$

Where,

$P_{tn-1-tn}$ = Total population production for the time interval Δt.

N$_{tn}$ = Number of individuals in the total population surviving at the moment t_n

$\frac{\Delta N}{\Delta t}$ = Number of individuals in total population that died during the interval Δt.

$\frac{\Delta W}{\Delta t}$ = Average weight (calorific) increment per survivors of each age group in the total population during the interval Δ t.

By interrelation between above two formulaes (I and II)

$$N' + N'' + ... = N, \frac{(\overline{W}' N') + (\overline{W}'' N'') + ...}{N' + N''........} = \frac{B}{N} = \overline{W}$$

Where `B' is the Biomass.

$\overline{W}$ is the average population increment

N is the survivors.

This formula will provide rough estimates of population production. But will be difficult to apply in population with overlapping generations.

Therefore, in overlapping population, the total population production can be calculated for a time interval to–tn which consists of more than one sampling interval Δt. Therefore, it will be necessary to sum of the period of growth of the population biomass for the whole time to–tn. Then the successive period of growth can be summed.

$$\overset{tn}{\underset{to}{S}} p \left|t\right| dt \; \Omega \; \sum_{to}^{tn} p$$

$$= \sum_{to}^{tn} \left[N_{tn} \cdot \frac{\Delta \overline{W}}{\Delta t} \right] + \left[\frac{\Delta N}{\Delta t} \cdot \frac{1}{2} \frac{\Delta \overline{W}}{\Delta t} \right]$$

2.5.3. Biomass eliminated

The total energy contained in the dead individuals together with their dead reproductive and other non-living organic matter products provides the informations on the sum of eliminated biomass. This represents the population energy which is transferred to other trophic levels by predation or decomposition.

Therefore, the eliminated biomass as production expressed in equation is as follow -

$$P_{tn-1} - t_n = \frac{\Delta N}{\Delta t}\left[\frac{\overline{W}_{tn-1} + \overline{W}_{tn}}{2}\right] + \Delta B_r$$

Where,

$\frac{\Delta N}{\Delta t}$ = Number of individuals in the population that died during the interval Δt.

W_{tn-1} and Wtn = Average weight (calorific) of individuals at the time t^{n-1} & t^n.

B_r = Biomass of reproductive output dead during Δt and other non-living organic matter products loss by all number of population during the interval Δt.

Hence, the total population production yielded to other trophic levels during longer period ($t_o - t_n$) will be –

$$\mathop{S}\limits_{to}^{tn} p \left| t \right| dt \;\Omega\; \sum_{to}^{tn} \quad p \sum_{tn}^{to} \left[\frac{\Delta N}{\Delta t} . \frac{W_{tn-1} + W_{tn}}{2} + \Delta B_r\right]$$

3. WATER QUALITY AND SOIL CONDITION OF FISH PONDS

Maintenance of a healthy aquatic environment and production of sufficient fish food organisms (plankton) in ponds are two factors of primary importance for successful pond culture operation. It is believed that water is the primary requisite to support aquatic life where as soil is essential to withhold it. To keep the aquatic habitat favourable for existence, physical and chemical factors like temperature, turbidity, colour, odour, pH value, dissolved gases-like O_2, CO_2 and also reducing gases like H_2S and CH_4 working lethal of fish life, will exercise their influence individually or synergetically, while the nutrient status of water and soil, play the most important role in governing the production of plankton organisms or primary production in fish ponds. The soil besides retaining water, depending on the texture, the bottom soil governs the storage and release of nutrients to the overlying water through various chemical and biochemical processes for biological production in the environment. On the otherhand, water in contact with bottom soil acquires nutrients from the soil, atmospheric gases and absorbs solar energy in the form of radiation essential for the activities of aquatic animals. This interdependence between the soil and water maintain a chemical equilibrium in the environment preventing wide fluctuation in environmental conditions deterimental to aquatic life and thereby influence the production, decomposition and consumption processes in the system. Rating of fish ponds on the basis of these factors is a difficult problem. Nevertheless, from a study of a large number of ponds under diverse physical and chemical conditions it is possible to arrive at some broad generalisation which can be gainfully used by aquaculturists. It may be remembered that different group of fishes behave differently as to the suitabilities of environmental condition and food habit. As the major carps like Catla, Rohu etc. are the widely cultivated fast growing fishes in India, these have been always used as standards except when otherwise mentioned.

3.1 Water

The physical conditions of water is greatly influenced with depth, temperature, trubidity and light. These constitute the more important physical parameters on which the productivity of a pond depends.

3.2 Depth

Depth of a pond has an important bearing on the physical and chemical qualities of water. Depth determine the temperature, the circulation pattern of water and the extent of photosynthetic activity. In shallow ponds, sunlight penetrates up to the bottom, warms up the water and facilitates increase in productivity. Ponds shallower than 1 meter get overheated in tropical summers inhibiting the survival of fish and other organisms. Generally a depth of about 2 meter is considered congenial from the point of view of biological productivity of a pond.

3.3 Temperature

Water termperature generally depends upon climate, sunlight and depth. That too, the intensity and seasonal variations in temperature of a water body have a great bearing upon its productivity. All organisms, including fish, posses well defined limits of temperature tolerance with the optimum laying some where in between. The temperature in fish ponds is generally minimum during the early hours of morning and reaches the maximum value in the afternoon showing diurnal fluctuations. The seasonal fluctuations of temperature of water influences greatly on the ingestive variation of fish. In winter months, the temperature of water compared with that of summer months are greatly different and variable. Such a wide fluctuations of water temperature in summer day does not have any direct adverse effect on the fish. But in ponds with high organic content in bottom mud causes large scale mortality of fish in summer months in early morning if the surface water is suddenly cooled by a shower of rain or cold wind. This is because, the bottom layer of anaerobic decomposition zone with reducing gases distributes itself throughout the volume of water and even in relatively oxygen rich surface layer of water suffer from oxygen depletion. That too, the increased rate of biochemical activity of microbiota during increased temperature are very significant. Moreover, the release of nutrients by decomposition of organic matter at bottom is more at higher temperature with consequent increase in the nutrient status of water.

Compared to the yields of fish in ponds in temperate zones, the natural water in tropical areas generally show a higher production due to more heat budget in the pond system. Apart from these, temperature plays very important role in physiological processes for breeding mechanisms in fish both under natural and artificial conditions.

All biological activities like ingestive variations, reproduction, movement and distribution are greatly influenced by water temperature. The chemical changes in both soil and water are greatly influenced by temperature. Decrease in dissolved oxygen is directly related to increase in temperature.

Fish display great variability in their tolerance to temperature. Indian major carps usually tolerate wide range of temperature and are called eurythermal. Fishes of temperate region can survive under ice in winter but those of tropics can not stand such low temperature. *Tilapia mossambica* can not stand temperature below 8.9°C. The Grass carp can withstand temperature up to 40°C. The silver carp shows sign of distress at 40°C. Indian major carps can thrive well in the temperature range of 18.3 – 37.8°C and temperature below 16.7°C and above 39.5°C prove fatal to them. The upper limit of temperature tolerance of Air breathing fishes like *Anabas, Channa, Heteropneutes, Clarias* etc. lies between 39-41°C.

3.4 Turbidity

The turbidity of water bodies may be either due to suspended inorganic substances like silt, clay and planktonic organisms.. Turbidity of water varies greatly with the nature of basin and inflowing sediments. Ponds with clay bottom are likely to have high turbidity. Turbidity restricts the penetration of light, therefore, reduces the photosynthetic activity hence acts as a limiting factor for productivity. That too, the suspended particles causing turbidity also adsorb nutrients like phosphate (PO_4), potassium (K) and nitrogen (N) in their ionic form making them unavailable for plankton production.

The turbidity tolerance of different cultivable species has not been studied systematically. Turbidity during heavy floods increases to the value of 2000 to 5000 ppm in river Ganga. The culturable plain dwelling fishes even tolerate to wide range of turbidity. Fishes dwelling in turbid waters have reduced eyes and mucus secreted by their skin possesses the property of rapidly sedimenting the suspended particles in water. The coagulating characteristics of mucus is important as a means of protecting the gills from choked by suspended particles. One or two drops of mucus secreted by the skin of *Pisodonophis boro* to half a litre of turbid water causes sedimentation in 20-30 seconds.

3.5 Light

Light is another physical factor of importance. Availability of light energy to a fish pond greatly influences its productivity as primary production or synthesis of carbohydrate is a photochemical process energised by light. Penetration of light of water phase is determined by turbidity which is measured optically and represent the resultant effect of several fctors such as : suspended clay and silt and disperson of planktonic masses. In shallow ponds, where light is able to penetrate up to the bottom, plants may occupy the entire basin and their luxuriant growth provides a cover to the water column. Such shading of aquatic vegetation is often advantageous in pisciculture for the control of algal bloom.

Chemical parameters

Among the chemical factors influencing aquatic productivity, the pH value, alkalinity, dissolved gases like Oxygen, Carbondioxide and dissolved inorganic nutrients like P, N are considered to be important.

The procedure generally adopted is to determine these chemical factors for a large number of productive, moderately productive and un-productive ponds. These indicate the ranges and average for the individual constituents. This obviously gives inconclusive result, because of the effect of one constituent is sometimes obscured by the synergetic influence of other chemical factors. Therefore, it is better to have a number of conventional ranges for different chemical factors and to determine the percentage of productive, un-productive and medium productive ponds in each of these ranges. With this procedure, the favourable ranges for the pH value, alkalinity, dissolved oxygen, dissolved inorganic nitrogen and phosphorous and N/P_2O_5 ratio have been suggested.

3.6 The pH value (Hydrogen ion concentration)

The pH value of the water is defined as the logarithm of the reciprocal of hydrogen ion concentration. It may be expressed mathematically as pH = $Log \frac{1}{(H)^+}$ where (H^+) is the amount of hydrogen ions in a solution in moles per litre. There is 0.0000001 of hydrogen ions in a litre of pure water. The pH value of neutral water is therefore 7. Increase in H^+ ions results in higher acidic pH value. The pH value is not affected by neutral salt (NaCl) but is determined by the absorbed CO_2. The pH value of pond water undergoes a diurnal change, it is being alkaline in mid afternoon and acidic just before day break.

According to some limnologists the largest fish crops are usually produced in water which is just on the alkaline side of between 7.0 and 8.0. The limit above or below which pH has a harmful effect is given as 4.8 and 10.8. Others have said that a weak alkaline reaction (pH 7.0 to 8.0) has been found in most productive ponds and highly acidic waters are undesirable.

In a study, it was observed that the ponds in the acid soil zones of Manipur and Assam which had a pH value below 6.0 and the ponds in Andhra Pradesh is alkaline soil showing water pH 8.5. Generally, the normal range of pH in freshwater ponds are between 6.5 to 8.5, unless contaminated by acidic or alkaline water. Average fish production is expected when pH value of a fish pond ranges between 7.5 - 8.5. Swingle (1967) stated that acidic water reduces apetite and growth of fish. The toxicity of H_2S, CH_4 and other heavy metals to fish is increased in increasing pH. Acid water influences other aquatic plants also. Fish gets prone to attack of parasites and diseases in acid waters.

3.7 Alkalinity, Carbonate, Bicarbonate and Free Carbondioxide

Alkalinity or acid combining capacity of natural fresh water ponds is generally caused by carbonate (CO_3) and bicarbonate (HCO_3) or hydroxides of calcium, magnesium, Na, K, NH_4 and Fe, calcium being form the major constituent. These along with dissolved CO_2 in water form an equilibrium system which is of primary importance in the ecology of the environment.

The source of CO_2 in water is from (a) Atmosphere, (b) Respiration (c) Bacterial decomposition, (d) Inflowing ground water, (e) within the water itself in combination with Ca and Mg. The free CO_2 which is necessary to retain calcium in solution in the form of Ca H CO_3 is called the equilibrium of free CO_2. This free CO_2 is contained in half bound state as HCO_3 and bound state as CO_3. Both HCO_3 and CO_3 are being together called combined CO_2.

CO_2 in natural water are reciprocally related to acid base relationship in the medium. Rain water contains nearly 0.6 mg/l of dissolved carbondioxide and only a small fraction of CO_2 forms carbonic acid which dissociates into bicarbonate and carbonate ions with consequent release of hydrogen ions.

$$CO_2 + H_2O \text{ - - - - - } H_2\,CO_3$$

$$H_2CO_3 \text{ - - - - - - } H^+ + HCO_3$$

$$HCO_3 \text{ - - - - - } H^+ + CO_3$$

Thus, an equilibrium between the reactants and products is always maintained in water and any change in the concentration of any one of the component cause either transformation of one form to the other or precipitate to restore the equilibrium and thereby prevents wide variations in pH value of water and does not allow it to drop below 4.5 and rise above 8.3.

Photosynthesis during day and respiration at night cause a net gain and loss of CO_2 respectively in the environment and thereby cause a diel fluctuation in the CO_2 concentration in water.

Bicarbonate and carbonate are the major constituent of pond water in their concentrations are expressed as total alkalinity. The change in the concentration of CO_2 results in the alternation in the proportion of HCO_3 and CO_3 concentrations in water. Swingle (1967) suggested total alkalinity as universal adoption. He suggested the titration and points to correspond to the following pH values. pH 5.1 when total alkalinity is 30 ppm as $CaCO_3$, pH 4.8 when total alkalinity is 150 ppm as $CaCO_3$, pH 4.5 when total alkalinity is 500 ppm $CaCO_3$. Usually free CO_2 and HCO_3 is encountered in water of pH 4.5-8.3. HCO_3 and CO_3 alkalinity is encountered in waters of pH ranging from 8.4 - 10.5. Hydroxide alkalinity generally occurs in polluted water.

In general, calcareous water with alkalinities more than 50 ppm are most productive. Waters with an alkalinity less than 10 ppm rarely produce large crops, water intermediate between these 10-50 ppm may produce useful results. Alikunhi (1957), reported that in highly productive water, the alkalinity is ought to be over 100 ppm. However, the range of alkalinity as 0.0 - 20.0 ppm for low production, 20-40 ppm for medium production and 40 - 90 ppm for high production is considered. All the ponds above 90 ppm of total alkalinity have been found to be productive. Some have reported that the range of alkalinity may 20-50, 50-100, 100-200 and above 200 ppm in different ponds studied. About 60-70% of the ponds were productive and the rest being unproductive. This shows that even above 20 ppm of total alkalinity can not work as a factor of good productivity. Its influence is probably masked (obscured) by the other more important factors such as dissolved nitrogen and phosphorous.

3.8 Dissolved Oxygen

Among the chemical substances in natural waters, oxygen is probably one of primary importance, both as a regulator of metabolic processes of plant and animal community and as an indicator of water condition. The pond water receives oxygen mainly through (1) interaction of atmospheric air on the surface water of pond, (2) by photosynthesis. The volume of dissolved oxygen diffused in water independent with (1) temperature (2) partial pressure of oxygen in contact with water surface and (3) concentration of dissolved salts. Further oxygen liberated during photosynthesis is influenced by temperature, light, concentration of dissolved salts, turbulence of water including the abundance of plants. Transfer of oxygen from air to water and vice versa occurs with the unsaturation and saturation of this element in water respectively. The intensity of turbulence regulate the net transfer of this element both ways. Thus, photosynthesis, respiration and slow rate of diffusion cause a diel fluctuation of dissolved oxygen in water and accordingly remain minimum during early morning and gradually increase to attain maximum in the afternoon and declines thereafter during night. It is possible that below 3.0 ppm of DO_2, asphyxia from low O_2 can be expected and to maintain a favourable condition for a varied warm water fish fauna, 5.5 ppm of DO_2 is required. Sometimes fishes congregate near the surface for respiration in such low DO_2 ponds. For average of good production, ponds should have DO_2 concentration above 5.5 ppm. It may be mentioned that, very high concentration of DO_2 leading to a state of supersaturation sometims lethal to fish fries during rearing of spawn in nursery ponds.

3.9 Total Hardness

In principle hardness is defined as the total of soluble Calcium and Magnesium salts present in the water medium, which is expressed as its $CaCO_3$ equivalent. In most natural waters, usually HCO_3 anions are associated with Ca, Mg, Na and K cations. Waters contaminated with ocean salts or from dryland area containing SO_4 and chloride ions are associated with Ca and Mg. Therefore, the total hardness of water includes SO_4 and Cl besides CO_3 and HCO_3. Usually bicarbonates of Ca and Mg cause temporary hardness. Permanent hardness of water is due to soluble Ca and Mg carbonates and salts of inorganic acids ($CaSO_4$). Swingle (1967) has suggested a total hardness of 50 ppm $CaCO_3$ equivalent to be the dividing line between soft and hard water. He reported that, the pond water having a hardness of 15 ppm or above are satisfactory for growth of fish and do not require addition of lime, but water having hardness less than 11 ppm require liming for higher production of fish. Water having hardness less than 5 ppm $CaCO_3$ equivalent cause slow growth, distress and eventual death of fish.

3.10 Dissolved Solids

In natural aquatic systems, water contain both inorganic and organic dissolved solids. Inorganic solids in solution contains anions of CO_3, Cl, SO_4, PO_4, NO_3 in combination with Ca, Na, K, Mg, Fe etc. Organic compounds are in the organic state of P, N, Sugar, acids and vitamins. Both organic and inorganic dissolved solids in pond water varies qualitatively and quantitatively with the season. The edaphic relationship that contributes to the productivity of the water is dependent to the total concentration of dissolved solids in a water. Sreenivasan (1967) reported that the electrical conductivity above 400 micro mhos did not limit productivity but productivity did not increase proportionately with conductivity.

3.11 Dissolved Nitrogen and its Compound

The importance of dissolved nutrients especially nitrogen is well recognised. It is an important element influencing the growth of phytoplankton in aquatic environment. As a constituent of protein. Nitrogen occupies a highly important place in aquatic ecosystem. Ponds having dissolved nitrogen below 0.1 ppm does not indicate productive condition, while the range of 0.1 - 0.2 ppm an average production is expected but above 0.2 ppm is considered favourable. However optimal limit of nitrogen can be in the range of 0.3 - 1.3 ppm.

The elemental nitrogen in water is derived from the atmosphere. Besides solubility of atmospheric nitrogen, fixation of atmospheric nitrogen through biological and meteorological processes and mineralisation of organic matter are the other sources of nitrogen enrichment in water. In water nitrogen is soluble to the extent of 12 mg/litre at 25°C. Therefore, the solubility of nitrogen in water varies with temperature and pressure. Supersaturation of nitrogen can occur at air-water interface which at time cause gas bubble disease in fish, a physiological condition caused by supersaturation of oxygen.

Nitrogen compounds in natural waters may be derived from outsides source (allochthonous) or within the body of water itself (autochthonous). In allochthonous source, nitrogen compounds like nitrates. Ammonia etc. are carried by rain, surface run-off, inflow of ground waters, seepage, springs etc. Nitrogen compounds are produced by plants and animal tissues. Many blue green algae, such as Anabaena, Nostoc etc. secrets extracellular nitrogenous compounds including polypeptides, amides and amino acids. The dissolved nitrogen in water vary seasonally indicating the intensity of productivity of a pond.

3.12 Phosphorus

The occurrence and abundance of phosphorus depends largely on geochemical conditions of the basin and surrounding areas. The main source of phosphorus in natural water bodies come from the phosphorus bearing rocks (e.g. apatite) and from the leaching of soils of the catchment area by rain. The other source of phosphorus in water body is liberated from dead plankton (animalcules) by the bacterial action. These planktons are broken down by autolysis. Hence, the phosphorus bound protein (nucleo protein) in the cell gets disseminated in water by the action of phosphatase enzyme.

Ecologically phosphorus is often considered as the most critical signle element in the maintenance of aquatic productivity. Moyle from a study of a large number of lakes and ponds gave the phosphorus fertility range as 0.00-0.02 ppm (low productive), 0.02-0.05 ppm (fair productive), 0.05 - 0.10 ppm (good productive) and above 0.20 ppm excessive.

Some workers have emphasized that besides the absolute concentration, the ratio of nitrogen and phosphorus concentration likely to influence aquatic productivity. Swingle and Smith (1938) estimated that nitrogen and phosphorus were being utilised in the plankton growth at the ratio of 3 : 1 to 6 : 1. However, excess of phosphorus in open water is a sign of heavy organic pollution.

3.13 Redox-Potential

Redox potential is an expression of the oxidising or reducing power of solution. This power is dependant on the nature of the dissolved substances which is considerably influenced by temperature and pH. Redox-potential generally lie between 0.4-0.5 volt. The value of 0.52 at 25°C is the best value of aerated water at 1 atmospheric pressure.

Calcium, Magnesium, Sodium and Potassium occurs in natural water bodies in combination with CO_3 ions, especially (Ca or $MgCO_3$) and chloride or sulphates especially (NaCl or Na_2SO_4).

The importance of Na and K is well recognised. As well SO_4 is ecologically important for growth of plants and its short supply may inhibit the development of plankton.

Silica is most abundant in sedimentary rocks and occurs in higher concentration in water located in such region. Silica is utilised by most of diatoms, cysts of yellow brown algae and spicules of sponges.

3.14 Trace Elements

Trace elements such as manganese (Mn), copper (Cu), zinc (Zn), aluminium (Al), gallium, molybdenum, nickel, cobalt and also uranium are of great significance in influencing productivity. The deposition of Magnanese in some cell walls of algae e.g. desmid and theca of Trachelomonas indicate the necessity of algae to utilise these trace elements. Copper, Zinc, Aluminum and Gallium remain either, ionic or collidal form gainfully import to the productivity of water. Molybdenum an important element for nitrogen fixing in Cyanophyta (Ex. *Anabeana* species). Similarly, Nickel and Cobalt are required by various blue green algae (euglenineae, Chlamydomonas etc.). Cobalt is used as a growth promoter as it is present as a constituent of vitamin B_{12}.

3.15 Organic Matter

The organic matter found in natural water bodies are in organic phosphorus, organic nitrogen, carbohydrates and vitamins. The organic matter may be derived in water bodies either allochthonously or autochthonously. The autochthonous source of organic matter is derived by the decomposition of plankton. Plankton contain 24% crude protein with a C : N ratio of about 12 : 1 and does not impart brown color to the water. The allochthonous source of organic matter is derived either from the basin or in the catchment area of influent streams, contain about 6% crude protein with C : N ratio of 50 : 1 and imparts strong brown color to the water.

Soil

Soil plays an important role in regard to the fertility of fish ponds. Types, characteristics and chemical conditions of soil influences the pond productivity. The Geographical boundary of India consist of 8 major heads of soils.

(1) Alluvial (2) Black (regur)

(3) Red (4) Laterite
(5) Forest (6) Desert
(7) Saline alkaline (8) Peat.

Alluvial soil is of two types : (a) Khadar and (b) Bhangar types. Khadar is newer alluvium of sandy, light coloured and less kankary composition where as Bhangar is older alluvium of clay, dark coloured and full of Kankar. Alluvial soil though deficient in nitrogen, contain adequate quantity of alkalies and phosphoric acid.

Black soil is called regur. It is derived from two types of rocks, Deccan and the Rajmahal trap and ferruginous gneisses and Schists. These soils contain fine grains, dark in color with high Ca and $MgCO_3$, Fe, K, Mg, Al but poor in P, N and organic matter. Black soil is highly imperivious to water and become sticky when wet. The soil contain high quantity of mineral.

Red soil are generally poor in N, P, Iron oxide, Lime and humus with high mineral. On the morphological point of view, red soils are either (1) Red loam or (2) Red earths. Red loam is characterised by agrillaceous soil having cloddy structrue. Red earth is characterised by loose and friable top soil, rich in secondary concentrations.

Laterite and Lateritic soils are compact and consists of vesicular rock. These soils are poor in phosphorus, potassium, calcium and nitrogen.

Forest soils are very rich in organic matter as derived from forest growth. These soils are either formed under acidic conditions with the presence of humus of slightly acid or neutral conditions with a high base status.

Desert soils are mostly derived from brown sand. These are poor in organic matter but rich in soluble salts.

Saline soils are rich in Nacl and Na_2SO_4 where as alkaline soils are rich in Na_2CO_3. Bacterial activity and organic matter are insignificant in such soils.

Peat soils can retain large quantities of water soluble alkali salts. These soils impart blue or blue-black color due to ferrus iron. It also contain considerable amount of organic matter.

The physico-chemical properties of pond water is more or less a reflection of the properties of the bottom soil. Productivity to waters are produced by chemical and biochemical process from raw materials consisting of organic matter and mineral components of clay fraction of the soil. Not only the total and available quantity of raw materil but also the requirement of the organisms of nutrients are essential for productive pond soil.

In this respect the major chemical factors of importance are pH, total nitrogen, total phosphorus, Organic carbon, C/N ratio, available N_2, available P, K and exchangeable calcium.

3.16 Hydrogen ion concentration (pH)

The pH value of soil depends on various factors. In pond muds, the decomposition of organic matter is slow due to lack of oxygen. The product formed of decomposition (in anaerobic condition) are H_2S, CH_4 and short chain fattyacids. These compounds make the condition acidic and leads to less productive. The release of essential nutrients at soil-water interface is greatly hampered due to low pH. The pH range of 5.5 is highly acidic, 5.5-6.5 moderately acidic, 6.5-7.5 nearly neutral, 7.5-8.5 moderately alkaline, 8.5 above highly alkaline. However, moderately alkaline pH for soil has been considered favourable for fish ponds.

3.17 Phosphorus

The importance of available phosphorus in soil for increasing productivity is well recognised. The total phosphorus in soil is not so important owing to the fact that PO_4 ions in the form of calcium phosphate is insoluble in alkaline condition and (Ferric Iron phosphate) Fe_2 $(PO_4)_3$ and Al_2 $(PO_4)_3$ are insoluble in acidic conditions, rendering the phosphorus ion unavailable to the water. The phosphorus in soil is both inorganic and organic forms. The inorganic phosphorus in the soil can be classified in to four groups : (1) Calcium phosphate, (2) Aluminium phosphate, (3) Iron phosphate and (4) reductant soluble phosphate. Calcium phosphate remain as apatite but dicalcium, monocalcium and octacalcium phosphate also exist in small amounts or as transitional form.

The organic form of phosphorus compound in the soil occur in three groups : (1) Phytin and Phytin derivatives, (2) Nucleic acids and (3) Phospholipids. The organic form constitutes about 35-40% of the total phosphorus content of the soil.

However, the available soil phosphorus (P_2O_5) below 3 mg/100 gm (30 ppm) as poor productivity, 3-6 mg/100 gm (30-60 ppm) as average, above 6-12 mg/100 gm (60-120 ppm) as high productivity and above 12 mg/100 gm (120 ppm) as excess are indicated.

3.18 Nitrogen

Nitrogen in soil is present mostly in organic forms as aminoacids, peptides and easily decomposible proteins where as the inorganic forms N^{+}_{4} and NO_3 are utilised by green plants. The conversion of complex organic forms of nitrogen to simple inorganic forms are carried out in the bottom mud by anaerobic microorganisms. Hence, it is important to know the available nitrogen than the total nitrogen in soil. The range of available nitrogen 50-75 mg/100 gms of soil is relatively more favourable for pond productivity. Loss of nitrogen also occurs in ponds through volatization of ammonia. The cause of volatization of NH_3 are high pH and high temperature in pond environment.

Besides organic form of nitrogen transformation in to inorganic nitrogen and loss of nitrogen in pond environment, some microorganisms, blue green algae, aerobic and anaerobic *heterofia-ophic* bacteria present in the soil and water fix atmospheric nitrogen in to organic nitrogen. The process of mineralization helps in the release of fixed nitrogen in available forms.

3.19 Organic carbon and C/N ratio

Compared to the mineral constituents of the soil, organic compounds are more varied and complex. Microbiologists believe that the bacterial activity depends not only on the carbon content but also on the ratio of C/N in the parent substance. Bacterial activity is low when the ratio falls below 10 : 1 and good when 20 : 1 or higher. The importance of carbohydrates and C/N ratio in NItrogen fixation has been indicated by Neess (1949). Studies indicate that very high organic content is also not desirable for a pond soil. However, organic carbon less than 0.5% may be considered poor, 0.5-1.5% as average while 1.5-2.5% appeared to be optimal for good production.

3.20 Calcium

Calcium is generally present in the soil as carbonate. The deposition of $CaCO_3$ in freshwater are referred as marl. The amount of exchangeable phosphorus in pond mud is inversely related to the marl organic matter. It was however, noted that no marked influence of exchangeable calcium upon productivity could be noticed.

4. NUTRIENT DYNAMICS

Aquatic ecosystems are complex entities. There exists dynamic equilibrium between the tendency to change and the capability of homeostasis. These two conflicting properties of system results in an environmental steady state under natural conditions.

4.1 Biotic structure and dynamics

In all aquatic systems there is a link between producers and the consumers which are of two physiologically different groups of organisms. One is plant life (producer) which builds-up organic matter from inorganic matter and release oxygen during photosynthesis. The second group (animal life, bacteria and fungi) feeds on these products. Both plant and animal life take part in the cycling of biological substances. The total production in terms of biomass in the system depends on the concentration in inorganic nutrients as these nutrients are easily available by the organisms. Basically if there is introduction or withdrawal of any available nutrients, the amount of matter produced will change. This will consequently change the composition of biological communities. Therefore, nutrient dynamics necessiates to establish the nutrient concentration and effect relation on biotic communities in aquatic ecosystem. The relationship can hardly be established by mere knowledge on exact concentration of the substance in water or from the chemical structure alone. Therefore, synergetic action and interaction informations on detailed chemistry assumes greater importance in ecosystem management. This will enable for setting the level of concentration of nutrients as well as the ecological criteria for aquatic habitat for their proper management.

The sustainable fish yield agreeably appears to be based on a dual food web : autotrophic (photosynthetic) and heterotrophic (bacteria and protozoa) Fig. 17. In this process, the inputs and management play an essential factor of production. Besides this, the autochthonous factors play some role in pond ecosystem. However, in aquatic ecosystems plant growth depends on the availability of some 20 nutrient elements of which the majority are required in trace. Oxygen, carbon, nitrogen and phosphorus are considered to be most essential nutrients for plant growth. Short supply of any one of these hampers the growth of plants.

Chlorophyll bearing organisms convert the solar energy into chemical energy in the form of carbohydrates, fats and proteins which get transferred to herbivores. Similarly the nutrients are transferred as well. Therefore, there is a transfer of both energy and nutrients from herbivores to carnivores and finally to the decomposers. Although there is a progressive diminishing of energy in the feeding chain (Trophic level - Fig. 18), the nutrient component is not diminished. In any event, nutrient content of plant and animal biomass is eventually subjected to decomposer activity and the nutrients are released to the environment for reuse. These two ecological processes of energy flow and mineral cycling constitutes the ecosystem dynamics.

Among nutrients, phosphorus and nitrogen are well recognised for pond productivity. Exchange of phosphorus between dissolved (water) and particulate compartments (bottom mud) is a dynamic process. The phosphorus supply and their exchange controls the size of biological communities. If phosphorus is present above growth limiting levels, it may lead to luxury up-take in algae and be stored as polyphosphates. The ratio between cellular carbon and phosphorus are more or less constant although Redfield *et al.,* (1963) reported that the ratio is around 60 : 1 for phosphate limited phytoplankton. The phosphate recycling in the water column is known as internal or metabolic P-cycle and recycling through the sediment (bottom mud) is known as geochemical P-cycle.

Nitrogen budget calculation in nutrient cycling is more complicated as nitrogen may enter the cycle as atmospheric gaseous nitrogen through fixation and leave the cycle through denitrification. Secondly, the organic nitrogen in the form of amides, polypeptides and aminoacids provide an important part of energy available to bacteria and to zooplankton that assimilate algae. Therefore, cyanobacterial nitrogen fixation is generally accepted as the major fctor controlling primary production in N-limited environment. This fixed nitrogen oxidised into Nitrite and nitrate, rapidly by nitrobactor reducing the nitrite toxicity to fish. The up-take of inorganic nitrogen is dependant on carbon sources generally formed from active photosynthesis. The algal nitrogen content is estimated approximately as 10% of its dry weight. After death of algae a rapid mineralisation occur by bacterial action and nitrogen is liberated in the form of NH_3. Hence, NO_3 and NH_3 serve as a nitrogen source for phytoplankton, although NH_3 toxicity is recorded in fish.

4.2 Plankton dynamics Vs. Nutrient availability

Primary productivity is related to nutrient concentration, light and termperature. These light and temperature are treated as exogenous factors which are otherwise called as driving variables. However, nutrient concentrations are linked dynamically for growth. With increasing growth in the system, the nutrient depletion is recorded accordingly. Therefore, addition of nutrients like carbon and nitrogen might greatly stimulate the rate of phytoplankton multiplication. This can not go beyond to the extent of carrying capacity of the ecosystem as the limit set is also dependant on available phosphates. This will result actual biomass production than the predicted biomass production. This differences between the predicted maximal sustainable biomass and the actual values found can be due to nutrient limitation or biomass loss due to grazing food chain or sedimentation. Hence, both phosphorus and nitrogen are required to produce large standing crop of phytoplankton, zooplankton and benthos. Although phosphorus alone can produce some increase in standing crop, nitrogen alone may or may not elicit a response depending on other factors.

4.3 Practical utility of Nutrient application

For practical purposes the difference between total phosphorus and soluble orthophosphate concentration in-water may be used as an index of phosphorus contained in plankton and detritus. Release of phosphate from sediment is an important source of this nutrient in unfertilised ponds. Iron phosphate, Aluminium phosphate and calcium phosphate in aerobic sediments are slightly soluble and a dynamic equilibrium exists betwen phosphate in sediment and the overlying water. Phosphate in mud is also released in sufficient amounts to water when Iron and aluminium phosphate dissociate under reducing conditions. Thus unfertilised ponds with high phosphorus content in mud are expected to be more productive than ponds with low phosphorus in mud. However, phosphorus fertilizer usually be added to pond to maintain high productivity. Calcium phosphate remains in insoluble form in alkaline water making it unavailable to plant growth. For this reason, ammonium phosphate is used as phosphorus fertilizer in ponds with hard and alkaline water. The rate of phosphate uptake by phytoplankton in fertilized ponds is estimated to be 4.5-5.5 times more than the amount of phosphorus released from muds. Thus high rate of primary propductivity necessary for intensive fish culture depends upon frequent application of PO_4 fertiliser to keep the orthophosphate concentration above the equilibrium concentration.

Inorganic form of nitrogen is available for pond productivity. This inorganic form of nitrogen contributes even less than 2% of total nitrogen in aquatic system. Most of the soil nitrogen occurs in organic form hence, not readily available to plankton until mineralised. Therefore, nitrogen fertilizers are advised for new ponds, where pond soils are devoid of humus. The humus is a colloidal state of mud which would produce some inorganic nitrogen itself by photochemical reaction from organic sources. The nitrogen levels indicate the carbon budget of the system. In pond fertilization programmes, the ratio between phosphorus and nitrogen is held to be very important. The most favourable P/N relationship is 1/4. The deficit of phosphorus affects nitrogen utilisation. A sustained level of about 0.25 - 0.5 ppm inorganic nitrogen is considered favourable for fish ponds and application of nitrogen fertiliser in split dose is advisable. Ultimately application of N-fertilisers must be avoided, which may lead to relatively high ammonia release and depress fish growth.

Organic manures favour multiplication of bacteria resulting favourable production of zooplankton. Hence external organic carbon source is apparently essential for high fish yield. Organic inputs serve both as a fertilizer and a potential direct feed. A general concentration of 1.5 - 3.0 mg/litre of carbon in water is a fair indication of productive potential. Addition of huge fresh organic wastes to aquatic ecosystem can create problems due to high oxygen demand from wastes. Hence, accordingly the dose of manuring should be adjusted.

A comprehensive pond fertilisation programmes must aim at maintaining inorganic nitrogen concentration between 0.2 to 0.4 mg/1, dissolved carbon between 1.5 to 3.0 mg/1, and soluble inorganic phosphate between 0.03 to 0.1 mg/1, through appropriate fertiliser and manure preferably at weekly intervals. The basal manural dose or starter dose could be some what in higher side. At this rate of nutrient loading a desired level of phytoplankton standing crop (25-30 mg/m^2 as chlorophyll-a) could be maintained for achieving a good growth rate of fish. As per the above estimate the quantities required per hectare per metre of water spread would be 2-4 kg as N, 15-30 kg as C and 0.3 - 1.0 kg as P for weekly doses.

5. MICRO-ORGANISMS AND NUTRIENT CYCLES

Micro-organisms particularly bacteria play an important role in recycling the organic wastes in the system. With their physiological activities, associated with transformation and regeneration of nutrients, the bacteria emerges as the first link joining the living world with abiotic factors. Hence proper understanding of bacterial population, decomposition of organic wastes, mineralisation nutrient circulations, sediment-water interaction, oxygen consumption and associated environmental conditions are necessary.

5.1 Bacterial community in ponds

Bacteria in a pond ecosystem constitute four distinct populations such as :

(1) Free floating planktonic bacteria

(2) Periphytic bacteria, those adhering to suspended particles

(3) Epiphytic and epilithic forms, those attached to submerged solid supports

(4) Benthonic or sediment bacteria.

However, on ecological point of view bacteria are of two major types such as : (1) autotrophic and (2) heterotrophic.

5.1.1 Autotrophic bacteria

Autotrophic bacteria derived their nutrition solely from inorganic sources. These include : (1) chemosynthetic autotrophic bacteria, *Example* - Iron and Sulphur bacteria. These bacteria obtain their energy by oxidising inorganic compounds without sunlight. (2) Photosynthetic autotrophic bacteria, *Example* - green and sulphur bacteria. They obtain their energy in the presence of sun-light.

5.1.2. Heterotrophic bacteria

Heterotrophic bacteria require organic matters for atleast a part of their nutrition. These include saprophytic bacteria and parasitic bacteria.

Based on oxygen, bacteria can be aerobic, anaerobic, facultative anaerobic and micro-aerophilic. On temperature preference, these are psychrophiles, mesophiles or thermophiles. On the basis of morphology, these are Bacilli, Cocci, Coccobacilli, Filamentous, rods and spreaders. However, on structural basis, they are spore bearing, spore less, capsulated, Bare, Flagellates or aflagellate.

5.2 Role of micro-organisms in nutrient cycling

5.2.1 Carbon cycle

Free CO_2 and bicarbonates become a part of phytoplankton and other aquatic vegetation during the process of photosynthesis. Many chemosynthetic autotrophic bacteria involve in direct fixation of CO_2. The dead organisms and organic wastes undergo aerobic decomposition leading to the formation of CO_2 and aquatic humus. The humus settles at the bottom as a part of silt and a fraction undergo further bacterial decomposition resulting in release of CO_2. The organic matter in the sediment undergo anaerobic decomposition with the formation of methane (CH_4). Hence, the atmospheric CO_2 cycles through phytoplankton and organic matter in the silt decomposes to form CO_2 resulting in repetition of carbon cycle in the system.

5.2.2 Nitrogen cycle

The important steps involved in the nitrogen cycle are: ammonification, nitrification, denitrification and nitrogen fixation. The insoluble organic nitrogen is mineralised by species possessing proteolytic enzymes, that include aerobic and anaerobic bacteria, fungi and actinomycetes.

(1) Bacterial activity cause deamination of protein and ammonia-N is produced. This process is called ammonification.

(2) After ammonification, nitrification is carried out by nitrosomonas in which ammonia is oxidised to nitrite via hydroxyla-mine, dioxyammonia, nitroxyl or hyponitrous acid. Nitrite is oxidised to Nitrate by nitrobactor.

NH_4 Nitrosomonas NO_2^-

NO_2 Nitrobactor NO_3^-

(3) The denitrification of NO_3 to free nitrogen is direct by saprophytic bacteria. The aerobes like *Pseudomonas, Achromobacter, Bacillus* and *micrococcus* act to reduce NO_3 to NO_2 and further to free nitrogen.

(4) The nitrogen input through biological nitrogen fixation is significant, carried out by free living nitrogen fixing bacteria (*Azotobacter* and *Closteridium*) and Blue green algae (*Anabaena, Nostoc* and *Cylindrospermum*).

Dalton and Mortenson (1972) reported that the annual estimates of biological nitrogen fixation is in order of 9.1 x 10^{10} kg and the chemical nitrogen fixation is in the tune of 2.2 x 10^{10} kg. That too, through Haber's and Bosch process, the chemical nitrogen fertilisers required high energy input at a temperature of 300°C and 200-1000 of atmospheric pressure. But through biological nitrogen fixation, much of the energy can be saved. These qualities implies the significance of biological nitrogen fixation in aquatic system for aquatic productivity. In 1862 *Closteridium pasteruneum* is the first reported biological nitrogen fixing bacteria and was confirmed again in 1894. Biological nitrogen fixation is done by Symbiotic nitrogen fixing bacteria such as *Rhizobium,* nodule baceria and non-symbiotic fixing bacteria such as *Azotobactor, Closteridium* etc.

The biological nitrogen fixation is relatively more in water than the sediment and nitrogen fixing bacteria is more in surface sediment than the sediment below 6-8 cm. Even seasonal distribution of nitrogen fixing bacterial populations are reported. It has also been reported that as many as 12 molecules of ATP is required to fix I molecule of Nitrogen by nitrogen fixing organisms. Energy required for fixing, comes through photosynthesis. In aerobic nitrogen fixation, the ATP is generated through oxidative phosphorylation and in anaerobic nitrogen fixation, the energy is generated through glycolysis. However, the factors that effect on nitrogen fixations are temperature, organic matter, ammonia and dissolved oxygen.

5.2.3. Phosphorus cycle

Major steps involved in phosphorus cycle comprise: (1) Increase in solubility of inorganic phosphorus, (2) formation of orthophosphate, (3) transformation of assimilated inorganic phosphate ion into organic phosphorus as a protoplasmic ingradient leading to immobilisation of phosphorus and reduction of orthophosphates. Microbial species such as *Pseudomonas, Mycobacterium, Micrococcus, Flavobacterium, Aspergillus* and *Penicillum* convert tricalcium phosphate to soluble secondary products. Microbial decomposition of organic matter release orthophosphate which participate in equilibria established by various Iron, Aluminium and Calcium phosphate compounds.

5.2.4. Sulphur cycle

The important steps involve in the sulphur cycle are: reduction of SO_4 to H_2S and oxidation or reduced compounds to molecular sulphur and sulphates. Nearly 80% of putrefying bacteria break down proteins producing H_2S mostly in the anoxic sediment layers. Sulphate reduction is carried out by vibrio species and *Desulphovibrio* species. The oxidation of reduced sulphur compounds yield considerable amount of free energy. Purple sulphur bacteria (*Thiobacillaceae, Thiobacteriaceae* and Thiorhodaceae), Purple non sulphur bacteria, and Green sulphur bacteria (Chlorobacteriaceae) are involved in the oxidation of reduced sulphur compounds.

The oxidation of H_2S takes place in stages. Beggiatom species oxidise H_2S to S which is retained in the cell (intracellular sulphur) and further oxidise when external supply is stopped. Extracellular sulphur is produced by *thiobacillus* species. Photosynthetic sulphur bacteria can use thiosulphates and oxidise H_2S up to sulphate (SO_4). But green sulphur bacteria carry on the oxidation till sulphur.

5.2.5 Iron and Manganese cycles

Iron and Manganese bacteria viz., *Leptothrix, Gallionella, Metallogenium, Siderococcus, Gaulococcus* and *Ochrobium* can precipitate. Iron and Manganese compounds at pond humus as hydroxides. Heterotrophic bacteria possessing slime capsule like *Cladothrix* spp. deposit Iron and Manganese while other like *Siderococcus* and *Pedomicrobium* utilise the organic moiety of the Iron humus.

6. MANAGEMENT OF SOIL AND WATER FOR AQUACULTURE

For successful cultivated fish raising, proper management of soil and water quality is essential. Maintenance of nutrient status, pH quality, dissolved oxygen and control of turbidity as a practical guidance to achieve the index of success are necessary.

6.1 Nutrient status

Primary producers (phytoplankton) requires several mineral elements like nitrogen, phosphorus, potassium, calcium, magnesium, sulphur, iron, manganese, copper, zinc etc. in addition to CO_2 and water for growth. In order to maintain a steady supply of nutrients for the growth of plankton, application of these elements in intervals are essential. Nutrients are derived either autochthonously or allochthonously. However, the nutrient releases after bacterial decomposition in case of organic manures, it is advisable to apply organic manures in heaps under water in split dosages to avoid the chances of oxygen depletion and pollutional problems.

In contrast to organic manures, inorganic or chemical fertilizers are highly soluble in water and releases nutrients soon after their application in pond. Usually the quantities of nitrogen applied vary from 100-200 kg N/yr, where as phosphorus application is 50-100 kg P_2O_5/ha/yr depending upon the available nitrogen and phosphorus contents of the soil. It is always advisable to apply the total quantity of fertilisers in split dosages at periodic intervals during culture period. In order to achieve better results, application of both organic manure and inorganic fertilisers are desirable. Besides micronutrients like manganese, cobalt, zinc, boron, molybdenum found to enhance plankton production or fish.

6.2 Maintenance of pH value

Application of lime raises the pH value of soil as well as that of water. Liming establishes a strong buffer system and stimulates release of nutrients through microbial decomposition. The amount of lime to be applied depends upon the pH, texture and organic matter content of the soil. The most common liming materials are agricultural limestone $CaCO_3$ or $CaMg(CO_3)$, hydrated lime Ca $(OH)_2$, and quick lime CaO. Of these, ground agricultural limestones are extensively used in fish ponds. Neutralising effect of $CaCO_3$ is 100%. Considering this the value for $CaMg(CO_3)_2$ is 109, Ca $(OH)_2$ is 136 and for CaO is 179.

Soil pH	Dose of lime $CaCO_3$ (kg/ha)
4.0-4.5	2000
4.5-5.5	1000
5.5-5.5	500
6.5-7.5	200

Liming can be applied to the dried pond soil or broadcasted over water surface either in single instalment or equal split doses. In certain fish ponds the pH value found to be 11.0 or 12.0 indicating highly alkaline water. In this case the pH of pond water may be reduced using agricultural gypsum ($CaSO_4$, $2H_2O$). The rate of application of gypsum depends upon the total alkalinity and calcium hardness of water. It is calculated as Gypsum (ml/h) = (Total alkalinity – Calcium hardness) x 2.15 x 2.

6.3 Maintenance of dissolved oxygen level of water

Dissolved oxygen in pond water should not be less than 3.0 ppm and for normal growth it should be more than 5 ppm. The dynamics of dissolved oxygen is strongly influenced by phytoplankton density. To maintain a desired level of oxygen, the organic rich sediment of pond should be removed and also limed. Proper regulation of manuring and fertilization application resulting in required phytoplankton cells per litre of water without causing bloom is necessary. During oxygen depletion, the pond water should be agitated smoothly by splashing. Aerators or small water pump may also be used for aeration to increase dissolved oxgen level in such waters, bannana trunks cut in to pieces is used to aerate the pond water by swimming.

6.4 Control of turbidity

Most undesirable type of turbidity is that resulting from suspended clay or silt particles. Turbidity due to plankton is somewhat desirable, if it does not lead to bloom. Turbidity caused by suspended silt and clay particles can be controlled by application of hydrated lime. Boyd (1979) tested the effectiveness of four coagulants : filter alum, hydrated lime, ferric sulphate and agricultural gypsum. However, most effective chemical to remove high turbidity is filter alum $Al_2(SO_4)_3$, 14 (H_2O). Application of filter alum, at the rate of 25-30 mg/l will remove turbidity from most waters. But alum application reduces pH value and alkalinity for which hydrated lime should be applied to the pond before or simultaneously with alum.

3

PREDATORY AND WEED FISHES

1. INTRODUCTION

There are numerous predators which pray on smaller fish than on larger fishes. It is customary to note that, culture fish during its life history is liable for predation at every stage. Predators are either vertebrates or invertebrates. However, among vertebrates most important groups are catfishes belonging to siluriformes.

Besides predators in unmanaged ponds, some varieties of uneconomic small fishes that naturally occur or accidentally introduced along with riverine carp spawn are termed weed fishes. Both these weed fishes and predatory fishes compete for food and space among the cultivable varieties of carps. Certain fishes are known to feed directly on carp hatchlings. That too, certain predatory fishes breed prior to the breeding of major carps in pond waters and go on feeding the available planktons with fast increasing in growth and size. When carp spawns are introduced, the predators are large enough to feed on them. Weed fishes have relatively good fecundity and attain sexual maturity in summer and breed even without rain prior to the monsoon. So their young ones are abundant in number during monsoon.

2. SOME EXAMPLES OF PREDATORY AND WEED FISHES (Figs. 19-23)

Predatory fishes	Weed fishes
1. *Clupisoma garua*	1. *Ambassis ranga*
2. *Silonia silondia*	2. *Aplocheilus lineatum*
3. *Rita rita*	3. *Aplocheilus panchax*

4. *Nandus nandus*
5. *Bagarius bagarius*
6. *Pangassius pangassius*
7. *Pama pama*
8. *Glossogobius giuris*
9. *Wallago attu*
10. *Clarias batrachus*
11. *Heteropneustes fossilis*
12. *Mystus seenghala*
13. *Mystus aor*
14. *Mystus cavasius*
15. *Notopterus chitala*
16. *Notopterus notopterus*
17. *Mastocembelus armatus*
18. *Pseudociana coitor*
19. *Channa striatus*
20. *Channa marulius*
21. *Channa stewarti*
22. *Channa punctata*
23. *Channa gachua*
24. *Anabas testudineus*
25. *Ailia coila*
26. *Ailia berg*

4. *Barilius barila*
5. *Barilius bola*
6. *Barilius vagra*
7. *Chela cochins*
8. *Chela laubuca*
9. *Esomus danricus*
10. *Gadusia chapra*
11. *Gonisola manmita*
12. *Neomochellus zonatus*
13. *Osteobrama cotio*
14. *Oxygaster bacaila*
15. *Oxygaster phulo*
16. *Setipina phasa*
17. *Puntius ticto*
18. *Puntius conchonius*
19. *Puntius sophore*
20. *Rasbora daniconius*
21. *Xenentodon concila*

Fig. 19.1. *Glossogobius giuris* (Ham.)

Fig. 19.2. *Esomus danricus* (Ham.)

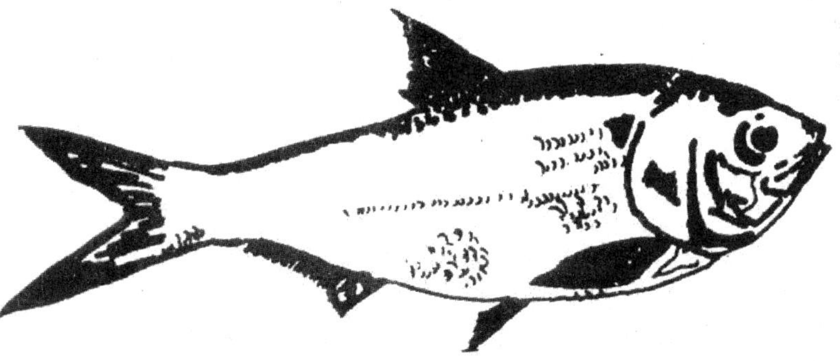

Fig. 19.3. *Gadusia chapra*

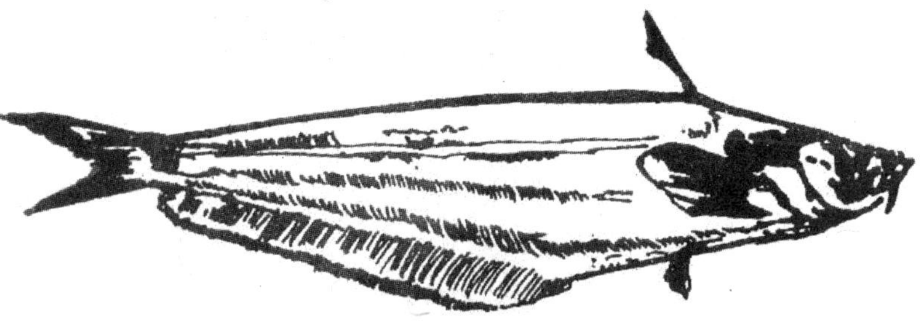

Fig. 20.1. *Ompok bimaculatus* (Bloch)

Fig. 20.2. *Channa marulius* (Hamilton)

Fig. 20.3. *Osteobrama cotio* (Ham.)

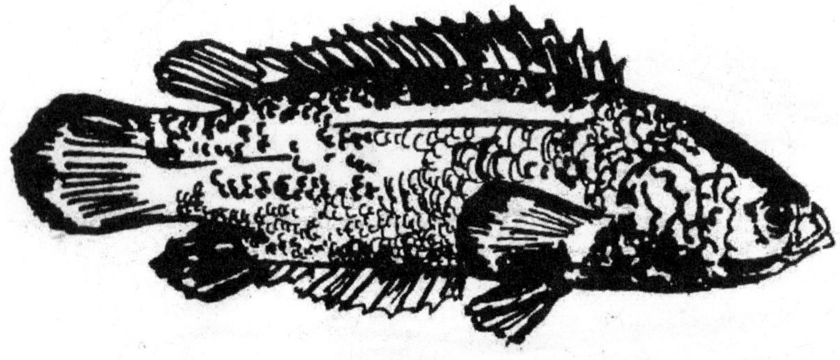

Fig. 21.1. *Anabas testudineus* (Bloch)

Fig. 21.2. *Puntius* spp.

Fig. 21.3. *Oxygaster bacaila*

Fig. 21.4. *Mastocembelus armatus* (Lacepede)

Fig. 22. *Pangassius pangassius* (Hamilton)

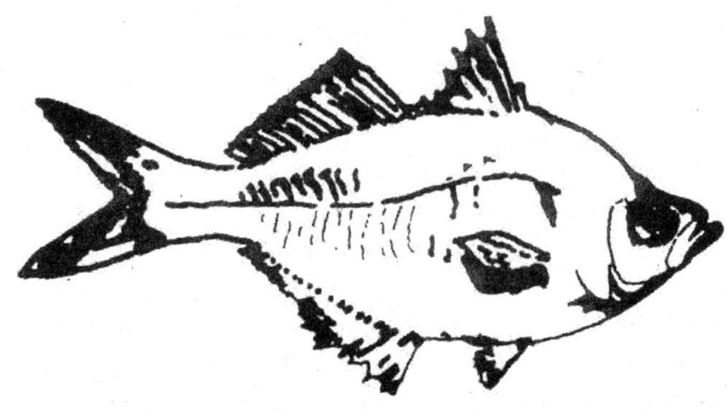

Fig. 23.1. *Ambasis* sp.

Fig. 23.2. *Amblypharyngodon mola* (Hamilton)

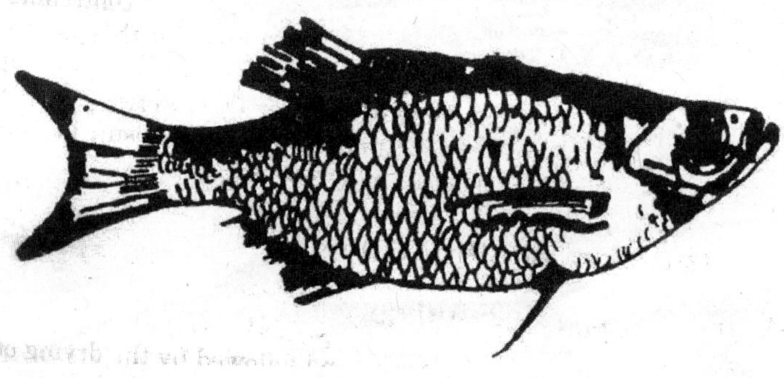

Fig. 23.3. *Laubuca laubuca*

3. ROLE OF PREDATORY AND WEED FISHES IN PONDS

Predatory fishes especially the Piscivorous varieties feed on the minnows and culturable species of fishes. Minnows being prolific breeders and profuse feeders on planktonic masses, they consume alomst all natural fish food organisms and occupy quite good space in the system. Therefore, these minnows not only compete for food but also compete for space with the culturable varieties of fishes. This reason necessiates complete removal or the control of predatory and weed fishes in carp culture tank. Besides this, the stronger and larger sized fishes show cannibalistic behaviour and may feed on the weeker and smaller fishes. Hence, it is desirable to completely eradicate the predatory and non-predatory minnows from culturable tanks through management.

Second reason necessiates about the consumer's preference on minnows and predatory fishes which are very low than the carp varieties. Therefore, carp culture is always preferable over predatory and weed fish culture due to above reason.

4. ERADICATION OF PREDATORY AND WEED FISHES

The eradication or control of the predatory and weed fishes from the culturable pond is suggested to be an important measure in the techniques of pond fish culture. In old days, eradication of these unwanted fishes were done mainly through hook and line with baits. From large water bodies, eradication of these fishes were next to impossible as it is uneconomical. However, control and eradication has been suggested for nursery, rearing and stocking ponds. Few scientists believe that, a small population of small mouth cleft fishes like *Notopterus chitala, Eutropichthys vacha* and *Notopterus notopterus* in the composite fish culture is beneficial. It is because that, predatory fishes feed on minnws as their food controlling the population of weed fishes. Therefore, in support of these views, in large water bodies, there should be strict control over the bigger sized predatory fishes like *Wallago attu* or *Ophiocephalus*, so as to control the population size of minnows. This may result favourable situation for cultivable varieties of fishes to grow.

5. METHODS OF ERADICATION

5.1 Dewatering and desiliting

Complete dewatering of the pond followed by the drying of soil bottom is suggested as a success method in eradication of unwanted fishes. It is economical only in small ponds. In heavily silted ponds, desilting after dewatering is advisable. Certain species of fishes like murrels, catfishes, Anabas and cuchia remain alive even in deep muds. Hence desilting helps in removing these fishes. Usually dewatering is suggested in summer, when the depth of water in ponds goes down to the minimum. This method is very convenient and economically applied in the seasonal or short perennial nursery and rearing ponds.

5.2 Netting operations

It is rather a primitive one and not a sure method. However, by repeated netting in a small water area with the help of a fine meshed dragnet ensures successful removal of the predatory fishes and minnows to a large extent. For more effective netting sinkers are used with the bottom rope. For the removal of predatory fishes from the pond subsurface soil, it is suggested to stir the bottom soil by means of bamboo poles or by human being or by any other means. Dragnet should be operated for effective removal of more numbers of predatory and weed fishes. The indegenous practice adopted by the fishermen for the removal of the fishes like cuchia, murrels, rata, singhi, Anabas etc. is to spread over a fine meshed net over the bottom soil of the pond. Sometimes in shallow ponds, such fishes were caught by hand picking also.

5.3 Hook and lines

Hooks with long lines are used for catching mostly the piscivorous fishes lik *Wallago attu*, murrels etc. Sometimes rods and lines are used with live baits to catch such fishes. Earthworms are used as baits in some cases.

5.4 Poisoning

Poisoning is a sure method of eradicatng the weed and predatory fishes from any sized water body but its use has been kept restricted because of its high toxicity effect to the aquatic and sub-aquatic animals. It affect on planktons also. Therefore, a more care is necessary in the selection of poison, dose calculation and application in carp nurseries and rearing ponds.

(a) Calculation of dosage

Chemicals are used in ppm (parts per million) and thus it is necessary to calculate the water volume of the pond to be treated. Different formulae are applid to find out the quantity of water of different sized ponds.

In case of square or rectangular shaped ponds the area is calculated by multiplying the length and the breadth. In case of circular pond, area is calculated by II r^2. In case of Triangular pond, the water area is calculated by applying the formula (Area = 1/2 X base X perpendicular to base). In case of different irregular ponds different compartments of square, triangular, circular, rect-angular shall be convincingly made. Then the water area of each of these compartments should be separately findout by application of different formulae mentioned above and finally adding the area of all such compartments. Then the water area is multiplied with average depth to give volume of water. To have more accuracy in the average depth, usually more numbers of possible representative points are taken in to account. The quantity of chemicals to be used can be worked out by the formulae.

$$\frac{wt.\ of\ water \times wt.\ of\ applicable\ dose\ of\ chemical}{1{,}000{,}000} = kg\ of\ chemicals$$

Chemical methods are suggested for eradication of unwanted species because dewatering, desilting and drying is difficult in perrenial ponds. It is better to suggest plant originated fish poisons because they affect on their respiratory system and the fishes die usually of Asphyxia. As a result, the flesh of these fishes may be used for human consumption. Moreover, the toxicity effect of the plant derivated fish poisons does not last for a long time as compared to chlorinated hydrocarbon or organo phosphates. Specially the endrin formulated fish poison chemicals are very strong poisonous and a little high dose of the poison may affect on the human beings and other animals if the treated waters are used. Further, the toxicity effect of the chlorinated chemicals is very much prolonged in to the water and as a result the treated water can not be used for a considerable period for fish culture. Some workers have therefore, suggested not to use frequently the hydrochlorinated chemicals as fish poisons. However, ICAR has suggested, the use of detoxifying materials like charcoal powder, sulphuric acid, potassium permanganate etc. in the chemically treated ponds to remove the toxicity within a short period. On the other hand the plant originated fish poisons example, mohua oil cake aids to the fertility of the pond as a fertiliser after their application. For eradication of weed fishes, CIFA, Kausalyagang has suggested the application of Urea at the rate of 100 kg followed by addition of bleaching powder (30% chlorine) at the rate of 175 kg per hectare meter of water. Urea acts as fertiliser in the latter days. Sometimes Bleaching powder alone at the rate of 350 kg/ha meter of water is applied to eradicate the unwanted fishes.

The fish poisons are of three types :

(1) Chlorinated hydrocarbon

(2) Organo phosphates

(3) Plant originated fish poisons

5.4.1 Chlorinated hydrocarbon

Chlorinated hydrocarbon like Aldrin ($C_{12}H_8Cl_8$), Dialdrin ($C_{12}H_8Cl_8O$), Endrin ($C_{12}H_8Cl_6O$) and Taffdrin-20 are used for the eradication of predatory and weed fishes. All the endrin formulated chemicals are very strong poisons and therefore, little doses are used for fish eradication. Its action in shallow ponds are quicker during sunny days but may have hardly any action in deeper water exceeding the depth of about 20 ft. Its action takes about 2-3 hours to affect Indian major carps and 4 to 8 hours to hardy fishes like *Tilapia, Anabas, Amphipneus cuchia,* Singhi, magur etc.

Planktons are also affected, specially zooplanktons but they may come up again after the poison effect is over.

Aldrin @ 0.2 ppm, endrin @ 0.01 ppm and dialdrin @ 0.01 ppm were found effective in eradication of many predatory and weed fishes. The toxic effect remain for 2 weeks in 0.01 ppm and for 2 days at 0.001 ppm concentration. Taffdrin contains 20% endrin as its effective poison. Taffdrin-20 at the low dose of 0.008 ppm can kill minnows, weed fishes etc. The higher dose of Taffdrin @ 0.02 ppm may kill all the aquatic and subaquatic animals within 8 hours.

Detoxification of Endrin formulated chemicals

The endrin formulated chemical gives long period of toxicity. For an example, when Taffdrin-20 is used in low dose (0.008 ppm), its toxicity remains up to 15-20 days. But, when used in high dose (0.02 ppm) its toxic effect may remain up to 40-50 days. But the endrin formulated chemicals can be detoxified by detoxifying agents such as charcoal powder, H_2SO_4, $KMnO_4$, lime and organic manures. The charcoal powder @ 20-25 ppm, H_2SO_4 @ 100 ppm, $KMnO_4$ @ 5 ppm when used in chemical treated ponds the pond will be detoxified within 4-5 days. This number of days can be further decreased to 2-3 days by application of charcoal powder @ 100 ppm. It is believed that the ponds having rich organic soil and algal blooms have rather quicker method of detoxification rather than the ponds without those too. Raw cow dung @ 18,000 kg/ha is also found effective in detoxification of the treated water.

5.4.2 Organophosphate

The organophosphates used in India are DDVP, phosphamidon, Thiomatin etc. The dose of phosphamidon and Thiomatin @

3 to 30 ppm can be applied for complete eradication of predatory and weed fishes. The dose of 0.0032 to 0.5 ppm is effective to the aquatic insects.

5.4.3 Plant originated fish poisons

There are quite good number of plants which could be used for fish poisoning. Various parts of plants such as stem, barks, seeds, leaves, roots can be used as the materials for fish poisoning. Some of the common plants are:

(1) *Barringtonia acutangula*

(2) *Crotton tiglium*

(3) Derris root powder

(4) Mohua oil cake (*Bassica latifolia*)

(5) Sugar cane Jaggery

The stem bark seeds or root of *Barringtonia acutangula* at the rate range of 10 to 20 ppm are effective. However, *Crotton tiglium* is effective at the rate range of 4 to 20 ppm.

In Assam, *Milletia piscida* and *Milletia pachycapra* are used to eradicate unwanted fishes. Derris root powder containing ingra-dient as rotenone is found to be effective in the rate range of 1 to 10 ppm in different group of fishes. Even for hardy fishes the dose was given at the rate of 20 ppm. The poison acts on the respiratory system of fishes and ultimately cause death. The poison is very effective up to a depth of 1.5 meter on hot sunny days when the temperature is more than 25°C. In deeper water its effect reduces gradually and beyond 6 to 10 meter, its effect is insignificant. The poisoning effect of Derris root powder lasts for 4-12 days when applied at the range of 4-20 ppm. In order to avoid risk factor, it should be applied one month before stocking of fish fingerlings in pond.

Bassica latifolia or *Madhuca indica* is very commonly known as Mohua. Oil cakes of these plants are available in market as Mohua oilcake. The active ingradients are mowrin or resparin or even saponin. It is recommended to be applied @ 200-250 ppm to fish ponds. It contains saponin, which on dissolving enters to the blood stream and through blood circulation to gills, buccal tissues. This results in hemolysis of RBC and cause death of fish. The sign of poison effect to fish is exhibited through distress, in-active and finally lose their balance. Mohua Oil Cake acts as a manure when its poisoning effect is lost. Poisoning effect remains for 21 days from the day of application. Hence, its application should be done atleast a month before stocking the carp spawn in nursery ponds.

Sugar cane Jaggary is used at the 1% concentration in fish ponds for eradication of unwanted aquatics. The effective ingradient is saponin.

The tea seed cake is made of the residue left after oil has been pressed out from the seeds of `oil tea' a sort of plants belonging to camellia family such as *Camellia sasargua, C. semiserrata*. It comprises saponin, a hemolytic toxic. The concentration of spaonin at 10 ppm may cause fish to lose their balance in 9-10 hrs or to die in 11 hours. The virulence of tea seed cake disappears quickly in higher temperature. Seed meal of tamarind, *Tamarindus indica* is used @ 1750-2000 kg/ha meter of water for eradication of unwanted fishes. Ammonia (NH_3) is also successfully employed for eradication of unwanted fishes and acts as fertilizer afterwards.

Other miscellaneous materials like calcium hypochlorite commonly called as Bleaching powder contains about 30% chlorine, or Urea and Bleaching powder is applied successfully for removal of unwanted aquatics as the price of mohua oil cake has increased. Even torpentine at the rate of 250 litre/ha is applied for eradication of unwanted fishes. It affects on the gills and checks effective gas exchange through gills causing Saffocation and eventually death to the organism.

Other biological agents like Amphibians are controlled by repeated netting, traps and scoopnets. Especially, frog tadpoles are eradicated by application of Tea seed cake @ 25 ppm. Reptiles are controlled by fencing and scaring. Those inside the pond are removed by repeated netting. Various fish eating birds like king fishers, Herons and ducks are removed by scaring, trapping and even by shooting.

The most important mammal for feeding on fishes are *Lutro lutro*. Other mammals feed on fishes are Beavers and Racons. These mammals can be controlled by fencing, scaring and shooting.

4

AQUATIC INSECTS AND THEIR CONTROL

1. INTRODUCTION

Aquatic insects constituted about 4% of the total insect fauna existing in the world. Any water area is invariably explored by large number of insects either in their adult or larval stage feeds directly on carp spawn or indirectly competing for food with carp spawn. Therefore, the pond culture technique includes the control and removal of harmful aquatic insects. Such eradication of harmful insects from ponds play a very important role in increasing fry survival rate. Insects are distinguished from other aquatic arthropods by their three thoracic segments, bearing 3 pairs of legs, 2 pairs of wings in the adult stage. The piercing, cutting and sucking type of mouth parts of the harmful, insects cause direct or indirect death to tiny fishes. However, common insects found in the culturable ponds being smaller in sizes, can not make any harm to rather bigger sized fishes including fingerlings and yearlings. Thus removal or control of insects in the stocking tank is not a compulsory.

2. AQUATIC INSECTS (Fig. 24)

Out of 11 orders of class Insecta, 3 orders namely Hemiptera, Coleoptera, and Odonata are relatively more common in fresh water ponds.

Order Hemiptera which includes waterbugs are relatively more dangerous as they are completely aquatic in both larval and adult stages. They have very strong piercing type mandibules and most fierce in praying upon fry.

Other common species belonging to Hemiptera are Belostoma, Lithocerus (Giant water bug), Nepa (water scorpion), Ranatra (water stick insect), Notonecta (Back swimmers), Anisopes, Geris (water spider) etc.

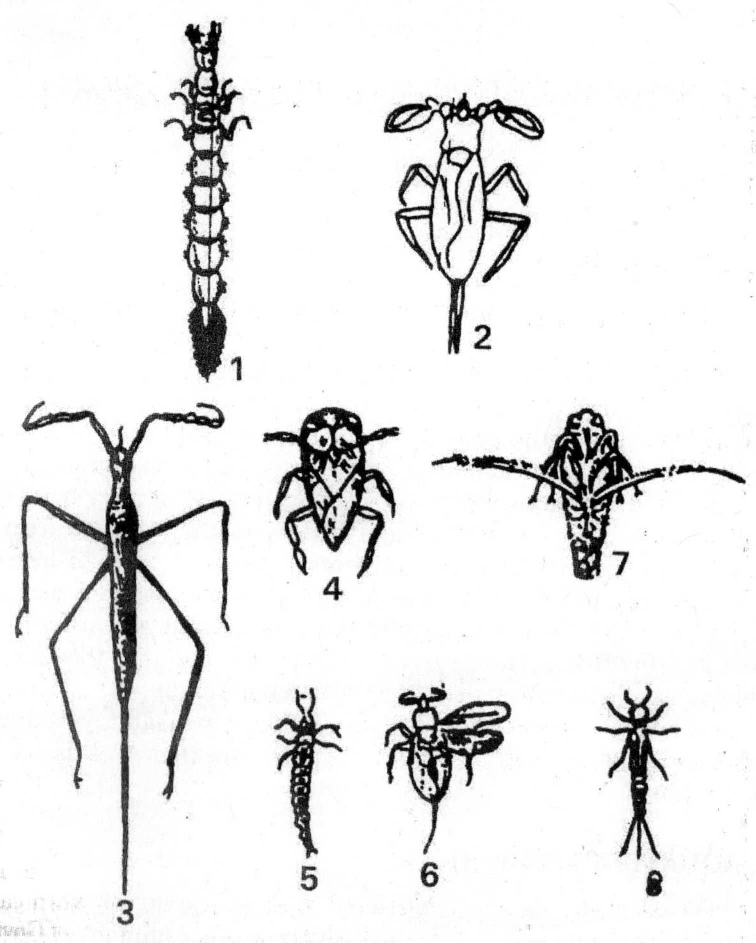

Fig. 24. Insects
1. Cybister larva, 2. Water stick insect, 3. Water scopion, 4. Water botaman, 5. Water tiger, 6. Diving bectle, 7. Back swimmer, 8. Damsel fly nymph.

Order coleoptera includes cybister (water beetles), dytiscus (diving beetles), Hydrophilus (Scavanger beetles), Gyrinus (whirling beetles), etc. These beetles lead their both larval and adult life in pond. The larval stages are extensively aquatic and in adult stages, they were very often found flying away out of water. Hence difficulties arise for controlling them. The strong cutting mandible may cause death of the fry by their attack.

The common insects belonging to the order Odonata in ponds are Dragon fly and damsel fly. The larval forms are aquatic while adult forms are terrestrial. The sucking type of mouth-parts of nymphs cause death of fish fry etc.

3. CONTROL OF PREDATORY INSECTS

Between initial poisoning of prestocking management and to the time of stocking spawn in the pond, predatory aquatic insect multiply rapidly. Many aquatic Insects fly from one pond to other pond. Hence, the pond which has been cleared of from insects is again repopulated with insects. For good survival of carp spawn, the nursery pond should be devoid of insects prior to stocking of spawns. Repeated dragging of pond with fine meshed dragnet does not eradicate completely the predatory aquatic Insects. Selective insecticides which kill selectively the aquatic insects without killing the fish food organisms is therefore necessary. However, spraying oil to kill the insects, by checking their respiration is a well known principle and commonly used among leading fish culturists. Mustard oil or coconut oil along with cheap washing soap at the ratio of 56 : 18 kg/ha is a well known technique developed by CIFRI for control of predatory insects within few hours of its application. The notonecta are killed within half an hours. Higher doses however kill water boatman (family corixi-dae), beetle larvae and bugs. Sometimes the substitution of soap is done by Teepol-B-300, a detergent manufactured by Burmashell in the emulsion. The recommended dose of Teepol is 560 ml with 56 kg of mustard oil.

The oil of Alexandrain laurel and water dispersible gamm-exan commonly known as Hertax U.P. are effective to kill insects. Gammexan only when applied @ 0.6-1 ppm, can safely eradicate insects. Pure gammaisomer of benzenhexachloride soluble in ethanol can kill insects within 6 hrs at 0.01 ppm. Carp spawn can tolerate chemicals up to a dose of 0.05 ppm. Fisheries Deptt. of Govt. of Maharahstra has described a dose of light speed diesel oil (1 litre), Hyoxid (0.75 ml) and water (40 ml) for every 200 m^2 of water surface for control of insects. The diesel oil at 50 litre/ha with 1/3 to 1/4 of washing soap can be applied to control insects.

Kerosene oil is also practiced among piscicultrists for effective eradication of aquatic insects. To kill water centipedes, quick lime can be applied before fru stocking. A drug namely Dipterex crystal (concentration 90%) may be scattered into the whole pond to create a concentration of 0.4 – 0.5 ppm in order to kill water centipedes in nursery tanks.

5

COMMON FRESHWATER AQUATIC WEEDS

1. INTRODUCTION

The menace of water weeds is reaching alarming proportions in many parts of the world. Weeds in short can be said as unwanted and undesirable plants which are adopted to grow and reproduce under aquatic conditions. The presence of these undesirable plants are definitely cause threat to the productive potential of the water body. The problem of weeds in fishery waters is more acute in tropical and subtropical countries than the temporate countries, although aquatic weeds now pose a global problem.

It is believed that the presence of certain plants in small numbers are frequently desirable in ponds. Any slackness given to check them would lead to their excessive growth which reduces productivity of the water body by utilising the nutrients or check the penetration of sunlight by shading. Heavy weed infestation restricts the living space for fish, fishing operation, gradual siltation and unwanted animals and parasites make their appearance and harm the existing fish population. Moreover aquatic weeds provide harbour place to predatory and weed fishes which are prolific breeders and profuse feeders and over populate themselves. These factors limit the available foods to the cultivable carp species resulting into less fish production.

Apart from fishery water, heavy growth of aquatic weeds also create serious problems in canals, irrigation channels, drainage, ditches, and large water bodies, interfering with hydroelectric production, wasting water in evapotranspiration and so on. More over, the decaying masses of weeds and algae pollute the potable water and environment. Furthermore, the problem of aquatic weed is worsened by increased enrichment of natural water bodies by fertilizer run-off and the nutrients from human and agricu-ltural wastes.

In India, over 140 species of plants have been reported to act as aquatic weeds both within and around various kinds of water bodies. The states comprising West Bengal, Orissa, Assam, Tripura and Manipur, have the maximum incidence of weed infestation ranging from 40 to 70 per cent, where as in other states, it may range between 20-50% (Philipose, 1968). West Bengal fisheries in 1951 had surffered a loss of 45 million kg of fish per year due to the overgrown of water hyacinth in 1.5 lakh ha of water area, 56 species of aquatic weeds of which 9 were floating, 11 emergent, 16 submerged and 20 marginal were identified (Philipose, 1972).

Problem of weed infestation in Bangladesh, Burma, Srilanka and Indonesia are almost similar to India. In most of the African Countries, the problem of weed infestation is encountered in large water bodies like reservoirs and certain slow moving rivers.

2. AQUATIC WEEDS

Hickling (1962) divides aquatic weeds according to the Softness and hardness of the stem of the plants. Pruginin and Zipshitz classify the aquatic weeds according to the type of association with the fish pond (Pruginin, 1968). From fisheries stand point the most useful classification based on habit and habitat of plants are as the following.

(a) Floating Weeds (Fig. 25)

These are free floating with their leaves above the water surface, roots are within water but not attached with the soil, e.g. *Eichhornia, Azolla* and *Pistia*.

(b) Emergent Weeds (Fig. 26)

These are rooted in the bottom soil but have all or some of their leaves, leaf lamina or shoots above the water surface. This group includes *Nymphaea, Trapa, Myriophyllum, Otellia, Vallisneria* etc.

(c) Submerged Weeds (Fig. 27)

This group of weed can be further divided in to 2 types i.e., submerged weeds which are completely submerged within water but may be rooted in the bottom soil for instance *Hydrilla* and *Najas* etc. and free floating submerged weeds such as *Ceratophyllum* and *Utricularia*.

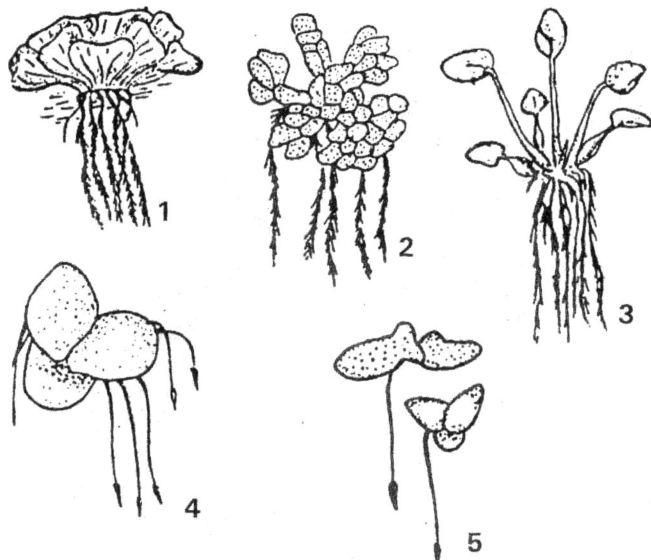

Fig. 25. FLOATING WEEDS
1. *Pistia stratiotes*, 2. *Azolla pinnata*, 3. *Eicchornia crassipes*, 4. *Lemna minor*, 5. *Lemna polyrhiza*

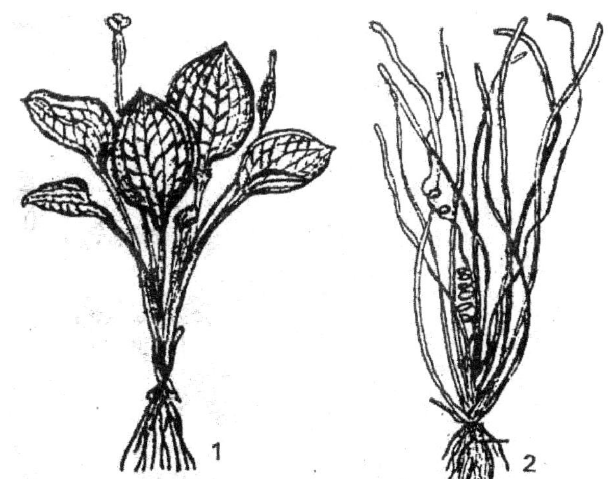

Fig. 26. EMERGENT WEEDS
1. *Otellia alismoides*, 2. *Vallisneria spiralis*

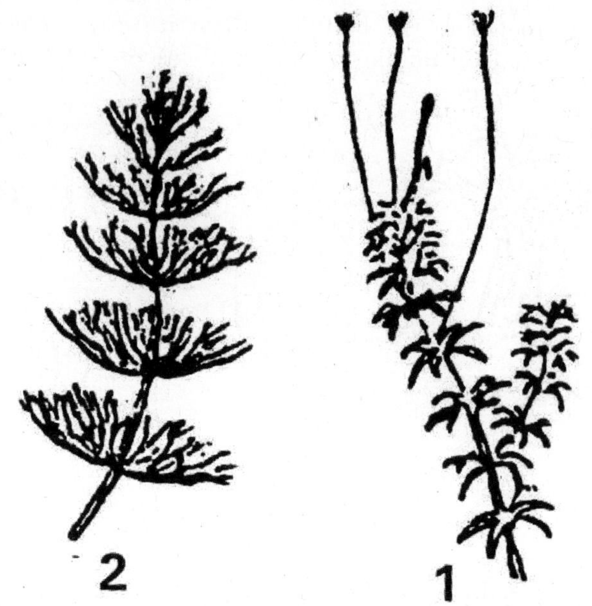

Fig. 27. SUBMERGED WEEDS
1. *Ceratophyllum submersum*, 2. *Hydrilla verticillata*

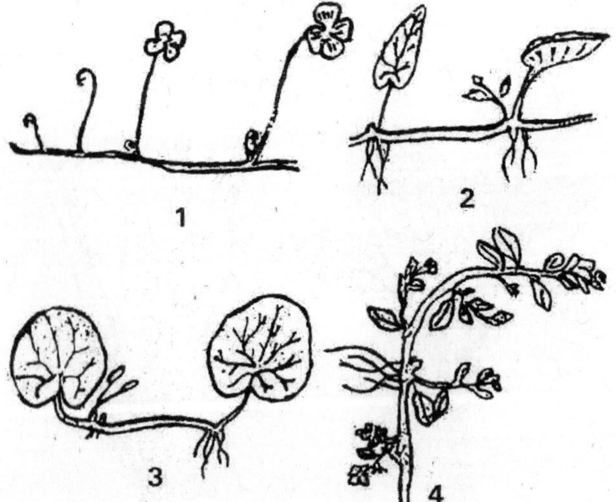

Fig. 28. MARGINAL WEEDS
1. *Marsilia quadrifolia*, 2. *Ipomoea aquatica*, 3. *Jussiaea repens*, 4. *Limnamthamum cristatum*

(d) Marignal Weeds (Fig. 28)

These grow on the margins or on the shore line of the water body. They are mostly rooted in water logged soils, e.g., *Typha and Phragmites.*

(e) Filamentous Algae

These algae form ``algal mats" in and around main water body. Most prominent filamentous algae in ponds are *Spirogyra* and *Pithophora.*

(f) Planktonic Algae

These are planktonic and their rapid proliferation results into algal blooms. The well known examples of this group are *Microcystis* and *Anabaena.*

Whatever manner the aquatic weeds be classified the most important are the methods of control which can be formulated only after their identification, habit, adoptive capacity, modes of multiplication, period of dormancy of the reproductive stages and such other ecological aspects. Further more, utilization of certain Aquatic weeds, in a sense constitute a free crop of great potential value - a highly productive crop that requires no tillage, fertilizer, seed or cultivation. Aquatic plants have potential for exploitation as animal feed, human food, soil additives, fuel production and waste water treatment.

3. CONTROL MEASURES

The measures adopted for the control of aquatic weeds can be grouped into following major catagories :

(1) Manual and management method

(2) Mechanical methods

(3) Chemical control methods

(4) Biological control of Aquatic weeds methods

(5) Control by Inter specific Competetions

(6) Control through utilisation methods.

3.1 Manual and Management methods

Before the discovery of mechanical, chemical and biological methods of weed control, manual and management methods alone were used to clear the weeds. In manual methods, human labour is employed with or without the aid of simple implements. In developing countries, manual methods are the oldest cheapest and frequently used for aquatic weed control. Manual measures necessiates constant vigil because of temporary effect and the weeds may even grow faster than they can be removed by hand.

The emergent and marginal weeds are removed by pulling them with hand or can be kept under check by cutting their floating leaves repeatedly. The floating weeds can be either hand picked or removed by wire, coir or nylon nets. For removal of rooted submerged weeds simple devices like handpulled bottom rakes, bamboo poles with toothed prongs can be used.

Management Measures

Management measures include :

(i) Construction or errecting barriers for checking the entry of floating weeds.

(ii) Periodic draining and drying of ponds.

(iii) Controlling the submerged weeds through turbidity. Algal blooms can be controlled temporarily by spraying liquid cow dung etc.

3.2. Mechanical Methods

Mechanical methods employed for control of aquatic weeds are expensive and require manual labour to complete follow up operations. Hence used in limited scale. Moreover, transportation, installation and operation of heavy machinery in swampy areas pose problems. However, at present a number of mechanical equipments like under water weed cutter, weed harvesters, weed harvester combined are available for physical control of the aquatic weeds. Therefore, mechanical control measures include cutting of weeds followed by secondary operation of collection and disposal. Other methods like dragging or chaining, dredging, draining, drying, ploughing, raking and burning etc are employed.

Devices used for cutting includes Hand Scythe, Double Scythe, Jointed Scythe and weed saw. Other equipments include turbine compressor and hydraulic pressure machine to dislodge the weed and collect them. In USA an ``aquatic Scavenger" for cutting and removing submerged weed was used.

3.3. Chemical control method

Chemical control of aquatic weeds has been reviewed by Blackburn (1968), Lawrence (1968), Philipose (1968, 1972), Patnaik (1972), Patnaik *et al.* (1988) and Srinivasan and Chacko (1952).

Some chemicals popularly known as weedicide can effectively control the noxious aquatic weeds. But some of their adverse effect on fish food organisms and the residual effect on the aquatic ecosystem has been felt with aquaculturists. A large number of chemicals which can be used as weedicides for the control of aquatics can be classified in to 6 groups.

(1) Inorganic chemicals - Sodium arsenite

(2) Organic chemicals - Xylene

(3) Auxin type regulater - 2, 4 - dichloropenoxy acetic acid (2, 4-D)

(4) Other growth regulators

(5) Contact herbicide/weedicide

(6) Algicides-rosin-amine D acetate

Hundreds of weedicides are available in the market in different trade names with wide range of chemical structures and functions in their killing action. These may be sprayed over the foliage, water or in the soil where root penetrates. Some herbicides are species specific and some work well on all types of aquatics.

For chemical control of floating weeds like *Pistia* and *Salvania,* foliar spray of paraquat @ 0.02 kg/ha, 2-4 D ester (MCPA) @ 25-40 kg/ha, Mineral oils (kerosene and diesel @ 775-1100 litre/ ha), Kerosene and urea at the ratio of 5 : 1, suing Diguat @ 1 kg/ha or paraquat mixed with aquous ammonia is applied. Water hyacinth (*Eichhornia crassipes*) is chemically controlled by 2-4 D (2-4 dichlorophenoxy acetic acid) @ 4.5 to 6.7 kg/ha. The dose of chemical application is also dependant upon the intensity of growth, age and time of application, environmental factor and spraying equipment employed for the purpose. Taficide 80 (2, 4-D sodium salt 80%) and Simazine are also effectively used.

For chemical control of marginal weeds, foliar spray of sodium salt of 2-4 D @ 5 kg/ha, 2, 2 dichloropropionic sodium @ 10-12 kg/ha, Dalapon @ 25-30 kg/ha, Amitrole @ 8 kg/ha respectively can be applied. Dalapon @ 15 kg mixed with 30 litre diesel and 1 kg detergent can effectively check these weeds.

For chemical control of emergent weeds, 2-4 D @ 1.5 kg/ha mixed with 1% wetting agent can be effectively used.

For chemical control of submerged weeds, Sodium arsenite @ 5-6 ppm, Urea with range of 50-300 ppm, $CuSO_4$ with Ammonium Sulphate, Ammonia @ 18 ppm in aquous solution, acrolein mixed with Xylene, Simazine @ 3-6 ppm, Aquathol (199 technical Endothaldisodium 3, 6 endoxohexa hydrophathalata) can be effectively used.

For chemical control of algal blooms and filamentous algae application of copper sulphate ($CuSO_4$) @ 0.025 to 0.5 ppm, Simazine @ 0.5 - 1.0 ppm is needed.

3.4 Biological control of Aquatic weeds

The biological methods considered the most convenient and least expensive methods for fish culture without any need for equipment or chemicals. A number of herbivorous fish species are being tested as aquatic weed control agents. Fish that control aquatic weeds are classified as -

(1) Grazers - if they eat stems and foliage like grass carp.

(2) Mowers - if they devour the lower portions of aquatic plants and thus cut them down.

(3) Roilers and rooters - if they stir the bottom sediments while foraging for food and preparing nests, there by creating turbidity for cutting off the light that the plants need for photosynthesis and growth or they root out the plants and destroy them.

(4) Algae feeders - if they consume filamentous algae.

(5) Plankton feeders - if they filter microscopic alage from the water.

Biological control through fishes viz., *Ctenopharyngodon idella, Puntius gonionotus* (Swingle, 1957), *P. javonicus* (Jhingran, 1983), *P. pulchellus* (David *et al.*, 1970), *Osphronemus gorami* (Villadolid, 936), *Tilapia mossambica, T. melanopleura* (Jhingran, 1983) have been used.

Some of the important herbivorous fishes are common carp (*Cyprinus carpio*), Grass carp *(Ctenopharyngodon idella*), the Golden carp (*Carassius carassius)*, the gold-fish (*Carassius auratus*), the Tawes (*Puntius javonicus*), the Nilem *(Osteochilus hasseti*), *Tilapia,* Gourami, *Osphranemus gorami, O. olfax,* Sepatsiam *(Trichogaster pectoralis)*, Milk fish (*Chanos chanos), Puntius dobsonii, P. pulchellus,* Silver Dollar fish *(Metynnis roosevelti)*, (*Mylossoma argenteum*), Brazilian fish (*Mylossoma bidens*), American Flag fish *(Jordanella floridae)* and Channel catfish (*Ictalurus punctatus* and *Barbus gonionotus*).

The grass carp is a truely phytophagus fish provided with powerful pharynageal teeth to tear and macerate plant material. Of the different species of Tilapia, *Tilapia melanopleura, T. zilli, T. rendalli, Tilapia quineensis, T. heudeloti, T. mossambicus* and Nile tilapia (*Tilapia nilotica*), are known to control weeds. Even silver carp *Hypophthalmichthys molitrix* feeds on phytoplankton there by controlling planktonic algae.

Control by Snails

A tropical freshwater snail *Marisa cornuarietis* (Seaman and Porterfield, 1961) and *Pomaca australis* have been found to feed voraciously on some submerged aquatics. The former is found in Colombia and the latter in Brazil. In India, *Pila globossa* (Thomas, 1975) has shown a promise to control *Salvinia molesta* from the waters of Kerala state.

Control by Mammals and other animals/Birds

Aquatic rat, *Mycocaster coypus* is a large sized rodent which is a native of South America has been found to be a powerful bioagent. In Israel and Poland, these aquatic rats are introduced in to fish pond for controlling weeds like *Typha, Phragmites.* But its burrowing habits and non-availability in large numbers are the constraints in its use as bioagent. In U.S.A. adult Manatee (*Trichochus manatus*) was tried experimentally for controlling aquatic vegetation. However, the availability of the animals in adequate numbers is a problem. Marginal grasses and marsh plants are effectively controlled by grazing cattle. Cray fish (Dean, 1969) and Turtle (Shah & Tyagi, 1985) have also reported to control aquatic weeds.

Swans, Ducks and Geese feed on algae and small sized weeds like *Wolffia, Lemna* and Marginal grasses (Ross, 1971).

Control by Insects

Several insects are known to forage upon aquatic plants. It was reported that, about 58 insect species were either phytophagus or used tissues of hydrophytes as food under certain conditions. Three species of insects viz., (1) *Agasicles hygrophila* (Flea Beetle), (ii) *Amynothrips andersoni* (alligator trip) and (iii) *Vogtia mallio* (Stem borer moth of Alligator weed) which are native of South America, have been very successfully employed in U.S.A. Insect enemies of water hyacinth *Neochetina bruchi* and *N. eichhorinae* (weevils), *Acigona ignitalis, A. infusella* (Pyralidae), *Epipagis albiguttalis* (Stem borer), *Cornops longicorne* (aquatic grass hopper), *Orthogalumna terebrantis, Leptogalumna* spp. (mites) and *Thrupticus* spp. (Diptera) were found effective. Larvae of *Arzama densa* feed on water hyacinth causing death of some plants and prevent seed formation in other. In India, *Gesonula punctifrons* was found to attack on water hyacinth. *Nigritulus,* fruit eating weevil is reported as good biocontrol agent for *Ludwigia adscendens.* The insect species *Neochatina eichhorinae* released in Florida was found for control of water hyacinth. Hussain and Jamil (1989) have reported the control of water hyacinth through Insects.

Salvania (Sankaran, 1976) and pistia (Chaudhuri and Janakiram, 1975) are also known to be predated by insects. These include *Samea multiplicaulis, Cryptabagous singularis, Paulina acuminata,* Noctuid (*Namengana pecticornis*).

The other insect agents are *Galerucella nymphaeae* (Smirnov, 1960) and *Simyra conspersa* (Sar, 1991) on water lilly, Aphis species *Nymphula diminutalis* and *Hydrellia* species, *Bagous* spp. on Hydrilla.

A moth *(Paraponyx stratiotata*) and weevil (*Litodactylus leucogaster)* were effective for biological control of *Myriophyllum* in Yogoslav.

Control by Fungi, Bacteria and Viruses

Attempts have been made to isolate pathogenic organisms to employ them as biological agent to control the weeds. Species of *Pythium, Curvularia, Verticillum, Aspergillus, Pencillium, Eunotium* and *Torula* were reported to control effectively Hydrilla. Fungal parasites viz., *Cercosporium, Cephalosporium* species, and *Myrothecium* species are to be attacking water hyacinth. Cyanophages are being explored for the control of blue green algae.

A virus is noted to be affecting *Pistia stratiotes* in West Africa (In Foletter - 1971).

3.5 Control by Interspecific Competetions

This is the simplest and well known method. It is practiced by fish culturists in parts of India to destroy algal blooms caused by microcystis by keeping a cover of *Lemna* (duckweed) for about a week or so. Floating weeds like water hyacinth, *Pistia, Salvania* and *spirodella* are sometimes used to suppress the growth of submerged weeds by cutting sunlight. But this method has certain limitations, because introduction of water hyacinth is very risky and secondary transporation of weed is difficult. The method is not suitable for large water bodies, but quite suitable for manageable small ponds.

3.6 Control through Utilisation

This is the new concept which has originated due to heavy costs entailed on the control of aquatics by manual, mechanical and chemical methods. The negative economic role which the aquatics play can be reversed through utilisation of various weeds.

Usefulness of Aquatic weeds can be categorised as,

(1) Weed as fertiliser

(2) Weed as feed for Animal/bird and fish.

(3) Weed as leaf protein

(4) Weed as food crops

(5) Weed as a source of energy

(6) Weed as wastewater treatment

(7) Aquatic weeds for pulp, paper and fibre

(8) Other uses.

3.6.1 Weeds as fertiliser

3.6.1.1. Compost

Compost is the decomposed plant matter for use as fertiliser in crop field as well as in fish culture ponds. The compost made from water hyacinth is superior to town compost and farm yard manure. That too, mineral value of some aquatic plants are very rich. The chemical composition like organic carbon, available N_2, available phosphorus/100 g of aquatic plant are higher than garden soil. In developing countries where commercial fertiliser is expensive and labour is comparatively cheap, compositing the weed to supplement as fertiliser is an ideal proposition.

Besides compositing, fresh aquatic weed can be used as green manure. Water hyacinth are used as bedding material for cultivation of mushroom after drying. Weeds are now a days being used as biological nitrogen fixer. Azolla and some blue green algae viz., *Anabaena, Nostoc, Rivularia* are known to fix atmospheric nitrogen and can be utilised for increasing nitrogen level of fish ponds also.

3.6.2 Weeds as animal feed

Many types of vegetation that human find inedible can be converted into meat, milk and wool by ruminants like cattle, sheep and goats. It is therefore reasonable to consider aquatic and semi-aquatic plants as potential ruminant feed.

The leafy parts of the aquatic plants such as duckweed, water hyacinth and some submerged weeds contain 25-35% protein, which is exceptionally high. The individual aminoacid constituents are present in about the same proportions as in land forages of similar crude protein content, however, the methionine and lysine amino acid levels are considered the limiting amino acids in plant proteins.

The weeds differ widely in fibre content. The fibre content of such aquatic plants make them potential substitutes for hay, cotton seed hulls and other roughages eaten by ruminant animals.

Some submerged aquatic weeds, such as Hydrilla and watermilfoil, are particularly rich in carotenes and Xanthophylls. These valuable pigments are being added to poultry rations in many countries. Moreover, the nutrient content and digestibility of some aquatic weeds are high indicating that for a roughage, the nutrients are satisfactorily available to ruminants.

In China, pig farmers boil chopped water hyacinth with vegetable wastes, rice bran, copra cake and salt to make a suitable feed. In Malaysia, the fresh water hyacinth is cooked with rice bran and fish meal and mixed with copra meal as feed for pigs, ducks and fish.

Aquatic weeds are used for silage. Silage prepared by mixing chopped and wilted water hyacinth with ground maize and molasses stored for 1-2 months in pits are well accepted as feed by Pigs.

Freshly harvested aquatic plants contain enormous quantities of minerals depending on plant type. Phosphorus, magnesium, copper, zinc and manganese were present in the same concentrations as they are in land forages, however, sodium was 10-100 times higher, Iron 4-19 times higher and potassium 3-6 times higher. Part of the reason why animals refuse to eat large quantities of such aquatic weeds is thought to be the over abundance of these minerals as animal getting only 10-20% aquatic plants in its diet would be receiving its full requirement for these minerals.

3.6.2.1 Feed for farm birds

Ducks, geese, swans and other water fowl forage on vegetation, controlling weeds on the banks of water ways and often clearing aquatic weeds and algae from small lakes, ponds and canals. Dried duckweeds (30-50 g) can be added to the daily poultry ration for their weight increase and also make up vitamin deficiencies. Chopped water hyacinth leaves are also fed to poultry.

3.6.2.2 Feed for fish

There are number of herbivorous fish which directly consume aquatic weeds. Aquatic macrophytes have been known to have potential value as human food, livestock fodder, fertilizers and food for herbivorous fishes (Edward, 1980). The growth performance, food conversion ratio and protein efficiency ratio alongwith nutritional qualities of weeds that are encorporated in pelleted feeds in fish have been reported by Tan, 1970, Hajra and Tripathi, 1985, Hajra, 1987, Patra and Ray, 1988 and Das and Ray, 1989.

3.6.3 Weeds as leaf protein

Various attempts are being made for the extraction of protein from aquatic weeds. Leaf protein from aquatic plants are similar in chemical composition to leaf protein from crop plants.

Certain freshwater algae such as *Chlorella pyrenoides* synthesise consideration amounts of protein (40-60% dry weight) constituting all essential amino acids as well as vitamins.

3.6.4 Weeds as food crops

Aquatic plants can provide 3 types of food : foliage for use as green vegetable, grain or seed that provide protein, starch, oil and fleshy swollen roots that provide carbohydrates.

Some common aquatic plants and their parts used for human consumption are -

(1) *Ipomea aquatica* - young leaves and stem

(2) *Marsilea* spp. - leaves

(3) *Nelumbo nucifera* - flowers, leaves, rhizome

(4) *Nymphaea* sp. - leaf, flower

(5) *Euryale ferox* - seeds

(6) *Trapa* - fruit

(7) *Ottelia* - leaves and fruits

(8) *Colocasia* - petiole, rhizome and tubers.

3.6.5 Weeds as a source of energy

In this era of energy crisis, all efforts are being made to find out unconventional sources of energy. One such source is biogas. The vast biomass of aquatic weeds can be utilised for the production of biogas by anaerobic fermentation process. According to an estimate, water hyacinth can produce more the 70,000 m^3 of biogas. Each kilogram of water hyacinth (dry weight basis) yields about 370 litres of biogas with an average methane content of 69% (Gupta, 1979). Recent trials in Kerala, for production of biogas using salvania have also given encouraging results. Water hyacinth and cowdung mixed in 1 : 1 ratio yielded 25% more gas than with cowdung alone.

3.6.6 Waste water treatment

Many aquatic plants are able to scavenge inorganic and some organic compounds from water. The water hyacinth is found to be very efficient in this regard. According to an estimate, the plants can remove about 1726 kg of Nitrogen/ha/year and about 387 kg of Phosphorous/ha/year from the sewage effluent. The water hyacinth is also capable of removing BOD up to 90%, COD up to 85% and 80% of fecal bacteria from sewage.

Other aquatic plants such as *Scirpus lacustris*, *Ceratophyllum demersum, Spirodella polyrhiza* and *Lemna minor* are used in waste water treatment.

The industrial and domestic effluents containing heavy metals, phenols, phenolic derivatives and pesticide resideues are removed to certain extent by water hyacinth. The water hyacinth can remove about 300 gm of heavy metals and 53 kg of phenol from 1 ha of polluted water per day. The alligator weed is also capable of removing some heavy metals.

It is hoped that the water hyacinth could provide safeguard to absorb radio-active isotopes, should they be accidentally released into the wastewater from a nuclear plants.

The Maxplank Institute of West Germany has developed emergent plant method for waste-water treatment through aquatic weeds. Similarly submerged plants method, Duckweed methods are adopted for waste water treatment. At the Louisiana State University, duckweeds are now being grown on dairy farm waste water. The harvested duckweeds are substituted for alfalfa in dairy and swine rations. In the NASA, waste water treatment project, water hyacinth uptaken heavy metals are fermented for methane gas.

3.6.7 For pulp, paper and fibre

The common weed *Phragmites communis* is extensively used in Romania, for making printing paper, cellophans, card board and various synthetic fibres. The row weeds and pulp mill wastes yield a variety of products like compressed fibre board, insulation material and fertiliser. Other aquatics like *Typha* (Cat tails) and *Cyperus* are also a source of pulp, paper and fibre although these are cultivated as fibre crop.

3.6.8 Other uses of Aquatic weeds

Rough and tough marginal weeds can be used for making huts, thatching roofs, traditional fishing rafts, fishing rods, musical instruments, writing pens and floats. These can be used for making screens and mats.

6

ARTIFICIAL PROPAGATION

1. INTRODUCTION

Artificial propagation includes series of processes of maturation, ovulation, spawning and incubation of fry under control condition. This necessiates to know morphology and physiological processes that are involved in artificial propagation. In this regard considerable literatures are available in the morphology of the teleost overy. It was Cavolini (1792), who for the first time gave an interesting account of certain morpholigical aspects of fish gonads. Earlier authors such as Brock (1878) and Calderwood (1891) confined their study to the general structure of the ovary. Afterwards the reproductive structures and the reproduction in fishes has been studied by Hoar, 1955; 1957; and 1969; Ball, 1960. The gonads have cyclic variation in development, controlled by endocrine and nervous systems and also by external ecological conditions. These Eco-physiological factors are associa-ted and the external may restrict the internal.

2. STRUCTURE OF OVARY

The female reproductive organ consists of a pair of ovaries, ther oviduct and any accessory gland. The ovaries are located in the posterior body cavity immediately ventral to the trunk kidney and swim bladder. The ovaries are symmetric, elongated sac like structure and in some brownish yellow in colour. These are attached to the air-bladder by mesovaria. Each ovary at its posterior end gives rise to a wide oviduct. The two oviducts from the ovary get united to form a common oviduct with an opening to the exterior of body. There are blood vessels and nerve branches on the ovary tissue.

Histologically ovary consists of ovarian wall and the oocytes at different stages of development. The ovarian wall could be distinguished in to an outer layer and an inner layer which are termed as the tunica albuginea and germinal epithelium respectively. The tunica albuginea consisted of connective tissues, cells and fibres. The blood vessels were seen along the periphery of the wall as well as within the connective tissue itself. The lumen of the ovary was obliterated by the development of ovigerous lamellae. Each lamellae was formed of connective tissue fibres, blood capillaries and germinal epithelium. The ovigerous lamellae consisted of oocytes at different developmental stages.

2.1 Development of Ovum

Development of ovum passes through

(1) Multiplication stage

(2) Growth stage

(3) Maturation stage

Multiplication stage is characterised by increase in number of reproductive cells through mitosis.

In growth stage, the oocytes increase in size from immature one to mature one. During growth stage yolk deposition starts and oocyte grows to primary oocytic stage.

In maturation stage, secondary oocytes are formed through meiosis and 2N oocyte (primary oocyte) becomes N oocyte (Secondary oocyte). Each primary oocyte (2N) produce one functional secondary oocyte and one polar body. Then by mitosis, one ovum and 3 polar bodies are formed.

2.2 Structure of Ovum

Fish's ova possess the common features of body cells. It consists of ovum cytoplasm, ovum nucleus and egg membrane. At the initial stage of development, the sphere shaped egg cells only have nucleus and cytoplasm. In further development the egg will accumulate by and by egg yolk material exogenously and endogenously. Polarisation appears in their distribution, thus the egg is called polarised yolk egg. The egg is topped by animal pole and yolk concentration increases at vegetal pole because of yolk's larger granule and specific gravity. Egg yolk contains protein, fat carbohydrates and vitamins etc. which are the nutritional material for developmental process. The nucleus is sphere shaped, but often leaf shaped. The nucleus consists of nuclear membrane, nucleoli, chromatin fibres, nuclear fluid and genetic material. Its function is to maintain genetic material (DNA) and cell metabolism, passing the genetic material to other generation.

Egg membrane is a membranous structure covering the exterior of an egg cell. The primary cell membrane is made of cell's own plasma and secondary cell membrane is secreted by folicle cells in ovary. The seondary cell membrane of fish is normally adhesive. Sticky eggs of common carp have both primary and secondary cell membranes. But in India major carps (IMC), Grass carp and silver carp, possess only primary cell membrane.

3. DEVELOPMENT STAGES OF OVARY

The development stages of ovarian maturity can be judged by visual observation on the morphology or through histological survey. Morphological staging of ovary is generally done on the basis of its appearance color, size, weight, intensity of blood vascularisation and the maturity of ova. Various types of morphological staging of ovary has been described by different authors. Rath and Mohanty-Hejmadi (1984) has described six morphological stages in *Channa punctata*. On the basis of histological changes, oocytes are also staged into various types by different authors. Rath (1980) describes 9 oocytic stages in *Channa punctata*. That too the classification of the developmental stages of ovary in carps adopted by each country is not identical. Five stages are admitted in India, Japan and U.S.A. Six stages are admitted in China while other several countries admitted seven stages. Wood (1930) recognised 7 stages of ovarian maturity in carps. However, the stages of ovarian maturity described below are purely arbitrary. Morphological staging as stage I, (Immature), Stage II (maturing), Stage III (mature), Stage IV (spawning) and V (spent) stages can be described (Fig. 29). Several others have described stage I as resting phase, stage II as early maturing phase, stage III as advanced maturing phase, stage IV as mature or prespawning phase, stage V as spawning phase and stage VI as spent phase. Oocytic staging on the basis of histological changes as chromatin nucleolus stage, early perinucleolus, late perinucleolus stage, yolk formation stage (primary yolk, secondary yolk and tertiary yolk stage), prematuration, maturation, mature and spent stages are described in different fish groups.

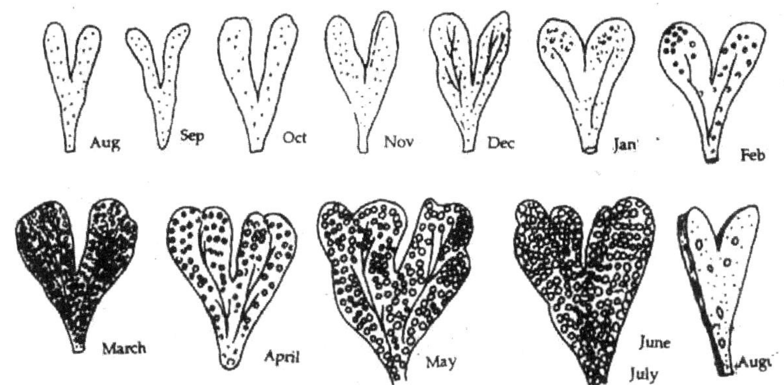

Fig. 29. Morphological staging of ovary
1. Resting phase, 2. Early maturing phase, 3. Advanced maturing phase, 4. Mature or prespawning phase, 5. Spawning phase, 6. Spent phase

A comparison on maturity stages of ovary in carps admitted by Chinese and also by Wood (1930) are given below.

Chinese pattern of staging ovary in carps	*Wood's pattern of staging ovary in carps*
Stage- 1	Stage-1. (Immature)
(a) Linear thread like in shape, transparent fresh white in color. Eggs are not visible to naked eye.	(a) Ovaries small, colorless and translucent, ova-invisible to naked eye.
(b) Cell tiny, 12-22 μ in diameter, nucleus big, nucleoli in the centre of nucleus.	
Stage II	Stage II. (Maturing)
(a) Ribbon shaped, fresh white or slightly pinkish, semi-transparent, small eggs are visible under a magnifying glass.	(a) Ovary occupy 1/2 length of body cavity, egg small, no yolk deposition, visible to naked eye.
(b) Cell diameter 90-300 μ, folicle layer surrounds the oocyte, nucleoli closely attach to the nuclear membrane.	
Stage III.	Stage III. (Maturing)
(a) Ovary is conspicuously enlarged, greenish grey in color. Egg are visible to naked eye. Blood vascularisation is clear.	(a) Length of ovary more than 1/2 of the body cavity. Greenish/brownish, ova opaque, coloured, yolked and visible.

(b) Follicular membrane is bilayer, egg yolk start accumulating, cell size 300-500 μ, nucleus oval and centrally placed, yolk vacuoles are present.	
Stage IV.	Stage IV. (Maturing)
(a) Ovary is long sac like. Eggs are greenish grey or light yellow in color, intense blood vascularisation. Easy to separate eggs.	(a) Ovary occupy 3/4 of body cavity, eggs opaque, yolked and larger in size.
(b) Cell size 800-1580 μ Egg yolk granules fill almost all the sapce outside the nucleus. Little cytoplasm. Nucleus edge looks like wave in shape, few nucleoli.	
Stage V.	Stage V. (Mature)
(a) Eggs are in the stage of flowing, ovary and belly are very soft. Egg come out with slight pressure on belly.	(a) Ovary extends entire length of body cavity, ovary coloured, eggs large.
(b) Yolk granules fuse in lump, polarisation of nucleus, nuclear membrane dissolved through peforation, Nuclei concentrate at the centre of nucleus.	
Stage VI.	Stage VI. (Spawning)
(a) Ovary is slack and shrink conspicuously. Major portion of egg has been laid.	(a) Eggs transparent central yolk mass
(b) Undischarged egg degenerate and get absorbed in the oocoel. Some eggs have irregular structure. Intrinsic oocytes develop in the ovary.	
	Stage VII. (Spent)
	(a) Ovaries shrunked and flaccid.

3.1 Ovary as an endocrine organ

Several experiments led to establish that there is an ovarian endocrine function but does not tell any thing about the chemical nature of ovarian hormone or hormones. Experiments indicated that the hormone or hormones of the fish ovary may be different from those of mammals (Hoar, 1955; Pickford and Atz, 1957). Various studies allow to say definitely that the fish ovary produces substances that are chemically and biologically estrogens. However, the literature devoted to fish steroids has expanded greatly. The exhaustive bibliography from Bern and Chieffi (1968) and Fostier *et al.* (1983) are valuable informations on gonadal steroids. The study on steroids in fish were done by using various techniques such as ultracytochemistry, radio immunoassay, mass spectometry and chromatographically. Steroidogenic activities have been localised in interstitial, thecal and granulosa cells according to the species and the stage of oogenesis with the exception to certain species such as *Sarotherodon aureus, Mugil capito, Cyprinus carpio* and *Perca fluviatilis.* Such differences between species can be attributed because all the studies have not been performed at every stage of the ovarian development and infact variations throughout the reproductive cycle have been observed. The steroid hormones so far postulated to be originated from fish ovary are progestine, corticosteroids, estrogen 17β - estradiol and estrone.

4. STRUCTURE OF THE TESTES

The male reproductive organs are suspended length wise by mesorchia and are situated ventral to the kidneys and the air bladder. The testes are paired, ribbon like in structure and dull white in colour (Fig. 30). Each testis at its posterior end is connected to a short seminal duct with an opening to the exterior of the body. Sometimes, accessory gland is present in association with the male reproductive organs in fishes. There are lot of seminiferous tubulues (ampullae) arranged irregularly inside. The spaces between seminiferous tubules (ampullae) are full of connective tissues. Each tubule is surrounded by a thin tunica albuginea composed of fibrous connective tissue. Formation of spermatozoa from the germinal epithelium appears similar to that in other vertebrates. The seminiferous tubules (ampullae) are composed of many spore sacs. These spore sacs are separated by a thin layer of follicular cells (Sertoli cells). In one spore sac there is plenty of synchronously developing germ cells (spermatocytes and spermatids. After the formation of sperms, spore sacs dissolve and sperm enter the cavity of seminiferous tubules (ampullae). Sperms mix with fluid secreted by testis forms so called milt which passes out to the exterior of the body through seminal duct, at the climax of brooder's estrus. The anterior part of testis differs histologically from the posterior parts. The posterior part generally consists of coiled tubules in some species of fish but lack of the seminiferous tubules. Spermatozoa are usually not present or present in some species in the posterior region of testis.

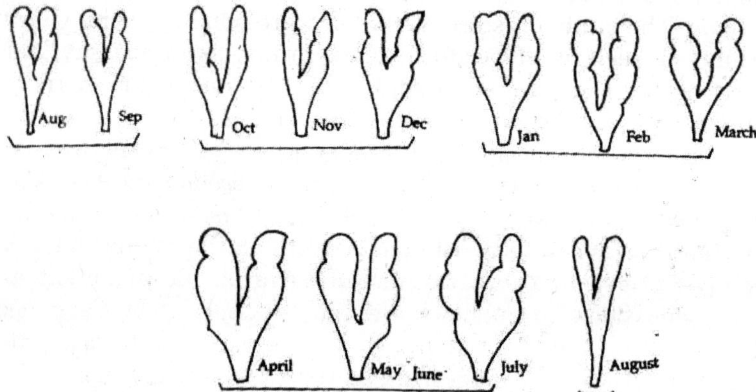

Fig. 30. Morphological staging of testes.
1. Resting phase, 2. Immature phase, 3. Maturing phase, 4. Mature phase, 5. Spent phase

4.1 Development of sperm

The process of development of sperm is known as spermatogenesis. Like that of ova development, development of sperm passes through :

(1) multiplication stage in which spermatogonium increases in number through mitotic division.

(2) Growth phase is characterised by increased number of primary spermatocytes and

(3) maturation stage in which secondary spermatocytes are formed through meiosis and then through mitotic division, spermatids are formed.

The differences between oogenesis and spermatogenesis are that, in oogenesis, primary oocytes produces one ovum and 3 polar bodies and cell size increases through development. In spermatogenesis, primary spermatocyte produces 4 spermatids and cell size decreases from large to small.

4.2 Structure of sperm

The morphology of spermatozoa has been studied in several species of teleosts reported by Nagahama (1983). The teleost spermatozoa can be morphologically sub-divided into head, neck piece, mid piece and tail. They lack an acrosome which occurs in all other vertebrate groups, this may be related to the presence of an egg micropyle in teleost eggs. The head is generally spherical or oval in shape, however in eels, the head is sickle or crescent shaped. Mid piece consists of a central flagellum and a surrounding mitochondrial sheath. Even a flagellated spermatoza and biflagellated spermatozoa occur in some teleosts. The sperm morphology appears to reflect the mode of fertilization, for example in medaka *Oryzias latipes* with external fertilisation, sperm morphology is primitive with a rounded nucleus and short mild piece. In guppy *Poecilia reticulata* with internal fertilisation, an elongation of nucleus and mid piece is observed.

Spermiogenesis – Transformation of spermatids into mature spermatozoa is known as spermiogenesis. Spermiation in teleost involve a hormone dependant thining and hydration of semen. The precise physiological significance of hydration is not clear although it seems that by increasing interlobular pressure that allows the sperm to migrate to the vasdeferens where they are stored. However, the morphology of spermiation in teleosts with tubule type (poecilidae) or lobular type testes has not yet been studied in detail.

A sperm cell of cultivated carps consists of a head, a neck and a tail (Fig. 31). Head is almost spherical in silver carp. It consists of an apex and a nucleus. The apex is the front part of the head and also called as penetrator for its function of penetrating into egg. The neck is very short and is situated between the head and the tail. The tail is narrow and long. It is the metabolic centre and motor organ.

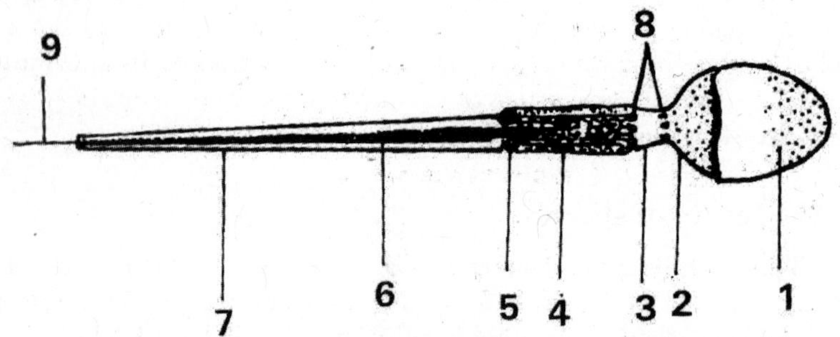

Fig. 31. Ultrastructure of a sperm of silver carp.
1. Apex, 2. Nucleus, 3. Neck, 4. Middle part, 5. End ring, 6. Spindle filament, 7. Plasma sheath, 8. Knots (front and rear), 9. Flagellum.

5. DEVELOPMENT STAGES OF TESTES

Just as the ovary, the testicular maturity can be judged by visual observations on the morphology or histological survey. However, many have classified development stages of testes as stage I as resting phase, stage II as late immature phase, stage III as maturing phase, stage IV as mature phase and stage V as spent phase. In the present context on carp maturity stages of testes admitted by Chinese and also by Wood (1930) are given below although classifications are purely arbitrary.

Chinese pattern of staging testes in carps	*Wood's pattern of staging testes in carps*
Stage- 1	Stage-1
(a) Testes are linear in shape and transparent.	(a) Testes thin thread like, Colorless and translucent.
(b) Scattered spermatogonia of 16 μ diameter with round nucleus of 9 μ dimaeter and no clear arrangement of sperm cells.	
Stage II	Stage II.
(a) testes are lace shaped either translucent or opaque.	(a) Testes elongated, slightly lobed and whitish.
(b) Multiplication of spermatogonia, seminal vesicles are arranged in	

bundles and ampullae are solid and separated by connective tissue.	
Stage III.	Stage III.
(a) Testes are rod shaped, pink or yellowish in color. Clear blood vascularisation.	(a) Testes elongated and wide lobed and whitish color.
(b) Hollow cavity appears in the middle of solid ampullae.	
Stage IV.	Stage IV.
(a) Testes are milky white, blood vascularisation conspicuous.	(a) Testes convoluted and creamy colored.
(b) Large primary spermatocytes, small secondary spermatocytes and smallest spermatids are observed.	
Stage V.	Stage V.
(a) Testes are white in color and full of milt.	(a) Testes convoluted, soft, creamy and milt extruded with pressure on the abdomen.
(b) Large number of sperm cells and sperms are seen inside ampullae.	
Stage VI.	Stage VI.
(a) Testes are greatly decreased yellowish white or pink in color.	(a) Testes creamy colored and freely oozing milt on gentle pressure.
(b) Spermatogonia, primary spermatocytes and connective tissue remain in seminal vesicle. After exudation of milt, the testis turns back to stage III and will develop on.	
	Stage VII.
	(a) Testes reduced in size and looks blood shot or pale.

5.1 Endocrine elements in the testes

Although no general agreement has yet been reached on the precise distribution and nature of the endocrine elements within the piscine testes, it is studied that two different endocrine arrangements occur in the testis of fishes.

(1) Intrestitial gland (leydig cells)

(2) Walls of the seminiferous tubules and at called as time boundary cells. It is stated that the boundary cells have the same endocrine function as interstitial leydig cells.

The testes of teleosts are thought to be the source of androgen important in regulation of reproduction, secondary sex characteristics and reproductive behaviour. Moreover, the alternative sources of androgen within the testes are the sertoli cells (Hoar, 1969).

The 3 β, 3 α and 11 β HSD (hydroxy steroid dehydrogenase) activity was observed in the interstitial cells and it was maximum in mature testis. This activity was gradually decreased after spawning took place, reaching to minimal level in the testis. Positive reaction for 3 β HSD was observed in the lobules containing spermatozoa as well as in sertoli cells of testis.

6. AGE AT MATURITY

Under different geogrpahical and ecological conditions, the maturity age of the same species is widely different. The maturity age of common carp is one year, for grass carp and silver carp 2 + years in India. In case of Indian major carps such as for Mrigal and Rohu maturity age is 1+ years while for Catla 2 + years in India. However, in the same region maturity age varies some how rather with ecological conditions and other intrinsic factors. The induced breed species reared in pond culture practice gets early maturity than the natural riverine stock. Generally males mature little earlier than the females, in cultivable carps.

Reproduction can be generalised, as the process of maturation, ovulation and spawning. Although these terms seem to be alike but scientifically these are different. Maturation can be defined as the development of primary gonadal cells to the formation of final germ cell, while ovulation is defined as the flowing or running state of the germ cells within the gonad but are not released to the exterior of body. But spawning is defined as the release of sexual products such as ova in female or sperm in male to the exterior of body through the process of estrous. There is also another terminology called artificial insemination in which reproductive germ cells are released to the exterior of body. But in the later case by pressure on the abdomen, the sexual products were forced to be released out, where as in spawning, sexual products are released out due to courtship behaviour under controlled conditions.

7. FECUNDITY

The fecundity of a female has a considerable bearing on the reproductive potential of a population. It is defined as the number of ripening ova in the female gonad prior to spawning period. This is in contrast with fertility, which is the number of eggs shed. Rate of fecundity has been determined for many fishes which provides informations on population dynamics, racial characteristics, production and stock recruitment problems. Due to variations in reproductive habits of fish, various terminologies that are acceptable in all circumstances have not yet been devised. Welcomme (1967) used fecundity for the number of eggs produced by a mouth breeding cichlid and fertility for the number of young produced. Raitt and Hall (1967) use three definations with the viviparous red fish *Sebastes marinus* as (1) Prefertilized fecundity, i.e., number of eggs before fertilization in ovary, (2) fertilized fecundity i.e. number of fertilized eggs in ovary (3) Larval fecundity i.e., number of larvae after hatching before extrusion. However, Gerking (1967) suggest that in a mouth brooding *Sarotherodon,* conveniently one should categories: (1) Ovarian fecundity and (2) Brooding fecundity. The temperate species which shed eggs in batches, the sum of all ripening eggs during the season can be called as fecundity. Contrast to this, in tropical species fecundity must only include total of ripening eggs shed continuously in one batch only. For fecundity study, Gilson's fluid is used as fixative for separating the ripe eggs from the ovarian wall which constitute 100 ml-60% alcohol, 880 ml water, 15 ml-80% Nitric acid, 18 ml-glacial acetic acid, and 20 gm-$HgCl_2$. This mixture hardens the eggs and helps in fecundity study.

In multispawner fishes like Bleak *Alburnus alburnus* and Sprat, *Sprattus sprattus* the only difference is the degree of egg release in a season than that of a single spawner which have only one spawning time in a season. In most marine fish, the mature eggs prior to spawning swell considerably to such a size that the ovary can not contain them all and the eggs must be shed in batches although interval of shedding is short. The fishes like angler *Lophius piscatorius* and the perch shed their eggs in gelatinous mass batch by batch and are truely non-multiple spawners.

7.1 Importance

Fish fecundity is associated with studies of natural mortality, racial characters and total population estimation. This study have assumed considerable economic importance with certain species. Another application of fecundity is concerned with productivity studies (Allen, 1951).

Some workers have calculated total number of pelagic plaice eggs and divided this by the average fecundity to give the number of female parents. Again the technique was refined by taking in to account the spawning ground of plaice.

The differences in fecundities of herring between autumn and spring spawners are probably due to racial differences (Jenkins, 1902 and Ferran, 1938). Lyamin (1956), showed in Icelandic Summer and winter spawners were morphologically and physiologically similar, but their fecundities differed markedly. The fecundity rate in flounder, dab, turbot and plaice were higher in the Baltic sea than in the north sea reported by Kandler and Pirwitz (1957). The Anadromous *salmonids,* spawning in different rivers have shown racial fecundity differences. The fecundity differences in sockeye salmon between river systems is believed to be due to racial differences (Rounsefell, 1957). Although many variations are reported on fecundity between different fish populations, still it is yet to establish wheather these are governed with genetics.

Absolute fecundity is called as the total ripe eggs in the ovary of an individual fish. It is usually related to length, weight or age of the fish. It is found that fecundity increases with size and age of fish. So inorder to compare fecundities of fish of different size and from different places, many prefer to calculate the relative fecundity which is the number of eggs per unit weight of ovary. Bagenal (1963) reported that in most fish the number of eggs dose not change significantly as the season progresses but the gonad weight increases due to an increase in water content, organic matter content or organic matter transferred from somatic tissue. Only in later case total weight will remain constant making the calculation of relative fecundity more meaningful. It has been reported that in Pike, gonad increases five fold in weight from October to March and this is apparently balanced by a somatic loss and so the relative fecundity remain constant through the season. Hence the use of regression coefficient in relative fecundity assumes to be in Log F = Log a + b Log W which does not differ significantly from unity.

7.1.1. Fecundity and age

The fecundity differences due to variations of age necessiates the importance of age specific fecundity for each age group. In this context, the population fecundity which is defined as the sum total of the absolute fecundities of all brooding females and the number of eggs laid by the population in one season is important. Zawisza and Backiel (1970) after thorough analysis conclude that age after eliminating length does not explain the fecundity relation as well as does length alone. Gerking (1967) concluded that variations in individual fecundity is so great that it hides any effects of age may have. However, ludwig and Lange (1975) used age and age length interrelation in his statistical model for estimation of predictive fecundity than length alone.

7.1.2. Fecundity and Length

A close relationship is usually found between fecundity and length. These two parameters have been plotted by several workers as a scarttered diagram (adopted by Bagenal 1966) and they have concluded that the relationship is $F = aL^{b}$. The logarithmic transfermation gives the straight line regression of, Log fecundity on Log length as Log F = Log a + bLogL, where a is constant and b is an exponent.

7.1.3. Fecundity and weight

It is often thought that fecundity is closely related with weight than length. However, in many fish the somatic weight changes significantly towards spawning due to nutrient flow from somatic tissue to ovarian tissue. Hence, the relation between fecundity and weight will differ as breeding season approaches

7.1.4. Fecundity and egg size

Fecundity and egg size are often correlated. It is generally assumed that fish that lay many eggs must lay small ones. Rounsefell (1957) has reviewed that the salmonids with the longest freshwater lives have the largest eggs such as *Oncorhynchus* and smaller egg in *Salmo genus.* within the genus, *O. nerka* lay smallest and *O. tshawyscha* the largest. The winter - spring spawners of herring have large eggs and low fecundities while the summer -autumn spawners have small eggs and high fecundities (Parrish and Saville, 1965).

The size frequency diagrma of ovarian egg gives the idea of the spawning cycle of a fish. The fish group with same egg size frequency in ovary indicate about a recruit spawner (single spawner) and the fish group with different egg size frequency in ovary indicate about the multi-spawners as in case of Bleak, *Vimba vimba* and Sand goby (*Pomatoschistus minutus*). In sand goby, the peculiarity observed is that a part of ovary containing ripe eggs get constricted and then separate off. The eggs contained in this part are shed as a batch and the second lot begins to bud off in the ovary.

Macer (1974) reported in his observation on some horse mackerel (multiple spawner) that there could be a substantial differences between fecundity and fertility because in such case of multiple spawner, the ovary goes through part spent to recovering state with many eggs unshed. However, nearly all spent fish have some residual eggs after spawning but in most species this is not a significant number compared with the total shed.

7.1.5. Fecundity and spawning time

Some had postulated a spawning sequence of herring in which fish travel to the spawning ground with the sequences of gonad development. The spawners which first spawn are less fecund due to transfer of somatic reserve to gonadal maturity. This loss of somatic reserve is build-up through compensatory feeding. This means higher the fecundity, longer must be the period of compensatory feeding. In carps, the same brooder can be breed more than once by intensive compensatory feeding under controlled conditions. Therefore, in high fecundity group of fish, the maturation takes longer period as in carps and in less fecund group of fish as in tilapia, maturation takes short period. It is also clear that high stocking density leading to overcrowding and subsequently food shortage results in a lower fecundity.

7.1.6 Fecundity and environment

It must be realised that any change in the environment may result in significant changes in fecundity and so alter production estimates. An increase in food supply with increase in food in take not only increase the growth rate but also increase the gonad weight due to exogenous flow of yolk (vitellogenin) for a given age. Therefore, the fecundity for a given length increases. Many examples of these changes are given by Nikolskii (1963, 1969).

7.1.7. Fecundity and egg quality

The eggs of most species vary in size (Bagenal, 1971) and chemical composition. Some of these variations will be important from the view point of fish production. Many attempts in this regard has been under taken and in the present context it is only to draw attention to the necessity of considering this aspect as a meaningful aspect of fish production studies.

8. FACTORS AFFECTING FECUNDITY

Various factors attributes to the variations of fecundities in each species of fish. As fecundity is correlated to ovarian development, the factors related to ovarian development have considerable effect on fecundity also. Among these factors the important ones are (1) Biological, (2) Environmental, (3) Nutritional, (4) Diseases and, (5) Physiological. Biological factors include stocking density, length, age, size at first maturity, Gonadosomatic Index, Ponderal Index, Relative condition factor, Ova size and seggregation to develop sexual urge for better fecund. Environmental factors include light, temperature, running water, pH and physico-chemical qualities of Soil-water interaction which could affect the food intake and keep the fish under stress are found to be less fecund. Nutritional requirement and nutritive qualities of diet enhance the fecundity in fishes. In the opinion of Chinese, Grass carp when fed with wheat sprouts, Soyabean, and Peanut dredy had better fecund of 57500 eggs/kg body weight compared to fish fed with corn and rice bran had 38214 eggs/kg body weight. The fish fed with Rice-bran, barnyard grass, and rice minced had 27236 eggs/kg body wt. The studies have also demonstrated that essential fatty acid deficient diets greatly affect the spawning of rainbow trout and red seabream. Hence essential fatty acid play an important role in reproductive physiology. Essential fatty acid deficient casein diet containing methyl laurate as the sole dietary lipid in rainbow trout caused maturation but the eggs produced had low hatching rate. However, total egg production and hatchability were significantly influenced in red seabream by essential fatty acid status in the diet and were quite low in fish groups given the essential fatty acid deficient diet. Similarly running water flow before spawning and slight running water all round the year in both Grass carp and Silver carp had resulted better fecund, spawning rate and fertilization rate than the fish subjected to no running water condition.

Jhingran and Sehgal (1978) emphasised that fecundity in cold water species are related to the green egg diameter. The fecundity is low with increase in diameter of egg. The number of green eggs per kg body weight ranging from 1417-1542 when fed with nutritive diet as against 1234-1342 in control. In rainbow trout the number of eggs was 1943 per kg body wt. in seggregated female with nutritive diet against 1649 in control. These aspects clearly indicate the impact of nutritive diet and seggregation on sexual urge that enhance considerably the fecundity in this species. Parasitic infection and physiological disbalance due to extrinsic or intrinsic factor affect greatly to fecundity as the progress of maturation is also severely affected due to above reasons.

9. FECUNDITY OF CULTIVABLE CARPS

There are considerable variations in fecundity in regard to each species, age of the fish and ecological conditions of habitat including the climatic factors prevailing on the locality of species occurrence. In several species, differences in fecundity is dependant with size of ova, length of fish, availability of food and other factors suggested earlier.

9.1 Catla catla

The fecundity range was found to be 20-246 numbers per gm body weight of pond reared fish and the number of eggs vary from 604 to 1330 per gm of ovary weight as reported by Sukumaran (1969). In a 5 kg female catla, a total number of 40,00,000 eggs were found which gives the fecundity to be 78 eggs per gram body weight. However, Jhingran (1966) reported the fecundity of different length groups of catla as follows -

Total length of fish	Wt. of fish	Wt. of ovary	No. of eggs	No. of ova per gm wt. of body	No. of ova. per gm. wt. of ovary
78.3 cm	113 kg	3000 gm	2300,831	20	767
84.0 cm	13 kg	3118 gm	2963,125	228	950
92.5 cm	17 kg	4222 g	4202,250	246	950

9.2. Labeo rohita

The absolute fecundity of several specimen of rohu, ranging from 51-75 cm (1.5-7.6 kg wt) was found to vary from 2,25,600 - 27,94,000 eggs. The average number of ova per gram of ovary wt. was found to be 1258. The number of eggs per gm. of body was 271. The fecundity range was found to be 109-413 numbers per gm. body wt. of fish and the number of eggs vary from 747 to 1528 per gm. of ovary weight as reported by Sukumaran (1969). Such variations may be due to age, length, nutritional requirements and ecological conditions of species occurrence.

9.3 Labeo calbasu

The data available for calbasu shows that a female weighing 500 gm (34 cm length) had 2,19,450 ova. The average being 438 eggs/gm body wt. and 1463 per gm ovary weight. It was also found that in a ripe female weighing about 2 kg had 7,39,400 eggs which works out to about 400 eggs/gm body weight.

9.4 Labeo bata

The fish get mature in 9-10 months (Alikunhi, 1956). He reported that a specimen weighing 1050 gm had total eggs of 3,00,000 giving an average of 286 egg/gm body wt. It was also reported that in specimens weighing 1.38 kg to 1.78 kg from Bhakra reservoir had the total fecundity range from 3,01,961 to 5.76,251.

9.5 Cirrhina mrigala

Both the sexes mature when they are two years old. The induced breed specimen become sexually mature in 1+ year. In a ripe female of 1.47 kg, a total number of 2,16,800 eggs were reported giving an average of 147 eggs/gm body wt. The fecundity ranges from 1.44-1.52 lakhs per kg body wt. of female has been also reported. The specimen weighing 2.5-7.5 kg and inhabiting in river Yamuna had the fecundity range of 800-1800 eggs/gram of ovary weight. Sukumaran (1969) reported that the fecundity ranges from 32-280 eggs per gram body wt. and 800-1836 eggs/ gm ovary weight.

9.6 Silver carp

Alikunhi *et al.* (1963) reported the fecundity of silver carp ranging in weight from 3-8.5 kg was found to vary 1,45,000 to

30.44,000 giving an average of 171 eggs per gm of body wt. However, the study made by Sukumaran (1969) suggests that the fecundity ranges from 46-313 per gm body wt. and 318-1768 per gram ovary weight. Chinese have reported that the average relative number of egg per gram body wt. of silver carp was 141 in case of pond reared specimen and it is only 131 in specimens studied from natural riverine condition. They are also in opinion that, running water condition provided before spawning or all year round had increased fecundity rate than that of fish without subjected to running water condition.

9.7 Grass carp

Males mature when 2 years old and the females are reported to mature a year later in relation to ecological conditions. Among the induced breed specimens, males started oozing by the end of first year, but females mature when about 2 years old. In a specimen weighing 4-7 kg. Alikunhi *et al.* (1963) found total number of 3,08,800-6,18,100 eggs. In other case of female fish weighing 3-4 kg were found to possess 1,80,000-4,92,750 eggs. However, Sukumaran (1969) reported that on an average, the fecundity was 82 eggs/gm body wt. and 610 eggs/gm of ovary weight. But Chinese have reported that the average relative fecundity per gm. body wt. in grass carp is 120 in pond reared specimens as against the fecundity of 90 eggs/gm body wt. in natural riverine conditions. They are also in opinion that running water condition provided before spawning or all year round had increased the fecundity rate than that of the fish without subjected to the running water condition. The size group and fecundity variations of Grass carp is greatly influenced with size of egg, age, length and other ecological parameters. Chinese are in opinion that the feed composition has considerable impact on the fecundity of grass carp. They have suggested that the Grass carp fed with wheat sprouts, rice sprouts, Soyabean and Peanut dred, gives more eggs (57500) per kg body wt. than those fish fed with corn and rice bran only (38214) or rice bran, barnyard grass and rice minced (27236 eggs).

9.8. Cyprinus carpio

The absolute fecundity is reported to vary from 39400-16,59,000 eggs. It is very difficult to compare the fecundity of common carp from different regions as it breed more than once in a year in tropical countries and only once in temperate country. Moreover, the absolute and relative fecundity of common carp in India varies with length group being reported by (Alikunhi, 1966). According to him a fish length range of 15-25 cm had absolute fecundity from 6360-25942 and relative fecundity per gm of ovary wt. range from 517-1388. In the length range from 25 to 45 cm, the number of eggs/gm of ovary wt. varies up to 1793 and after that with increase in length group, there is decrease in number of eggs/g of ovary weight suggesting that age and age-length interaction to be taken in to consideration for statistical interpretation of fecundity than length alone.

Length alone or age alone in relation to fecundity study does not provide much meaningful statistical interrelationship. The variations in fecundity studies in some group of fish species reported by each of the authors are necessarily to be computed statistically to get better insight in the subject.

9.9 Big head

Chinese have reported that the average relative number of eggs per gram body weight is 124 in pond reared species as against the fecundity of 96 eggs/g body weight in natural riverine condition.

9.10 Black carp

It is cultivated in China and the fecundity is 114 per gram body weight in pond reared species as against the fecundity of 93 eggs/g body wt. in natural condition. Variation in fecundity is also reported in this species due to ecological and physiological factors.

9.11 Mud carp

The average relative fecundity per gram body wt. is 240 in pond reared specimen.

In other group of fishes, like *Notopterus,* the absolute fecundity is low ranging from 1000-2000 and measure about 3.5 - 4.5 mm diameter. In Singhi, fecundity varies between 1100-37,000 in species of 120 to 285 mm in length weighing 11 to 112 gm. Fecundity of *Clarias batrachus* is size dependant which varies from 1000 to 35000 in the length range of 15-40 cm. on an average, such fish produces 53 and 511 number of eggs per gram of its body weight and ovary weight respectively. However, the report of coordinated project on air breathing fish culture in Swamp suggests that the fertility i.e. the number of hatching produced in *Chana striatus* through artificial propagation ranges from 1304-23180, in *C. puncctatus* ranges from 3560-13960, and in *C. gachua* ranges from 3059-8089. In *Anabas testudineus* fecundity varies from 6000 to 29000 in specimens ranging from 91 to 140 mm. Rath and Hejmadi (1976) reported that the fecundity of *C. punctata* ranges from 733 to 9225. The fecundity in *C. punctata* was found to be linear with respect to its body length, body weight, ovary weight and length. Some have reported the curvilinear relationship with length in some other species i.e., *M. cephalus* probably due to ecological differences. In *C. punctata* the lower range of fecundity is an indicative of parental care as in case of cold water species and in certain other livefish groups. Such low fecundity in *C. punctata* is correlated with a shorter development time and lower mortality rate in natural condition which means higher percentage of survival. It is accepted as a generalised statement in Airbreathing or live fish groups. In contrast high fecundity of carps is a generalised selection character for aquaculture.

10. ECOLOGICAL CONDITIONS ON GONAD DEVELOPMENT

Several abiotic factors such as light, temperature, water and metereological conditions play an important role in transmitting the informations through neural signals to the hypothalamus and then to hypophysis for the release of pituitary gonadotropin within the organism and thereby control reproduction of fish.

The range of environmental factors required to stimulate gonadal maturity varies from species to species. Beyond the optimum range, these factors act adversely, inhibiting reproduction in fishes. The main ecological factors that control and effect sexual maturity and sexual cycle of fish are nutrition, water temperature, water current, dissolved O_2 etc.

Nutrition affects greatly the development of gonad. In cultured fish, the gonadosomatic Index is generally 3 to 6% before February and after April, it rapdily increases to 14-22%. This indicates that with increase in temperature and when fish are better fed, gonad begin to grow rapidly. Among the required nutrients (proteins, fats, carbohydrates, vitamins and minerals) vitamin E is essential for gonad development. In order to ensure proper gonad development, proper nutrition must be combined with other favourable ecological conditions so that the gonad reach maturity. As fat deposition retards gonadal maturity, the supplementary feed should have least amount of fat (1-2%) and there by ensure least fat deposition.

10.1 Water temperature

Water temperature is an important factor affecting metabolic rate, age at first maturity and maturation rate of gonad. The differences in water temperature restricts the growth period of fish. The growth period is counted generally when the monthly average water temperature is above 15°C. In tropical climates usually the growth period of fish is in the range of 10-12 months in a year. In subtropical or temperate climate the growth period of fish is in the range of 8-11 months and 5-5.5 months respectively in a year. Moreover, grass carp the native of white Amur river matures in 2+ years in India but it even takes about 9 years to mature in Russia. Similarly silver carp which matures in 2+ yrs. in India, takes about 5-6 year to mature in Heilong Jiang province of Northern China. But silver carp matures in 2 years in Guangxi province of South China and 3-4 years in Jiangsu province of Central China due to differences in water temperature and growth period. Nevertheless, the accumulated temperature required for maturity is around 18,000-20,000 degree days. However, Horvath (1978b) has reported that it may be assumed 8,000 to 10,000 degree days are necessary for entire cycle of stage I oocyte to VII oocyte in carps of central Europe. The carps usually take 4 years to attain maturity. He also reported that the number of degree days between two ovulations may be even less than 2,000 degree days. This may be due to the presence of large number of stage VI oocytes in the ovary after ovulation. Warm temperature plays a primary role in stimulating the maturation of gonads in a number of fishes and also accelerates spermiation (Ashan, 1966). The Indian major carps are observed to spawn within a temperature range varying from 24-31°C. Observations show that there are optimum temperature ranges for induced breeding of cultivated fishes. However, above or below the critical temperature limit, fish do not reproduce. However, at a uniform temperature throughout the year, one female can theoretically propagate, 4-5 times and can serve as a basis for industrial propagation method, independently of the season. The scientists of CIFA have practically demonstrated the multiple breeding of cultivated carps in Kausalyagang fish farm, Orissa, with 45 days of gap between two successive breeding of same individual carp feed with nutritional enriched diet. However, fish raised in relatively low ater temperature (i.e. below the critical temperature of water) can be ripened by adminstration of gonadotrophic hormone. Thus it is apparent that temperature has direct effect on gonad maturation as well as indirectly stimulates the pituitary for release of gonadotropin for gonadal maturation. With reference to the spawning, fishes if given an effective dose of pituitary, spawn successfully even when there is a substantial increase or decrease in water temperature. According to Clemens (1967), pituitary injection to some extent bypass the environmental variables of temperature, light and rain.

10.2 Water current

Allowing mild water flow at a definite or indefinite time keeps water quality relatively fair which is beneficial to growth. Moreover, the maturation of gonad is enhanced in confined water system by maintaining a mild flow of current towards the onset of breeding. Such mild flow of water current can regulate the composition of the natural food as well as nutrient cycle to raise the nutritional level of the brood fish. Running water accelerates metabolism to the gonad. Water current has direct as well as indirect effect (stimulating for release of gonadotropin) on the gonad.

Slight running water year round maximizes the spawning rate of carps, particularly silver carp. Slight running water year round and before spawning have similar positive effects on the spawning rate of carps particularly grass carp. If there is no running water, the spawning rate drops significantly.

10.3 Dissolved Oxygen (DO_2)

Oxygen is essential for survival. When dissolved oxygen is 2mg/litre, normal physiological activities are drastically reduced and fish gasp for air. At this level excessive energy consumption negatively affects gonad development. Usually 5-5.5 mg per liter dissolved oxygen (DO_2) of water is necessary to meet the demand of fish. Horvarth and Peteri (1980) reported that oxygen concentration also conditions the success of hypophysation.

10.4 Light/Photoperiodicity

Intensity of light and time of exposure are very important factors in controlling reproduction in fishes. Early maturation and spawning of fish as a result of enhanced photoperiodic regime have been reported in number of fishes. Contrary to these some fishes in northern latitude attain early maturity in short light period and delayed maturation in long photoperiod. For example *Salvelinus fontinalis, Oncorhynchus rhodurus* and *Plecoglossus altivelis* (Shiraishi and Fukuda, 1966) attain early maturity in short light period under experimental condition and delayed maturation in long light period. But in case of rainbow trout long photoperiod is necessary to stimulate gonadal development. Varghese (1967) reported that, *Cirrhinus reba* attains early maturity when subjected to artificial day length longer than natural day even in winter months of low temperature. Resorption of gonad can be inhibited and the spawning condition of this species was maintained up to November. Photoperiodic sensitivity passes through retinal route and helps in gonadal maturation. In certain fish groups retinal route were checked through blind, still maturation of gonads were observed. Therefore, it is believed that such photoperiodic sensitivity passes through some extraretnal route and enhances maturation.

These studies reveal that the requirement of light for stimulation of gonad cycle varies from species to species, from latitude to latitude, with variations of accumulated heat budget and exposure of day length.

10.5 Social factor

This includes the presence of opposite sex. Yamazaki (1965) and Yamamato *et al.* (1966) indicated in female goldfish that the presence of sexually active males induce ovulation. Kyle *et al.* (1985) reported that the presence of a receptive female or stimulus pairs of goldfish induce an increase in expressible milt volume and plasma gonadotropin (G t H).

Nikitina and Godovich (1983) have deomonstrated the dynamics of sex hormones in the blood of male common carp, depend on the presence or absence of spawning in the females placed with them. In China, mud carp is induced breed by the presence of active male spawner.

Besides food availability, water current, dissolved oxygen concentration, light (Photoperiodicity), temperature, other factors like pH (Hydrochemistry), water space (density), benthos, age, rain fall/monsoon, floods (turbidity) and presence of the opposite sex affect gonad development and spawning. Although these factors are responsible for gonad development, still no critical factor was assigned to the triggering of spawning in nature or in induced breeding. However in a natural spawning in riverine condition a sudden drop in the electrolytic level in the water by flood or heavy monsoon rain induced hydration of the gonad and those which were ripe started spawning are reported (Sinha *et al.,* 1974).

Further the total water content of the ovary, the fluid space content between the follicular cells, loosening of follicular cells and increase in diameter of egg just before spawning suggests about the absorption of water into the ovary and ovarian egg accompany ovulation process. Although fish get mature in confined water, but they do not breed because of mechanical barrier i.e. chorion which may mechanically oppose or support the entry of ions and water to the egg for its final maturation. Contrary to this males do spermiate in ponds because the sperm has no mechanical barrier as chorion in the case of egg. Wheather endogenous or exogenous inflow of water results hydration still greater possibility may exist to manipulate extrinsic factors and synchronising physiological intrinsic factor so as to provide better success in induced spawning.

Swingle (1956) observed that fish secretions released into water constitute what he called `repressive factor'. These secretions are NH_3 and other metabolites which inhibits gonadal maturation and spawning. When such highly metabolite water is diluted by fresh rain water run off or flood water in bundh or tanks results in spawning. According to Lake (1967), the factor stimulate the fish to spawn is produced when water comes in contact with dry soil and called it ``petrichor" which he identified as an oil obtained by steam distillation of silicate minerals and rocks. According to him aroma of petrichor stimulate the fish to spawn as fishes possess a remkarkable sense of odour perception.

Therefore, it is essential to know the environmental factors like photoperiodicity, temperature etc. which act as selective factors for the reproductive strategies in fish. The role played by each of these factors and combined effects of all these factors need to be understood. Moreover, nutritional requirements, water exchange and related factors responsible in order to govern the fecundity and physiological development of spawners are to be clearly understood. It is most important to get the fish physiologically fit to breed for which procedure for systems's approach have to be finalised. Mulispectral appraoch like nutritional assimilation rate, metabolic rate, conversion rate, enhancement of formation of oocytic stages, time mapping of plasmagonadotropin after inducing has to be systematically studied so as to make the fish physiologically suitable to breed successfully. Moreover, hydro-inflow in relation to conductivity of pond water, permeability of ovarian wall, follicular cell and circadian rhythm relationship after inducing by hormone in carps can provide new tool for successful spawning.

The studies reveal the functional approach to the physiology of hormone secretory glands, type of hormone and hormonal effects on maturation, ovulation and spawning of fish. The effective dosage of hormones to be administered for spawning of carp at present is roughly arrived at by actual experimentation in each particular instance depending on the species, stage of maturity, potency of hormone adminstered and the physiological state of the recipient. The studies indicate two possible hypotheses in regard to functional approach of reproductive physiology of fish through direct gonadotrophic route (Hypothalamo hypophysial-gonadal axis) and indirect corticotrophic route (Hypohalamo hypophysial-ovarian Interenal axis). However, side effects of hormones on metabolism, genetic and sexual behaviour of fishes has given better insight in this aspect.

11. RELATION BETWEEN ENDOCRINE SYSTEM AND REPRODUCTION IN FISHES

Like other chordates all the physiological activities of lower vertebrates particularly fishes are regulated, mediated and controlled by the nervous system and the endocrine system (Fig. 32). External stimulus mediated through the neural signals to the hypothalamus a part of central nervous system is the principal system of receiving nervous impulses and mediated the signals to organ concerned in response to such stimulus (Rath, 1987a). The secretions of hypothalamus being mediated to organ concerned is called neurosecretions. Ramaswami (1978) has given an account of review articles on vertebrate neurosecretions. He has cited the relation between the hypothalamus and hypophysis in vertebrates particularly fishes.

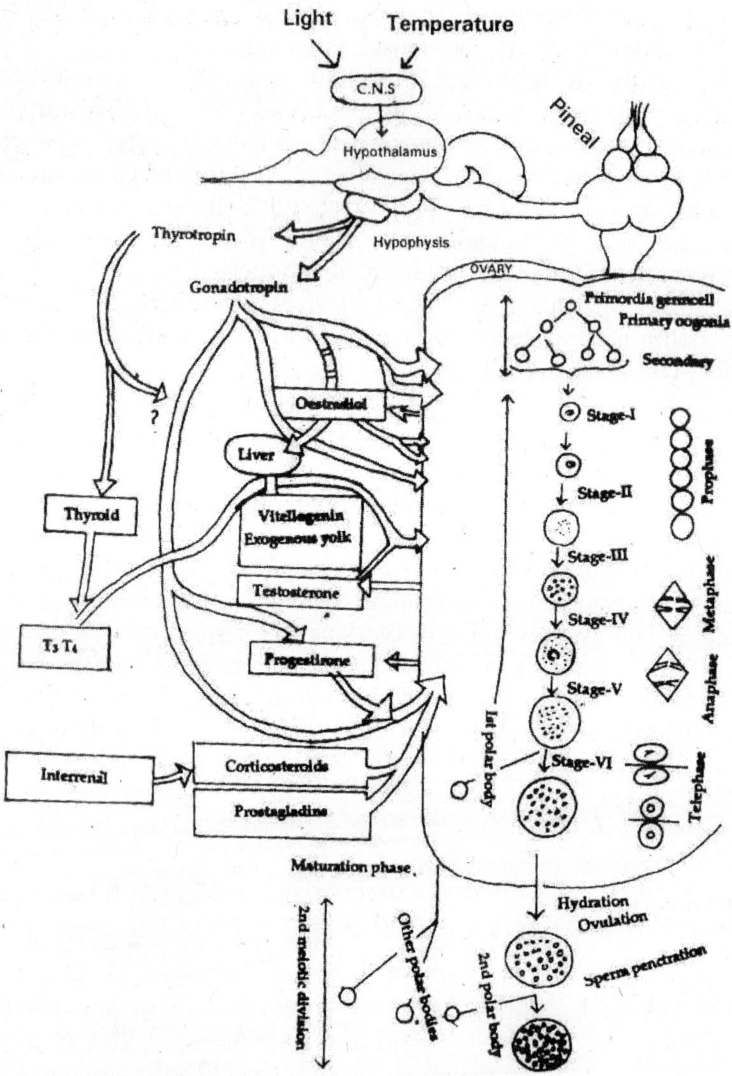

Fig. 32. Endocrine control of reproduction.

11.1 Hypothalamo-hypophysial relationship

Dahlgren (1914) is the pioneer in the study of neurosecretion by describing that certain glandular cells in spinal cord of sharks exhibit neurosecretory functions. Further it is described that certain glandular cells from the preoptic nucleus (PON) of the hypothalmus of the minnow (*Phoxinus laevis*) and speculate upon their endocrine function. In 1940s, it was established that these cells were found ending up in the neurohypophysis (nervous part of hypophysis) and by neural mechanism, monitors hormonal secretions of the hypophysis. It was stated that hypothalmus is primarily concerned with the functions of light, heat, sleep and apetite regulations. But subsequently, hypothalamus was recognised as an independent endocrine gland. Tranducer role of neuron passes neuron-neural information from the hypothalamus to the hypophysis. It is stated that the vascular connection plays as important role and that the hypothalamus secretes a substance (hypophysiotropin chaemotransmitter, neurohormone, neuro transmitter) which controls the secretion and release or inhibition of the hypophysial hormones in a masterly way.

11.1.1. Cyclostome (Myxinoidae)

Olsson (1959) identified a vascular region in the ventral hypothalamic floor in myxine into which neurosecretory material appeared to be released. Further it was described that a prehypophysial vascular plexus in the region of optic chaisma in *Polistorema stoutii* from which blood drains into the neurohypophysis and there are few connecting vessles between neurohypophysis and adenohypophysis in myxinoides. Although inter connections are few, still the hypothalamic influence is mediated through these connecting vessels to the adenohypophysis. So in general for myxinoids (hag fish), hypothalamic regulation in adenohypophysial function is set at a very poor level.

11.1.2 Cyclostome (Petromyzon)

Ozton and Gorbman (1960) reported that in Petromyzon, the floor of the infundibulum is thin and connected with the underlying pituitary gland by a set of blood vessels. The preoptic nucleus send bundles of fibres posteriorly of which one set goes to the hind brain, other one to neurohypophysis and the third set of fibres passes close to the floor of the infundibulum and enters the neurohypophysis. Further it was found that the pituitary receives its vascular supply from the internal carotid and then it takes the form of a series of arches. However, Lampreys do not have median eminence and the pars distalis of the Lamprey can maintain normal reproductive physiology, when transplanted into the eye or into a pharyngeal muscle (see Larsen and Rothwell 1972). Larsen and Rothwell (1972) believes that the central nervous coordination may not be required for sexual maturation in Lamprey.

11.1.3. Elasmobranch

The hypothalamo-hypophysial system in the Indian Shark, *Scoliodon sorrokowha,* has been studied and in this case the hypothalamus has a anterior thicker and a thin posterior portion. Further more, there is some evidence that the hypothalamus may control activity of the isolated ventral lobe via the systemic circulation.

11.1.4. Teleosts

It was recognised in teleosts, that the hypothalamus functioned as an intermediary between nervous and endocrine effector systems and thus between the animal external and internal environment. Besides this, there are data suggesting that fish reproductive cycles are controlled by this interplay. The preoptic nucleus (PON) was first identified in bony fishes as long as 1891 (Herrick, 1891) and its neurosecretory function and direct connection with neurohypophysis were established by Scharrer (1930) and Palay (1945) respectively. Stahl (1953) believed that this PON acts as an intermediary between the external environment and pituitary in controlling fish reproductive cycle.

Peter (1970) demonstrated the importance of nucleus lateralis tuberis (NLT) by the use of lessioning techniques. Through this technique, it is found that the external stimuli can pass through neurons in NLT and affect the pituitary. Both nervous and vascular routes are possible between hypothalamus and pituitary and evidence shows that infact both routes are present. Breton *et al.* (1971) and Breton and Weil (1973) have shown that extracts of carp hypothalamus as well as Synthetic LH/FISH-RH, when injected into carp cause a rise in plasma-gonadoropin in 2 to 4 minutes following the injection. So there exists a hypothalamic control on pituitary gland (hypophysis) of teleosts.

11.2 Neuro secretory nuclei in lower vertebrates

In the fish generally there are two principal nuclei (1) the Pre-optic nucleus (PON) and (2) the nucleus lateralis tuberis (NLT) in the neurohypophysial region which contain several neurosecretory neurons (Fig. 33). The type A fibre (neuron) is stainable called peptidergic where as Type B fibre non-stainable is called aminergic. The peptidergic fibres arise from the pre-optic nucleus and aminergic fibre from nucleus lateralis tuberis (NLT). Some still believe that besides aminergic fibres from NLT, some peptidergic fibres may be present. Further some are in view that these neurons or fibres are arranged in distinct groups called as nuclei.

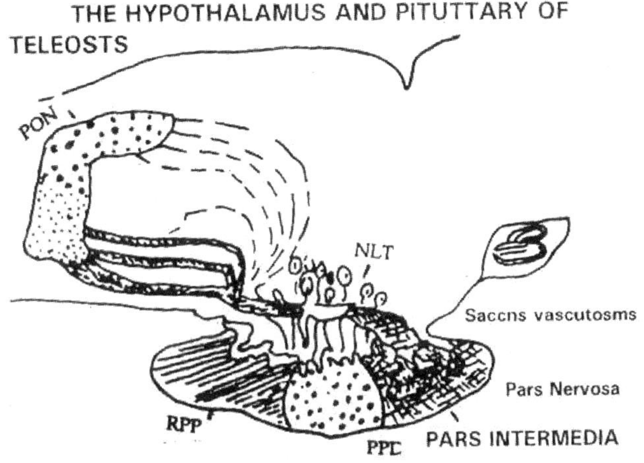

Fig. 33 : Nuclei locations (NPO and NLT) in Hypothalamus and pituitary of a Teleost.

12. PITUITARY GLAND

There have been several excellent review articles on the reproduction of fishes (Hoar, 1969; Donaldson, 1973; Dodd, 1975 & Matty, 1985) indicating that pituitary regulates both gametogenesis and steroidogenesis in fishes. That too, Tripathi and Khan (1990) gave an exhaustive account on carp seed production technology and Nandeesha *et al.* (1990a) gave an account of alterante inducing agents for carp breeding in India. However, besides hypothalamus and pituitary, hormone secretory glands in fishes are : (1) Gonad (ovary and testes), (2) Interrenal/corpuscles of stannius, (3) Thyroid, (4) Adrenal, (5) Granulosa cells, (6) Urophysis and (7) Pineal.

Khan (1938) has described on the effect of administration of extract of anterior lobe of pituitary gland on the ovulation of fish. Other workers like Ball (1960), Chaudhury and Alikunhi (1957),

Ramawamy (1962), Sahu and Biswal (1988), Rath (1980), Chondar (1985), Kaul and Rishi (1986), Nandeesha *et al.* (1990a) have described different chemicals and hormones on the gonadal development and ovulation. Studies by Pickford and Atz (1957), Atz (1973), Ball and Baker (1969), Kuo *et al.* (1973), Sundararaj and his group. Hirose (1973), Ishida *et al.* (1972), and Yamazaki (1969) have given detailed account of reproduction by endocrine glands.

12.1 Evolution of Pituitary glands in lower vertebrates

12.1.1 Protochordates

In protochordates, the head regions have a ciliary organ e.g., Saccoglossus. In hemichordates, the head region has a pre-oral ciliary organ e.g. Branchiostoma. In amphioxus, the head region has a Hatschek's pit and wheel organ. The ascidians have neural complex. All these animals are highly specialised. So it is unlikely that any of these ciliated organs are direct forerunners of the pituitary. However, Hatschek's pit and the wheel organ develop from the pre-oral pit of the larval amphioxus which opens into the most anterior of the coelomic pouches. Rathke's pouch, which gives rise to the adenohypophysis during vertebrate ontogeny, has closely similar relationship with the anterior coelomic pouch. Barrington (1964), accounts that Rathke's pouch must be homologous with Hatschek's pit and the wheel organ.

12.1.2. Cyclostomes

The class cyclostome comprises two very different orders (1) Myxinoidae or hagfish with only 2-3 living species and (2) Petromyzontidae or lamprey with 31 living species. Studies have been made by scientists to test the effect of mammalian pituitary extract, preganancy urine, testosterone and oestrone on the secondary sexual characters of intact adult and larvae. Report suggest that some stimulation of the secondary sexual characters, especially the cloacal labia was produced by all the substances tested.

The striking characteristic in pituitary gland of myxinoidea was the absence of contact between neurohypophysis and adenohypophysis. Fernholm (1972) stated that although complete separation of neurohypophysis and adenohypophysis was usual, still modified adenohypophysial bridging tissues were encountered in

17 of 264 animals investigated. This suggests that probably neuron elements may penetrate the adenohypophysis in some cases (Fernholm, 1972).

In petromyzontidae (Lamprey), the pituitary is highly organised than that of hagfish. Roth (1957) and Evenett (1963), suggested the presence of gonadotrophs in pituitary which enhance maturation of gonad in petromyzontidae.

12.1.3. Elasmobranch

The elasmobranch pituitary show great degree of sub-division into regions (Norris 1941, Dodd, 1963, Dodd 1972, Ball and Baker 1969). In elasmobranch the bioassay data and fine structural study reveal that ventral lobe as well as median lobe play role in reproduction and the importance of this ventral lobe in the reproductive physiology of the Dog fish was established. This ventral lobe has come subsequently as the main if not the only source of gonadotropin in the elasmobranch pituitary (Ball and Baker, 1969). This ventral lobe is vascularised directly from the internal carotid arteries and the importance in reproduction has been well established.

12.1.4. Teleosts

Interest has been focused on the study of the pituitary gland of teleost fishes and several workers have studied its structure, morphologically, histologically and histochemically. Attemps have been made to know its physiological significance.

The pituitary gland of mature fish is a creamy white in colour. It is situated on the underside of the brain anterior to the lobi inferiors. The gland is lodged in the depression known as Sella turcica in carps and is covered by a thin membrane called duramater. The pituitary gland in teleosts has been categorised in to two groups on the basis of its attachement with the brain by infundibulum or hypophysial stalk. If the pituitary is attached with a hypophysial stalk, then that type is called as Leptobasic and if it is without stalk, then that type is called as platybasic type (Bretschneider and Dewit, 1947). This corresponds to `B' and `A' type of Kerr (1942) respectively. The leptobasic type has again been differentiated into cranio, caudo and dorso basic types according to the position of the hypophysial stalk entering the pituitary gland from either of the anterior, posterior and dorsal regions of the gland respectively.

The fish pituitary gland possesses two distinct parts (1) adenohypophysis (glandular part) and neurohypophysis (nervous part). The neurohypophysis is directly connected to the hypothalamus with its nerve fibres and blood vessels planted deep into the adenohypophysis. The adenohypophysis could be divided in to anterior (proadenohypophysis/Rostral pars distalis), transitional (mesoadenohypophysis/proximal pars distalis), posterior (meta adenohypophysis/pars intermedia).

The teleostean pituitary, the orchestra of all endocrine glands contain gonadotrophic cells which are variously named by different workers. However, these have long been classified as ``Basophils", or ``Cyanophil-I". These gonadotrophic cells are generally found in the proximal pars distalis and in sexually mature bony fishes like eel and pacific salmon, these cells also invade the rostral pars distalis. Cyanophil-II are recongised as thyrotrophs. However, the cytochemistry techniques are now going to be replaced by immunocytochemistry inorder to study the different cell types in adenohypophysis and their functional relation to gonadal maturation, ovulation and spawning in fishes. With reference to the histochemical studies one type of gonadotrophic cells could be recognised favouring the existance of a single gonadotropin in teleosts. Knowles and Vollrath (1966) in eel pituitary, some of the gonadotrophic cells contain granules of 190 nm (nanomicron) diameter may be LH secretory while some other gonadotrophic cells contain granules of 130 nm (nanomicron) may be FSH cells. Some with fluorescein labelled antibodies identified the gonadotrophic cells in teleostean pituitary. This hormone stimulates growth, development, maturity and ovulation of eggs. The hormones secreted from pituitary are protein or peptide. That too the number of hormones secreted by the adenohypophysis of fish pituitary is disputed, but majority of the endocrinologists are in view of following important hormones such as :

(1) Somatotrophic hormone (STH)

(2) Adrenocorticotrophic hormone (ACTH)

(3) Prolactin

(4) Thyrotrophic hormone (TSH)

(5) Gonadotrophic hormone (FSH and LH)

The STH and ACTH are proteins or peptides with no carbohydrate moiety but TSH, FSH and LH are glycoprotein or mucoprotein in nature.

The hormone cells present in rostral pars distalis have been identified as ACTH and prolactin cells. The proximal pars distalis contains somatotrophs (Growth hormone) thyrotrophs and gonadotrophs and the pars intermedia contains MSH and PAS+ ve cells.

12.2 Neurohypophysis

Two chemically related hormones are usually secreted by the neurohypophysis of vertebrates. These are peptides containing eight aminoacids.In most teleost fishes these two hormones are Arginine vasotocin and Isotocin. Isotocin is otherwise known as Ichthyotocin. So far nine such homologous peptide hormones have been identified in other higher vertebrates such as :

(1) Arginine vasopressin (AVP)

(2) Lysine vasopressin (LVP)

(3) Arginine vasotocin (AVT)

(4) Oxytocin

(5) Mesotocin

(6) Isotocin

(7) Glumitocin

(8) Valitocin

(9) Aspartocin

The endocrine elements in the gonad and their role in reproduction has been described earlier.

13. INTERRENAL AND CORPUSCLES OF STANNIUS

Corpuscle of stannius are very small in size, more or less spherical or oval in shape and lie usually as a single pair on the dorsal aspect of the mesonephros. Idler and Truscott (1972) gave detailed account on corticosteroids in fish. The role of steroid hormones in reproduction in fish is well documented in the International Congress held in Delhi by Indian National Science Academy in 1978.

Donaldson (1973) concluded that in some catfish species, gonadotropin may stimulate ovulation by induction of corticosteriodogenesis in extra ovarian tissues, possible the interrenal (corpuscles of stannius) while in other species such as medaka *Oryzias latipes,* ovulation may be caused by gonadtropin induced ovarian corticosteroidogenesis. Robertson and his coworkers and McBride and Overbeeke (1969) reported progressive hypertrophy of the interrenal cells during sexual maturation and increase in the plasma concentration of 17-OH. Corticosterids (Pickering and Pottinger, 1983; Hane and Robertson, 1959; Hane *et al.,* 1966). A similar observation was made concerning the plasma corticosteroids by Schmidt and idler 1962 in *Onchorynchus nerka.* Hane *et al.* (1966) reported that the possibility of hypertrophy of interrenal is due to the influence of gonadal hormones. Several workers have demonstrated that change of the teleost interrenal may induce alternation in the corticotrophic cells of pituitary (Ball and Olivereau, 1966; Hanke *et al.* (1967). McBride and Overbeeke (1969) supports the view of Hane *et al.,* 1966 that the gonadal hormone exert their effect directly on the interrenal and hyper-trophy of interrenal is caused by gonadal hormones which may act on the adrenal homologue without mediation by the pituitary gland.

Assem and Hanke, 1981; suggested that the corpuscle of stannius were concerned in Osmoregulation. But Fontaine and Leloup-Hatey (1959) extracted significant amount of corticosteroids from the corpuscles of Salmon.

Bara (1968, 1972) reported that no histochemically demonstrable reaction occured for 3 β, 3 α, 11 β and 17 β Hydroxysteroids in the Corpuscle of stannius of *Fundulus heteroclitus* and *Pseudopleuronectes americanus.* It was indicated that also no corticoides are produced by corpuscle of stannius of *Salmo gairdneri.* Buss and Larsen (1975) reported the absence of known corticosteroids in blood of *Lampetra fluviatilis* after treatment with mammalian corticotropin. The absence of corticoids activity of corpucle of stannius in many species seems to negative the idea of a universal function in fish.

14. THYROID

The ventral lobe of hypothalamus contains the thyroid stimulting hormone which may be involved in vitellogenesis (*Lewis* and Dodd, 1974). It was also reported that gestation in *Torpedo* (*Rivulus marmorata*) (an elsmobranch in which self fertilization takes place) is accompanied by high thyroid activity and hyperaemia of the gland. Oliverau (1949) found that the onset of sexual maturity in *Scyllium canicula* is associated with signs of increased thyroid activity, further in mature females the thyrosomatic index (TSI) is approximately twice than that of the male. This was further supported by thyroidectomy experiment. It was found that in partially operated animals, inaddition to previtellogenic follicles, vitellogenic follicles of varying sizes were found, where as in completely thyroidectomised animals none of the follicles was undergoing vitellogenesis. These results appear to indicate that thyroid hormones are implicated in vitellogenesis.

Thyroid activity in relation to spawning in bony fishes are reported by Buchmann, 1940; Matty, 1960; Dodd and Matty, 1964 and Leatherland 1982. Hoar (1955) stated that thyroid activity during spawning may be casually related to the physical activity involved for migration and spawning behaviour rather than gonadal function as such. Experimental data confirm that lowering the thyroid activity in fishes (by antithyroid drugs a-radioiodine) results, inter alia, and inhibition of gonad development. Scientists are inview that thyroid indirectly helps in gonad development by controlling metabolism which is acceptable. Triiodothyronine (T3) influences positively the maturational effect of gonadotropin and steroid hormones.

15. ADRENAL GLAND

The adrinal tissue is situated in the wall of the posterior cardinal vein and its branches. The major part of this tissue is found on the right head kidney. This gland composed of only chromaffin or both chormaffin and cortical-tisues in fishes. In Elasmobranch, the chromaffin and cortical-tissues are spearated. Oguri (1960) described the distribution of chromaffin cells and interrenal cells in the head Kidney of fishes. The adreno-cortical tissue were assayed histo-chemically for 3 β, 3 α,11 β and 17 β hydroxysteriod dehydrogenase (HSD) activities at certain intervals during sexualcycle. Typical adrenocortical steriods have been identified in a wide variety of bony fishes (Chester Jones *et al.,* 1959) Several group of workers have demonstrated histochemically the 3 β HSD activity in the adrenocortical tissue of some teleost fishes by using 3 β-hydroxy-androst-5-en-17-one (DHA) or 3 β-hydroxypregn-5-en-20-one (pregnenolone) as substrate (Chieffi and Botte *et al.,* 1964, Hanke and Chester Jones, 1966. Bara (1968) also demonstrated the presence of 3 β-HDS in *Fundulus heteroclitus* when 3 β-hydroxy-5 β-androstan 17-one and 3 β-hydroxy-5-α androstan-17-one i.e.3 β-hydroxy steriod with 5β or 5α configuration were used as substrate. The similar result was obtained with the adrenocortical tissue of *Pseudopleuronectes americanus* (Bara, 1972). The incubation of the head kidney of *Fundulus heteroclitus* in media containing added progesterone gives cortisol and aldosterone from this precursor. Nandi and Bern (1960) found that incubation of head kidneys of Sable fish *Mugil cephalus, Tilapia mossambica*, revealed the presence of cortisol and cortisone, their yields of steroids were too small to allow of the identification of aldosterone. In fishes there is probable existance of mutual overlapping and memicking in the endocrine function of the adrenal cortex and the ovary as a case of mammals (Chesters Jones, 1957). The possibility of gonado-adrenal relationship was evident from the hypertrophy of adrenocortical cells and this hyperactivity is associated with gonadal development and spawning (Robertson and Wexler, 1957).

16. GRANULOSA CELLS

In some teleosts. investigation have been directed to the possible endocrine function of granulosa cells of oocytes in steroid hormone production.

Yaron (1971) demonstrated the presence of 3β-hydroxysterpid dehydrogenase (HSD) activity in cytoplasm of granulosa cells. and thecal cells of *Tilapia nilotica*. The enzyme activity was more prominent in large follicles. But 3β-HSD activity was not demonstrable in granulosa invaded atretic oocytes of either Tilapia or Acanthobrama. Therefore, the source of ovarian hormone in the teleost ovary has been controversial. However, the granulosa in some fish may have specialisd in hormone production to the point where granulosa invaded oocyte (corpora atretica) became functional corpora lutea. Bretschneider and Duyvene' de wit (1947) described granulosa invaded ooccyte as "Pre ovulatory corpus luteum". They suggested that this structure function as an endocrine gland in *Rhodeus amarus*.

17. UROPHYSIS

At the caudal end of the spinal cord, urophysis is present in both Elasmobranchs and teleosts. Urophysis has neurosecretory function and plays a physiological role in osmoregulation. The teleost caudal neurosecretory cell have two types of aminergic neurons and two biologically active peptides are isolated from the urophysis. These two peptides are urotensin I and urotensin II and were first separated by Lederis (1977) and his colleagues. Although both urotensin I and urotensin II appear to be present in all fish, still the presence of two other peptides urotensin III and IV in the urophysis is being claimed (see Matty,1985).

Although the function of urophysis is mainly concerned with salt balance, but its resultant product play a physiological role in the reproduction of fish.

18. PINEAL

Pineal is supposed to have photoneuroendocrine function. Although photosensory function of pineal of fish is known, still, there is some physiological evidence that pineal of fish functions as an endocrine gland (see Matty, 1985). The pineal gland of bony fish contain melatonin, serotonin and number of free amino acids, along with the enzyme hydroxyiondole-o-methyl transferase (HIOMT), which are believed to have some physiological role. Daily rhythm in these secretions have been noticed in the pineal and plasma of bony fish. Wheather such diurnal changes are associated to influence the release of pituitary or other hormones is yet to be established, though endevours are in progress in many research institutes.

19. ULTIMOBRANCHIAL GLAND

The ultimobranchial gland in teleosts is the site for synthesis and storage of the polypeptide calcitonin and usually exhibits a hypocalcemic role as in other vertebrates. However, other studies have shown that the ultimobranchial gland in teleosts is more involved in reproduction, especially in ovarian maturation, rather than the calcium regulation.

The ultimobranchial gland remains in close approximity with the ventral surface of the oesophagus and just lies dorsal to sinus venosus. However, in females the glandular mass is comparatively larger than males. The follicle cells of gland consist of both granular and agranular cells.

The ultimobranchial gland of matured females exhibit higher activity than those of males, as in females the glands are relatively large and display hypertrophy and increased vascularisation.

The architectural details of ultimobranchial gland of mature females are suggestive of being in a more active state than those of males of the same age group as evidenced by the small size of the follicles, reduced cellular height of follicular epithelium and poor vascularisation in the males. Similar difference in the ultimobranchial gland has been reported in different fish groups *Onchorhynchus niasu, Brachydanio rerio, Puntius sophore, Crossocheilus lutius, Esomus danrica.* Several authors not only have noted higher concentration of calcium and intense cellular activity associated with ovarian maturation and reproduction as indicated by higher degree of cellular activity during spawning phase in females but also opine that ultimobranchial gland in female teleosts is more closely associated with ovarian maturation and reproduction as indicated by higher degree of cellular activity than in other sex.

20. PANCREAS

Although pancreas in vertebrates is regarded as part of the alimentary tract, still the endocrine pancreas is concerned with functions of hormone secretion for metabolism and growth. The pancreas of teleosts are found in three forms such as compact, lobulated bodies spread to various parts of the body cavity and scattered in small bodies over the whole body cavity. Brockmann 1846 (See Matty, 1985) described these bodies contain islet tissues. These Brockmann bodies are derivatives of the dorsal epithelium of the embryonic gut. As these bodies contain largely endocrine tissue, they have been isolated and examined biochemically and physiologically. The histology of the teleost islet shows A, B and D cell types. D cell is some times called as A_1 cells. In teleosts using the immunohistochemical demonstration, insulin has been located in B cells and glucagon in the A cells in *Xiphophorus* (Red Sword tail). In addition to A, B and D cells , two other types of amphiphillic cells have been described as occurring in teleost pancreas but their function is not known. The function of D (A_1) cell in teleost is believed to produce a third pancreatic hormone, similar to gastrin like peptide analogous to the D cells of the mammalian islets of langerhans. Other authors have suggested that D (A_1) cells produce somatostatin (the release- inhibiting factor for growth hormone in *Xiphophorus* as seen in Mammalian D cells. The teleost endocrine pancreas shows cells with characteristic granules. The B cell granules show marked variation from species to species and are crystalloid in nature. Such variation may be due to differences in the macromolecular arrangement of the insulin or the type of binding protein. In *Xiphophorus*, B granules are circular or oval in shape with few rod shaped while the granules in D cells are much smaller than islet cell granules. Heavy nervous innervation is seen in the islets suggesting the presence of direct nervous control.

A number of teleost fish insulin have been isolated, crystallized and their amino acid sequences determined. Although insulin appears to play little part in glucose homeostasis in fish there is evidence that it is linked to amino acid metabolism. It is not known whether insulin inhibit the physiological role of somatostatin in fishes. The biosynthetic pathway for insulin in the Brockmann bodies of teleosts has been studied and insulin precursors identified. Pro-insulin has been identified. Two fish glucagons have been isolated and characterized.

21. TYPES OF HORMONES

The hormones related to maturation, ovulation etc. in fishes are broadly of three different types such as : (1) Gonadotrophic hormones, (2) steroid hormones and (3) Neurohormones.

21.1 Gonadotrophic hormone

In earl;y 1960's the presence of gonadotropin in agnatha and chondrichthyes was detected by means of bioassay technique. Burzawa-Gerard detected the gonadotrophic activity in a saline extract of the pituitary gland of the lungfish (*Protepterus*) species. The positive response in mid 50's through bioassay technique represent the first detection of pituitary gonadotropin. Isolation of gonadotropin from the pituitary gland of the teleost have been under investigation for a long time. Robertson and Rinfret (1957) extracted gonadotropins from *Onchorynchus tshawytscha* pituitary gland using a solution of acetone and acetic acid. Other bioassays for piscine gonadotropin were frog spermiation assay (Fontaine and Chauvel, 1961), Goldfish testicular hydration assay (Clements and Grant, 1964) and weaver finch assay (Witschi, 1955). However, one day old chick testicular radiophosphate, uptake (p^{32}) assay was employed as it has advantage of rapidity, sensitivity precision and ability to respond to mammalian LH as well as Salmon gondotropin (Donaldson *et al.,* 1972). The process of GTH radioimmunoassay has allowed the study of gonadoropin secretion during the sexual cycle in different cyprinid species.

Methods of isolation of gonadotropin involves essentially two steps : (1) Crude extraction and (2) Purification of crude extract. Ultrafiltration or pressure dialysis using membranes of specific pore size are successfully employed for crude extraction of hormones. Various absorbants for direct adsorption of hormones Ex, HCG from urine are in use. Benzoic acid and Kaolin are some of the adsorbants reported.

By employing the method of immunoelectrophoresis and by ultracentrifugation or coloumn chromatography on a carboxymethyl sephadex and gel filtration on Sephadex G-200 or G-100, the hormone can be purified.

Purification of piscine gonadotropin and Salmon gonadotropin are done by the process of Burzawa Gerard (1971), Burzawa-Gerard and Fontaine (1972) and Donaldson *et al.,* 1972.

The physical characteristics of gonadotropin includes the molecular weight of piscine gonadotropin, rate of velocity of sedimentation and subunits. Physical characteristics of gonadotropin have been studied by (1) Molecular exclusion chromatography (2) Preparative ultracentrifuge sucrose gradient and (3) Gel filtration. However, variations in the molecular weight, sedimentation rate within same species and other species of fishes may probably be due to different techniques employed at different pH's. That too the state of maturation of the species at the time of study probably be dissimilar or the polysaccharide moiety of the hormone content for encountering such variations.

The chemical nature of both carp and Salmon gonadotropin suggests that they are acidic in nature and thus similar to mammalian FSH than LH (Fontaine and Gerard, 1963; Donaldson *et al.,* 1972). In relation to amino acid composition, the carp gonadotropin resembles more with FSH rather than LH. It was described that carp gonadotropin is a glycoprotein with 8.6% hexoses, 4.9% hexosamine and 0.35% sialic acid. That too the biological activity of FSH is affected by the removal of sialic acid when treated with neuraminidase but the biological activity of LH is not affected. The Salmon gonadotropin lost much but not all biological activity after removal of Sialic acid and thus more closely resembles with ovine FSH which lost 90% of biological activity than LH which lost 50% of its activity.

Immunological studies of piscine gonadotropin suggest that circadian rhythm of plasma-gonadotropin exists on fish. Its concentration in blood is high in the evening and low in the mid-night followed by spawning. Release of gonadotropin at the time of ovulation corresponds with the degranulation of pituitary gonadotrophs. Bioassay studies indicated an increase in pituitary gonadotropin content during maturation followed by a drop after spawning.

With reference to the number of gonadotropins in fishes, it is believed that, biological cycle of lower vertebrate exhibits differential stimulation for gametogenetic and steroid producing tissues, suggesting the existence of FSH like and LH like hormones. The other argument in support of two gonadotropins in due to gonadotrophic cells and Luteotropes or corticotropes, both related to gonadotrophic function. This has led to support the FSH like and a LH like hormone. However, biological and immunol-ogical properties of carp gonadotropin indicates the existence of a single protein homologous with thyotropin (TSH) but have 2 classes of receptor in ovary. All these views keep the hypothesis in dielemma regarding how many gonadotropin actually are present in fish.

Crude extracts of gonadotropin from teleost pituitaries have long been used to induce spawning and are crucial to the development of fish culture. However, effective dosage of hormone of pituitary gland for ovulation differs from species to species, stage of maturity, potency of the hormone and physiological state of the recipient (Rath, 1987b). Timing of injection which influences circadian rhythm of plasmagonadotropin plays significant role for ovulation. Autosynthesis and heterosynthesis of protein in to ooccytes are influenced by hormonal action for enhancement of maturation (Rath and Mohanty Hejmadi, 1991a). Follicular envelope is an indispensable for hormonal action and hydro-inflow resulting hydration for ovulation.

21.2 Steroid hormones

The site of steroid hormone production in teleosts are: (1) adrenocortical tissue, (2) adrenal gland, (3) corpuscles of stannius/interrenal and (4) Granulosa cells of gonad.

Usually glucocorticoid steroid (Ex-cortisol and corticosterone) and mineralcorticoid (Ex-aldosterone) steroids are reported. The presence of oestriol, oestrone and oestraidol 17- β are reported in the fish ovary. The presence of progesterone was identified in fish ovary of African Lung fish *Protepterus annecten.* Some believe that oestradial is the primary product of ovary where as oestrone and oestriol are the secondary products.

The general biosynthetic pathway of steroids on the whole is cholesterol or the probable pathway are acetate and chloesterol-Pregnanolone-Progesterone and so on.

Steroid hormones are effective for inducing maturation and ovulation in fishes. Cortisol is the most potent steroid for *in vitro* ovulation of *Oryzias latipes* (Hirose, 1976). Progesterone is effective when follicles are treated with steroid solution and then washed before incubation in *O. latipes* (Iwamatsu, 1974). The 17 α - 20 β progesteron seem to be strongest inducer *in vitro* maturation of trout (Fostier *et al.,* 1973). Goldfish (Jalabert 1976), Carp (Epler *et al.,* 1980). Jalabart and coworkers reported that on trout *S. gairdneri* the control of maturation is through direct gonadotrophic route and indirect corticotropic route. Others reported that the androgen injection resulted in spermiogenesis on *Lebistes reticulatus, Fundulus heteroclistus* and *Poecilia reticulata.* Sundararaj and his group have reported that Deoxy corticosterone and hydrocortisone was effective in inducing maturation and ovulation in *Heteropneustes fossilis* under *in vitro* condition. In goldfish ovulation was induced invivo by prostaglandins (Stacey and Pandey, 1975) and in vitro in carp (Epler, 1978; Epler *et al.,* 1985) as well as by adrenaline (Epler, 1978). Spermiation seems to be dependant on androgen via gonadotrophic hormone (GtH). Different steriods are active in fish such as 11-keto testosterone, testosterone, methyltestosterone, testosterone propionate and progesterone (reviewed by Billard *et al.,* 1982).

Excretion, osmoregulation, protein synthesis and various other metabolic activities are controlled by hormones. Hormones also suggests the practical application of sex manipulation, monitoring of growth by genetic control and has opened new horizons for aquaculture. That too the reproductive process is marked with sexual responsiveness of gonadal steroids and pheromones. Colombo *et al.,* 1982 reported that the control of reproduction, fertility, sexual behaviour and parental behaviour of fish are governed by pheromones. These are indications that the female fish may utilise urine as well as ovarian fluid as pheromonal vehicle for sexual behaviour. Thus it is believed that gonadal steroid exhibits both hormonal regulation and bioregulatory function in fishes. Therefore ovarian hormone, central sensory motor mechanism for sexual response and gonadotropin involved in maintaining physiological state of ovary is a key for success in spawning in fishes.

Tavolga (1956) demonstrated the presence of a chemical compound produced by gravid females in *Battygobius soparator,* which when added to a tank containing sexually active males, induced courtship response in them. Since then numerous reports are available on the presence of pheromones in fishes (Amouriq, 1964; Liley, 1966; Todd *et al.,* 1967; Wallace, 1970; Stacey and Liley 1974; Partidge *et al.,* 1976).

Timms and Kleerekoper (1972) reported that the males of *Ictalurus punctatus* were attracted by a source of water which had previously held gravid females of that species. Some reports are also available on the involvement of pheromones on the recognition of partners at the spawning site, nest building and care of eggs (Reviewed by Kapur, 1981). Moitra and Sarkar (1975) reported that the breeding of non injected carps along with injected ones in usual practice of bundh breeding appears to be due to the release of pheromones. Such type of breeding in un-injected cases is referred as sympathetic breeding being reported by several workers.

21.3 Neurohormones

Neurohormones are produced by some specialised neurons. The axon terminal is specialised to store and release the hormone. This axon terminal is ended with blood vessels and constitute neurohaemal organ. The other neurons of the hypothalamus or brain has the capacity to produce neurotransmitters. But neuro-transmitters differ greatly from neurohormones with reference to the source of production, site of action, transport system and neurohaemal organ.

The differences between neurohormones and neurotran-smitters (Neurohumours) are as follows :

	Neurohormone		Neurotransmitter or Neurohumours
1	It is released to the blood stream.	1	It is not released into the blood stream.

2	The site of production is either far or near from the site of action.	2	THe site of production is very near to the site of action.
3	Secretions are released into axon terminal that is neurohaemal organ from which it is secreted into the blood stream.	3	It act on the post synaptic junction of another neuron.
4	Neurohaemal organ (the storage place) is present.	4	No such organ is present.

The aminergic neurons produce neurotransmitters like (1) Adrenalin, (2) Non-adrenalin, (3) Dopamine and (4) Serotonin. All these are catecholamines. Similarly the peptidergic neurons produce 10 peptides. These are as follows :

(1) Thyrotrophin releasing hormone (TRF)

(2) Leuteinizing hormone-releasing hormone (LHRH)

(3) Growth Hormone releasing inhibiting hormone (GH-RIH) or somatostatin

(4) α - endorphin

(5) β - endorphin

(6) Enkephalin

(7) Neurotensin.

(8) Substance-P

(9) Vaso-intestinal polypeptide (VIP)

(10) α-MSH

21.3.1. *Neurosecretion in Fishes*

It was reported that in fishes, the neurosecretory cells send their axons mostly to the posterior part of the neurohypophysis, but a few axons terminates in the aniterior part correspond to median eminecne (ME). However, in teleosts, as seen in *Anguilla,* (Knowles and Vollrath, 1966) and *Leuciscus,* the fibres of the preoptic nucleus (PON) terminate in the metaadenohypophysis (pars intermedia) and those of the nucleus lateral tuberalis (NLT) in the proximal pars distalis. In *Phoxinus* where an NLT is absent, the preoptic nucleus fibres innervate the meso and metaaden-dypophysis.

Sathyanesan (1969) studied the neurosecretory tract of *Clarias* by staining technique. His group of workers described the neurosecretory tract and also the hypothalamo-hypophysial system of the *Heteropneustes.* These workers reported that the axonic fibres carrying neurosecretion from the nuclei may end near the cyanophils of the pituitary sinusoids thus carrying the chemical to the cells. Several authors including Sundararaj and others reported that there is no portal system in the catfish for the connection of ventral hypothalamus with that of pituitary gland. So the hypothalamoneurohypophysial system in teleost enrich the scientific study that portal system is poorly developed and neuro-secretion is directly carried to the adenohypophysial cells. Hence, neurohypophysis controls the activities of adenohy-pophysis.

Peter (1970) reported that NLT has a stimulatory regulation of gonadotropin (GTH) secretion via a gonadotropin releasing hormone (GnRH). Peter and Paulencu (1980), through experie-ments of brain lesioning have dmonstrated the presence of gonado-tropin-release-inhibiting factor (GRIF) in the anterior-ventral preoptic region.

Chang and Peter (1983) found that dopamine has GRIF activity in goldfish modulating spontaneous release of GtH as well as LHRH-analogue stimulated release of GtH (Gonadotrophic hormone).

21.4 Pheromone

Fish possesses chemosensing and chemical signaling systems that in general controls the reproductive behavior of fish. Such chemosensing and chemical signaling system are controlled by pheromones.

Hence, pheromones are defined as the chemical substances which are secreted to the exterior of body by an individual and recognized by a second individual of the same species, in which they produce one or more specific reactions.

21.4.1. Sources

Pheromones are believed to be present in different parts of body of fish *viz.,*

1. Epidermal cells of skin:-Ex ***Phoxinus phoxinus***

   Urine: Ex-***Poecillia reticulate, Icatalurus nebulosus,***
2. Mucus cells: ***Ictalurus natalis,***

3. Ovarian fluid: Ex- Goby, Loaches, Goldfish

The pheromone reception organs are mainly

1. Olfaction: as seen in African cat fish and gold fish,
2. Gustation or by tasting

Very little is known about brain mechanism responsible for pheromonal recognition (see Matty, 1985). Further pheromones are applied for fishing and fish breeding.

21.4.2. Function

Pheromone systems has various implications in different aspects of fish behavior by producing.

1. Alarm substances: for examples- *Phoxinus phoxinus* releases pheromone when there is cell injury and *Schereckstoff* releases pheromone when there is danger.
2. Social behavior: for example- female zebra fish would not ovulate in the absence of male zebra fish (*Brachidanio rerio*). In female of shrimp, *Paratya*, ovarian develop-ment is delayed without males. Similar situation is also observed in *Macrobrachium rosenbergii* & *M. malcolmsonii.*
3. Sex and individual reorganization: for example- male species of guppies, zebrafish, white catfish are attracted to their respective female species through odour perception.
4. Social hierarchy: for examples- the dominance of yellow bull head (*Ictalurus natalis*) is communicated by pheromones.

II. HYPOPHYSATION

1. INTRODUCTION

Since the pioneering experiments of Houssay (1931) of Argentina, the hypophysation technique was successfully applied in Brazil in 1934 by Von Ihering (1937) and his collaborators. In India, the first attempt to induce breeding was made in 1937 (Khan, 1938), using mammalian pituitary extract. The hypophysation technique in major carps was established however in India in 1957 (Chaudhuri and Alikunhi, 1957). Since then several workers have used this procedure in different parts of the country. Later by adopting same technique, Chinese Silver carp and Grass carp reared in ponds were successfully breed (Alikunhi *et al.,* 1963) in India. The first hypophysial technique in China was adopted in 1958 by aquaculture researchers of Guangdong province by injecting the hypophysis of Common carp into brood fish of Silver carp and Bighead cultured in ponds. In the same year, researchers of Zhejiang province also succeeded in induced spawning of Silver carp and Bighead by applying human chorionic gonadotropin (HCG). In 1960, artificial propagation of Grass carp in China succeeded by using pituitary gland of Common Carp. In 1963, in China, the artificial propagation of Black carp was done similarly by use of Common carp pituitary.

However, at present most of the fingerling produced in ponds come through artificial propagation so that it plays an important role in dveloping aquaculture in most parts of the world.

2. PRE-REQUISITES FOR INDUCED SPAWNING OF CARPS

2.1 Brooder maintenance and care

An adequate brood stock is an essential prerequisite for the successful induced breeding programme. Brood fish means the male and female fish which are used for artificial propagation. Only after they reach sexual maturity, can be induced to spawn. Therefore, stocking and rearing of brood fish is very important link in the chain of spawning. The brood fish may be grown in the farm or collected from outside sources and reared in proper brood fish ponds. Transportation of brood fish involves transport cost, injury and physiological stress. Therefore, it is always advisable to rear them in farm tanks itself. The stocking rate for brood fish is usually in the range of 2,000-2,500 kg/ha with usual management to keep them healthy and in condusive condition. The selected brood fish should be examined periodically for their condition, gonad development and progress of maturation. Freedom from parasitic infection is necessary to keep brooders in most suitable condition. Filling and creating running water condition to the brooders are essential for better fecundity, spawning and fertilization rate. That too, age of brood fish and sex are important considerations in induced breeding.

The potential brooder weighing 1.5 Kg to 5.0 Kg are chosen and daily fed with supplementary feeding of 35-40% protein range.

2.2 Parent fish transportation

General care should be taken during transportation of brooder fish from distant places. Various designs of transport carriers for livefish have been developed by Indian workers, of which Mammen's model (1962) is a compressed air aerated carrier tank of splashless design to check physical injury to fish. The record of successful packing density is approximately 1 : 2.5 (fish weight : water weight) for a haul of 1/2 hour duration. The other improvised Parto's model (1969) is similar to a fuelgas tank type. It consisted of two barrels, smaller one is inverted inside a bigger one, Oxygen under pressure from a cylinder is introduced through the water column in to the gas tank. However, transporting the brood fish individually in separate hand nets by keeping inside fish carrier with water has yielded considerable good results. That too the use of anaesthetics (drugs or chemicals) have solved this problem to certain extent. Although there exists potential danger from these drugs being misused, on live fish as their limit between safe and fatal doses is very narrow. Hence trained and experienced persons are required for handling and treating the fish before transport. The brood fish after transport should be disinfected and conditioned before released in the pond. The most common way in China is to catch mature brood fish such as Grass carp, Black carp, Silver carp and Bighead from large water bodies as parent fish and rear them in ponds for a short period of time for inducement. By this way, the time is shorter than that of rearing them from fry to parent fish in ponds. The most proper time of collection is in autumn and winter when the temperature of water is low. At that time the activeness of fish is rather weak, so not to cause serious injury. That too shortage of dissolved oxygen in cold water in winter month is unlike that of summer months.

During selection of parent fish, the male or the female fish from different blood relations will improve the vitality of the next generation. Usually in China, canvas tub or wooden barrel are used for short distance transportation. In 100 litres of water, one parent fish of about 10 Kg in body weight is put for transportation. The quantity of fish in transportation is determined according to water temperature, size of fish and time tentatively taken for transportation. For long distance transportation, oxygenated plastic water bags ae used. Plastic bags made up of vinyle-film cylinder 30-35 cm diameter and 50 cm longer than the fish body can contain one parent fish after being filled with some water. The water level is as high as fish body. Transportation with a perforated boat is convenient if there is water way. As the water can come in and out through the holes of the hull of the boat, the density of fish, can be higher. Some time net screens can be provided, so as to check the occasional jumping of fish. Transportation after tranquilization is done when the fish are in a state of unconsciousness. Sodium barbital solution at concentration of 13.3 ppm with a short period of time has also over-come the difficulties of transportation of brooder fishes.

2.3 Differentiation of the male and the female fish

It is often necessary to maintain the ratio of the male and the female brooder fish when induced to spawn. The correct discrimination between male and female is important. The methods of distinguishing the sex of carp is generally the same. The sex is mainly identified in accordance with the characteristics of pectoral fin. In case of males when the abdomen near the vent region is pressed slightly, milt would ooze out easily. Care should be taken only that the least loss of milt could be allowed during selection. The mixing of milt with water is to be judged. Sometimes, thick milts do not mix thoroughly in water, causing less percent of fertilization rate. Such male brooders with less mixing capacity of milt is usually avoided in induced breeding programme. The pectoral fin of the male is rough, coarse and thorny due to row of fine scale serration on pectoral fins during breeding season.

Females with soft and bulging abodmen is considered. Slightly swollen belly and pinkish vent are considered for good female brooders. When the female brooder is kept ventrally upward, the belly on both the sides are swollen due to ripe ovary with a median depression. The abdominal keel region at the median depression of belly is very soft and form a distinct groove clearly indicating the good brooder condition. However, care should be taken for Grass carp, because the bulging of belly may be due to excessive feeding and in Catla due to considerable deposi-tion of abdominal fat within the body cavity. In a ripe female the eggs are slightly loose inside the ovary. By introducing gently a catheter or a canule through the vent and suck carefully to get a sample of ovarian eggs. The egg can be put in teleost saline to observe uniform shape and size. Besides this, the position of nucleus in the egg can be judged. If the nucleus is about 1/3-2/3 above the central part of egg, then it is considered as a good female brooder ready for induced spawning as reported in Black carp (Rath, 1991b). However, some have reported that, if 90% of withdrawn eggs show the diameter more than 1.0 mm depending on (Carp species) the female is suitable for induced breeding.

3. INDUCING AGENTS

3.1 Pituitary gland

It can be collected from freshly dead but not spoiled male and female matured carps. Collection can be done either by scale scrapping or through foramen of magnum. Sex specificity and species specificity of pituitary gland homogenate is observed in fish although pituitary glands from both sexes are equally potent in carps. However, the homoplastic fish pituitary extracts are in general, more effective than the heteroplastic pituitaries. In contrast to this, common carp pituitary serves as a common donor of pituitary gland for breeding of different species of cultivable carps.

Freshly taken pituitary gland can be preserved, dehydrated and defatted by absolute alcohol or acetone without any perceptible loss of potency. The Inland Fisheries Research Institute has opened up facilities of pituitary banks for private and public sector fish breeding enterprises. The absolute alcohol preservation method is widely followed in India whie acetone-dried method is largely practised in USSR and USA. These acetone dried pituitary glands are stored inside air tight sterile phials and placed inside a desiccator for dehydration which is kept inside the refrigerator for maintaining its potency. The extract of pituitary gland prepared out of homogenization can be stored in glycerine solution. Sometimes Triglyceraldehyde is used to store the extract of pituitary gland although these are not commonly practiced in India. The preparation of glycerine extracts is tedious and the administration of more than 10% glycerine has a harmful effect on the recipient fish (Clemens and Sneed, 1962).

3.1.1 *Dose and Frequencies of injection*

Till 1960 the pituitary dose was calculated in terms of a part or as a whole gland. But their after, a dose in miligram of pituitary gland per kilogram body weight of fish has been developed by Indian workers. This dose expressed in mg/Kg body weight is widely accepted among the growing aquaculturists. Other dose systems like ``Fish Units" (FU), ``Loach Unit", ``Vyun Unit", ``Frog Unit" or ``Magur Unit" which were suggested based on biologically assaying fish gonadotropin are no more in use in Indian system. In India, the female brooder fish receives two split dosages, the preliminary dose consisting of 2-3 mg/Kg body weight and a second dose of 5-8 mg/Kg after about 4-6 hrs. The amount of dose is also valuable with the Eco-physiological state of brooder female. However, sometimes a single ``Knock-out" dose or ``booster dose" is given to ripe females for successful spawning. At the time of last or second dose, the case may be the male receives a single dose of pituitary extract at the rate of 2 to 3 mg/Kg body weight.

Similar method is also followed in Exotic carps like Grass carp and silver carp cultivated in India. The female is usually given two split doses. The first dose of 3-4 mg/Kg body weight and after 4-6 hours of first injection, the second injection at the rate of 7-10 mg/Kg body weight was given to the recipient female. The males are injected at the time of second injection to the female, the dose being 3-6 mg/Kg body weight of the recipient fish. The dose may vary depending on sexual maturity, and ecological conditions prevailing for induced breeding of the recipients. Sometimes, over dose causes plugging in carps thereby checking the pass out passage for eggs. These unshed eggs due to overdose are characterised by the presence of multiple oil globules and histological abnormalities suggesting artesia. In certain cases, third injection of higher dose is administered when the injected female does not spawn within 8 hours of time of 2nd injection. In such cases the spawning under the influence of excess hormone, results poor yield of fertilisable eggs.

3.1.2 *Breeding set and mobility of gamete*

Usually a breeding set consists of one female with two males. The weight of one female is more or less similar to the weight of two males. It is practised that, two males of equal size are usually preferred because, during estrous and courtship behaviour, the egg released by females due to abdominal muscular movement and the milt secreted by the chasing male make the eggs to contact with milt for fertilization. As the active mobility of milt in water is only for few seconds, if within that time, sperm does not come in contact with egg, that sperm dies. Although more number of sperm attachment in egg is seen, only a single vital sperm penetrate through micropyle of chorion or vitelline membrane to fertilise the egg by fusion of two nucleus. The average life span of grass carp sperm in fresh water is 112 seconds, that in 0.6% normal saline is 649 seconds (Pearl River Research Institute, China). The vitality of eggs differ obviously in the body fluid from those in freshwater. Decreasing relation between prolonged fertilization time and fertilization rate of the matured eggs of carp has been noticed. The Pearl River Research Institute, China has mentioned that the released eggs in freshwater if fertilised just after 30", the rate of fertilization is 30% and if fertilised after 1'30", the rate of fertilisation drops to 6% only in Bighead carp.

4. METHODS OF INJECTION

Usually two methods are adopted for injecting the brooder fish. These are intramuscular and intraperitonial (coelomic).

4.1 Intramuscular

This method is very effective and convenient. Therefore, it is very popular in Brazil and India. The intramuscular injection is given on the dorso-lateral muscle towards the caudal peduncle region or the muscles in the humeral region.

As two split doses are given for carp spawning, the pricks are made on either side (right and left) alternatively to reduce the physical strain to fishes. While inserting the needle, care has to be taken that the needle is inserted under the scale parallel to the body of the fish and then turned to 45° angle to pierce quickly the muscle, and inject the fluid. By this way absolutely no damage is caused to any of the scales because scales are arranged in an imbricate fashion on the body of carps. Intracranial method of injection used earlier by Russians have switched over to intramuscular injection, finding its conveniency in carp spawning.

4.2 Intraperitonial/coelomic

This method of injection is mostly adopted in USA and China although not uncommon in India. During injection, a brood fish is placed over a foam rubber cushion laterally. In carps, there are soft regions at the base of pectoral or pelvic fins through which the injection needle is inserted. When intraperitonial injection is given, point the syringe needle towards the head at an angle of 45° to the body's longitudinal axis, insert the needle. Care has to be taken not to damage the internal organs and inject the fluid.

5. SPECIFICATION OF NEEDLE AND PROCEDURE FOR ADMINISTRATION OF INJECTION

A 2 ml hypodermic syringe with 0.1 ml graduation, preferably with interlocking arrangement is convenient for use. The specification of needle comprising thickness and the length depends on the size of the recipient fish. Usually with increase in weight and length of fish, the specification number of the needle proportionately decreases. Generally B.D.H. No. 22 needle is used for fishes weighing between 1 to 3 Kg and No. 19 for the larger ones. For smaller fishes (below 1 Kg) No. 24 needle is considered.

The procedure followed for handling brooder fish during injection is very important. Care has to be given for less physical strain to the brooder. Usually the recipient fish is taken out from the breeding enclosure by hand net (scoope net) and are gently put laterally on a rubber cushion of suitable size. Before inserting the needle, the syringe loaded with required quantity of extract usually not exceeding 1 ml is taken. The piston of the syringe is hold firmly with ring finger on the ventral alongwith thum finger on the top. The middle finger and the finger placed on the right side of it can rest on ventral side of the graduation cylinder of syringe. Before inserting the loaded syringe, one person places his hand on the head of fish and the injector holds the caudal peduncle and inserts the needle in suitable way and injects the fluid with thumb finger pressing the head of movable piston.

6. SEASONS FOR INUDCTIONS OF SPAWNING

The most suitable season for spawning is dependant on the weather and the gonad development of fish. These are the key links in a chain of artificial propagation in carps. Although experimental initial propagation period in Aquaculture Research Institutes of India starts prior to late April or mid May in some cases, still, the major initial propagation period in India is from June onwards with the onset of South-West monsoon. The sequence of induction is Mrigala and Rohu first, followed by Catla, Silver carp and Grass carp although there is no hard and fast rule. The sequence of estrualisation practised in South China is Grass carp first, Silver carp and Bighead the second, Black carp the last. However in India, the propagation lasts till August-September in some cases, but majority fin-fishes propagation is done by August with reference to usual natural monsoon and meteorological conditions.

7. ALTERNATIVE INDUCING AGENTS

Pituitary forms a major inducing agent for breeding (a) although several problems such as varied potency of pituitary glands (b) difficulties involved in collection and storage and non-availability in adequate quantity and quality were encountered. Besides this sacrificing brood fish for pituitary gland hampers to achieve the target seed production due to non-availability at the time of requirement. Therefore, scientists have felt an urgent need to reduce the use of pituitary gland by suitable substitutes which means saving the brood fish for utilising them to go far up for seed production. Considerable progress has been achieved in the use of releasing hormones in combination with dopamine antagonists i.e. ovaprim (Nandeesha *et al.* (1990 b).

7.1 Freshwater catfish pituitary

Pituitary obtained from freshwater catfishes such as *Bagarius bagarius, Pangassius pangassius, Silonia silondia* and *Mystus seenghala* when injected at a slightly higher dose also effect spawning (Bhowmich *et al.,* 1986).

7.2 Marine catfish pituitary

Varghese *et al.,* 1975 reported for the first time the induced spawning of *Labeo rohita* and *Cirrnina mrigala* with marine catfish *Tachysurus* with higher dose. The subsequent study by Varghese and Rao (1976) indicated the effect of marine catfish pituitary gland extract of *Tachysurus jella* and *Tachysurus thalassinus* on induced ovulation of Silver carp and Catla. The effective dose reported by them were 40 mg/Kg for female and 30 mg/Kg for male in Silver carp which responded 50% to spawn. The dose required for Indian major carps were 30 mg/Kg of female and 20 mg/Kg of male for 100% response to spawn. Marine catfish pituitary is being successfully employed in many fish farms of Karnataka with relatively good success. However, the collection of pituitary-from catfish at right time of the year is the major constrain faced.

7.3 Mammalian gonadotropins

A large number of workers have studied the effect of several crude and purified mammalian gondotropins (LH, FSH, LH+ FSH), pregnant mare serum (PMSG) and Human chorionic gonadotropin (HCG) on various reproductive functions of teleosts (Devlaming, 1974; Jhingran, 1975; Fontaine, 1976; Sundararaj and Goswami, 1977; Kapur, 1978; Chondar, 1985; Rath, 1988 a).

The discovery of HCG is traditionally ascribed to Aschheim and Zondex who in 1927 demonstrated in the blood and urine of pregnant woman, a substance which induces ovarian hypermia, corpus luteum formation and vaginal estrous in immature female mice. However, ethanol as precipitant for the precipitation of HCG from the urine was employed. Various absorbants for direct absorption of hormones from urine are used. The crude extract prepared by Infar (India) Ltd. is available in the trade name as `Summach'.

HCG is a protein hormone which is produced by the placenta during pregnancy. HCG is also called as sailo protein or glycoprotein because the carbohydrate molecules are attached with the protein molecule. The molecular weight of HCG hormones are in the range of 30,000-40,000 depending on the carbohydrate moiety in the hormone. Besides protein, HCG contain carbohydrates such as mannose, galactose, glucosamine, galactosamine, fructose and N-acetyl-neutraminic acid (NANA). The presence of NANA is essential for their biological activity. HCG and Synahorin (a mixture of HCG and mammalian anterior pituitary extract) have been found to be successful in the induced spawning of *Labeo rohita* at the rate of 25 rabbit units/Kg after priming with 2-4 mg/Kg of carp pituitary extract (Barrackpore, 1968; Jhingran 1975; Bhowmick *et al.,* 1976) and they reduce the total carp pituitary requirement by about 50-60%. HCG efficiency depends on the species. It is alone effective in induced spawning of silver carp (Chondar, 1985) and ovulation in goldfish (Yamamoto and Yamazaki, 1967). Chinese studies report that it is weakly effective in Grass carp and Mud carp, the relatively low efficiency of HCG and synahorin can be improved by using them in combination with a lower dose of pituitary extract (Chaudhuri, 1976). The rate chart released by Infar (India) Ltd., company manufactures crude HCG indicates tht HCG can be administered in two split dosages totalling to 13-16 mg/Kg in Silver carp breeding. In China for both Silver carp and Bighead, the dose, 4-5 mg PG or 800-1000 IU HCG/Kg is administered for spawning. However for Indian major carps 70-80% of pituitary gland can be replaced be HCG, although HCG has played some role in reducing considerably the demand of pituitary gland which is not yet totally replaced. In China, fish which were spawned for several consecutive years with HCG developed a resistance to the hormone and become `immune' to it (Cooperative team for hormonal application in pisciculture, 1977).

The dose required may vary in different species depending on how closely related the endogenous gonadotropin is to HCG (Lam, 1982). However, the dose response relationship has not often been studied.

7.4 Purified fish gonadotropin

In recent years, purification of gonadotropin of several species of fish have been done (Burzawa-Gerard 1971; Banerji *et al.*, 1989). Some attempts have been made to use some of these preparations to induce ovulation in fishes. However, only salmon gonadotropin SG-G 100 has been extensively tested for induction of ovulation or spawning in cultured species (Lam, 1982; Donaldson and Hunter, 1983). The induction of ovulation with salmon gonadotropin (SG-G 100) in Indian major carps has been achieved (Chaudhuri *et al.*, 1977). One of the three fractions of carp pituitary extract obtained by Shina (1971) proved to be a potent ovulating agent when administered into common carp, silver carp and *Puntius gonionatus*.

Hypothalamic extract and L.R.H. - Analogues

The functional activities of hypophysis of fish are directly controlled by hypothalamus which secretes LHRH. The first report of Breton *et al.* (1972b), indicated that the hypothalamic extract (LH-RH) stimulated the gonadotropin release from the pituitary glands *in vitro*. This has led to a number of valuable investigations in this field. It was further observed that the hypothalamic extract from sheep stimulated the gonadotrophic secretion in carp pituitary *in vitro* and that the similar extracts from the carp and trout source effected the sheep pituitary (Breton *et al.*, 1972 a). LH-RH has been successfully used for inducing ovulation in a goldfish at 12°C (Lam *et al.,* 1975) and in combination with fish pituitary in grass carp, silver carp, bighead and black carp (Conference on Application of Hormones to Economic fish 1975). LH-RH caused germinal migration and GVBD in common carp but not ovulation (Sokolowska *et al.,* 1978). In 1974 the experiment of synthetic luteinizing releasing hormone (LRH) proved its effectiveness. In 1975, the high effective luteinizing release hormone analogue (LRH-A) was synthesied. About 2000 analogues of LRH is prepared. LRH was refined out from the hypothalamus of sheep which is a polypeptide consisting of 10 amino acids. These are :

1. Pyroglutamic acid
2. Histidine
3. Tryptophan
4. Serine
5. Tyrosine
6. Glutamic acid
7. Leucine
8. Arginine
9. Proline
10. Glycine amide

The molecular weight of LRH is 1182 (IFF, China). LH-RH, a synthetic decapeptide is effective in inducing gonadotropin release and ovulations in fish but its superactive analogue (LRH-A) are more effective in teleosts (Peter, 1980). In 1975, LRH-A, a nonopeptide hormone containing nine aminoacids was synthesised. These are :

1. Pyroglutamic acid
2. Histidine
3. Tryptophan
4. Serine
5. Tyrosine
6. D-Alanine
7. Leucine
8. Arginine
9. Proline and acetylamine

Its (LRH-A) molecular weight is 1167. The sixth glutamic acid and the tenth glycine-amide are replaced by D-Alamine and acetylamine respectively. The biological activeness of LRH-A is about 100 times higher than LRH to fish. It is a high effective estrualising agent and is a white powder. The LRH-A available in the market is often added with mannite as a filler, that is 10 times as much as its weight. It is easily dissolved in water. Hence, stored in a dry and shady place under airtight condition. The advantage of analogue is that it has got high potency, higher receptor affinity and longer duration of action. However, the recent investigations on obtaining most potent LRH-A by modification of three aminoacids i.e., 6th, 7th and 10th are in progress. D ($alan^6$) LRF is 4 times more potent than LRF, D (phe^6) LRF is 15 times more potent that LRF, D(Trp^6) LRF is 36 times more potent than LRF, Gly^{10} ($DAla^6$, pro^9) LRF is 15 times more potent than LRF, GLY^{10} (D-Trp^6 - Pro^9) LRF is 144 times more potent than LRF.

They were Chinese who made the first claim on the successful use of a mammalian based LRH analogue (D-Ala^6, Pro^9 NET) for breeding carps. This has showed 78.5%, spawning success (Cooperative team for Hormonal application in pisciculture, 1977). However, silver carp and bighead, have been breed with LRH-A (D-$alanine^6$) alone or in-combination with HCG. Grass carp have been breed with LRH-A (D-ala^6) alone, where in Black carp, LRH-A in primed followed with LRH-A and pituitary extract for successful breeding. (Training manual IFF, China, 1987). Breeding trials with the purified gonadotropin releasing hormone from Catla have yielded positive results in Rohu and Mrigala (Haldar *et al.,* 1989). Peter *et al.,* 1988 indicated that in practice, Chinese farmers always use LRH-A in combination with pituitary or human chorionic gonadotropin. But LRH-A alone does not always result in successful spawning of many species of carps. The reason for unsuccessful spawning with LH-RH or LRH-A alone become clear with the investigation of dopaminergic inhibitory regulation of pituitary gonadotrophin release in common carp (Billard *et al.,* 1983) and goldfish (Chang and Peter, 1983; De-leeuw *et al.,* 1983). De-Leeuw *et al.,* 1989 observed that the goldfish pituitary contains two classes of GNRH binding sites, a high affinity low capacity site and a low affinity high capacity site. Injection of domperidone 40 μmol/Kg body weight of dopamine angatonist, resulted in a dose and time related increase in capacity of both high and low affinity GNRH binding sites. The effects on GNRH receptor capacity correlated very closely with changes in serum gonadotrophin concentration. These investigations showed the blocking action of dopamine with dopamine receptor antagonists, potentiates the action of LH-RH-A in gonadotropin release. Such investigations led to the development of ``Linpe method" (Peter *et al.,* 1988), where in the LH-RH-A is combined with dopamine antagonist for successful spawning. Sherwood *et al.,* 1983 achieved a major break through in isolating and characterising the (S-GnRH) Salmon gonadotrophic releasing hormone.

Among the various LRH-A analogues, D-Arg^{6} -Pro^{9} -NEt-S-GnRH was more potent, because of its higher affinity to binding site in pituitary (Little and Dawson, 1989). Peter *et al.* (1985, 1987) have demonstrated that this hormone is 17 times more potent than D-Ala^{6} -Pro^{9} - NEt analogue. The usual dopamine antagonists like pimozide or domperidone are used in combination with LRH-A for spawning the carps. As domperidone or pimozide are not dissolved in water, its application encountered certain problems in fish breeding. Ovaprim, a ready to use drug in solution form, manufactured by Syndel laboratories in Canada has overcome such problems. Ovaprim is now available in India by Glaxo company. With reference to the chemical constituent of ovaprim, it contains calibrated amount of S-GnRH (Salmon gonadotropin releasing hormone) and domperidone, a dopamine antagonist dissolved in an organic solvent. It is potent even in room temperature and does not require freez storage. Ovaprim has been tested successfully with IMC and Chinese carps (Nandeesha *et al.,* 1990 b). Similarly the effectiveness of pimozide with LHR-A (D-Trp^{6}-des Gly^{10} LHRH, Peninsula laboratories Europe, merseyside, England) proved successful in ovulation in Gudgeon, *Gobio gobio* (Kestemont, 1988). Reddy and Thakur (1998) described ``Ovatide" having the base of synthetic peptide structurally related to the naturally occurring hormone (GnRH) is used for successful fish breeding. The advantage of ``Ovatide" over other commercially available spawning agent is that ``Ovatide" a low viscosity injectable solution, highly active and cost effective inducing hormone. The dosages of Ovatide varies from 0.20 to 0.40 ml/Kg for female brood fish of Rohu, Mrigal and Calbasu and 0.10 to 0.20 ml/Kg for male brood fish of Rohu, Mrigal and Calbasu. Similarly the dosage of Ovatide varies from 0.40 to 0.50 ml/Kg for female brooders of Catla, Grass carp and Silver carp and 0.20 to

0.25 - 0.30 ml/Kg for male brooders of Catla, Grass carp and Silver Carp. However, the dosages are adjusted depending on the maturation of brood fish and environmental conditions. It is M/s Hemmopharma, Thane, Mumbai and CIFE, Versova, Mumbai, those have made the endeavour of evolving Ovatide. M/s Hemmopharma has taken commercial production of Ovatide and also its marketing in India.

Similarly, WOVA-FH, a synthetic hormone developed by Central Institute of Fisheries Education (CIFE), Mumbai and marketed by Wackhard India Ltd. is used for induced breeding of fishes. It constitutes a calibrated amount of synthetic GnRH and a dopamine antagonist. It is potent even in room temperature and being used widely.

7.5 Trials with Alternative inducing agents in India

Successful spawning of *Cirrhinus mrigala* with LRH-A (desGly^{10}-D- Ala^{6}) LH-RH ehtylamide was achieved (Kaul and Rishi, 1986). Similarly Jose *et al.,* 1989, successfully spawned *Cirrhina mrigala* and *Labeo fimbriatus* with LRH-A (Des-Gly^{10} D-Ala^{6}) LH-RH when applied in two split doses. Parameswaran *et al.,* 1988 achieved successful spawning of *C. mrigala* with LH-RH-A, buserelin acetate (D-Ser (Bu)^{6} - des-Gly^{10}) LH-RH ethylamide in combination with progesterone.

7.6 Steroids

Several workers have reported the effects that mediated the action of gonadotropin or steroid hormones on maturation, ovulation and induced breeding in a number of teleosts (Jalabert, 1976; Jalabert *et al.,* 1977, Sunderaraj and Goswami, 1977; Moses in Baraj and Shaider, 1988). However, a high level may cause an inhibition of gonadal functions because of their feed back effect at the hypothalamo-hypophysial level (Donaldson, 1973; Devlaming, 1974).

7.7 Progestogens

There is a good evidence that progestogen or/and corticosteroids mediate gonadotropin action on oocyte maturation and associated oocyte hydration may also be under their control.

Fostier *et al.*, 1973 reported that 17 α - 20 β progesterone seems to be strongest inducer of *in vitro* maturation of *Salmon giardneri.* Maturation of oocytes can be induced by steroids belonging essentially to the progesterone group in trout, pike and goldfish (Jalabert and Breton, 1973; Jalabert, 1976; Jalabert *et al.*, 1976, 1977). Progesterone seems to induce final maturation characterised by germinal vesicle migration and break down in teleost oocytes (Jalabert, 1976 and Goetz, 1983). At these stages of oocyte development, the blood gonadotropin level has risen considerably compared to the previous stages (Jalabert *et al.*, 1976). This increased gonadotropin level could have triggered the synthesis and storage of presumptive ovulation mediator in preparation for steroid-induced completion of oocyte maturation Jalabert *et al.*, 1978). This hypothesis is supported by the findings that, if fish (trout) with oocytes close to maturity (Tertiary yolk granule stage) and have high endogenous levels of plasma-gonadotropin, final maturation and ovulation can be induced with the injection of 17 α - 20 β progesterone, but in the fish (trout) with less mature occytes which have low level of endogenous gonadotropin, 17 α - 20 β progesterone should be supplemented with gonadotropin to bring about final maturation and ovulation (Jalabert *et al.*, 1978). Similarly in common carp, priming dose of pituitary extract followed by 17 α - 20 β progesterone caused induced ovulation. This shows that gonadotropin use is in some extent replaced by 17 α - 20 β progesterone. However, in Indian major carps, the pattern by which oocyte maturation due to progesterone is not known, as 17 α - progesterone seems to be ineffective (Kapur, 1978). Moreover these studies also do not clearly demonstrate the nature of the active steroid, site and mechanism of its action. Recently Breton *et al.*, 1983 investigated the possibility of using a low dosage of LHRH or LRH-A to replace pituitary extract or gonadotrophic hormone (GtH) as a primer before 17 α - 20 β progesterone injection in the carp, but only limited success was obtained.

7.8 Corticosteroids

High dose of corticosteroids have been shown to induce oocyte maturation and ovulation in several species of fish. The finding of Donaldson (1973) in some catfish suggests that the gonadotropin may stimulate ovulation by induction of corticosteroidogenesis in extraovarian tissue, possibly the interrenal (corpuscles of stannius) while in medaka (*Oryzias latipes*), ovulation may be caused by gonadotropin induced ovarian corticosteridogenesis. Hence, corticosteroids are largely of interrenal origin which may also be synthesized in gonadal tissue.

The increase in plasma corticosteroid concentration during reproductive development and spawning have been reported in teleosts. As in the case of 17 α - 20 β Pg, the efficacy of corticosteroids as inducer of ovulation may be improved by a priming injection of pituitary extract of GtH (Hirosa and Ishida, 1977). Although *in vitro* effectiveness of corticosteroids (cortisone) in common carp, *Cyprinus carpio* have been initiated but cortisone were slightly effective at their highest concentration tested (Moses in Baraj and Shaider, 1988). Hence use of corticosteroids have not been met with success in Asiatic carps as a whole.

7.9 Estrogens and Androgens

After the completion of vitellogenesis, it is seen that the plasma estrogen level in female declines. This indicates the unsuitability of using estrogen as an induction of final maturation and ovulation in fish. This has been demonstrated in grass carp and silver carp by using a synthetic estrogen, stilbestrol dipropionate, in combination with PMSG as a primer which could not induce ovulation (Brandt and Schoonbe, 1980).

Informations are available on *in vitro* spermiation by administration of Androgen and methyltestosterone in number of teleosts. The presence of androgen in female teleosts, which could induce *in vitro* maturation has been reported by Goetz and Gergman (1978). Jalabert (1975) observed that androgen can reduce the quantity of dose of gonadotropin required for *in vitro* maturation. However, the attempt made to use androgen, androsterone and testosterone by moses in Baraj and Shaider (1988) in *Cyprinus carpio* is ineffective. Hence, virtually no attempt has been made to use androgens to induce ovulation in Asiatic carps.

7.10 Prostaglandins

Prostaglandin have been shown to stimulate follicular contraction and rupture, hence ovulation in teleosts (Jalabert and Szollasi, 1975; See Kapur, 1979; Kapur and Toor; 1979; Stacey and Goetz, 1982). It also stimulates gonadotropin secretion (Singh and

Singh, 1976; Stacey and Goetz, 1982) and in the regulation of spawning behaviour (Stacey and Peter, 1979). Kapur (1981) reported that PGE^1, and PGF^2 failed to bring about ovulation and/or spawning in Indian major carps. The reason may be due to incomplete gonadal maturation. Such failure can be over-comed probably priming these fishes first with pituitary followed by prostaglandins (Tripathi and Khan, 1990).

7.11 Synthetic drugs/Antiestrogens

Antiestrogens are synthetic compounds that are capable of competing with estrogen for binding sites on estrogen recepter (Donaldson and Hunter, 1983). Clomiphene citrate and tamoxifen have been used for inducing gonadotropin secretion in teleosts (Breton *et al.,* 1975, Billard and Peter, 1977. Success was first reported for clomid with increased plasma gonadotropin level in goldfish when implanted, which is considered responsbile for induction of ovulation (Pandey and Hoar, 1972). The effect of clomid on ovulation and gonadotropin stimulation in *Cyprinus carpio* was reported (Breton *et al.,* 1975; Kapur and Toor, 1979) and in *H. fossilis* (Singh and Singh, 1976). Clomid has been found to be unsuccessful in inducing ovulation in *Labeo rohita* (Bhowmick *et al.,* 1976, 1986, Kapur, 1978), though it stimulates gonadal steroidogenesis (Kapur, 1981).

The other synthetic non-steroidal chemical compound, sexovid which may also stimulate gonadotropin release and ovulation in teleosts (Singh and Singh, 1970; Kapur, 1978). Cyclofenil Sexovid F6066 treatment in *Labeo rohita* has been found to stimulate ovulation in 50% of the cases. This ovulation percentage was raised to 100% in the gonadotropin primed individuals followed by sexovid administration. The use of Hoe 766 vet, a synthetic drug have given positive results in spawning of *C. mrigala* and *Labeo rohita* (Datta *et al.,* 1989). Gupta *et al.,* 1988 have reported the effect of an anabolic steroid (Winstrol) in induction of maturation in *Labeo rohita.* Those treated anabolic steroid contained the highest percentage of mature oocytes with germinal visicles in the centre or slightly eccentric. Chowdhuri (1989) reported on the use of homeopathic medicines like *palsittella* and *Natrum muraticum* on ovarian maturation of teleosts. However, these are not standardised for commercial fish seed production.

7.12. Metal and metal salt

Khosa and Chandrasekhar (1972) observed accelerating effects of cupric acetate and asphalt on the vitellogenesis in two species of teleosts. In case of *Labeo rohita*, cupric acetate stimulated steroidogenesis, where as cupric sulphate and cupric chloride were either ineffective or inhibitory in their action. None of these salts could bring about ovulation and spawning in these fishes.

8. RESPONSE TIME

Under normal condusive conditions, the injected brooder take 6-8 hours after the last injection to appear in the state of spawning, although a while after last injection, parent fish appear to be in a state of courtship, chasing and sexual play comparable to the state of estrous. This period of time of sexual play resulting in spwaning is called as the response time. Factors responsible for sexual responses is dependant on water temperature, the kind of estrualising/inducing agents, the frequency of injection and the species of brooder. The response time to ovaprim, ovatide and WOVA-FH is shorter than the response time to pituitary gland injection. Similarly, the response time to pituitary (Pg) injection is shorter by one to two hours than that of HCG injection, while the response time for LRH-A injection is longer than that of PG injection being reported. Hence it is suggestive to record that the success on spawning not only depends on the potency of hormone administered but also the physiological responsiveness · that synchronises the final result.

The brooders after injected with inducing agents are stocked into the breeding enclosures to release eggs or exude milt to complete fertilisation process. This is called as induced spawning or natural spawning under induction and fertilisation. The spawning pools or enclosures adopted in India are of two types : (1) Traditional spawning enclosure (2) Modern spawning enclosure.

9. SPAWNING ENCLOSURES

9.1 Traditional spawning enclosure

In India, the immediate environment for induced breeding is considered to be the breeding happa, inside which the injected brood fish of both the sexes consisting of one female and two males are released to spawn. Breeding happa is a box-shaped cloth enclosure with an opening on the upper breadth wise provided with cloth loops and button holes to close the opening securely after releasing the brooders inside. The happa should be fixed at a depth where it is easily accessible to the aquaculturist. Happa is fixed in bamboo poles in such a way that, 3/4 part of its remain immersed in water without touching to the pond bottom mud. Usually 20-30 cm above the bottom mud, the four lower sides of the happa is fixed in respective bambo poles. SImilarly 1/4 part of net (usually 15-20 cm) which is above the pond water with its four sides fixed with respective bamboo poles tightly so as to keep the breeding happa in set right condition. The cloth usually used is strong, fine meshed mosquito net type, cheap and durable. The mesh size of net cloth should be such that the ovarian eggs do not pass out of the breeding happa. The happa should be tanned with preservatives and washed throughly before use. The same breeding happa can be put for subsequent operation after thorough washing with detergents, immersed in $KMnO_4$ solution and completely dried under sun light. Various sizes of breeding happa are in use. For a breeding female weighing over 4 kg, the size of breeding happa is 3.6 meter X 1.8 meter X 1.0 meter. For brood female weighing between 1.5 and 4 kg, the size of breeding happa is 2.4 m X 1.2 m X 1.0 m. For the brood female weighing under 1.5 kg, the size of breeding happa, 1.8 m X 1.0 m X 1.0 m is usually practiced.

9.2 Modern breeding enclosures

Now a days modern, hatcheries are operated for mass production of carp seeds. The essential components of a modern hatchery proper, are : (i) ante-tank or storage tank. (Brooder holding tanks), (ii) breeding tanks or ward tanks, (iii) Incubators or hatching jars and (iv) larval rearing tanks. However, the modern breeding tanks are usually made of bricks with concrete cover. These are called as ward tanks. With te passage of time in carp aquaculture, the breeding tanks have under gone modifications in their shape, design and size. Even automation has been installed for self transfer of fertilised eggs into incubation/hatching tanks.

A breeding tank may be rectangular in shape. The size of the breeding tank varies with the brooder holding capacity. For holding 4-6 brooders, weighing 3-6 kg each, the size of breeding tank may be 2.5 m X 1.5 m X 1.0 m. For holding 8-10 brooders weighing 3-6 kg each, the size of breeding tank, may be 4 m X 2 m X 1 m. For still larger brooders of 12-20 kg each, even 7.5 m X 2.5 m X 1.0 m, breeding tanks are in use. However, in India the convenient size of brooder weighing 1.5 to 5 kg each is taken for induced breeding. Besides induced natural breeding, artificial insemination for seed production is practiced in India. Therefore, it is convenient to construct a side tank of 20-25 cm deep and 50 cm wide along the long axis of the breeding tank. The fish which are to be stripped for artificial insemination can be pushed conveniently into the ditch for separating without disturbing the other brooders. It is advisable to fix a breeding happa inside a breeding tank which is much easier to handle, procure, isolate or release the brooder as per requirements.

In modern times, circular breeding tanks are in use with additional facilities. Breeding tanks of circular shape are of Chinese origin. A convenient size of a circular breeding tank is 2 m diameter and 1 m deep which would hold about 1800 litres of water. These breeding tanks essentially need a continuous supply of filtered, clear-clean and well oxygenated running water of optimum temperature. A common feature of modern breeding enclosure is that, they have a sloping bottom leading to the outlet, centrally located in circular or sideway located in rectangular for water drainage. The other common feature of breeding tank is that to maintain their water level. In circular design of breeding tank, the water inlets are set at a 40° angle target to the tank wall inorder to impart a circular motion to water. The rate of flow of water in a 1800 litre circular tank may be 6-8 litres/sec. The outlet is usually installed in the centre of the bottom into which would fit a uniformly perforated straight pipe and a screen covering the per forations, and serves as a passage for the collection of eggs. The pipe which is hidden underground, passes directly to an incubation raceways. It is possible for a breeding tank to function also as an egg incubator, hatching tank or fry rearing tank, by installing extra facilities like air diffuser or air blower.

Elliptical spwaning ponds are also in use for breeding of cultivated carps in China. The inlets and outlets are placed on same straight line. The water depth is 1m. The bottom of the pond is sloped toward the outlet. There is an adjacent egg collecting chamber. This is more or less comparable to the type of Bangla breeding adopted in West Bengal.

Rectangular	*Circular*
Advantage	
1. Provides some aspects of riverine condition.	1. Condusive to a continuous flow system and provides reverine environment.
2. Adoptable to large scale operation.	2. Centrifugal flow of water makes the operation of inlet and out let more effective.
3. Water quality can be controlled.	3. Adoptable to large scale operation.
4. Additional aeration is possible but not very effective because of shape.	4. Dead spaces are absent and makes uniform distribution of oxygenated water. Hence hatching is more effective.
5. Provides oxygenated water but not uniform because of dead areas at curves or angles.	5. Water quality can be controlled.
	6. Condusive to protect eggs from being washed out due to screen surrounding the centrally located out-let.
	7. Additional aeration is easier.
	8. Combine the function of hatching tank and fry rearing tank.
Demerits	
1. Required large quantity of water.	1. Requires large quantity of water for effective functioning.
2. Any defect in drainage system requires dismantling to repair.	2. Defects in under outlet pipe or drainage system require dismantling to repair.
3. Separation of egg shells from hatchling not possible.	3. Separation of egg shells from hatchling is combersome and not very effective.

10. ESTROUS AND SPAWNING

Under the influence of exogenously introduced pituitary extract or hormones, brood fish will generally begin to chase each other excitedly. This phenomenon is called estrous. This chasing can be apprehenced soon after seeing the irregular rippels that appear on the water surface. Sometimes, brooders emerge on the surfce of the water and the male hits the abdomen of female with its head. The female lies on either on the surface of water or in under water. The abdomen and caudal fin of female intensely constricted and then eggs flow out like a jet. At the same time, the male come close to female and discharge milt. Sometimes male and female swing their pectoral fins closely to spawn or discharge milt. Usually two types of spawning are noticed. When the brooders spawn at the surface layer of water, the spawning is called as floating or surface spawning. Mostly Grass carp and Silver carp shows floating spawning. The under water spawning is called as muffled spawning being exhibited by Bighead and Black carp. Although Bighead is a surface feeder but because of sluggishness they show muffled spawning (IFF, China).

In India, spawning behavior in ecohatchery is categorized mostly in to three types based on the chasing movement of brood fish. These are 1. Hypogyne 2. Subgyne 3. Laterogyne. In Hypogyne the male brood moves below the female brood as seen commonly in Silver carp but occasionally in Indian major carps.

In Subgyne, the male chase behind the female brood as seen commonly in IMC.

In Laterogyne, both male and female brood move parallel to each other in the surface level of water. Besides these three types of spawning behavior, some times solitary movement of the male brood on hormone induction is seen when female brood is not receptive or vice versa. Such type of spawning behavior is termed as Ungyne.

During spawning of carps three types of courtship locking mechanisms are noticed in the spawning pool of ecohatchery . These are 1. Pecto-caudal 2. Dorso-caudal and 3. Isolateral.

In pecto-caudal locking system, both male and female brood fish inter lock their pectoral fins together followed by caudal peduncles. This phenomenon is observed when strength and size of brood fish are mostly similar.

In dorsal caudal locking, the male brood fish encircles the female brood at the end of the dorsal fin followed by coiling towards the caudal peduncle keeping their vents to maximum closeness for release of gametes. Such type of phenomenon is seen where the male is smaller and slender than the female brood fish.

In isolateral ,although there is no locking but both the male and female brood come nearer to take a parallel position lateral to each other. At this posture, both releases their gametes but the possibility of fertilization rate of eggs are less. This type of posture is observed when brood are too fatty and bulky

10.1 Collection of fertilised eggs

Eggs should not be collected or transferred out of breeding enclosures until they are water hardened. A water harden egg does not rupture or get spoiled easily. In traditional breeding enclosure, the bottom corners of the breeding happa are untied first and lifted slowely from the back to the other open end of the happa. Then brood fishes are skillfully pushed back and carefully removed by scoope nets without disturbing much of the eggs. The happa with the eggs is then lifted above water and transferred to a bucket by a graduated known capacity enamel mug. However, in modern breeding tanks the water flowing stimulation not only create a riverine condition, but also raise egg laying rate and fertilisation rate. After collecting the brooders, the central outlet pipe is opened, so with mild gravitational force, automatically eggs are collected at the egg collection chamber.

10.2 Method of counting eggs

In India usually volumetric method is widely accepted. Eggs after their removal from the breeding enclosure are kept suspended over a rectangular piece of cloth, till all the water has drained out. The eggs after water hardening are then measured by a mug of known volume. From the number of mug, the total volume of eggs are find out. Then average number of eggs in a known volume of sample are found out. From this and from the total volume of eggs obtained, the total number of fertilised eggs can be calculated out. The number of eggs per ml. of water also is size dependant. The diameter of eggs after water hardening in Catla is (5-6.5 mm), Mrigala (4.5-5.5 mm), Rohu (4.0-5.0 mm), Silver carp (4.2 - 5.0 mm) and Grass carp (4.5 - 5.5 mm).

However, there are about 1000 eggs in 100 ml. of eggs of grass carp and silver carp after water hardening. Generally 8-13 eggs per ml. for fully water absorbed eggs in carps are noticed.

11. ARTIFICIAL FERTILIZATION (STRIPPING)

When the brood fish are in sexual play and exhibit courtship behaviour just before the release of eggs and milt by the female and male respectively, they are captured immediately. Egg collection and milt collection are done so as to bring the matured egg and sperm together what is called as artificial insemination. The time of egg collection and milt collection are dependant to various inducing agents, species of carp induced and water temperature. This is the key to the success of artificial insemination as per the vitality of gametes in water is very short.

11.1 Composition of gamete fluid

11.1.1 *Seminal fluid*

The mineral composition of the seminal fluid in carps varies widely (Morizawa *et al.*, 1983; Kruger *et al.*, 1984). Morizawa *et al.*, 1983 showed that the concentration of carp seminal fluid was 74 ± 23 mM Na and 77 ± 10 mM K. Large amount of aminoacid (36 mM) composition has been reported in seminal fluid of carp (Menezo *et al.*, 1983). The osmotic pressure is around 300 m osm/kg. Variations in lipid content of the seminal fluid and spermiation process of various cyprinids was studied by Belova 1982, 1983. The pH of seminal fluid in carps was in the range of 7.6-8.6 (Zhukinskiy and Bilko, 1984).

11.1.2. *Ovarian fluid*

Scanty informations are available on the composition of ovarian fluid. The osmotic pressure is 306 m osm for Carp and 218 m osm for Silver carp. Plouidy (1982) and Ghosh (1985) reported large differences in mineral and organic contents between common carp and grass carp. Protein, Amino acid, Urea, Hydrolases, Dehydrogenases, glucose amd lactic acid were identified in carp ovarian fluid. The pH values are high 8.5 ± 1.1 for carp and 9.1 ± 0.5 for Silver carp (Billard *et al.*, 1986).

11.1.3. *Gamete survival*

The rate of deterioration of gametes after exude from respective female and male spawner are directly related to temperature. That too, the vitality of gametes are reduced in water with passage of time. Therefore manipulation of gamete is very essential in artificial insemination.

For *in vitro* fertilisation, eggs should be collected as soon as possible after induced ovulation and if they are not fertilised immediately can be cooled and stored in 2-4°C. Collection of eggs should always be in ovarian fluid but not in water. Ovum fertilizability is very short in freshwater or various saline solutions, even at isoosmotic strength. This phenomenon is due to autoactivation without any sperm contribution was reported (Yamamoto, 1961) in goldfish and in carps. The fertilisation rate of carp begins to decline immediately after dilution either in freshwater or in saline solution and become nil within few minutes (Sjafei, 1985). However, various attempts have been made to inhibit autoactivation. By adding an esterase inhibitor in goldfish the autoactivation of egg was delayed (Hamano, 1957), by adding soldium oxalate to *Oryzias latipes* (Medaka) ova (Yamamota, 1954), or by adding ficoli to zebra fish ova (Harvey, 1982). After dilution in a pooled ovarian fluid, autoactivation is delayed by few hours (Christen, 1985, P.C. see Billard *et al.,* 1986) suggesting that this medium includes a factor(s) which prevent autoactivation.

Cold preservation of milt for considerable hour without deterioration are reported (Belova, 1981). He indicated that the survival of Silver carp, Grass carp and Bighead carp sperm cells declined after 12 hours storage at 6-8°C. Sperm may be diluted with an extender with a chemical composition of the milt (pH, Na^+, K^+, Ca^{++}, Mg^{++}, Osmolarity etc.) to study the sperm count, spermatocrit, milt volume and motility. Withler (1982) showed that frozen spermatozoa of *Labeo rohita* after thawing were as effective as the fresh control. The best percentage of fertilization even recorded has never exceeded 50% verses control, although more cryo-preserved spermatozoa were used.

Usually the method followed for cryopreservation of milt is to dilute the milt with an extender (Distilled water 100 ml, sucrose 4.28 gm, $KHCO_3$ 1.00 gm, Reduced glutathione 0.20 gm). The extender solution also contain a cryoprotectant and freezed to around –45°C and reduce ice crystal formation. Glycerin, ethylene glycol and propylene glycol have been used as cryoprotectants for fish sperm but the most widely effective is dimethyl sulphoxide (DMSO). However, for cryopreservation, the freezing rate of 30 to 160°C per minute have been successful.

At present, mini straws of fine plastic tubing with 250 ml aliquots of extended milt, plugged at both the ends and freezed in liquid nitrogen are used in experimental basis. An appropriate cooling is achieved by keeping the straws of milt on a metal plate 4 cm above the surface of the liquid nitrogen by a polystyrene float. After 10 minutes, the straws should be immersed in the nitrogen and then stored until required. Once in liquid nitrogen, the frozen milt can be stored indefinitely since the rate of deterioration is negligible.

12. METHODS OF DEGUMMING AND STRIPPING

Usually in sticky eggs of fishes, especially of common carp, the stripped eggs and milt mixture are degummed. Various degumming methods are available for the purpose.

The mixture of urea and sodium chloride was initially used in 1960's for removing the sticky mucoid envelope of common carp eggs. Other method of degumming includes salt-carbide method in which 4g Nacl and 3g urea/liter of water is put in the mixture of stripped eggs, milt and milk for 5 minits with constant stirring. Then these eggs are again treated with a solution containing 4g Nacl and 20g urea/ liter of water for 30 minits. There after, the eggs are washed thoroughly in pond water to get degummed free eggs. Often human urine is used to degum the common carps eggs.

Similarly cream milk method is employed for degumming the sticky eggs of fishes in which the eggs, milt mixture is treated with 20g of full cream milk powder (fat content 26-28%) dissolved in 1 liter of water for 1 hour. By these degumming methods, the sticky eggs get free from each other.

13. METHODS OF STRIPPING (ARTIFICIAL INSEMINATION)

Dry method and wet methods are followed in artificial insemination in carps. In some regions semi-dry method is followed, the procedure of which are given below.

13.1 Dry method

Press out the eggs into the basin in which the seminal fluid is added immediately . Stir the eggs and sperm gently with bird's quill so as to mix them. After that, add a clear water, stir the mixture for one minit, wash the eggs with clear water 3-4 times. Then the process of fertilisation is finished.

13.2 Semi-dry method

Suck the seminal fluid with a pipette or a syringe without needle and dilute it with little normal saline and then add it to the eggs and stir the mixture.

13.3 Wet method

Put little clear water in to basin. Press out eggs and sperm in to the basin immediately and then stir the mixture.

For better success the process of artificial insemination should be carried out skillfully and completed as quickly as possible, otherwise fertilisation rate will reduce.

13.3.1. Catfish breeding

Government of India has also identified catfish farming as the National priority and has emphasized on the diversification of cultured practices. The major bulk of catfish comes from capture resources that include air breathing and non airbreathing varieties. It is well know that magur (*C. batrachus*) is the most preferred indigenous catfish species in India. It breeds in nature during monsoon. For breeding, maintenance of healthy brood stock is a pre-requisite for successful seed production in captivity. This species takes about one year with 100-150g in weight for maturity. Fish can be stocked in cemented tank with soil base of 5-10g thicknesses with mild continuous flow of water. A mixture of GNOC, fish meal, soybean meal, rice bran with vitamin and mineral mix containing around 30% protein is fed to stocked fish at 2-3% of body weight. Male and female are distinguished by secondary sexual characters. These are bred either with hormonal administration or through environmental manipulation. Females are induced bred through synthetic hormones that is Ovaprim, Ovatide, WOVA-FH @ 1-1.5 ml/kg body weight. Heteroplastic pituitary gland @ 30mg/Kg body wt can induce bred the magur. Single injection is given to female and male do not get any injection, as sperm suspension is required for fertilization. Sperm suspension is made with 0.9% sodium chloride solution, which can be used for 24 hours. Ecohatchery is used for small scale and large-scale seed production. The yolk sac of newly hatched larva gets absorbed in 3-4days. The hatchlings are reared in plastic containers in indoor system by giving zooplankton, artemia nauplii, molluscan meat, tubifex or eggs custard (41-65% protein). Vitamins and minerals can be added in eggs custard preparation. During rearing in indoor system, continuous aeration and water exchange is made for better retrival rate. A stocking density of 2000-3000 larvae/m^2 is considered optimum during indoor rearing. On attaining the size of 10-20 mm and 30-50 mg size within 12-14 days of rearing , fry are transferred to other cemented tanks for fingerling raising. Floating weeds are given to provide shelter to magur fry. Fish fry reared at a density of 200-300 fry/m^2 and fed with formulated feed containing 30-35% protein grow to 0.8-1.0 g during 30 days of rearing. These fingerlings are now ready for stocking in culture ponds.

Another important catfish is *Heteropneustes fossilis*. It attains sexual maturity when one year old with length range from 8 – 12 cm. Usually female brooder has swollen belly, prominent round genital papilla where as male is slender, streamline body with pointed genital papilla during breeding season. It breeds during monsoon months (June to August) with peak in July. Brood fishes of 50-100g body weight are reared in cement cisterns before breeding operation with continuous flow of water. A mixture of fish meal, Soyabean meal, groundnut oil cake and rice bran at 2 – 3% of body weight is fed to stocked fishes. Hormones like Pituitary gland extracts @ 15 – 20 mg/ Kg body weight or Ovaprim @ 0.6 – 0.9 ml/Kg body weight is used for induce breeding of the female fish. No injection is given to male. After 14 – 18 hours of hormone injection, female broods were taken out for stripping and the male brood fish is cut open for preparation of sperm (Milt) suspension in 0.6% saline solution for viability of sperm. The milt suspension is added in to the stripped eggs and mixed thoroughly by bird's feather for 2 to 3 minits. The fertilized eggs are washed repeatedly in freshwater. Fertilised eggs are greenish blue in colour and settles down to the bottom of the container. The unfertilized eggs are whitish and kept floating above the fertilized eggs. The fertilized eggs hatch out in to hatchlings within 16-19 hours. The yolk sac of hatchlings get absorbed in 3–4 days after which spawn feed on natural foods. These spawn are reared to fry for 10–12 days in shallow containers with live feed, molluscan meat suspension, egg custard etc., so as to attain a size of 10–12mm. These fry are further reared for about 20–22 days at a stocking density of 3000–5000/ M^2 in large cement cisterns to harvest advanced fry . These advanced fry are reared in earthen ponds at the stocking density of 300–500/ M^2 for production of fingerlings. They can be fed with molluscan meat and rice bran @ 5–10% of body weight during evening time. Fingerlings are then cultured in stocking ponds.

Stripped cat fish (*Mystus vittatus*) is also reported to be a valued cat fish in wet land of Kalkota. It matures when 6 cm in length.

13.4 Methods of calculating eggs

Both weight and volumetric methods can be adopted for measuring the egg. In weight method, weigh the eggs before absorbing water, multiply the weight by the egg number per unit weight, then the total egg quantity can be calculated. In volumetric method, measure the volume of eggs before absorbing water, multiply it by the egg numbers per unit volume and then the total egg quantity can be calculated.

14. COMPARISON BETWEEN TWO FERTILIZATION METHODS

Natural Induced spawning	Artificial Insemination or stripping.
Advantage	
1. Eggs can be collected at proper time and egg quality is good.	1. Fertilised eggs are clean without pests and miscellaneous matter.
2. Brood fish are not easily injured.	2. Convenient to carryout cross breeding.
Disadvantages	
1. Requires more installation and less limited by conditions.	1. Brooders are easily injured.
2. Fertilised eggs are mixed with pests and miscellanous matter.	2. If proper time of egg collection is not made, fertilisation rate goes down.
3. Difficult to carryout cross breeding.	

15. BUNDH BREEDING

The potential for freshwater fish culture in India indicate a need for more than 16 billion fish fry every year. The present supply is much less than the target requirement of fish seeds. The question arises how to bridge this gap in shortest possible time.

The answer lies in ``Breeding carps in bundhs". Breeding of carps in bundhs was known to have originated from West Bengal state, especially from the districts of Midnapore and Bankura. Yet Madhya Pradesh was the pioneer state where the bundh breeding was undertaken in the public sector on scientific lines for the first time in the year 1958. With the expansion of fish culture industry in India, the bundhs have been established in several other states namely Bihar, Uttar Pradesh, Andhra Pradesh, Punjab and Orissa.

In recent years, in India, the production of fish seed through hypophysation contributes 1.5% of total spawn production of the country, the rivers still continuing to be the main source. Bundhs contributed about 5.3% towards the total spawn production (Government of India, 1966).

During the year 1976, the total spawn production in West Bengal and Madhya Pradesh was 11.5 and 54% respectively by bundh breeding.

15.1 Types of bundhs

The bundhs are principally of two types. (1) Dry bundh and (2) Wet bundhs. However, successful adoption of a circular breeding chamber and a specially designed rectangular masonary structures with an arrangement of water pipe systems led to a hatchery system popularly known as ``Bangla Bundh" (Chondar, 1990), which is a dry type of cemented bundh tank with specified slopes at sides and bottom, low cost devise developed in India in early 1980's to breed IMC, Grass carp and Silver carp on a mass scale.

15.1.1. *Dry bundh*

A dry bundh is a shallow depression having bundhs on three sides and an extensive catchment area on the other side. Bundhs get flooded with the monsoon water during the South West monsoon, but remain completely dry for a considerable period during the remaining part of the year.

The topography of land plays an important role in the location and distribution of the dry bundhs. Usually undulated land is preferred because it provides large catchment area. This topographical structure favours in quick filling of bundh even with a short rain, at the same time quick and easy drainage is possible due to gravitation. In West Bengal, a catchment area of more than five time the bundh area is considered suitable. In Madhya Pradesh, the relation between actual bundh and its catchment area of Sonar dry bundh of Nowgong works out to be 1.25 and is considered essential (Dubey and Tuli, 1961). In Bankura district of West Bengal, most of the dry bundhs are fed with water from storage tanks constructed in the upland areas.

15.1.2. *Construction and Design of a dry bundh*

Seasonal bundh as such with little improvement can be used for dry bundh breeding, provided, it has vast catchment area so that the bundh may accumulate sufficient water in the first rain. These seasonal bundhs can be improved by creating facilities for breeding grounds at different levels on both the sides of the water current to give the fish an opportunity to breed on accumulation of water at different levels. The cleaning of the bed may also be done so that the egg can be collected without difficulties.

Soil has not been found to play any significant role in the breeding of carps in bundhs. Spawning having been observed on laterite, rocky and even clayey soil. Hence, any suitable site on a small rivulets can be selected for construction of a dry bundh. The bundh should be provided with a weir guarded with fine mesh iron netting through which the surplus water can flow over. An outlet can be provided at the bed level to dewater the bundh proper. Provision shall be made to hold the male and female brooder separately in near by cemented tanks (Ante tank) along with hatchery facilities for incubation of eggs collected through bundh breeding.

15.1.3. *Wet bundh*

The wet bundh is a perennial pond located on the slope of the vast catchment area of undulating terrain, with proper embankments. The inlet face towards the upland and an outlet towards the opposite lower ends. A greater portion of this bundh dries up during summer and is cultivated while the actual pond always contain some water harbouring mature fish. After a heavy shower, a major portion of the bundh gets submerged with water due to rain run off from the catchment area. The excess water flows out of wet bundh through the outlet. The fish starts spawning in such a stimulated natural condition in shallow areas of bundh. The outlet is protected by fencing to prevent escape of brooders. The wet bundhs are comparatively much bigger in size and modified Chittagong wet bundhs are reported suitable for the spawning of fish.

The wet bundh could be of any size with a catchment area ranging from 20-100 times of the bundh to as large a body of water (reservoir) have come to be recognised as wet bundh. However, large reservoirs where the fish breeds in their upper reaches after migrating some miles in the rivers or streams or fields should not be considered as wet bundhs.

16. TECHNIQUES OF BREEDING OPERATION

Matured brooders are collected from perrenial ponds or reservoirs before the onset of monsoon or during the first showers and are kept separately sex wise in ponds preferably on a rainy day. The inlets and outlets were kept properly quarded and sufficient inflow and outflow of water was permitted during rain.

Sufficient quantity of freshwater in the dry bundh is allowed to accumulate before the brooders are released. A ratio of one female to two males is followed preferably on a rainy day; however, the ratio need not be strictly followed. The stocking density of brooder varies from 3000-3500 kg/ha. After stocking, the brooders are allowed to remain for 10 to 12 hours in order to get acclimatised. In Bankura district of West Bengal, a few sets are injected with pituitary gland extracts. This has now become a common practice. But in Madhya Pradesh injection of pituitary extract is not practiced. The injection dose is 3 to 4 mg/kg body weight for males and 7 to 8 mg/kg body weight for females. Even sometimes, second injection of a higher dose is followed when brooders do not breed. After pituitary injection, water is released from the storage tank to dry bundh by means of a pump or sometimes by cutting the bundh of a storage tank. Water flows in to the dry bundh for about 3-4 hours and excess water is passed through the outlet. In dry bundhs of madhya pradesh, the current is created by the rain water entering into the dry bundh directly from the catchment area.

Consequent with the effect of pituitary injection and the artificial current due to flow of rain water stimulated the brooders to chase and after 4-5 hours the brooders spawn in shallow areas of the bundh. Depending upon the availability of brooders and water source, five to six breeding operations can be carried out in one dry bundh. Before starting the second operation, the dry bundh is completely drained through out let and spent brooders were removed. The bundh bed is dried.

In wet bundhs, the brooders stock may be maintained throughout the year or replenished prior to monsoon. The brooders are generally not injected with pituitary hormones but are stimulated to spawn due to water current, entering from catchment area like the case of dry bundh.

17. COLLECTION AND HATCHING OF EGGS

After lowering the water level, eggs are collected by dragging a piece of mosquito net cloth. Collection of all the eggs are impossible especially in wet bundhs due to its larger area. In Madhya Pradesh, after collection of eggs, these were hatched in traditional hatching happa fixed in bundh itself. Sometimes hatching is carried out in rectangular cement cisternaes of portable sizes. In West Bengal, hatching of eggs were carried out in specially dug out earthern pits with mud plastered walls.

18. FACTORS INFLUENCING CARPS IN BUNDH BREEDING

It is observed that strong current influence the breeding of carps in bundhs. Others have also considered the heavy monsoon flood to be the primary factor responsible for the triggering mechanism of spawning. Das and Dasgupta (1945) believed that the molecular pressure of water particles and silt on the body of the mature fish has got a stimulating effect for spawning in conjunction with rise in temperature. However, they believe that monsoon flood from the hills, having special physical, chemical and certain electromagnetic properties apart from a peculiar smell or fragrance induces spawning in bundhs. However, Khan (1942) considered shallow spawning ground is a factor for spawning in bundhs. According to Saha *et al.* (1957) temperature has no specific influence on spawning but cloudy days accompanied by thunder strom and rain seems to influence the spawning. Mookherjee *et al.* (1944) mentioned that pH and oxygen content of water do not influence the spawning in fishes. The breeding of carps in bundhs took place without inflow and outflow of water (Alikunhi *et al.,* 1964) without rains and even at fall water level

(Dubey and Tuli, 1961). These observations clearly indicate that current, floods etc. are not an essential requirement. Such spawning can be attributed to the pheromone action (Moitra and Sarkar, 1975) and (Ranganathan *et al.*, 1967).

19. ADDITIONAL FEATURES ADOPTED IN DRY BUNDHS

By adopting additional features in recent year, the dry bundh breeding and hatching recovery has been improved.

19.1 Improvements in breeding facilities

(i) Selection of shallow spawning grounds in undulating terrains of laterite/sandy soils with maximum catchment areas.

(ii) Retaining of water for longer period by encorporating additional facilities.

(iii) An overflow weir, protected by wire netting can be provided to maintain desirable depth of 2 meter of water on the bundh.

(iv) Preparation of spawning grounds at different levels so as to get them flooded at different water levels in the bundh.

(v) Few Ante tank (storage tank) either cemented or earthen ponds can be provided adjacent to the bundh to stock the brooders temporarily prior to their introduction in to the bundh.

19.2. Hatching facilities

(i) Construction of cement hatcheries or modern hatcheries adjacent to bundh with water requirement facilities for its operation.

(ii) Facilities for store and shelter can be created for vigilant operations.

20. PROBLEMS ENCOUNTERED IN BUNDH BREEDING

(1) It is difficult to synchronise the collection of large quantities of eggs at a time, particularly in the case of wet bundh breeding.

(2) Serious problems were encountered during egg collection from wet bundh, due to entry of unwanted fishes and predatory insects.

(3) Although fertilisation rate is high, still poor recovery is encountered. This can be improved by using modern hatcheries.

(4) Transportation of brooders, cause great physical strain and suffer injuries. Therefore, facilities for brooder stocking ponds can be provided.

21. PRODUCTION OF CATFISH SEEDS IN BUNDHS

In recent years, culture of some of the species of catfish especially *Clarias batrachus* has made good progress in India. Although it has great production potential, its culture have not made headway due to non-availability of adequate quantity and timely supply of fish seeds. This is evident due to exist of a wide gap between demand and supply of its fish seed. Greater and sincere efforts are made in their breeding in artificial ponds and bundhs.

Breeding of *Clarias batrachus* on commercial scale is not a problem. The fish is easily breed under the influence of stimulated ecoloical conditions (Thakur, 1984). The fishes are made to breed in special spawning ponds with holes at definite intervals.

After breeding, the hatchlings remain in the holes and after 9 to 11 days, good quantity of fry were harvested. In India, this catfish usually breed in paddy fields. This phenomenon is very common in South Bihar, Darjeeling district of West Bengal, Raigarh and Santna district of Madhya Pradesh. Therefore, creating suitable bundhs for major carps, remodelling of paddy fields and construction of special types of ponds for catfish breeding, it is possible to have mass production of fish seed of major carps as well as catfishes.

7

EMBRYONIC DEVELOPMENT AND INCUBATION

1. EMBRYONIC DEVELOPMENT

Embryonic development takes place soon after the egg is fertilised by a sperm. The eggs of bony fishes have a relatively large amount of yolk for which these are categorised under telolecithal eggs which is in contrast to alecithal (non yolk) and mesolecithal (moderate yolk) eggs.

1.1 Cleavage

The egg plasma which is seggregated from the yolk begin to concentrate towards the animal pole of the egg. Cleavage is confined to the superficial of the cytoplasm and is incomplete (meroblastic) (Fig. 34). In the earlier stages cleavage planes are vertical so that all the blastomere lie in one plane only. The blastomeres are separated from each other by furrows and lie over the yolk. In the later stages, cleavage occurs in horizontal planes, so that the blastomere becomes smaller and smaller and are arranged in more than one row. The disc of cells thus formed on the animal pole of the egg is called the blastoderm.

1.2 Morula

The blastomeres accumulate on the top end of the yolk and is called the morula stage.

1.3 Blastual stage

The central cells of the blastoderm divided to form a number of free blastomeres which becomes arranged on the top of the yolk, so as to form a layer of cells called the periblast (Fig. 35). The space beween the blastoderm and periblast is the blastocoel and the embryo is in blastula stage. The blastoderm gives rise to the embryo while periblast cells probably serve to digest the yolk and supply it to the developing embryo.

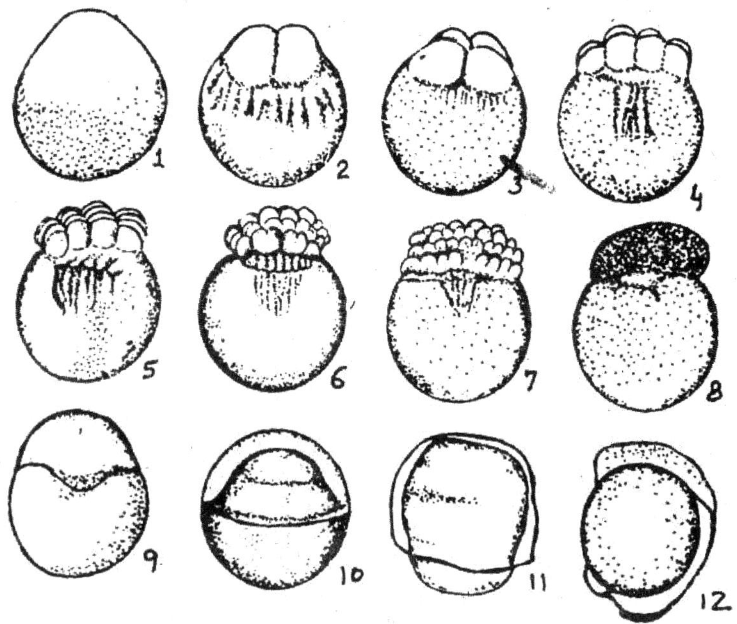

Fig. 34. Early Embryonic development stages (1-12)
1. protuberant blastoderm; 2. 2 Cell stage; 3. 4 cell stage; 4. 8 cell stage; 5. 16 cell stage; 6. 32 cell stage; 7. 64 cell stage; 8. morula stage; 9. mid-blastulla stage; 10. early gastrula stage; 11. late gastrula stage; 12. blastopore closing stage.

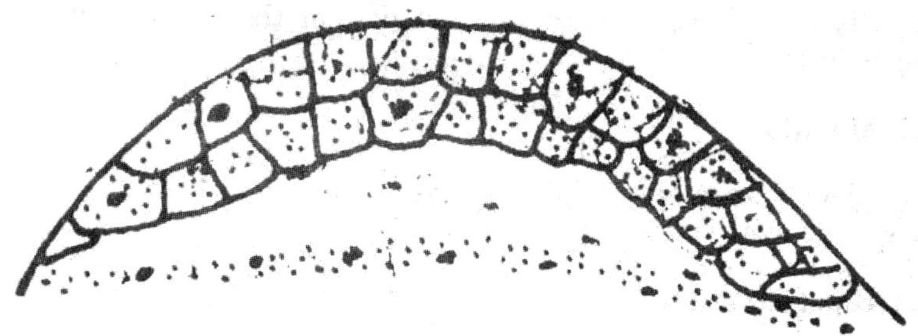

Fig. 35. Blastula showing periblast

It is possible to identify various regions of the blastula wall that are destined to give rise to specific organs of the embryo. Thus a fate map of the teleostean blastula can be constructed showing the presumptive ectoderm, mesoderm, endoderm, notochord, neural plate etc.

1.4 Gastrulation

Gastrulation is a process by which the three germinal layers (ectoderm, mesoderm and endoderm) are formed. Prior to gastrulation, there is a mass of cells in the shape of plano-convex disc which is a blastoderm formed at the blastula stage. The gastrulation in bony fishes in accomplished by (1) Epibody and (2) Involution.

1.4.1 *Epibody*

The edge of blastoderm thicken due to concentration of cells. This thickend edge takes a ring shaped appearance and is called germ ring. The germ ring keeps on extending down ward to surround the entire yolk. The downward extension of the periphery of the blastoderm is called epibody. However, the entire germ ring does not extend downward. At one point of this germ ring, there is a greater concentration of cells due to cell division and becomes thickned. This thickened area is called embryonic shield (Fig. 36). The embryonic shield is an elongated area.

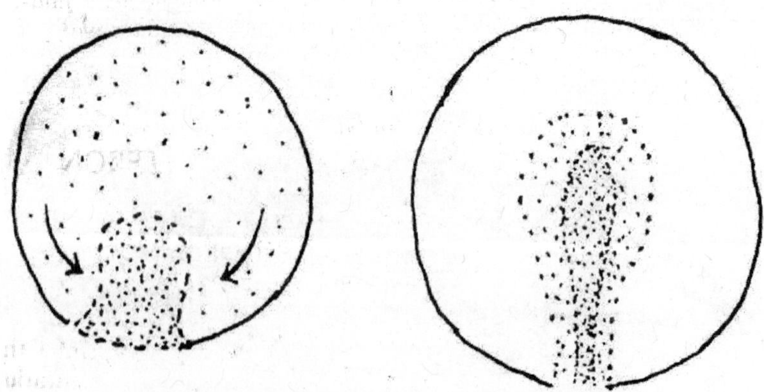

Fig. 36. Begining of embryonic shield formation

1.4.2 *Involution*

The cells of posterior end of the embryonic shield instead of going down wards, curl under the embryonic shield and start spreading as a layer of cells towards the anterior end. this incurling and spreading of cells is called involution. By involution, a new sheet of cells formed below the blstoderm. This sheet of cells is called the hypoblast (Fig. 37). Hypoblast spreads in the segmentation cavity dividing horizontally it in to upper segmentation cavity proper and other below the hypoblast which is called the archentreron cavity (Fig. 38).

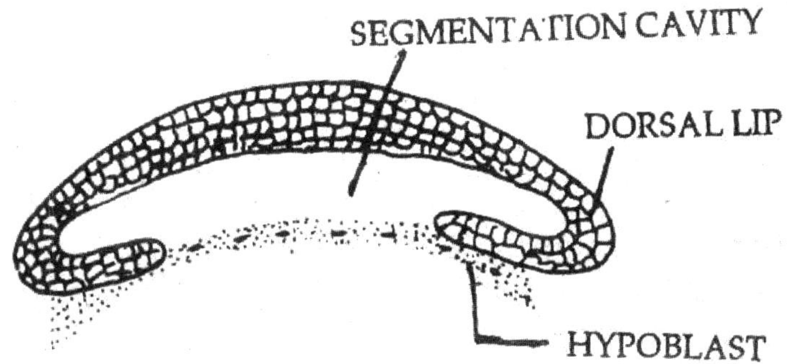

Fig. 37. Blastula showing hypoblast

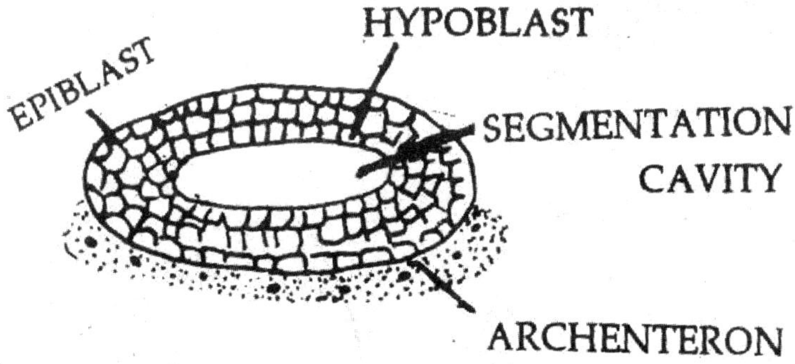

Fig. 38. Blastula showing Epiblast

1.4.3 *Blastoparal lip*

The place, where the incurling initially starts is called the dorsal blastoporal lip. The layer of cells above the segmentation cavity proper is called the epiblast. Both hypoblast and epiblast take part in forming the embryo at the embryonic shield region.

1.4.4 *Convergence and concrescence*

As the blastodermal cells multiply, some of the cells from the top and sides of the blastoderm move towards the dorsal blastoporal lip. This movement of cells is called convergence. At the same time these cells aggregate in the middle of cells is called concrescence. Some authors do not distinguish between convergence and concrescence. They refer to these by one name as convergence. When epiboly ends, the down wardly extending epiblast, cells form a sac which covers the yolk. This sac is called the yolk sac. When yolk sac is fully formed, the periblast become indistinguishable.

1.4.5. *Formation of notochord and mesoderm*

The hypoblast and the epiblast come in contact and the segmentation cavity between them disappear (Fig. 39). Some of the hypoblast cells along the mid length get separated from the rest of the hypoblast to form a medium rod of cells. This rod is called notochord. The remaining hypoblast split horizontally to form two sheets of cells. The upper sheet towards the epiblast is called the mesoderm. The lower sheet towards archenteron is called the endoderm. After the formation of mesoderm and endoderm, the epiblast is called the ectoderm (Fig. 40).

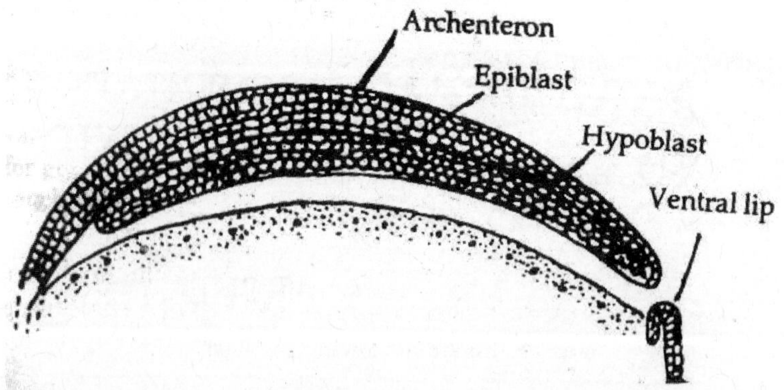

Fig. 39. Stage of Gastrula showing the formation of epiblast and hypoblast.

At the edge of embryonic shield, the ectoderm continues with the yolk sac ectoderm. Later the outer edge of mesoderm spread downward in between the ectoderm layer of yolk sac and the yolk proper.

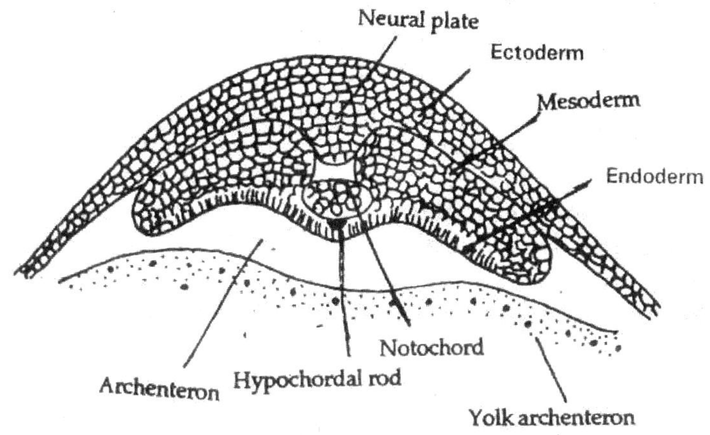

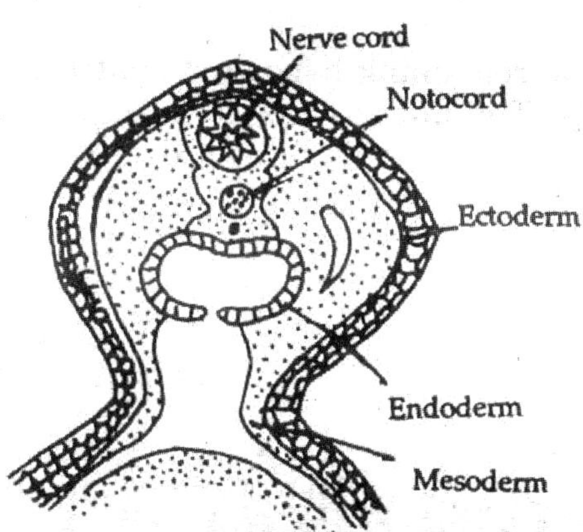

Fig. 40. Formation of the neural plate, notochord, hypochoroderm and endoderm

The endoderm curl downwards and fuses in the midline to enclose the archenteron inside it. The tube thus formed is the future epithelial lining of the alimentary canal.

1.4.6 *Formation of Neural Plate*

At the time of formation of notochord, the ectodermal cells of the mid length region of embryonic shield multiply to form a plate of congested ectodermal cells. This plate is the neural plate. In front of the neural plate appears the head position. Body segmentation occurs in the mid part of the embryo. Various kinds of organ primodia are differentiated from these blastodermal layers.

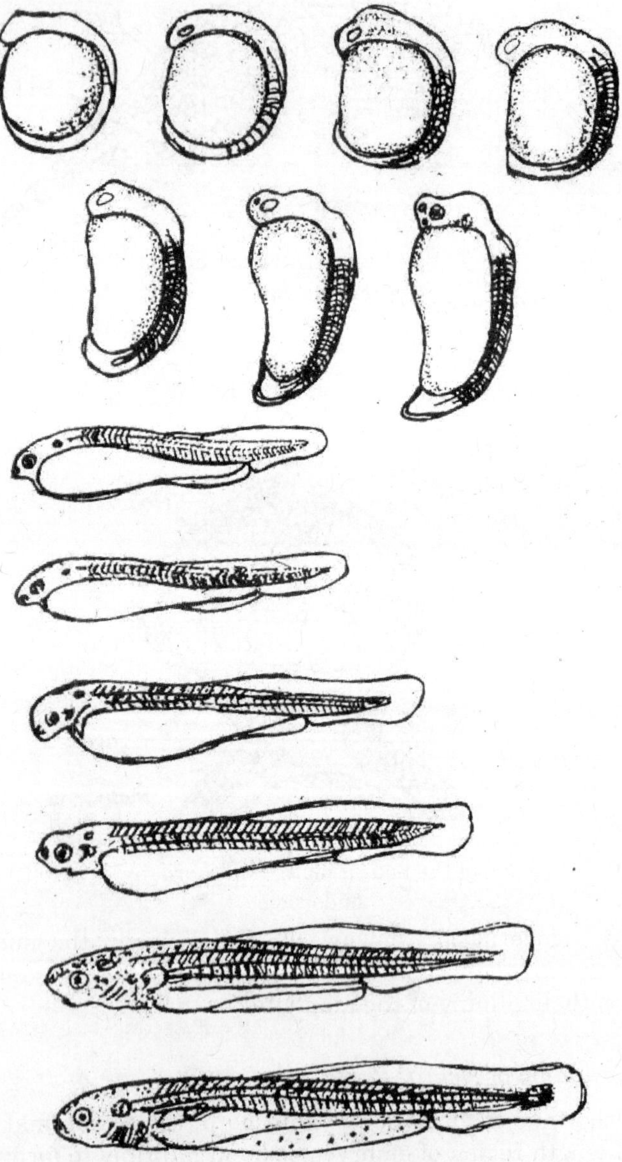

Fig. 41. Embryonic development

1.5 Organogenesis

After the completion of gastrulation various organs of the body are generally formed. These include appearance of optic vesicles on the two sides of the fore brain, then olfactory plate, primodia of forebrain, midbrain and hind brain appear successively. The body segmentation increases (Fig. 41). Gradually auditory vesicles appear on both sides of the hind brain. Caudal bud appears which grow to develop and elongate. With the extension of caudal bud, the yolk sac also elongate. The crystalline lens appear in the optic vesicle and oval gill plates appears below the optic vesicles. Segentation of body increases considerably showing wriggling and twistchling movement. Otoliths appear in the auditory vesicle and the wriggle become violent. Embryo begin to break the membrane and come out. The embryo is more or less cylindrical and bilaterally symmetrical.

To ascertain the ecological conditions of the embryonic development to the fish and then to provide them with the most satisfactory incubation condition are the most important steps in raising the hatching rate and survival rate. The important considerations are the water quality and optimum temperature upto 22°C–28°C. The rate of embryonic development is obviously affected by variations of water temperature.

2 INCUBATION

It ncludes the process by which the fertilised eggs obtained from the brooder develops through smooth embryonic development to hatchlings and subsequent stages. The duration of embryogenesis varies according to the species in degree days :60-70 for carps, 24-30 for grass carp and silver carp, 30-50 for bighead carp, and 60-70 for tench (Billard *et al.,* 1986). The optimal incubation temperature is 20-22°C for carp and 22-25°C for the Chinese carp and the tench (Horvath and Lukowioz, 1982), and 10-12°C for roach (Gulidov and Popova, 1979). The size of the hatched larvae is also influenced by the temperature during incubation for *Idus idus* the length was shorter at 9°C than at 13°C. Carp eggs can support termal fluctuation for short hour during synchronous segmentation (Jaoul and Rouband, 1982). However, the incubation period of IMC is about 14-18 hours when water temperature ranges from 26°C to 31°C. For the Chinese carps it is about 18-20 hours generally. The rate of development and the time of incubation vary according to the temperature of water. With extreme flutuactions of temperature between lower limit and higher limit of optimum temperature, the embryonic development stops or abnormalities are encountered.

During incubation, the water should be filtered and should have oxygen content not less than 5 mg per liter of water. Below 2 mg oxygen/litre, the embryo can not develop normally. The phosphate content should be low (0.12 mg/L) (Toor, 1983). That too literatures on problem encountered in hatching of carp eggs in hatcheries are well documented (Rath, 1988b) so as to take precautionary step for raising the survivality of carp hatchlings. Plankton free especially cyclopids free water is suggestive to use for incubation. The mild flow of water, create a running water system to keep the fertilised eggs in rotation. This allow maximum surface area for gaseous exchange. This will maximise the recovery of hatchlings. In this context the shift from earthen hatching pit (Hora, 1943); to earthen pot, plastered hatdchery pot and traditional double walled cloth hatching happa (Ahmed, 1948; and Chaudhury and Tripathi (1976) to various modern running water type hatching jars, Zoug Jar hatchery (Huel, 1971); vertical flow Jar of transparent polythene (Shirgur, 1972); glass Jar (CIFRI, Barrackpore), Aluminium and Plastic bin hatchery (Shirgur, 1973); and PVC hatchery Dwivedi, 1976) for incubation and circular Chinese hatcheries are more population in the field of induced breeding as well as incubation. However, a wide variety of incubators and other hatchery components are described by Woynarovich and Horvath, 1980; Jhingran and Pullin, 1985; Dehadrai, 1984 and Dwivedi *et al.,* 1988).

2.1 Double hatching happa (Fig. 42)

Such happas are used extensively all over the country. The component consists of an outer happa made of thin muslin cloth (2 X 1 X 1 m) and inner happa of round mesh mosquito cloth (1.75 X 0.75 X 0.5 m). These happas are tied in a series in ponds having clear well oxygenated water free from predators and crabs also.

Each unit of hatching happa hold about 75,000 to 1,00,000 eggs. The fertilized eggs are spread uniformly in the inner happa. When the hatching is complete, within 14-18 hours in case of IMC the hatchlings escapes to outer hatching happa leaving behind the egg shells. Then the hatching happa is taken out. The yolk sac get absorbed within 72 hours and the hatching is now called as spawn which are transferred to especially prepared nursery ponds for further rearing.

Fig. 42. Double hatching happa

2.2 Carp hatchery jar

In a carp incubation jar, the water inlet can be below or above but the exit is always at the top. Ordinarily incubation jars with water supply from below are more commonly used. The shape of incubation jars may vary from cylindrical contour to a funnel, conical or barrel shape and made of clear glass or polyethylene/ plastics (Fig. 43). Baked clay vessels can also be used which, though cheap and easily replaceable have the disadvantage of being opaque. The size of individual jar may vary from 1 litre to 200 litres and installations of jars is done as per requirement by a culturist. A one litre jar has the capacity to hold 1,00,000 fertilised eggs. However, the jars of 6.5 litre capacity are common and they are fixed vertically in two rows either in a wooden table or on an iron stand. The outlets of all the jars open in a common conduit leading to spawnery (spawn collection unit). Each jar is loaded with 50,000 eggs with the flow of water at 600-800 ml/minute for IMC and 800-1000 ml/minute for silver and Grass carp.

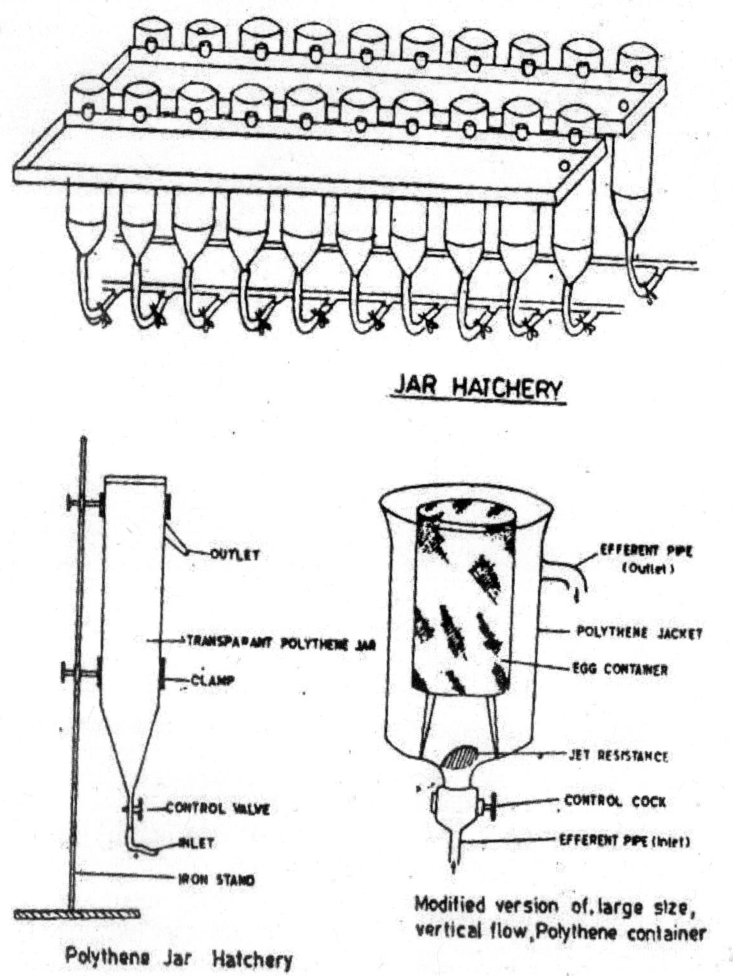

Fig. 43.

In a hatching jar with water supply from above, the water inlet duct goes down to the jar having round bottom. On hitting the round bottom, the water is reflected up creating a current to keep eggs bobbing up and down as they develop (Fig. 44).

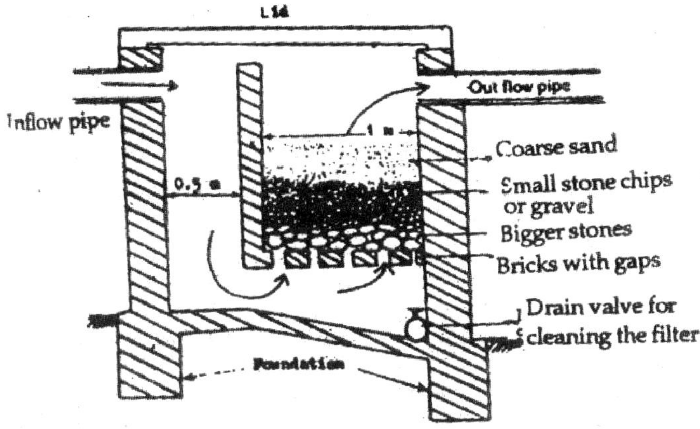

Fig. 44. Reverse flow inlet hatchery

2.3 Funnel shaped incubators

Sometimes, soft material like nylon or canvas is used to prepare funnel shaped immersed type of incubators. These are installed or hanged in rectangular water tanks. The funnel incubators have water inlet from the bottom, to keep the eggs in rotation for development (Fig. 45).

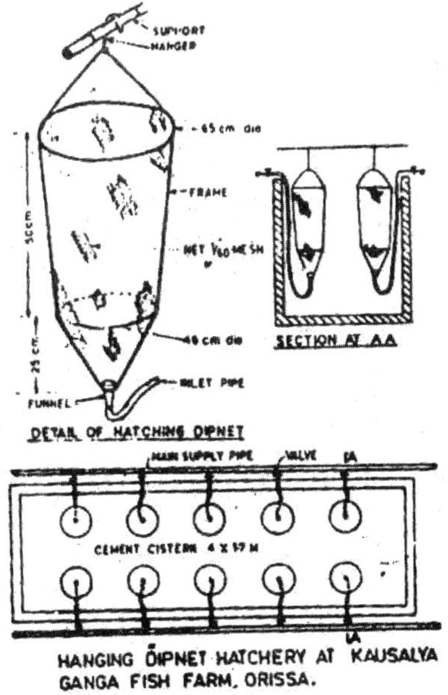

Fig. 45. Hanging dipnet

2.4 LDPE-Hatchery (Fig. 46)

This is known as ``Dwivedi designed" hatchery. All the water holding components of this hatchery are made of translucent low density polyethylene material called sintex (sinter plast container). The material is strong, unbreakable, poor conductors of heat, does not corrode, light weight and easy to clean. The hatchery unit comprises vertical hatching jars each of 40 litre capacity. Incubation capacity of each jar is one million eggs at a time where the water flow is maintained at 12-15 litre/minute. The model is portable and can be operated on a commercial scale.

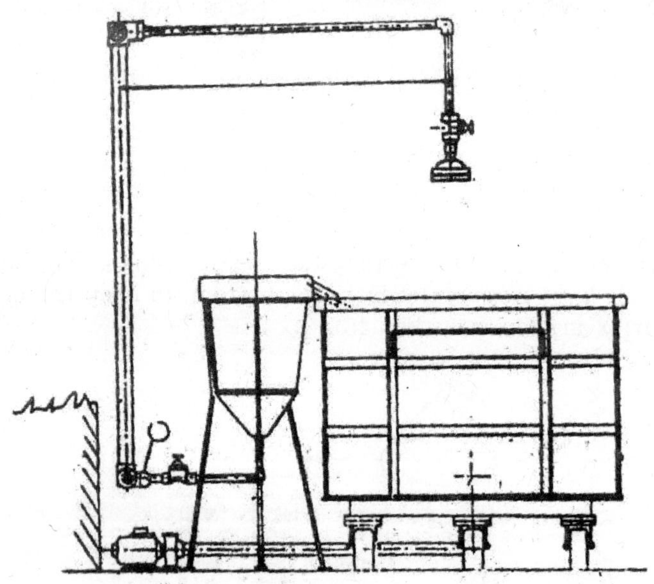

Fig. 46. Low density polyethylene (LDPE) hatchery

2.5 Circular hatching tanks (Fig. 47)

This is ring shaped lines of cement and bricks. It may be single or double. Various diameters ranging from 1.5 to 4.0 m of circular incubation tanks are noticed. The circular course is 1 meter wide, and 0.9 meter deep. This system is meant for incubating large number of eggs ranging from 0.7 to 1.0 million eggs per cubic metre of water. Jet pipes are fitted at the bottom for providing circular flow of water in the chamber which varies from 0.1 to 0.5 m/sec. The centrally placed drainage hole fitted with a vertical standpipe through which desired water level is maintained. This vertical stand pipe is encircled with a iron frame with fine nylon screen of 1/60" mesh size to prevent the escape of developing eggs and spawn. The out let fitted at the bottom leads to spawn collection chamber where a nylon happa is fixed for collection.

Fig. 47. Circular hatchery

3. INDICES USED IN INCUBATION

(i) Calculation of Fertilization rate

This is calculated by the following way. Fertilization rate (%) = Number of fertilised eggs/Total number of eggs x 100

Note that this calculation can be carried out in late gastrula stage. The eggs which are transparent with transparent nucleus are regarded as fertilised egg. The unfertilised eggs are white and opaque.

(ii) Calculation of hatching rate

Hatching rate (%) = Number of hatched fry/Number of fertilised eggs x 100

(iii) Survival rate/Retrival rate

Survival rate/retrival rate (%) = Number of individuals harvested/Number of fertilised eggs x 100

4. MANAGEMENT

Management adopted during incubation process stress upon regulation of flowing speed of water in the incubation jar, cleaning of filter or screen, periodical examination of the fertilised eggs and the embryogensis process. The nylon net with 6-8 mm meshed can be put across the circular hatching unit in a slanting position to collect the discarded egg shells. The central screen fitted should be periodically checked to remove the clogging caused by egg shell bits or can be carefully replaced by a new one.

4.1 Operational aspects of incubation pool

1. Use of water cushion is utmost essential while loading the incubation pool with fertilized eggs.
2. Duck mouth opening and speed of water must be maintained in such a way that it keeps the developing eggs away from both screen and water ,preventing them from mechanical injury.
3. The water speed of the incubation pool is to be regulated at 4-5 m/sec for first 1-2 hours, 1-2 m/sec for next six hours and 3-4 m/sec for rest of operation in order to avoid premature hatching of developing eggs.
4. Maintenance of suitable temperature and to avoid infection to developing embryos, potassium permanganate solution can be sprinkled at 2 hours interval in the incubation pool.
5. Egg loading in the incubation pool is in the range of 8-10 Lakh eggs per metre cube of water.
6. For better recovery and survival of spawn, it is required to clean the pool (after each collection) by effective cleaning devices. Such devices include sticks, coir rope with soft bristles tied at 10-15 cm distance. The dead eggs and spawn get stick to the bristles of the coir rope. These ropes are cleaned manually from time to time during operation of the pools.

5. CARP SEED TRANSPORT

The transport of carp seed (Spawn, fry and fingerlings) of all cultivable species of fish for stocking culturable waters posed problems for a long time ago. A large scale mortality of carp seed during transportation was a very common phenomenon in earlier days. With increasing demand of carp seed to the order of 1200 crores in India at present, the improved method of carp seed transportation is practiced.

Traditional techniques of transportation of fish seeds are largely impirical. However, the development of the technique of tranporting fish seed in sealed, carried under the enclosed oxygen is a breakthrough in this field. Still there is lot of scope for further improvement in the current transport techniques to work out optimum and economical packing load of fish seed without mortality.

5.1 Techniques of transport

There are mainly two kinds of transport techniques namely (i) open type and (ii) closed type. Before transportation, the seeds are acclimatised, conditioned and unfed in order to reduce the discharge of metabolites and excreta during transportation that may cause pollution to the water, in which these are carried in.

In open system, the surface of water, carrying the fish seed is kept exposed to atmosphere with or without water circulation, aeration and oxygenation. In Bengal, open seed transportation was carried out mainly by two types of hundiies having 23 and 32 litres of water holding capacity. These hundiies or pots are filled with water from spawning grounds and measured quantity of spawn 3/4 to 1/2 ``Kunka" is put in each of bigger (containing 23 litre water) and smaller pots (containing 14 litre water) respectively.

The measure ``kunka" is an arbitrary unit used in trade with a capacity of about 1 lakh spawn on an average. About 58 gms of pulverised red soil are then sprinkled over the water surface of each pot and the entire medium is subjected to occasional shaking during transit. Periodically of about 4-5 hrs. or even earlier if necessary, the water is changed and the red soil is again added. The traders are of opinion that transport of fish seed without typical red soil would cause heavy mortality. As earthen pots are liable to break while transporting by road or rail, other containers made of aluminium, galvanised iron, or ordinary tins are being used. But these metalic containers were not found to be as effective as earthen pots in keeping the water medium cool and better oxygenated. Transport of fingerling is rather difficult and needed much more care and attention in open system. However, use of plastic pools of 250 litre capacity have been successfully for the purpose.

Closed system: The technique followed in this system is that the water surface is exposed to compressed air or pure oxygen introduced to fill the enclosed atmosphere of the carriers, which are sealed air tight. The use of polythene bag measuring 650 mm X 450 mm inflated with oxygen has become very popular in transporting fish seed by rail and road or air. In this method, the bag is first put into up to 1/3 of its capacity with water (6-7 litres) and the required number of fish seed is put into it and then the bag is inflated with oxygen upto about 2/3 from a oxygen cylinder. The upper 10-15 cm of the bag is twisted, bent and tied securely with string. The spawn numbering 20,000-40,000 per bag depending on the distance are packed and transported. About 300-600 fry (30-40 mm), or 40-70 fingerlings per bag are transported in this manner. The mortality is found to nil to 5%. After the reaching of seed to the nursery site, it is acclimatised gradually in the water of the pond where it will be stocked and then released.

8

STRUCTURAL FEATURES OF A FISH FARM

1. CONSTRUCTION OF A FISH FARM

1.1 Site selection

The primary consideration for the construction of a fish farm is the slection of site. The selection of suitable site for construction of pond depends on principally two main guiding factors such as the water retentivity capacity of the soil and its inherent fertility and also respond readily to organic and inorganic fertilisation. Besides these, there should be dependable perrenial source of adequate water supply to fill in ponds at any time of the year. The site should preferably be gently slopping, so that terraced self draining ponds can be possible on higher elevation. In swampy and marshy areas. bundhs have to be piled up by depositing earth to construct ponds of required size and barrage ponds are made by constructing dams/barrages in accessible narrow flowing streams. That too, the site should be easily accessible by road or rail with market complex nearby the vicinity for easy dispose of farm products. The necessary raw materials as inputs like feeds, fertilisers and building materials for construction of store house and other accessories are available near the site. There is no possibilities of industrial, domestic or insecticide and pesticide pollution at the site. Other facilities include educational and medical in the vicinity of site. The farm proper should have additional scope for integration of aquaculture with agriculture and animal husbandry. This can be further expanded to develop processed products and other allied multistructural components at the site.

1.1.1. *Soil and water*

The propagation of natural fish food organism (Productivity) of a natural pond depends primarily on the characteristics of bottom soil. A satisfactory pond bottom soil is one which, apart from being impervious to water, permits rapid mineralisation of organic matter and release them slowly over a long period to maintain normal propagation of natural organisms.

Silty clays, clay-loams, loams etc. generally make good quality soils for a fish pond. Soil fraction should be about 90% of the whole soil, gravels not exceeding 10%. Most rocky, sandy, grave and limestone areas are to be avoided. In case of porous soil, pond bottom may be treated with bentonite, clay or other soil sealants. Sprayed on asphalt liners and plastic film liners can also be used to reduce the rate of seepage. But such treatment apart from being expensive, prevent exchange of soil water minerals and nutrients. Such locking of nutrients in the pond bottom to certain extent influence negatively the productivity of pond system. The chemical additives for leak proof as reported by Lopinot (1991) are as follow:

(i) Polyphosphates

This chemical acts as a dispersant to break down coarse soil particles into finer particles. The material is applied at the rate of 1 pound to 20 sq ft of soil and thoroughly mixed to a depth of 8 inches.

(ii) Perma-zyme

This multiple enzyme is made from organic materials. It lubricates the soil particles and produces a milk cementation action to control leaks. The lquid is applied at the rate of one gallon for each 15 cubic yard of soil to be compacted. It should be mixed into the top 12 inches of pond basin soil. If pond can not be drained, dry to do the above, then one gallon of the material can be applied to every 800 sq. ft of surface area.

(iii) Soluble salts

Salts have been used as dispersing agents to porous soil particles to rearrange the clay particles to reduce soil permeability. These soluble salts include NaCl, Sodium polyphosphate and Na_2CO_3. NaCl is applied @ 1 pound/3 sq ft of soil. Sodium polyphosphate is applied at the rate of 1 pound per 10 sq ft of soil.

(iv) Lime

It is used sccessfully to control leaks. The pond has to be dried and the top 12 inches of basin soil is removed. Then a 2" layer of lime is put down and the removed soil is placed over the lime layer. The lime increases the structural integrity of the soil making it impervious to water.

(v) Organic material

The organic manure such as animal manure, straw leaves, grasses etc. when added to the pond basin cause a reaction to take place with the soil particles which develops into a sticky layer that can act as a sealant. The precaution has to be taken that some materials used to control pond leaks may be toxic. However, application of organic manure over a long period of time, not only tends to reduce the rate of seepage by sealing the soil pores but also reduces water turbidity caused by suspended silt and colloids. If a polythene or plastic film liners, is to be installed to prevent seepage, it is desirable to install it about 200 cm below the pond bottom soil. This will create a water - soaked soil substratum at the bottom of the pond. Very little is known on water-submerged soils of ponds in relation to pond fertilisation. This is one area in which extensive research will play rich dividends in economising aquaculture.

Fish can not live for longer time without water. So availability of an adequate and dependable source of water is another prerequisite for setting up a fish farm. The pond should be filled with freshwater at regular intervals so as to adjust water depth. Control water quality, to check serious surfacing of fish. The usual source of water supply to fish farm are reservoirs. streams, springs, canals, surface run off (rain), wells, tube wells etc. However, the best thing to do is to have as far as possible natural, preferably rain water and other large water bodies, because in natural water bodies, dissolved oxygen content, pH, water quality and water temperature are more stable and suitable for fish to grow. On the contrary, the water from under-ground source is often deficient in dissolved oxygen. The waste water discharged from factories and mines usually contain harmful chemicals which is not suitable for fish farming.

The ponds should be away from rivers and reservoirs so as to avoid water table fluctuations. Proper elevated wide embankments are necessary so as to avoid flooding. Water should not be highly acidic nor alkaline. It should be corrected if acidic by the application of lime and if alkaline by adding organic manure.

1.2 Size and depth of the pond

Other factor involved for site selection, is the purpose of farm. The purpose of the farm depends how large is the size of the area. For commercial fish farm, which may comprise, nursery, rearing, stocking, brooder pond and breeding ponds etc., a large area is required.

With the possibility of self reliance in carp reproduction and hatchery for incubation, the embryonic development results in to the production of spawn which is a 3 day old larvae. To raise these to about 10-15 cm size, hatchlings take 2-3 months. During this period, they are reared in nursery and rearing ponds. The resultant product from the nursery pond is the fry (2-3 cm length) and the time period taken varies from 11 to 30 days in respect to different management practices adopted. The ideal size of each fry pond is 0.04 ha and their number vary in accordance to target production with water depth of 1-1.5 m. The shape of a nursery pond is regular as a rectangle and flat even bottom for easy netting operation. Similarly the resultant product from the rearing tank is the fingerlings (10-15 cm length) and the time period taken varies from 2-3 months. The size of each fingerling pond varies from 0.05 ha to 0.1 ha and their number vary in accordance to target production of fingerlings with water depth of 1.5-2 m. The shape of a fingerling pond is rectangular with flat even bottom for easy netting operation.

The stocking pond, each having an area of 1 acre- 1.5 acre is considered to be an optimal size for intensive culture of food fish with water depth of 2.5-3 metre. The stocking material in the stocking pond is the fingerling of 10-15 cm length and the resultant product is the table size fish.

The number of stocking ponds vary according to the target table fish production. The time period of rearing varies from 8-10 months. The stocking ponds are used as breeding pond or brooders raising pond as per the requirements by an aquaculturist. However, there is no hard and fast rule regarding the size of the ponds.

1.3 Dyke

Dyke should be compacted, solid, leak proof so as to conserve fertile water. Dyke should be stable in all weather conditions and not liable to collapse during heavy rains. The Dyke height should be more than 0.3 - 0.7 m above the maximum flood water level. For Dyke construction, a soil containing 15-30 per cent silt, 45-55 per cent sand and 30-35 per cent clay is most suitable.

A berm of sufficient width should be provided for stabilising the slope. Wider berm also helps in operating the net in the pond. The berm should be sufficiently wide and in no case should be less than 1 m. The slope of the embankment in horizontal to vertical axis should be 2 : 1 in good quality clay soil. The slope (base : height) in loamy silt or sandy soils should be 3 : 1. While raising the Dyke, the clay puddle (mixture of 1 : 2 of sand and clay) is deposited in 10-15 cm thick layers with a precaution that it should be adequately moistened before next layer of clay puddle is laid. The puddle may be at the centre or in the water side of the pond. The crest or crown of the Dyke should be sufficiently wide, so that allied farm activities can be takenup gradually and the minimum top width of the embankment should be about 1 m. Excess water outlet pipes can be provided on the embankment, as safety measure so as to safeguard against damage due to excessive rise in water level.

Short creeping grass is recommended to be grown in the top and sides of Dyke to check soil erosion. Short herbs and shrubs can be grown on the Dyke which can be made use as green manure when needed. On wider embankment, commonly banana and coconut trees are planted on east side so as to avoid shade in the morning hour when photosynthesis starts. Even in wider crest and slopes, terrestrial grasses like hybrid napier, gunny grasses and elephant grasses are cultivated to supply feed wholly or partially to herviorous fishes like grass carp reared in cultured tanks. Sluce gates or drain pipes are fitted to have desired water level in the pond. The diameter of drainage pipe depends upon the size of the pond and also on the volume of water.

1.4 Ratios between different ponds

Fish culture commences with the construction of ponds, Specific types of ponds are required for the species of fish and various life history stages. Various ponds like nursery, rearing, stocking and so on are required for a fish farm. The ratio of these ponds varies in relation to stocking density and survivality.

1.4.1. Stocking density and survivality

The usual stocking density in nursery, rearing and stocking pond are 6 million spawn/ha, 0.3 million fry/ha and 5000 finger lings/ha respectively.

In nursery pond depending on the pond's productivity the stocking density of spawn is determined from 2.5 million to 10 million/ha. Under scientific management an average survival of 50-88% is obtained in nurseries. However, in rearing pond muiltispecies rearing is practiced to make use of different natural food niches. The stocking density varies from 0.1-0.3 million per hectare in 1 : 1 : 1 ratio of catla, rohu and mrigal depending on the pond productivity. The survivality is very high to the order of more than 75%.

In intensive rearing system, the stocking density used is always above the normal rates. Compounded nutritive artificial diet in different feeding frequencies is given with optimal water quality. This water quality is maintained by recirculatory and reconditioning system. A high survival of above 80% is normally recorded. The spawn stocked at the rate of 50-60 million/ha recorded a growth of 12 to 18 mm in 14 and 19 days with a survivality ranging from 51 to 63%. In case of fingerlings, the stocking density varies from 0.4 to 0.6 million/ha with survivality of 83 to 98%. The size attained in each individual fingerling varies from 80 to 170 mm in 32 to 85 days of rearing.

In super intensive rearing system the stocking densities of 100 to 435 million spawn/ha with a survival rate from 79 to 85% in 15 days have been reported in laboratory conditions. With the improvement of traditional system with regulation of water, aeration, filtration and formulated diets, high rates of survival and growth are possible on commercial basis.

The ideal shape of the pond should be rectangular. That too the length and breadth in the proportion of 3 : 1 is ideal. In any case, the breadth should not be more than 30-40 meter as it is difficult to operate efficiently a net longer than 40 metre. The corners of the pond should be rounded, so that while operating net, fish may not escape.

Schaperclaus (1933) has given the following ratio of various ponds required for a fish farm under European condition.

Breeding pond – 0.25% of the area.

Nursery pond – 2.75%

Fingerling pond – 10.00%

Rearing pond for 2nd yr. fish – 23%

Over wintering pond – 3%

Third yr. rearing fish – 60.00%

Holding pond for spawners – 1.0%

According to Alikunhi (1957), under Indian conditions a 4 ha farm should be divided into 0.2 ha nursery pond, 0.8 ha rearing pond and 3 ha stocking ponds.

With the development of advanced technology such as enhanced rate of stocking and survival of spawn, fry and fingerling, the ratio of different types of ponds in a farm needs reconsideration (Barrackpore, 1973). According to them the ratio of nursery, rearing and stock ponds should be 1:40:1280 considering stocking density of 10 million spawn/ha with 80% survivality, 0.2 million fry/ha with 80% survivality and 5000 fingerlings/ha. However, Sinha and Ramacandran (1985) suggested that for a self sufficient farm, the ratio of nursery, rearing and stocking ponds may be 1 : 10 : 420 respectively, considering stocking density of 6 million spawn/ha with 50% survivality, 0.3 million fry/ha with 70% survivality and 5000 fingerlings/ha.

In adopting the intensive rearing system, the hectare ratio of nursery, rearing and stocking ponds may vary considering stocking density of 50 million spawn/ha with 63% survivality, 0.6 million fry/ha with 98% survivality and 25000 fingerlings/ha in stocking ponds under rotational stocking and harvesting.

However, if higher the stocking rate or low retrival rate, the requirement for successive tank is less. Hence, the requirement of rearing tank is directly related to the production of nursery tank. Similarly, the production of rearing tank directly related to the requirement of stocking tank. So a farmer with his own experience can reconsider the ratio of various tanks needed in the farm, so as to fulfil the farm objective and requirement.

1.5 Pond renovation

Good pond conditions are regarded as one of the key factors in achieving high and stable yields. One can judge the pond conditions on the basis of water quality, depth, area, dyke with or without inlet and outlets and other factors. Inorder to have high and stable yields, the old ponds should undergo renovations.

(1) *Shallow ponds are dug up to become deeper ponds.* The increase in water volume per unit area creat a favourable condition for poly-culture of different varieties of fish at high density. In deeper ponds, the water will not be turbid during harvest, but keeps fresh and clear. In addition, water temperature and water quality are more stable without any sudden change. It is beneficial for fish growth. If ponds are too deep it is not so good for fish to grow and results poor yield. Normally the optimum water depth is kept around 2.0-2.5 m all the year round.

(2) *Change inaccessible and stagnant water ponds to free exchanging ponds.* Inorder to facilities the drainage, adjustment of water level and quality, it is advisable to change the inaccesible and stagnant ponds to free exchange water ponds for better fish yield. This is because aeration can boost up fish production reasonably.

(3) *Change low Dyke pond to high Dyke pond.* In order to prevent floods, it is necessary to heighten pond dykes. At the same time, the widened dyke crest can be used for integration with other agriculture or animal components in line with local conditions.

(4) *Small ponds are combined to form reasonably larger ponds.* There is a saying by fish farmer that, broader the water body is the larger the size of fish in it. Big ponds could provide larger space for fish to live in and satisfy their ecological conditions. Thus the stocking ponds of about 1.5-2.5 acre are used.

1.6 Maintennce

In course of fish culture practice, the pond bottom accumulates considerable amount of silt and organic matter. This need to be removed periodically in every 3 to 4 years. If excessive pond silt is there, it should be desilted. It is better to expose the dry pond bottom to sunlight by dewatering. These pond bottom silt can be used as bed for agriculture crops which is undertaken in the farm itself. The pond dyke should be repaired annually after monsoon. The dyke crest has to be levelled, thoroughly rolled and grass planted to bind it. The holes created in pond sides or embankments should be filled up by clay or loam. The dyke can be reinforced by driving poles of wood or bamboo.

1.7 Placement of different type of ponds in a fish farm

The location of different kinds of ponds in a farm is of considerable importance for ease operation and minimisation of operational cost.

The detailed contour survey of the area is required before fish farm construction. This will minimize filling or digging and movement of earth, which all cause expenditure on labour or mechanical earth moving.

The deeper and lower area of the farm site should be developed in to stocking ponds. The appropriate higher areas can be developed in to rearing ponds and areas higher still in to nursery ponds. The highest of the areas are to be developed for office building, staff quarter, laboratory and hatchery proper. Nearest to hatchery proper, water treatment, filtration, sedimentation plants and the water tower should be placed. Inside hatchery, the fish breeding tank is placed which is connected to the hatching tank. The hatching tank is connected to outdoor nursery tank through suitable conduits. Brood fish ponds in the above stated arrangement would be farthest from the hatchery. However, antetank in the hatchery provide holdng space for brood fish close by the fish breeding tanks.

In the farm, where modern hatchery facilities are not there, some of the stocking ponds very close to rearing and nursery ponds are selected as brood ponds, breeding ponds and hatching ponds. However, care should be taken, how easy it would be for farm operation, management and operating costs.

9

POND FERTILIZATION

1. SIGNIFICANCE

Pond fertilization is one of the key factor in increasing the maximum carrying capacity. Farmers adopted the method of manuring to rear fry ages ago. In nursery and rearing ponds, fertilisation is aimed at developing natural food and saving artificial feeds. Satapathy (1990) has described a positive relation in nutrient dynamics and fertiliser application in aquaculture management.

Phytoplankton, the primary producer of pond, carry out photosynthesis, converting the inorganic materials into organic nourishment needed for their propagation and growth. This is a link in the food chain and food web process that occurs in the ecosystem. Therefore, the importance of pond fertilization lies in the cultivation and propagation of various fish food organisms for the cultured fish. That too, utilisation of energy is related to the length of the food chain. The shorter the food chain, the higher the rate of energy transfer. In other words, the higher the utilisation rate of energy, the higher the fish production.

In a pond ecosystem, the materials are in a constant rate of circulation, mainly through food chain. This is called ``pond material circulation''.

The fertilizers used in fish culture ponds are of two categories (1) organic and (2) inorganic.

2. ORGANIC MANURE

Increasing attention has been paid towards recycling of various agricultural and animal wastes through aquaculture production process for enhancing fish yield (Rath, 1989c). The major objective of various waste utilisation is to recycle different nutrient elements present in such wastes. Millions of farmers in developing countries require adequate resources for augmenting food production. Proper management ensuring continued maintenance and building up fertility of an ecosystem is indispensable for greater productivity. This necessiates the importance of organic and inorganic fertilisers for enhancing the productivity of an ecosystem. With the rising cost of fossil fuel an acute scarcity in the area of energy and production of chemical fertilser is felt with the enterprenures. That too availability of chemical fertiliser to the farmer at reasonable cost is fast declining. There is thus a need for utilising organic manure/wastes for supplementing chemical fertiliser in the present context. That too, the accumulation of wastes/products can generate pollution to the environment. If such great amount of excreta (waste) is not disposed of, it must pollute the environment. Therefore, these great amount of wastes can be recycled for food production.

2.1 Advantage

1. The advantage of applying organic manure is to improve the soil structure, water holding capacity, drainage, base exchange capacity and also check soil erosion.
2. The research worked out in All India Co-ordinated Project on `Decomposition of Organic matter in Indian Soils' showed that farm yard manure or compost was the best source for maintenance of soil organic matter at higher level. Therefore, regular application of organic manure is a sound practice for maintaining soil fertility.
3. Due to energy crisis, prohibitive cost of chemical fertilisers and poor purchasing power of marginal and small farmers. it is necessary to use organic manure/ wastes to its maximum potential with proper technology to meet the shortage of chemical fertiliser and also for improving the culture system.
4. The use of organic manure has the advantage of converting unusual surplus wastes into useful product for use in agriculture, aquaculture and also for certain livestock raising.

5. Besides, nitrogen (N), Phosphorus (P) and Potassium (K), Organic manures are a potential source of micronutrients.

2.2 Varieties of organic manure

Organic manures are mainly farm animal excrement. Generally the term refers to manures containing organic matter. Now a days organic manure, are applied as base and additive manures to fish ponds in most part of the country although inorganic fertilizers are applied in scanty. The manures often used in fish ponds are : (1) faeces and urine of livestock and poultry, (2) green manures, (3) compost and (4) silkworm faeces. However, in China, besides these above noted organic manure, human excrement (Night soil) is used not only in Agriculture but also in aquaculture practices. Only through microbial decomposition, the organic manure converted to nutrients that the plants can absorb.

2.3 Factors influence on the composition and nutritive value of animal wastes

All the animal wastes are a mixture of semisolid excreta (dung) and liquid excreta (urine). Their composition and nutritive value depends on several factors like age, breed variety, digestibility, composition of the feed ratio, physiological condition of the animal etc. These factors also influence the ratio between solid and liquid excreta. For example, more water the feed contain, the over liquid is excreted. The more readily digested the feed, the less dry matter is contained in solid excreta and more in liquid ones. The more concentrated the feed given to animals and richer in protein, the more nitrogen manure contain. The wastes of a young growing animal is poorer in manurial ingradients than that of an adult animal, as the growing animal retain much more nutrient than that of an adult. A working animal excretes wastes of richer quality than an animal at rest. Similarly the wastes of a milking cow is poorer than that of a dry cow.

2.4 Composition of solid excreta (dung)

The faeces of livestock chiefly consists of undigested food and also residue from digestive fluids, waste mineral matter, wornout cells from intestinal linnings, mucus, bacteria and foreign matter such as dirt consumed along with food. The nitrogen and phosphorus of the solid animal excreta form part of the organic compound and become available to plant only after their mineralisation while potassium is in a mobile form. The dung of most of the animals is rich in phosphorus and calcium.

2.5 Composition of liquid excreta

In comparison to dung, urine contains more water, nitrogen and potash but less phosphorus. All the nutrients in urine are in a readily soluble or easily mineralisable form. Sheep urine is richest in manural ingradient followed by horse urine. The liquid of most of the animal is rich in nitrogen and potash.

2.6 Nature and characteristics of organic manure

2.6.1. Faeces and Urine of Livestock Poultry

(i) *Pig manure :* Pig manure includes much organic matter and other nutritional elements such as N.P. and K. Pig manure is a fine complete manure (Table 1). Pig faeces are delicate with moisture content ranged from 70-77%, containing more nitrogen than other livestock faeces (C/N; 14 : 1), making them more susceptible for rotting. The major of pig urine is nitrogen in the form of urea. It decomposes easily.

Table -1. Nutritional elements in some of the organic manures (per cent in fresh matter).

Variety	Organic matter (%)	Inorganic matter		
		N	P	K
Pig Faeces	15	0.6	0.5	0.4
Urine	2.5	0.4	0.1	0.7
Cattle Faeces	14	0.3	0.2	0.1
Urine	2.3	1.0	0.1	0.4
* Buffalo Faeces	12.7	0.26	0.18	0.17
Urine	–	0.62	traces	1.61
* Sheep Faeces	33.1	0.7	0.51	0.29
Urine	9.3	1.47	0.05	1.96
* Horse Dung	21.0	0.50	0.3	0.33
Urine	8.0	1.29	0.01	1.39

Chick Dung	25.3	1.63	1.54	0.85
Duck Faeces	26.2	1.10	1.40	0.62
Goose Faeces	23.4	0.55	0.50	0.95
Night Faeces	20.0	1.0	0.50	0.37
Soil Urine	3.0	0.5	0.13	0.10

Source : Asian-pacific Regional Research and Training Centre for IFF, China, 1987.

* Organic manure, Gaur *et al.*, 1984.

The excretory amount of a pig is generally associated with its body weight and food intake. A pig per day discharges excreta 7-20% of its body weight. A pig excretes about 1000 kg faeces and 1200 kg of urine in the culturing period of 8 months from pigling to an adult.

(ii) *Cattle manure* : The elements of cattle manure are similar to those of pig manure. But cattle are ruminants and the food stuffs are repeatedly masticated, making the excrement quite delicate. Cattle manure contains less nitrogen than pig manure (C/N, 25 : 1), cattle urine contains more nitrogen than pig urine in the form of hippuric acid (C_6H_5 $CONHCH_2$ - COOH). Therefore, cattle excreta decomposes slowly. The average daily excreta of a cow of about 500 kg weight is 25 kg in which the ratio of faeces and urine is about 3 : 2. However, the quantity of annual dung and urine production per individual of cow are around 5400 kg and 3600 kg respectively (IFF, China).

Usually the cattle dung consists of about 75-85% moisture, 15-25% organic matter and 2-5% mineral matter. The organic matter of dung mainly comprised of 78-90% of total carbohydrates (Crude fibre + nitrogen free extract), 9-18% of crude protein and 2-5% of ether extract.

(iii) *Poultry manure* : Poultry manure include the faeces of chicken, ducks and goose and are rich in both organic and inorganic matters since liquid and solid excreta are excreted together resulting in no urine loss. Poultry manure ferment quickly and their nitrogen is mostly in the form of uric acid which can not be absorbed directly by plants. If the dropping come from the cages or dropping pits, superphosphate may be added at the rate of 1 kg/day/100 birds. This improves the fertilising quality and help the control of flies and odour. Poultry manures are more effective after decomposition. Although poultry excreted about 5%

of its body weight per day, an average the annual amount of excrement per fowl is as follows :

Chicken	5	to	5.7 kg
Duck	7.5	to	10 kg
Goose	2.5	to	15.0 kg

Although the quantity of annual amount of excrement per fowl is comparatively small, the number of poultry culture is often great, therefore, the total amount of faeces is significant.

With regard to natural value of poultry manure, the dry matter content of the fresh poultry excreta from an adult egg layer ranged from 20-30% and it comprises 20.2% ash. The organic fraction of the excreta contained 21.5% crude protein, 1.9% ether extract, 13.4% crude fibre and 42.9% nitrogen free extract (See Gaur *et al.,* 1984).

(iv) *Goat and Sheep manure* : The droppings of sheep and goat make a very good manure, which is applied in cultivated lands as common practice by millions of farmers. The chemical composition of their excreta has a dry matter content of 42-48%. The organic fractions of dung component comprised 52-93% crude protein, 14-19% ether extract, 27.8 to 36.4% crude fibre, 40-47% nitrogen free extract, and 0.35-0.77% ash. Sheep and goat manure contain about 3% N, 1% P_2O_5 and 2% K_2O.

(v) *Night soil* : This type of organic manures are used in China for terrestrial grass production in an integrated fish farming system. The composition of night soil (human excrement) is greatly dependent on the food consumed. Nitrogen is high (C/N - 3 :1) and 70-80% of it is in the form of urea. This facilitates for easy absorption. On an average, an adult excrets annual 90 kg faeces and 700 kg urine of waste material.

Night soil to be used as manure must be fermented before application. This is equally done by storing the manure in anaerobic conditions for 2-5 weeks. The decomposition of human excreta produces ammonia. Under air tight conditions and at certain concentration, ammonia can sterilise human waste. Quick-lime (1-2%) or/and formalin (0.1-0.2%) are effective in killing the harmful organisms and bacteria in night soil.

(vi) *Silk worm dregs* : Silkworm dregs are composed of silk worm faeces, moult residues and mulberry residues. These are rich in organic matter (87%) and nitrogen (13%). They make good fish feed and also used as manure.

(vii) *Green manure :* All wild grasses, cultivated plants of the composite family with some gramineous and leguminous plants which are nutritively rich, less fiber and are easily putrescible, if used as manure, are called as green manure (Table 2). This is commonly called as ``Dacao or Totsao" in China. Green leaf manuring is also common in India. Cowpea (*Vigna catjang*), Dhancha (*Seabenia aculeata*), Cluster bean (*Cyanopsis tetragonoloba*), Horse Gram (*Dolichos biflorus*) and Sun hemp (*Crotalaria juncea*) are the plants mostly used for green manuring purpose. The N P K in these plants varies from 0.34 - 0.84%, 0.12 - 0.15% and 0.51 - 0.58% respectively. With the rapid development of fish culture now a days vegetable wastes and other poisonless plants with soft stems and leaves are als used to fertilise the water. These manures rot and decompose easily, providing ideal environment for bacterial and plankton propagation.

Table 2. Nutritional elements in Green manure (% wet weight)

S.No.	Items	N	P_2O_5	K_2O
1.	Stems and leaves of broad been *Vilia faba*	0.55	0.12	0.45
2.	Rape (*Brassica napus*)	0.43	0.26	0.44
3.	Alfalfa (*Medicago falcata*)	0.54	0.14	0.40
4.	Wild grass	0.54	0.15	0.46
5.	Barnyard grass	0.35	0.05	0.28
6.	Water peanut	0.20	0.09	0.57
7.	Water hyacinth	0.24	0.07	0.11
8.	Water lettuce	0.22	0.06	0.10

Source : Asian Pacific Regional and Training Centre for IFF, China.

(viii) *Compost* : Compost manures are the decayed refuse like leaves, twigs, roots, stubble crop residue and allied refuge, which are decomposed along with animal wastes (Table 3). Mixing several manure together may easily produce a fertiliser that is more suitable for fish pond. Lime is included in the composted manure to neutralise the organic acids produced during rotting and decomposition. Compost is prepared either by heaping or soaking method. The compost made from water hyacinth is superior to town compost and farmyard manure as reported (Table 4). In addition rich mineral value of certain aquatic weeds are evident to influence the pond fertilisation.

Table 3. Chemical composition of various compost.

Name of the compost	pH	Organic carbon (%)	Available N_2 (mg/100 gm)	Av. phosphorus (mg/100 gm)
Pistia compost	7.5	1.311	38.08	10.05
Hydrilla compost	8.2	1.887	59.92	6.56
Najas Compost	7.6	1.229	47.60	13.55
Ottelia compost	7.5	1.422	48.72	10.05
Water hyacinth	7.3	1.140	35.84	11.80

Source : FAO, Singh, 1963.

Table 4. Nutritional elements in various composts (%) in wet weight

Constituents	Water hyacinth compost %	Town compost	Farm yard compost
Nitrogen	2.05	1.0	0.50
Phosphorous (Phosphoric acid)	1.10	1.0	0.25
Potash	2.50	0.8	0.30
Lime (CaO)	3.91	3.5	0.20
C/N ratio	13.0	10.0	12.13

Source : Basak, 1948.

VIII (b) Vermicompost

In ordinary compost preparation, composting of aquatic weeds and waste green matter with cattle dung is done by way of layer to layer distribution in a pit. Vermicomposting is done preferably in one side open cistern above the ground level. The open side of cistern is covered with strong bamboo gratings which allow air to flow from the side and exude excess water that is sprinkled on the cistern pit regularly. This exuded manure water is collected in a side drain and used for manuring. After 15-20 days of decomposition, adequate number and quantities of earthworms are released in the cistern pit. About 1800 earthworms per cubic meter of compost are required. These earthworms burrow and feed on the decomposed material and leave their excreta on the compost which enriches the nutrient content of the material. After 4 - 6 weeks, when decomposition is completed the earthworms are collected from the base of the pit and the same worms are used in another pit. Then the compost is dried and grinded in to the form of granules, packed and sent for marketing The nutrient composition of vermicompost (De, 2008) is as follows: Organic carbon (27-30%), Nitrogen (1.8-2.05%), Phosphorous (1.32-2.05%), Potassium (1.28-1.5%), Calcium (1.0-4.5%), Magnesium (0.5-0.7%) and Sodium (0.02-0.3%). Silicon, Manganese and Aluminium are found in traces in the vermicompost.

(ix) *Pond silt* : The effectiveness of manure is greatly dependent not only on the composition and quantity but also method of application. These applied manures in pond soil decomposes and forms humus and muck. These pond humus and muck are of high nutritive value and can be used for agriculture bedding.

(x) *Biogas slurry*: The waste components of cattle and buffalow offers great scope for exploiting a potential source of highly effective manure. The saving of 29% of dung, that is presently being burnt as fuel in villages and its use as manure is a consideration of great practical importance in the context of the new strategy for increasing food production. As burning of cow dung as fuel is in escapable in rural area, the alternative solution is to setup a biogas plant for energy and biogas slurry for fertilisation can be practised. Biogas slurry is standardized as a manurial input @ 30-45 tons/ha /yr to produce 3-5 tonnes of fish/ha/yr (Tripathi, 2008). It serves the purpose of nutrient input in duckweed culture, carp feed ingradient and preventing pond seepage. The manural policy of the farmer is to take every possible step to ensure the return of all organic waste material to the productive system for increasing protein food production.

(xi) *Horse manure* : Horse dung consists of 76% of moisture, 21% organic matter and 4% of mineral matter. The dung mainly comprised of .5% N, 0.3% P_2O_5 and 3% K_2O.

2.7 Methods of organic manure application

2.7.1 Application of livestock manure

Livestock manure as a base manure is applied by heaping it at a corner of the pond or in small piles in shallow water with a sunny exposure. These decompose and spread gradually into the water. If the manure is used as an additive it is added in small quantities every 10-15 days interval in pond water.

2.7.2 Application of mixed compost

After fermentation, the compost is diluted with pond water and the supernatant liquid part is sprayed into the pond around the dykes and the residue (manure dregs) can be used to fertilize crop fields. In the case of a large pond, the manure may be loaded over a raft, flushed in batches with pond water and sprayed evenly over the pond. The residue is used to fertilize the crop fields. As decomposition has already been done in earthen pits, these manures consumes less dissolves oxygen. The nutrients of the compost are quickly absorbed by phytoplankton.

2.7.3. Application of green manure

The application of green manure is like that of livestock manure. Green manure is applied by heaping it at a corner of the pond and turning the pile once in every two days. The rotten parts will spread into the pond water. The root and stems which decompose slowly are dredged out of the pond periodically. The decomposition of green manure in water consumes a great amount of oxygen. For this reason, green manure is applied as additive manure in small quantities with more frequencies in culture period of fish. That too avoid any chance of O_2 depletion to the critical level. Fresh oxygenated water can be drained in to pond or aeration can be provided.

2.7.4. Application of night soil

Application of night soil is prevalent in Jiangxi and Hunan province of China. Before application, the fermented night soil treated with lime for sterilisation is diluted with twice part of water. This dilution is then sprayed along the pond dyke. The quantity depends on the fertility of pond water.

2.8 Constraints in application of organic manure

The main constraint in the use of animal wastes for aquaculture is the problem of depletion of dissolved oxygen in water. The wastes undergo decomposition through bacterial action and the process uses dissolved oxygen creating biochemical oxygen demand (BOD). So in such cases, monitoring of dissolved oxygen and B.O.D. of the pond water is absolutely essential. However, addition of freshwater with aeration facilities to some extent solve this problem.

2.9 Biofertilizers

Biofertilization has been observed to be a potential biotechnological tool improving the contribution of biologically fixed nitrogen to the total nitrogen budget of fish pond ecosystem. In aquaculture practices, Azolla is used as a biofertilizer. Azolla is an aquatic fern with wide distribution all over the world having the capacity of assimilating atmospheric nitrogen through the cyno-bacterium *Anabaena azollae*, a symbiotic blue green algae present in cavities on the dorsal lobe of the leaf. It is not only used as a biofertilizer for wetland rice, but also used as a fodder for livestock, poultry and forage for fish. Its high nitrogen fixing capacity, rapid multiplication and also rapid decomposition rates resulting in quick nutrient release have made it an ideal nutrient input in farming system.

Azolla is a heterosporous fern belonging to the family Azollaceae (Salviniaceae) with 7 living and 20 extinct species. Composition of Azolla in dry weight percent is as follows.

Crude protein	–	13-30%
Crude fat	-	4.4-6.3 %
Cellulose	-	5.6-15.2%
Hemicellulose	–	9.8-17.9%
Lignin	-	9.3-34.8%
Ash	-	9.7-23.8%
Phosphorous	-	0.1-1.59%
Nitrogen	-	1.96-5.3%
Potassium	-	0.3-5.97%
Calcium	-	0.45-1.70%
Magnesium	-	0.22-0.66%
Sulphur	-	0.22-0.73%

For long run use of Azolla as biofertilizer, it is necessary to go for Azolla multiplication. Multiplication may be either small scale or large scale according to requirement or use in farming practices. Nursery plots with water depth of 5-10cm are needed for initial inocculation of Azolla for multiplication. 100 g of SSP and 3 – 5 g of Carbofuran (Furadon) is added and Azolla is inocculated at the rate of 1kg per 8 m^2 (2 X 4 m) of nursery plot. After a week, Azolla is thinned and multiplication procedure is repeated in new plots. Important factors for Azolla cultivation are

1. Water and lrrigation
2. Temperature
3. Light
4. nutrients like phosphorus.

It is reported that one kg of phosphorous application to Azolla result in 5-10 kg of nitrogen fixation. Common pests of Azolla are Insects (Pyralis species and *Nymphula* sp.), Snails, Algae and Fungi (*Rhizoctonia* sp.). So careful management is to be taken in Azolla cultivation plots.

Wastewaters from agro-based industries are processed in terms of microbial oxidation or nutrient removal through algae and other macrophytes like Lemna, Azolla, Spirodella, Wolffia etc. Further trials on application of heterocystous blue green algae like Anabaena, Nostoc, Calothrix are underway to serve as biofertilizer. Increase in heterotrophic bacterial nitrogen fixation rates through pond environmental amelioration measures is contemplated.

Biofertilization with Azolla has been standardized at 40 tones/ha/yr providing 100kg N_2, 25kg P, 90kg K and 1500Kg organic matter in aquaculture system (Tripathi, 2008).

2.10 Effect of manure on fish food organisms

The application of organic manures results in the rapid multiplication of plankton through bacterial decomposition and release of nutrients that are leached to the pond water. These bacteria and planktons are the food source for the aquatic animals and filtering fish. Sometimes, the dung is also directly consumed by some types of fish which are called coprophagus.

The nature and properties of manure depend closely on the predominant species of plankton production in pond water. If organic manure is applied, initially phytoplankton belonging to the group chrysophyta and pyrophyta appear but gradually the zoo-plankton production is dominant. After each manure application, the nutrient content of the water increases resulting in a planktonic peak. Phytoplankton peak is achieved within 4-5 days after manure application, followed by zooplankton peak in 5-7 days. Usually protozoans, rotifers, cladocerans and copepods reach peak subsequently. Protozoans multiply by binary fission, rotifers and cladocerans multiply by parthenogenesis reaching peak. Ideally the peak in plankton population should coincide with the dement of the fish fry.

2.11. Equivalency of organic manure in terms of chemical fertiliser

Very few reports suggest the equivalency of organic manure in terms of chemical fertilizer. The Chinese people consider the pig as a costless fertilizer factory moving in hooves. The total manure obtained from 20-30 pigs during a period of one year was found to be equivalent to one ton of $(NH_4)_2$ SO_4. They add that on an average, an adult human excrement of 90 kg faeces and 700 kg urine per year is equivalent to 22 kg $(NH_4)_2$ SO_4, 6.80 kg of calcium super phosphate and 3.5 kg of potassium sulphate (K_2SO_4) IFF, China).

2.12. Conversion rate in fish yield

It is reported by ASFA that, the conversion ratio between pig dung and fish biomass was studied to be 17 : 1. But Chinese reported that under experimental study of monoculture of common carp, 50 kg of pig manure can be converted into 1.25-1.5 kg fish. In polyculture system with common carp as amjor species, 50 kg of pig manure can be converted into 1.75 - 2.0 kg fish. In polyculture with silver carp as major species, the 50 kg of pig manure can be converted into 3 kg of fish. The probable explanation may be that, manure is not only used as fertilizer but also to some extent used as direct feed. The fecal matter discharged by silver carp provides fertilisation to pond for which production has enhanced. Schaper-clause (1959) reported that in Germany, the most advantageous dose of pig dung varies from 3-5 ton/ha. Each ton of pig manure resulted in an extra yield of 30-40 kg of fish. However, with pig manure in polyculture system, the production has been reported to be 7200 kg/ha/yr (Sharma *et al.,* 1979). The conversion factor of cow manure is 3.15 kg in dry weight or 21 kg in wet weight basis with filter feeder and omnivorous fish species, and it is 3.3 kg in dry weight and 26 kg in wet weight in only filter feeders. According to the experiments counducted at the Chinese freshwater fisheries Research Institute, the output of Silver carp, Big head, Common carp and Crucial carp in manured pond is 3.5 times, 2.8 times, 3.3 times and 2.2 times more than the output in unmanured pond. It is also found out that 8 kg of silk worm waste as manure can produce 1 kg of fish and the waste water during cocoon or pupae processing can additionally be used as manure. Every 200 kg of waste water can sustain and produce 1 kg of fish. In India different workers such as Jhingran and Ghosh (1988), Banerjee *et al.* (1989), Dutta and Goswami (1988), Patra and Ray (1988) and Tripathi (1990) have studied the use of organic manures such as the dung of livestock, birds, sewage, gobargas slurry and town compost, as fish pond fertilisers. All have shown that the output in manured pond is higher than the unmanured pond. Although some have recorded the daily increment in weight, but Dutta and Goswami, 1988 recorded that the total fish production in manured ponds without supplemental feed is 3 times more than the control. The biogas slurry when fed daily in fish ponds at 75-80 litres/ha/day gives high fish yield, as high as 6 ton/ha (see Tripathi, 1990). The prodction of 2227 kg/ha and 3270 kg/ha fish in fish culture tank in 10 months rearing was achieved by using water hyacinth input based biogas slurry, which when applied at the rate of 15,000 litres/ha/annum and 30,000 litres/ha/annum respectively (Tripathi *et al.*, 1989).

3. VARIETIES OF INORGANIC FERTILISERS

Inorganic manures are also referred to as chemical fertilisers, According to composition, chemical fertilisers can be (1) Nitrogenous, (2) Phosphorus and (3) Potash fertilisers.

3.1 Advantage

1. Exact composition of inorganic fertiliser is advantagious.
2. Mineralisation is very fast giving quick effect on pond productivity.
3. Lack of pollution. As they need not to undergo any bacterial decomposition, the abnoxious gases like CH_4, NH_3, SO_2 are least possible to pollute the environment.
4. No biochemical oxygen demand (BOD) is required for chemical fertilisers, or in other words their beneficial effect on oxygen content (requiring no decomposition).

5. Used in small quantities as additive manures, hence convenient for utilisation.

However, when chemical fertiliser is applied in ponds, the first link of the food chain is principally phytoplankton which are not as nutritious to zooplankton as are bacteria. Therefore, in chemical fertilizer treated ponds, the zooplankton propagation often lags far behind than that pond treated with organic manure. The other disadvantage is that the effect of inorganic fertiliser is rather short and difficult to control the water quality. Therefore, application of alone chemical fertiliser is no way better than alone application of organic manures. But it is always advisable to use organic manure as base manure and both organic and inorganic as additive manures with greater frequencies probably 10-15 days intervals or seeing the planktonic conditions on pond system.

3.1.1 Nitrogenous fertilisers

Liquid ammonia, Ammonium Sulphate and Urea are the nitrogenous fertilisers used in production systems. The nitrogen content in these fertilisers vary due to different manufacturers with different trade names. Nitrogen content in liquid ammonia is 12-16%, in $(NH_4)_2 SO_4$, 20-21% and in urea is 44-46%. Aquous ammonia is readily volatilised, hence should not be exposed to the air for a long time. This is also used in killing the unwanted predators during prestocking managment operation. Liquid ammonia can be applied sufficiently 7-8 days before stocking fish seed and can not be used till the culture is going on. In contrast, $(NH_4)_2 SO_4$ can be used as additive manure during culture period. However, urea does not ionize when dissolved in water. Hence, cannot be absorbed directed by plants. Only after decomposed, they are ionised and made available to plants and planktons.

3.1.2 Phosphorus fertilizer

Single super phosphate (SSP) or Triple Super phosphate (TSP) are used as additive phosphorous fertiliser in pond culture system.

3.1.3. Potash fertiliser

Potassium is also an essential nutritional element of plants. However, it is plentiful in the water as influenced by geochemical cycle of pond basin. There is no particular need to apply potash fertiliser, until and unless recommended by a technical expert.

3.2 Application method for inorganic fertilisers

While applying liquid ammonia, the container containing liquid ammonia should be put under water and open the lid to let the liquid ammonia diffuse slowly. In this way, volatilisation can be avoided. pH and temperature of water are to be considered while applying liquid ammonia, because in strong alkaline water, the NH_3 toxicity of water is noticed. At a water temperature of 25°C, the percent of N_2 as ammonia at various pH is as follows :

pH	N_2 as NH_3 in (%)
6	0.05%
7	0.49%
8	4.7%
9	32.9%
10	83.1%
11	98%

Source : Asian-Pacific Regional Research and Training Centre for IFF China (1987).

Ammonia at 0.3-0.4 mg/litre of water is fatal to the Juveniles of rainbow trout. Although silver, grass and common carp can tolerate ammonia concentration to some extent, but the maximum NH_3 concentration permitted for fish farming is 0.1 kg/litre. Therefore, special care is needed when liquid NH_3 is applied as nitrogenous fertiliser.

The application amount of nitrogenous fertilizer depends on its nitrogen content. For example, if it is required to apply total 20 kg of nitrogen in a specified water area throughout culture period in instalments in the form of $(NH_4)_2\ SO_4$, 100 kg of $(NH_4)_2\ SO_4$ is required for whole year as the nitrogen content of ammonium sulphate $(NH_4)_2\ SO_4$ is 20%.

Most water sources lack phosphorus. Therefore, it is important to apply a phosphoric fertiliser to accelerate the reproduction of azotebacteria and complement the action of nitrogenous fertiliser. The application amount is calculated according to the phosphoric acid (P_2O_5) content of the fertiliser. The method of calculation and application is usually same as that for nitrogenous fertilizer.

3.3 Doses of manure application

Dose of manure application is dependant on the inherent qualities of soil base and water. Hence before deciding the doses of fertilisers, the soil and water analysis has to be carried out.

Acidic water can be made alkaline by application of lime a week prior to manuring. In highly acidic soil (pH 4-5), the lime dose is 2000 kg/ha, in moderately acidic soil (5-6.5 pH), the lime dose is 1000 kg/ha and near neutral (6.5-7.5 pH), the lime dose is 500 kg/ha and in slight alkaline soil (7.5-8.5 pH), the dose of lime application is 200 kg/ha. In highly alkaline of 8.5-9.5 pH) lime application is not required.

The standard combination of N:P:K as 18:8:4 is commonly used to fertilise the pond, the dose of which is decided on the criteria of the soil as high, medium and low.

	Available N (mg/100 g soil)	Available P_2O_5 (mg/100 g soil)
High	6-12	50-75
Medium	3-6	25-50
Low	3	25-0

The application of phosphatic and nitrogenous fertilizers may be so that the phosphate and total dissolved inorganic nitrogen in water do not surpass 0.25 and 1 ppm respectively. Application of fertiliser should be made when there is a real need to boost the production of plankton. This must be based on chemical analysis of soil and water.

To bring Available Nitrogen to higher level as mentioned above in mg/100 gm soil, the $(NH_4)_2 SO_4$ @ 750 kg/ha/yr or calcium ammonium nitrate @ 750 kg/ha/yr or sodium nitrate @ 940 kg/ha/yr or urea @ 335 kg can be applied in instalments. This means 150 kg nitrogen is made available to the soil through these fertilisers having respective percent of nitrogen content. Similarly to bring available phsophorous to higher level as mentioned above i.e. 6 mg/100 gm soil, the SSP can be applied at the range of 375-470 kg/ha/yr and TSP at the range of 165-190 kg/ha/yr. This means 75 kg phosphorus is made available to the soil through these fertilisers having respective percent of phosphorus content in it, the SSP available commercially has 16-20% and TSP has 40-45% of phosphorus.

Depending on the organic carbon in soil the amount of organic manure can be applied. If the organic carbon in soil is low (0.5%), raw or compost farm yard manure should be applied @ 10,000-15,000 kg/ha/yr. However, pig dung can be applied @ 3-5 tons/ha. If poultry manure is applied as base manure, the rate may be 5000-7000 kg/ha/yr. However, the dose is dependant to the inherent biological productivity of the system. It is always advisable to treat the dungs with lime, bleaching powder or form-alin, so as to ensure that, these do not harbour any pathogenic or harmful organisms. With due fermentation, decomposition and treatment with sterilised agents, dungs are to be diluted and the supernatant should be sprayed in the pond for quick fertilisation effect. If dungs will be applied directly as heaping on a corner place, silting process will go on making the pond more shallower gradually which will need reclamation after some years of fish culture practice. Besides this, the harmful pathogens harbouring untreated dungs may cause severe O_2 depletion due to biochemical oxygen demand (BOD) and disease problem in intensive fish culture practice. Care is given that, the fertilisers are applied when pond water receives plently of sun light, optimum dissolved oxygen, condusive temperature, clear transparent, no wind, and adequate water level is maintained in pond. Hence, the pond water can be kept fertilie flexible and crisp with manuring and water filling. ``Fertilie" means that fish ponds have a fast material recycling and a great biomass of plankton, which are abundant as natural food for fish. ``Flexible", means the color of pond water is always changeable. There are diurnal and monthly variations of planktonic mass imparting brown color (due to zooplankton mass) in morning and green colour (due to phytoplankton mass) in night. ``Crips" means pond water is fertile but not turbid.

With the change of water color, the water can be divided as (1) Fertile (2) Sterile, (3) water bloom and (4) deteriorated water. With suitable managment like manuring, rational application of aerators, and timely water filling, the water can be utilised for fish culture for high out put.

10

FISH FEED

1. INTRODUCTION

Fish feed falls under the category of nutrition. This concern the nature of food, food nutrients and the needs for rearing aquatic organisms. The physical organisation and the anatomy of animal affects the nutritional needs. One of the objectives of nutrition is to direct the selection of food and to establish desirable eating habits. Food eaten by fish varies considerably according to the species and their developmental stages. Exact preferential food should be matched with different developmental stages of fish. There are distinct feeding phases for the fish according to the age. Such as :

(1) First phase-includes the resorption of yolk sac of the hatchlings.

(2) Spawn, fry and fingerling stage

(3) Fingerling to table size stage

In intensive rearing of fish spawn and fry, feeding constitute a major factor. Their growth mainly depends upon the quality of diet provided. Although some of the formulated feeds are nutritionally rich' in protein than the live food but spawn and fry prefer live food rather than formulated feeds. That too, the mineral and micronutrient needs of spwan and fry are little understood to be incorporated in formulated feeds. These mineral and micronutrients are supplied through live food thus compensate for these when formulated feeds are used. Therefore, it is always prferable to have a regular supply of live food. Hence, live food organisms serve as living capsules of nutrition.

Live foods are mainly phytoplankton and zooplanktons. For regular supply of live food, it is better to have species wise live food calture tanks. For aquaculture practices essentially prawn and other invertebrates, the most common phytoplankton species used for culture are :

(1) Melosira	(6) Nitzschia
(2) Cheatoceros	(7) Cylindrotheca
(3) Rhizosolenia	(8) Skeletonoma
(4) Tabellaria	(9) Thalassiosira
(5) Chlorella	

and the most common zooplankton species are: (1) Brachionus, (2) Moina, (3) Daphnia and also (4) Artemia.

The technique of culture includes perfect culture media and the material for inocculation. The phase of successive growth of plankton consists of : (1) Log induction phase (no increase in numbers), (2) Exponential phase (Geometric increase in concentration), (3) Constant or linear growth phase, (4) Stationary phase and (4) death phase.

Different types of live food cultures are practiced. For all microalgal culture, the nitrogen to phosphorus concentration should be 16 : 1. Several culture media are advocated. These are Schreiber's medium, Miquel's medium, Conway or Walne media. These include : (1) Batch culture, (2) Large scale culture, (3) Continuous culture, (4) Synchronus culture and (5) Dialysis culture. In USA, monoculture and stock culture of planktons are prevalent whereas in India, open culture is practiced Isolation of microalgae for pure culture is also adopted. These isolation methods are: (1) Pipette method. (2) Centrifuge method and (3) Social dilution method. However, the most common practice for isolation is the dilution mathod.

Considering the principal natural diets of cultivated fishes which includes not only phyto and zooplankton but also worms, molluscs, insects and other benthic faima. Red worm (chironomid larvae) and tubifix are very useful in rearing tender aquatic organisms. Among molluscs, gastropods are the most important as food for fish. These include Limnaea and Planorbis. Fish such as Black carp and Gourami preferably feed on molluscs.

With advancement of fish culture technology, the extensive carp farming method has gradually been shifted towards intensive culture method. The fish originally live solely on the natural products of the pond impairs growth, reproduction and health. In the farming habitat, feeding of stocked population with nutritionally balanced and quality test diet is of critical importance to ensure optimal biological and physiological processes as well as production. However, test diets are either dried diet, semi-dried, moist, encapsulated or particulate diets. Dried diets includes the diets of pure plant origin, animal tissue meal, compounded or formulated.

Besides the need for feeds for grow out systems, feeds are also required for hatcheries and nurseries to produce healthy stocking material. The larvae/Juvenile of most of the fishes, prawns and bivalve molluscs require microparticulate diets and in most cases suitable live food are being fed. The potential source of diets for larvae/Juveniles of fishes are either viable which include bacteria, motile gametes, spores, yeast, diatoms, unicellular algae, zooplankton or non-viable. These non-viable diets include detritus organic aggregates, micro particulate, micro encapsulated and tissue suspension. In microencapsulated diets, the liquid or particulate materials are enclosed by a membrane made of gelatin, gums, waxes, natural polymers or the synthetic polymer of ethyl cellulose or polyvinyl alcohol. Thus for larvae especially of crustaeans and molluscs, nutritionally adequate microparticulate or microencapsulate diets with appropriate size, texture, taste etc. is most important.

Akhtar and Alexander, 2008 emphasised on single cell protein in Aqua feed. The various single cell proteins used in aqua feed are

1 Yeast : Yeast cell wall contains mannan oligosaccharides and a-glucans. The levels of nutrients (dry weight) such as Protein (25 -29%), Carbohydrate (21-39%), Lipid (2.5 – 9%), Nucleic acid (8-12%) and Ash (4.7-14%) are in yeast also.

2 Bacterial Single cell Protein contains bacterial protein meal (BPM). BPM contains high levels of copper and nucleic acid. Eurolysine Fodder Protein (EFP) is a commercial SCP concentration obtained from the bacterium *Micrococcus glutamicus*. EFP was found to replace up to 40% of fish meal in the practical diet. Proximate composition of bacteria in dry weight includes Protein (25-49%), Carbohydrates (2.5-11%), Lipid (2.5-9%), Nucleic acid (8-16%) and Ash (29-40%).

3 Algal SCP: This includes spirulina and chlorella.

Spirulina in powder form serves as food for fish and shell fishes also. Spirulina is a blue green algae belonging to class Cyanophyceae, order Nostocales and family Oscillatoriaceae. Spirulina powder had excellent nutritional composition.

Protein	:	62.5 – 71.0%
Lipids	:	6-8%
Carbohydrates	:	15-17%
Crudc fibre	:	0.02-0.90%
Ash	:	7-9%
Aminoacids		
Isoleucine		4.13%
Leucine		5.80%
lysine		4.00%
Methionine		2.17%
Phenylalanine		3.95%
Threonine		4.17%
Tryptophan		1.13%
Valine		6.00%
Vitamins		**mg/Kg**
Biotin		0.4
Cyanocobalamine		2.0
Calcium pantothenate		11.0
Inositol		35.0
Nicotinic acid		118
Riboflavin		40
Thiamine		55
Tocopherol		190
Ash		**g/Kg**
Calcium		1.3
Magnesium		1.9
Sodium		0.4
Potassium		15.4
Iron		0.6
Phosphorous		8.9

Spirulina is a rich source of betacarotene (1.7 g/Kg) that is precursor of Vitamin-A and also linolenic acid that are not normally encountered in dietary items. With energy content of 346 K cal/100g, its digestibility is as high as 84% with a net protein utilization level of 61%. The USA has declared Spirulina as the "Best food for tomorrow". Spirulina contains three types of pigments

(1) Phycocyanin (blue)

(2) Carotenoides (orange)

(3) Chlorophyll (green)

Spirulina is used as (1). Dietary supplement (2). Food coloring (3). Cosmetics (4). Pollution control in wastewater and (5). Therapeutics.

Chlorella contains more than 20 vitamins and minerals, rich in a-carotene and powerful antioxidant, Vit-C and Vit-K. Chlorella minerals include Calcium, Iron, Magnesium, Potassium, Phosphorus and traces of Iodine, Selenium, Zinc and Copper. It contain 19 naturally occurring amino acids including 8 essential amino acids. It also contain omega-3 and omega-6 fatty acids. Chlorella Growth Factor (CGF) a complex combination of amino acids, nucleic acids and peptides stimulate the growth of valuable friendly bacteria which in turn has probiotic effect. CGF and an acidic polysaccharide in Chlorella's cell wall stimulate the production of interferon by the body. It helps in improving the activity of T-cells and B-cells which are principal immune system defenders against viruses and other invading microorganisms. CGF is identified as being responsible for the production of macrophages within the immune system which can assist in destruction of cancer cells and removal of cellular debris.

Compounded diets should contain adequate level of nutrients to meet the physiological requirements of the organisms such as energy, body building, repair or maintain cells, tissues, and regulate body processes. According to Halver (1976) any nutritionally balanced compounded diet must include an energy source with sufficient essential amino acids, essential fatty acids, and non-energy nutrients (Vitamins and minerals) to maintain and promote growth. Any imbalance of these nutrients may have sparing action that would affect the efficacy of conversion of food by the organisms. The specific nutritional requirements of fish vary greatly with species, size, physiological condition, temperature, stress, nutrient balance of the diet and environmental factors. Therefore, programming of nutrient constitutents must be done in order to have most economic compounded ration for fish.

Several species of fishes in the world have been studied in great detail for evolving standard diets. Among the carps, common carp is the most widely studied by nutritionists. The studies on the nutrition in Indian major carps and Exotic carps have been made in India in recent years. For the preparation of the formulated feed having the optimum dietary nutrient levels, the essential prerequisite is to have an understanding of the nutrient requirements of the species selected for culture. The quantitative nutritive requirements that have been derived for several species of fish are based on estimating the nutrient needs of their closest related groups.

2. NUTRITIONAL REQUIREMENTS OF FINFISH

A considerable progress has been made in the study of the dietary nutrient requirements in number of fishes (Halver, 1972; Millikin, 1982; Cowey and Sargent, 1979; Cowey and Tacon, 1983; Cho, Cowey and Watanabe, 1985; Pandian, 1989). Although fishes exhibit similarities in respect of qualitative nutrient requirements with that of terrestrial vertebrates, still marked differences exists in the quantitative nutrient needs (2-3 times higher than mammals) in them at the dietary level (Pandian, 1987). Such differences can be attributed in carnivorous and omnivorous fish which show their preference to use protein over carbohydrate as a dietary energy to maintain their body temperature and expend relatively low energy for reproduction compared to other vertebrates. Hence, fishes are considered to be better feed converters than the other vertebrates. That too, the biochemical composition of fish flesh reflects their nutritional requirement.

Fish requires 40 or more essential nutrients, among these the most important ones are :

(1) Protein

(2) Carbohydrates

(3) Lipids and Fats

(4) Vitamins

(5) Minerals

(6) Trace elements

Protein, carbohydrates and fats are regarded as energy food while vitamins, minerals etc. are regarded as non-energy food.

2.1 Proteins

Proteins are peptide molecules which consists of aminoacids. Although about 300 or so aminoacids are known to occur in nature, only 20 of these are present in proteins. The sequence of aminoacids in a particular protein is genetically controlled. Thus each protein differs from other in the sequence arrangement and in the number of aminoacid molecules in it. Protein is used as a source of energy when fat and carbohydrates are depleted; 1 gram protein yields 4 k cal (17 k joul) energy. Over 20 species of fishes have been studied for the dietary requirement of protein. These have shown a high dietary requirements ranging from 35 to 55% which is equivalent to 45-70%, of the energy content of the diet in the form of protein. Usually the recommended protein content in fish feed is 25-40% as the dietary protein need is dependant to species, size and environmental factor particularly temperature. Small sized fishes require higher levels of proteins for growth than the larger ones. Carnivorous fish such as rainbow, trout and eel demand feeds with high protein content; herbivorous fish require less protein. Since proteins are continuously being used by the animal either to build new tissues or to repair tissues, adequate protein in the diet is necessary, otherwise there is rapid reduction or cessation or loss of weight. In Indian Context, the first ever systematic study on nutritional requirements of Indian Major Carps has been studied by Sen *et al.,* 1978. They have suggested that the gross protein and carbohydtate requirements of carp, fry and fingerlings vary from 36-45% with variations in water temperature ranging from 26 to 32°C. Further, investiga-tions supported these views to a great extent (Singh *et al.,* 1986; Swamy *et al.,* 1989 and Mohanty *et al.,* 1990). The protein requirement is higher in catla fry (45-47%) than fingerlings (40%) (Singh and Bhanot, 1988). Such differences probably be attributed to various factors including eco-physiological variations within the organism during experimentation. However, Renukaradhya and Varghese (1986) estimated about 30% protein in the diet is enough to meet the dietary protein requirements of both Catla and Rohu.

The Chinese Grass carp, fed with diet consisting leaf protein concentrate (LPC) did not show variations in protein (36%) requirements with change in temperature (Das and Tripathi, 1979). The protein requirement in Silver carp fry ranges from 37-42% (Singh, 1989) at a temperature range of 23-28.7°C. The protein requirement in Common carp is to be 31% (Varghese *et al.*, 1976), 45% (Sen *et al.*, 1978) and 54% (Pandian, 1988) with different feed ingradients being reported. Such variations may be attributed to the size, age, temperature, sex, dietary energy level, meal size, water quality, pellet density, nutrients in the feed, feeding procedure, frequencies of feeding and management of feeding practice as a whole.

Finding such discripancies in protein requirements, the amino acid requirements had been studied which is found to be similar in Indian Major carps. Hence, by taking the common amino acid input balanced diet, a conventional, low cost feed can be formulated (see Tripathi, 1990). So, it is now recognised that the general requirement of protein in fish is the requirement of aminoacids. Nutritionally aminoacids are essential (indispensable) and non-essential (dispensable).

The dispensable aminoacids are those that can be readily synthesised in adequate amounts to support growth. Indispensable aminoacids are to be encorporated in the diets as these are not synthesised by the organisms at all. These essential aminoacids are : (1) Methionine, (2) Arginine, (3) Theronine, (4) Tryptophan, (5) Valine, (6) Isoleucine, (7) Leucine, (8) Phenylalanine, (9) Histidine and (10) Lysine.

Besides essential aminoacids, some other aminoacids like Tyrosine, Cystine, Glycine and Serine could not be synthesised by the organism in sufficient level and so need to be supplied in a lesser extent. Therefore, these aminoacids are called as semi-essential aminoacids. Glutamic, aspartic acids, alanine, proline and hydroxyproline are non-essential amino acids, since these can be synthesised in required levels. However, the food-stuffs under metabolic process illustrate the synthetic and interconversion of non-essential aminoacids. Though cystine from methionine and tryosine from phenylalanine could be synthesised, in the absence of cystine and tyrosine, the requirement for methionine and phenylalanine is increased.

Although, the different individual essential amino acid requirement of several of the fishes have been determined, the dietary requirement of all the 10 essential aminoacids are established only for four species of fishes. These are,

(1) Common carp

(2) Japanese eel

(3) Channel catfish

(4) Chinook salmon

The optimum requirement of essential aminoacids for common carp has been shown as 13.7% of the diet. The EAA requirements of rohu is estimated to be about 27% of protein which is almost similar to common carp on the basis of dietary protein.

It is important to note that both quantity and quality of protein are essential for the growth of fish.

Table 5. Essential aminoacid requirements (g/kg dry weight) at stated dietary protein levels of certain fishes.

Amino acid	Chinook salmon	Japanese eel	Common carp	Channel cat fish
Arginine	24	17	16	10.3–17.0
Histidine	7	8	8	3.7
Isoleucine	9	15	9	6.2
Ieucine	16	20	13	8.4
Lysine	20	20	22	15.0
Methionine + cystine	16	19	12	5.6
Phenylalanine + tyrosine	21	22	25	12.0
Threonine	9	15	15	5.3
Tryptophan	2	4	3	1.2
Valine	13	15	14	7.1
Protein in diet	400	377	385	240.0

2.2 Carbohydrates

Carbohydrates are available in the form of Saccharides or Polysaccharides. Through digestion, these are broken into monosaccharides, which are absorbed and utilised by the fish. 1 gram carbohydrate yields 4 k cal (17 KJ) energy and hence included under energy feed. Some monosaccharides are oxidised and some are stored in liver as glycogen or are converted into fat, which serves as a reserve energy source during food shortage. It is the least expensive form of dietary energy and helps in pelletising quality of practical fish diets. Dietary carbohydrates are known to be utilised by various fishes but only limited information is available on their digestibility and metabolism. Various metabolic intermediate like non-essential amino acids formed during carbohydrate metabolism which are necessary for growth of fish. Higher levels of dietary digestible carbohydrate may lead to death through accumulation of glycogen in liver. However, diets containing over 40% carbohydrate reduces the feed efficiency and retards growth in common carp.

The metabolisable energy of carbohydrates for fish range from near zero (for cellulose) to 3.8 k cal/gm for easily digested sugars. However, much of the carbohydrates entering into the diets of carps is of plant origin. Carbohydrates have a sparing effect on protein, since 0.23 gm of carbohydrates feed per 100 gm of diet will spare 0.05 gm of protein.

Utilisation of dietary carbohydrates in fishes differs with their complexity. Common carp has higher utilisation levels of dietary carbohydrates. This is also true for fry and fingerlings of rohu. Common carp has higher digestibility for rice and wheat bran, where as rohu fingerlings exhibit high digestibility for GNOC + wheat bran mixture, fish meal + GNOC, wheat bran, yeast (84-86%) and mrigala fingerlings for rice powder (88%). Starch is well accepted as dextrin by fry and fingerlings of rohu, mrigala and common carp. Studies on Common carp have shown that carbohydrates up to levels of 25% of the diet are effectively utilised as energy source. For fry, fingerlings of rohu and common carp, 26% is shown as optimal requirement and 28% for mrigala fingerlings. Lower levels of carbohydrates with increased levels of protein did not result in increased growth. This shows that protein had been used as a source of energy. On the other hand, a higher level of carbohydrates with lower dietary protein level (30%) spared protein in fry and fingerlings of rohu, mrigala, and common carp. Hence, the fish species that can readily adopt to high carbohydrate diets and convert the excess energy into fat are much more efficient in carbohydrate utilisation than those that lack such an ability.

Trouts show relatively high digestion of carbohydrate i.e. glucose 99%, maltose 92%, sucrose 73%, lactose 60%, starch 57%, and raw starch 38% (Phillips *et al.,* 1948). Buhler and Halver (1961) found that chinook salmon tolerated relatively high levels of dietary carbohydrates without the development of abnormal conditions. The carbohydrates assimilated by trout assumed the presence of sucrose, maltose, lactose and amylase enzymes in the digestive tract. The maltose activity was higher than sucrose and lactoses. Mc Geachin and Debnam (1960) demonstrated a relatively high amylase activity in the digestive tract of number of freshwater fishes. In predominantly herbivorous fish tilapia, amylase activity was distributed throughout the gastrointestinal tract, but carnivorous perch, the pancrease was the only source of amylase. Such differences in amylase activity in the digestive tract of fish can be attributed to different feeding habits.

Nagai and Ikeda (1971) have shown for the carps that, under starvation, lipid reserve was converted into glycogen in hepatopancrease and increase in blood glucose level was observed. During breeding, the carbohydrate content in diet increased or protein decreased, glycogen in hepatopancrease increased while lipid in hepatopancrease and blood glucose level decreased. This suggested in carp that carbohydrate is hardly converted to lipid but that protein is principally converted to lipid. Hence, protein and lipids are the main energy source for carps but certain amount of carbohydrate is necessary for the activity. This was experimentally again supported by them through C^{14} technique. The Glucose 6^{-14} C was incorporated into glycogen in all groups were oxidised to $14CO_2$. When more than 50% of protein was contained in diet, oxidation of glucose $6^{14}C$ decreased and at the same time blood glucose increased. From these observations it was suggested that carp possesses the active and reversible Embdon-meyerhof path way but that glucose is not a principal storage depot of energy in fish.

Cellulose, the constituent of plant cells are not digested by *Cyprinid* fishes, although in Tilapia and Milkfish cellulose is digested, still the utilisation rate is low. This indicates that *cyprinids* lack cellulolytic enzyme which is unlike to ruminents. Fibre in plant material serve not only as a diluent for other nutrients but also act as an extender in the ration. Levels as high as 21% reduces nutrient intake and impare digestibility (NRC, 1983). Fibre in small quantity (about 8%) may add structural integrity to pelleted feeds, but high amounts is unsuitable, as it may lead to ingestive variations, increase faecal waste and consequently pollute the culture water.

2.3 Lipid

Lipid are water insoluble biomolecules which play an important role in energy production of animal tissues and as a sources of essential fatty acids (EFA). Researches carried out during the past revealed the significance of dietary lipids in the nutrition of fish. Lipids are important source of energy (8-9 k cal/gm) and fatty acids are essential for normal growth and survival of fishes. Besides this, lipids have important dietary role as carrier of other fat soluble nutrients like vitamin A, D and K. The unsaturated fatty acids play an important role in the transportation of other lipids. Lipids, especially phospholipids and sterol ester (cholesterol) play vital role in the structure of biomembranes at both cellular and subcellular levels. Lipids are also important in the flavour and textural, properties of the food consumed by warm water fishes. However, excess or deficiency of lipids affects the growth as well as the body composition of the fish. Lipids are involved in many other aspects of metabolism, like steroid hormone biosynthesis. Besides, the long chain polyunsaturated fatty acids (PUFA) are precursors for prostaglandins in fish. Na^+, K^+, activated AT-pase requires phospholipids, phosphotidylcholine and phosphotidyl ethanolamine for its normal function. Particularly phosphatidylserine or phosphatidylglycerol are effective in activating this AT-pase enzyme. The possible role of PUFA of the W_3 series in brain and nerve activity is well known. About 85-90% of dietary lipids are digestible by fish. Cultured fish have shown that the optimal lipids intake is essentially similar to that for wild fish. The cold water fish (rainbow trout) at optimum temperature of water (12°C) fed with 24% dry wt. as herring oil results excellent growth rates. Even diets with 9% lipid was found to increase the weight gain and utilisation of protein is high. Channel catfish have shown best growth with diets cantaining 35% crude protein and 10-12% lipid. Lipid incorporation in the diet not only increase the dietary energy level but also improves protein utilisation. Cowey and Sergent (1979) reported that lipids not less than 10% and not more than 20% can be added to fish diets with excellent results. It has also been shown that the protein level could be reduced in marine fish diet, if the energy content is maintained at a high level.

It is now recognised that there are 3 essential fatty acids (EFA) i.e. Linoleic (18 : 2 : W6), Linolenic (18 : 3 W3) and Arachidonic. All fatty acids posses a hydrocarbon chain and a terminal carboxyl group. Fatty acid is saturated if no double bond exists between hydrocarbon chain (Palmitic acid). If one double bond exists, then it is a monounsaturated fatty acid (Oleic acid). If 2 or more double bond exists it is called poly unsaturated fatty acids (Linoleic acid), (Palmitic acid 16 : 0), Oleic acid (18 : 1W9) Linolenic acid (18 : 3W3). Corn oil is an example of Linolenic series. The requirement of (EFA) in fish is reviewed by Castell, 1979; Watanabe, 1982). Particularly linolenic series have been demonstrated in a number of fishes for achieving better growth rates, food conversion and to avoid certain pathological conditions. Inclusion of 18 : 2 : W6 in the diet resulted some improvement in growth and feed conversion compared with EFA deficient diets. So fish requires a mixture of W_3 and W_6 fatty acids. However, their requirements differ from species to species as the EFA requirement is found to be far less for channel catfish, eel and carp than those of rainbow trout. Even EFA requirements of fish may change slightly with temperature and culture conditions. As flexibility and permeability of cellular membrane is concerned, the importance of EFA in fish metabolism and physiology is well recognised.

Significant variations exists among various fish species in their ability to elongate and desaturate dietary 18 carbon fatty acid to 20 or 22 carbon fatty acids. Several marine species such as red sea bream, black sea bream and yellow tail and also turbot (*Scaphthalmus maximus*) appear to have lower enzymatic elongation desaturation capabilities than freshwater fishes. Results indicate that marine species have limited ability to elongate and desaturate 18:3W3 resulting in dietary essentiality of eicosapentaenoic (20:5: W3) or docosahexaenoic fatty acids (22 : 6W3). Consequently for these marine fishes, it is essential to supply highly unsaturated fatty acid (HUFA) in the diet. Hence the ratio of W3/W6 fatty acids is greater for marine species than freshwater species (Castell, 1979). Hence same trend in fatty acid composition and the ratio of W3/W6 is expected to be in fish that migrate from freshwater to marine or vice versa. Most cold water and marine fish require W3 or W3 HUFA fatty acids. But the requirement of W3 and W6 PUFA shows greater diversity in different warm water species than in cold water species.

Lipid requirement of common carp are met by providing soyabean oil (5%) in the diet. The requirement of fat for catla fry and fingerlings has been estimated to be 6%. But fat requirement for mrigala fry was estimated to be 6% and rohu fry 4% (Swamy *et al.*, 1988). Therefore, as per requirements vegetable oils and animal fats may be combined with marine fish oils to satisfy the PUFA-requirements.

Oxidised fat in the diet has adverse effects. Rancidity and autooxidation can be controlled by using one of any antioxidants cited else where.

2.4 Vitamins

Vitamins are complex and essential organic trace elements. These are distributed in food stuffs in small quantities and form a distinct entity from other major and minor food components. The importance of vitamins as essential constituents in the diets of animals have been well recognised. Water soluble vitamins except cholin and inositol (soluble growth factor) act as co-factors or co-enzymes and are necessary to be incorporated in fish diet as fish can not synthesise in its body tissues. But contributions to vitamin nutrition in aquatic species are relatively less, compared to mammals and poultry. Such slow progress in vitamin nutrition in aquatic animals can be attributed partly due to inherent problems posed by the aquatic medium. That too oxidation of vitamin particularly ascorbic and during feed preparation, storage and leaching when introduced into the water are some of the major constraints. The actual dietary requirement of vitamins in certain fishes also posed difficulties because its conribution to some extent is influenced by gut microbial flora. Therefore, vitamin requirements remain unknown and the formulated diets may still be deficient in certain vitamins even after supplementation.

Varieties of vitamins carry out various physiological function. Some vitamins participate in metabolic process : Vitamin B_1, (carbohydrate metabolism); Vitamin B_6, (protein metabolism); Vitamin C, (protein synthesis); Vitamin D, (calcium and phosphorus metabolism). Vitamin A is involved in calcium transport across some membranes in reproduction, embryonic development and in cellular and subcellular integrity. Vitamin-E seems to be related to its ability to product HUFA in lipids of cell membrane from oxidation in the presence of molecular oxygen. Vitamin-E also act as antisterility. Ascorbic acid is involved in erythrocyte maturation. Vitamin requirements are affected by size, age, growth and by environmental factors.

Two principal (1) water soluble vitamin and (2) fat soluble vitamin categories exist. So far four fat soluble (Vit-A, D, E and K) and eleven water soluble Thiamine (Vitamin-B_1), Riboflavin (Vit-B_2), Pyrioxine (Vit-B_6), Choline, Niacine, Pantothenic acid,

Inositol, Biotin, Folic acid, Cyanocobalmin (Vit-B_{12}) and Ascorbic acid, vitamins are known to be required by finfishes and shell fishes. Many of the water soluble vitamin functions either directly or as coenzyme for one or more enzymes. None of the fat soluble vitamin is known to function as coenzyme.

Although studies on vitamin requirements of warm water fishes are very limited still the survival and growth of spawn, fry, fingerling, growers and brooders of IMC are influenced with vitamin B complex when provided at 0.1% level in the formulated diet as feed. Hence vitamin premix is normally added in the feed for healthy growth of carp. Care should be taken to provide the supplemental level in excess so as to compensate for losses due to processing and leaching.

2.5 Minerals

Minerals are required in small quantities in the diets. Though minor in quantities added, the mineral nutrition is no less important. These are very essential to be supplemented in fish diet because fishes are not capable of synthesising them. Many of the minerals either through feed or from environment cause significant loss in unit of flesh production. Minerals not only provide basis skeletal structures but also are important co-factors of enzyme and other biological chemical involved in life process. Since fish need to maintain their osmotic balance through mineral ion exchange, therefore, mineral requirement is of paramount important. Mineral requirements in fish may be classified as bulk elements (such as calcium, phosphorus, potassium, chlorine, sodium and magnesium) and trace elements (copper, cobalt, iron, iodine, manganese, selenium, zinc aluminium, chromium and vanadium).

Bulk elements

Calcium and phosphorous are the most abundant mineral elements in fish, human and other living organism. Approximately 99% of the calcium and 85% of the phosphorous are present in bones as calcium phosphate and hydroxyapatite. The remaining part of the calcium and phosphorous are found in extracellular fluids, intracellular structures and cell membranes. Further calcium is responsible for a number of regulatory functions. Phosphorous is directly involved in energy producing cellular functions.

Thus phosphorous play an important role in overall metabolism as well as in various metabolic processes involving buffers in body fluids.

Magnesium is an essential element in bone of fish and its metabolic activity is interrelated with calcium and phosphorous metabolism. It is an important element in some enzymatic processes especially in the metabolism of carbohydrates. It is not only essential for maintenance of ribosomal structure but also helps in protein synthesis. It helps in biological function such as membrane transport, generation and transmission of nerve impulses, contraction of muscles and oxidative phosphorylation.

Sodium is important for acid base balance and osmoregulation. It is abundant with chlorine in ionic form. The concentration of sodium, potassium is higher in fish than in land animals, as fish is always surrounded with medium rich in potassium.

Trace elements

Zinc is essential for growth and reproduction, wound healing and normal functioning of the immune system and other physiological processes. It can regulate many metabolic processes of carbohydrates, lipids and protein metabolism and synthesis of nucleoproteins. Further iron as a component of haem in haemoglobin, myoglobin, cytochrome etc., play an essential role in oxygen transport, storage and utilization. Iron is also a cofactor for a number of enzymes. Similarly manganese acts as a cofactor for many enzymes to form meta-enzyme complexes.

Copper is involved in glucose metabolism, haemoglobin synthesis and phospholipid formation. Further selenium is involved in the activity of glutathione peroxidase. Similarly chromium play a role in maintaining the structural stability of protein and nucleic acid but the primary role of chromium is to potentiate the action of insulin. Molybdenum is essential for several enzymes including xanthine oxidase, aldehyde oxidase and sulfide oxidase. Further Iodine is needed for biosynthesis of thyroid hormone, thyroxine and triiodothyroxine. Similarly cobalt and fluorine are essential trace elements. The functional role of cobalt is known for the formation of Vitamin- B12.

Freshater fish show higher mineral requirements compared to the marine fish as sea water is enriched with most of the essential minerals. Fluorine and molybdenum have also been added of late to the list of essential mineral elements. Calcium is essential for blood clotting, muscle function, proper nerve impulse transmission, osmoregulation and as a co-factor in enzymatic processes. Phosphorous is involved in energy transformation, permeability of biological membrane, genetic coding, general control of reproduction and ovulation through hydration and growth. In addition, minerals have a role in hormonal regulation, respiratory pigments, structural elements, high energy bond and enzyme co-factors. Thus studies on mineral requirements assumes greater importance in nutrition of fish. Hence in fish feed preparation, mineral bulk elements (Ca^{++} and phosphorus) are to be added (1.5% in dicalcium phosphate form), sodium chloride (0.3%) and trace mineral mixtures (0.1%). However, it is reported that 6.5% dicalcium phosphate or monocalcium phosphate when encorporated in the diet, resulted better growth in intensive culture. But increase in phosphate level beyond optimum may lead to environmental pollution. Trace elements are growth stimulants. These trace minerals helps in improving protein assimilation and survival rate. Cobalt nitrate and cobalt chlorides are added in diets of carps to ensure survivality and growth. Manganese and cobalt promote growth, and enhanced survival rate in Asiatic carps. Besides Alumino-sillicate, natural zeolites, sodium bentonite, boron, zinc, and trivalent cromium are used as growth promoters in aquaculture production process. Shirgur and Indulkar (1999) described three different growth promoters for the growth of *Macrobrachium rosenbergii*. Their results indicated that 94.7 to 95.4% average percent gain in weight of the juveniles (30-35 mm length) were possible by incorporating 2.5% ``Bhimnsu", 0.2% ``Staffac-20" and 0.02% ``Bioboost Forte" in 30 days of culture experiment.

2.6 Antibiotics

Antibiotics are costly ingradients and should be fed only when medicated feeds are to be given for therapeutic reasons. Antibiotics such as sulphonamides, tetracycline, 4-Quinolones, Nitrofurans, Erythromycine and chloramphenicol are used as fish chemotherapy and chemoprophylaxis. Routine feeding may result in development of resistant strains of pathogenic bacteria. The use of certain antibiotics as growth promoters in fish are demonstrated. The addition of virginiamycin at the rate of 40, 80 and 100 mg/kg of high protein incorporated diet of carp resulted in length-weight increase significantly than the control. Terramycin also brought length-weight gain when encorporated in the diet at 10, 50 and 100 mg/kg of protein encorporated diet. However, best results have been obtained in carp by feeding 20,000 units of Terramycin resulted in 9.5% increase in weight and saves 10.5% of feed. Not all antibiotics such as Avoparcin and Tylosin show growth promoting effect (Matty, 1987).

However, use of antibiotics in aquaculture has raised considerable concern regarding

1 Their persistence in the environment
2 Development of antibiotic resistant strains of wild bacteria
3 Presence of residues of antibiotics in the cultured food product and
4 Influence the appearance of resistance in bacteria of other niches through transmission.

The aquaculture authority considered a list of 20 antibiotics, drugs and pharmacologically active substances not to be used in aquaculture practices. The expert group also noted that this ban would in no way hamper the farming practices as there were several safe alternatives like probiotics, immunostimulants and other compounds available to be used by the fish farmers during crisis.

Antibiotics and other Pharmacological active Residual substance level	Maximum permissible (in PPM)
1. Chloramphenicol	nil
2. Nitrofurans including Furaltadone, Furazolidone, Furylfuramide, Nifuratel, Nifuroxime, Nifurprazine, Nitrofuratoin, Nitrofurazone	nil
3. Neomycin	nil
4. Nalidixic acid	nil
5. Sulphamethoxazole	nil
6. Aristolochia spp. and preparations there of	nil
7. Chloroform	nil
8. Chlorpromazine	nil
9. Colchicine	nil

10. Dapsone	nil
11. Dimetridazole	nil
12. Metronidazole	nil
13. Ronidazole	nil
14. Ipronidazole	nil
15. Other nitroimidazoles	nil
16. Clenbuterol	nil
17. Diethylstilbestrol (DES)	nil
18. Sulfonamide drugs (except approved sulfadimethoxine, Sulfabromomethazine and sulfaethoxy pyridazine)	nil
19. Fluroquinolones	nil
20. Glycopeptides	nil

Source: aqua international January, 2008 PP 34.
(Banned by Aquaculture authority)

2.7 Hormone

Addition of hormones in fish diets not only ensure control of sexual development but also improve feed conversion efficiency thereby resulting enhanced growth. Hormonal control of sexual development is advantageous where either male or female has superior culture characters. Use of synthetic anabolic steroids as growth promoters are now common in aquaculture research. Pelletised feed encorporated with 17 Beta methyl testosterone at a concentration of 1 mg/kg (IPPM) in diet resulted better growth in catla, rohu and common carp while for mrigal better growth was obtained at a concentration of 3 mg/kg of pellated diet (Deb and Varghese, 1987). However, diets with combination of synthetic steroid hormones such as 17 alpha methyl testosterone and diethylstillbestrol at 4 mg/kg of diet has resulted faster growth and better food utilisation in silver carp and *Tor khudree* than with individual hormones (Shyama and Keshavanath, 1987). Nonsteroid hormones that have growth promoting effect in fish are thyroxine, triiodothyronine, insulin and growth hormone itself. Recently genetically engineered hormone bovine somatotrophin (growth hormone) has been put into trials to increase lactation (milk) yield in cow. It would be of much use, if this product can be used to grow larger and better fish. However, Sen and Chatterjee (1976) had used 15 growth promoters on the basis of potency, economy and ease of handling the substances. The substances fall under protein, carbohydrates, vitamins, minerals, antibiotics and hormones. These substances are as follows: (1) Proloid, (2) Eltroxin, (3) Serein, (4) Macrobin, (5) Vit-B complex, (6) Yeast, (7) Starch, (8) Selenium, (9) Molybdenum, (10) Boron, (11) Cobalt chloride (12) Enterocycline, (13) Chloromycetin (14) Hoestacycline and (15) Manganese.

2.8 Pigments

There are variety of pigments in many plants and animals which can impart coloration to skin, flesh etc. These pigments are called biochemics. The groups of biochemics are collectively called as biochromes. These biochromes are :

(1) Carotenoids
(2) Quinones
(3) Flavonoids
(4) Flavins
(5) Tetra pyrroles
(6) Pterins
(7) Indole pigments (melanin)

Carotenoids are fat soluble lipochrome and can be divided in to two groups, such as (1) carotenes (made of carbon and hydrogen only), (2) Xanthophylls (made of carbon, hydrogen and oxygen). Carotenoids are present in the organism either in carotenoproteins complex form or dissolved in lipoprotein or lipoglycoprotein components. Though the animals lack the capacity to synthesise carotenoids, still they contain varieties of carotenoid molecules acquired through food wed. Based on carotenoid biochromes, Fox (1979) divides the organisms in to (1) carotene selector, (2) carotenoid rejector, (3) xanthophyll accumulators, (4) Nonselector and (5) carotenoid innovators. A few fishes come under carotenoid innovators. The importance of carotenoids is ascribed only when a clear cut physiological function of these as precursor for vitamin-A was understood in animal kingdom. Apart from this though many a varied roles ascribed to them in the physiology of the organisms are highly speculative in nature. However, in order to have great acceptance by the consumers, these are incorporated in the fish diets to impart coloration to the fish flesh. As animals can not synthesise carotenoids, the deposition of carotenoids in hepatopa-ncrease of penaeids speculates that perhaps lipid or proteins are the carriers of carotenoids in to hepatopancrease of prawn. Their nutritional value is yet to be well understood.

3. PHYSIOLOGIL APPROACH TO THE NUTRITIONAL BIOEERGETIC IN FISHES

The ultimate source of energy comes from food consumed that are processed through kreb cycle and cytochrome oxidase to produce energy for various metabolic (catabolism and anabolism) process, that an animal carries out. The immeidate source of energy is the Adenosine triphosphate (ATP) which is produced through this process. The Adenosine diphosphate (ADP) traps energy that is formed during catabolism, forming ATP. Therefore before any growth can be achieved by an organism, there must be sufficient maintenance and any movement associated with food intake. It is desirable to examine the exchange of energy within an organism so as to proceed for specific considerations in fish, which is given as a flow chart.

Energy of food consumed

↓ Faeces (F)

Digestible (assimilated) energy (A)

↓ Non faecal loss (U) exogenous

Physiologically useful energy or metabolisible energy

↓ Heat increment

Net energy (energy available for growth and work)

↓ Metabolism (R)

Growth (production) P

A general balance sheet of total energy consumed (C) by an organism flows as biological uses and losses that are associated with the conversion and utilisation of energy. It is quite evident that, the losses of energy are faeces (F), Urine and NH_3 (U) and heat increment. Similarly the uses of net energy in to Growth (P) through metabolism (R) is evident from the flow chart (Fig. 48). Hence, it is possible to draw up a balance sheet or energy budget in a very simple form as $C = P + R + U + F$.

Where, C = Consumption

P = for growth

R = for metabolism

U = for non-faecal excretion.

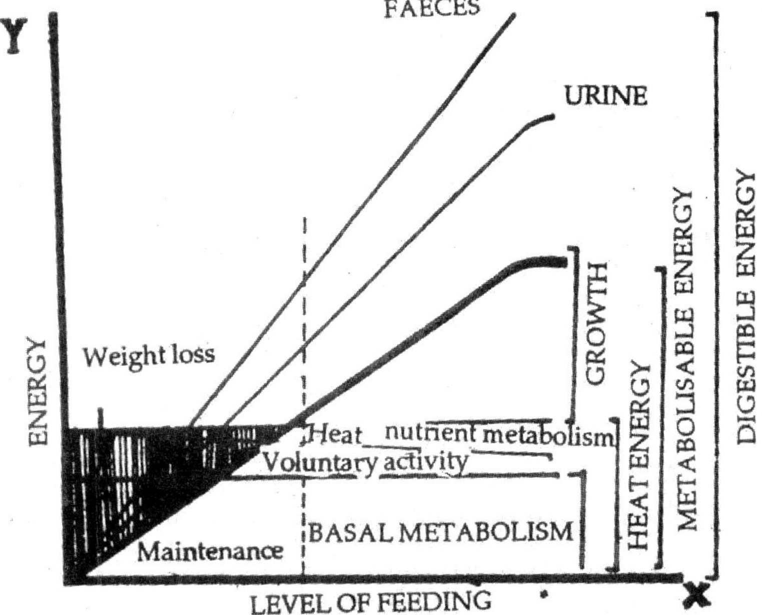

Fig. 48. Distribution of food intake gross energy in a growing fish at various levels of feeding

By understanding the environmental and biological factors that affect metabolic rate, it is possible to reduce these energy costs to a minimum and to optimise food conversion. Many factors that influence the metabolic rate and consequently energy requirements are broadly abiotic and biotic. The abiotic factors include temperature, salinity, oxygen, carbon dioxide, ammonia, pH, photoperiod, season and pressure. The biotic factors are activity, weight, sex, age, schooling, starvation and diet composition.

3.1 Energy value of feeds for fish

In nutrition, it is often necessary to know the calorific value of the feed, faeces and flesh. These values are necessary to compute the energy budget and to determine the assimilation and conversion efficiency. Generally the energy content is expressed in terms of calorie (cal) or kilogram calorie (k cal). The usefulness of calories is due to the fact that all forms of energy can be converted into calories, while they can not be wholly transfered into any other forms of energy. However, in the international unit system, instead of calorie, the Joule (J) a unit of mechanical energy has been introduced. The conversion of mechanical energy and heat energy is termed as Joule's law in which the equivalent of one calorie in units of mechanical energy is 4.187 Joules. A list of energetic equivalents and conversion factors are given below.

1 kilo calories (K cal. C)

= 1000 gm calorie

1 k cal. = 3.968 British thermal unit (BTU)

1 k cal. = 4187 Joules-at 60°F

1 Joule = 2.388 X 10^{-4} cal.

1 Btu = 0.252 k cal. at 15°C.

3.2 Conversion factor for determining energy content of feed

The major components of feed constitutes proteins, lipids and carbohydrates. Crude protein content of a food material is assumed by determining the nitrogen content of feed. As it is assumed that all crude proteins contain nitrogen 16% by weight, so the conversion factor used in 6.25 (100/16). This is not always so. Therefore, check need to be made for percentage nitrogen content before using the factor. For a few biological materials, the factors for converting nitrogen to crude protein is given herewith.

Cottom seed meal	18.86% N	Conversion factor	5.30
Soyabean meal	17.51%	-do-	5.71
Barley, Oats and Wheat	17.15%	-do-	5.83
Maize, Egg and meat Milk	16.00%	-do-	6.25
Milk	15.68%	-do-	6.38
Fish protein (ogino)	13.85%	-do-	7.22

However, in case of lipid, carbohydrates and protein if their quantitative values are known, calorific value can be calculated by putting appropriate calories conversion factors. This is called as component analysis of energy content. The calorie conversion factor for protein is 5.65 k cal/gm and for lipid is 9.41 k cal/gm and 4.1 k cal/gm of carbohydrates. Although component analysis for determining energy content of feed posed some constraints still it is advantageous because in nutrition, the proximate composition of the feed is always worked out. The other method for calculation of energy content in feed are wet oxidation, thermochemical method and Bomb calorimetry.

3.3 Calorie versus protein, as unit of measurement in nutritional bioenergetics

Although in animal kingdom, the study of energetics and energy is expressed in terms of calories or Joules, still for aquatic organisms partitioning (energy budget) based on protein as nitrogen units is of more suitable over energy units. This is because :

(i) Fishes are poikilothermic, hence use less or no energy for regulation of their body temperature.

(ii) Since shellfishes are bottom dwellers for much of the time, they need not to spend any energy for locomation and position, in water column.

(iii) Shellfishes need not actively maintain ventillation of gills for the purpose of respiration, hence relatively less energy is required.

(iv) In fishes and shellfishes, protein is preferred over carbohydrate as dietary energy source that serve as nutrient for growth.

(v) Flesh produced through protein assimilation is essential. That too lipids and carbohydrates is stored as glycogen and fat.

(vi) Unlike other land animal, the end product of nitrogenous metabolism in aquatic organism is in the form of Ammonia which get eliminated through passive diffusion. Thus energy need not to be spend in converting the toxic Ammonia into urea and uric acid.

Therefore, in feeds, the per cent of protein content is relatively essential than the total energy content which may be due to lipids or carbohydrates. In view of protein, the formulae and indices used in nutritional energetics are follows :

3.4 Formulae and indices used

1. Assimilation of protein

$$\alpha = \frac{\text{Protein consumed }(g) - \text{Faecal protein }(g)}{\text{Protein consumed }(g)} \times 100$$

(Assimilation efficiency of Protein)

2. Nitrogen balance (NB)

= N consumed – (N in faeces + N, excreted)

NB is measured in terms of mgN/100 g body wt/day.

3. Protein efficiency ratio (PER)

$$= \frac{gain \quad live \quad weight \quad (g)}{protein \quad consumed}$$

4. Protein conversion ratio (PCR)

$$= \frac{\text{Pr}\ otein \quad gained \quad (g)}{\text{Pr}\ otein \quad consumed}$$

5. Net protein retention

$$NPR \quad = \frac{loss \quad of \quad PGF \quad (g)}{protein \quad consumed} weight \quad gain \quad of \quad TPG \quad (g) + weight$$

TPG = group fed with test protein

PFG = group fed with protein free diet.

6. Productive protein value (%)

$$PPV \quad (\%) = \frac{gain \quad in\ body \quad protein \quad (g)}{protein \quad assimilate \quad d\ (g)} \times 100$$

7. Meat produced in assimilated protein (PAP %)

$$= \frac{Live \quad weight \quad gained \quad (g)}{protein \quad assimilati \quad on\ (g)} \times 100$$

8. Protein produced in assimilated protein (PAP %)

$$= \frac{\text{Pr}\ otein \quad gained \quad (g)}{\text{Pr}\ otein \quad assimilate \quad d\ (g)} \times 100$$

9. Gross protein value (GPV)

$$= \frac{A}{Ao} \times 100$$

A = (weight gain in group 2 – that of Gr1) + wt. gain of Gr2

Ao = (wt. gain in Gr3 - that of Gr1) + wt. gain of Gr. 3

Diet group = Gr. I - fed with basal diet

Gr. 2 - fed with basal diet + Cg of test protein

Gr 3 - fed with basal diet + Cg of casein.

Basal diet will have optimal crude protein.

10. Apparent biological value

$$= \frac{N\ consumed - (faecal\ N + Urinary\ N)}{N\ consumed - faecal\ N} \times 100$$

11. Biological value BV (%)

$$= \frac{N\ consumed - (faecal\ N - MFN) + Urinary\ N - EUN}{N\ consumed - (fecal\ N - MFN)} \times 100$$

MFN = Metabolic faecal nitrogen is that quality of N excreted in the faeces when the animal is fed with nitrogen free diet.

EUN = Endogenous urinary nitrogen is that quantity of nitrogen excreted through gills and as urine when the animal is fed with nitrogen free diet.

12. Daily protein requirement (% live wt/day) =

$$\frac{Optimal\ dieatry\ protein\ requirement\ \% \times Consumption\ of\ feed\ \%\ live\ wt.\ per\ day}{100}$$

13. Protein required for wt. gain (g/kg live wt.) =

optimal dietary protein requirement %
x Food consumed /wt. gained in dry basis x 10

14. Chemical score = Here essential Amino acid content of the source is consumed with that of a standard protein. The standard protein by the nutritionists is hens egg white. The chemical score is calculated as follows.

Suppose the Tryptophan in egg white = 1.7% and Tryptophan in sardine 1.2%. Then chemical score (cs)

$$\frac{1.2}{1.7} \times 100 = 70.59\ \%$$

15. Essential Amino Acid Index (EAA - I)

In this case, all 10 EAA are taken into consideration. It would be defined as the geometric mean of egg ratios of these acids. This can be expressed as

$$EAA - I = n\sqrt{\frac{100\ a}{ae} \times \frac{100\ b}{be} \times \frac{100\ c}{ce}, \ldots \frac{100\ j}{je}}$$

a,b. . . J = Per cent of EAA in the protein source.

ae, be. . . .Je = Per cent of EAA in the egg albumin.

n is the number of EAA entering into calculation. EAA-I has the advantage of predicting the effect of supplementation in combination of proteins.

3.5 Optimal dietary protein requirement

This can be calculated, when the actual values of indices like protein efficiency ratio (PER), protein conversion ratio (PCR), net protein retention (NPR), productive protein value (PPV), meat produced in assimilation (MPA), protein produced in assimilated protein (PAP) are plotted against percent protein in diet, whereby optimal requirement is ascertained. However, it has been observed that at high percent of protein level, feed consumption show a rise although assimilation for all nutrients is low. Thus the animal shows a super fluous feeding which is called as ``gluten-effect".

3.6 Daily protein requirement

In fishes the daily protein requirement ranges from 0.75–5.25 in terms of percent of body wt/day due to species variations in their size, age, temperature of water etc. However, Tacon and Cowey (1985) reported a linear relationship between daily protein requirement and specific growth rate for which it is evident that optimal dietary protein requirement and daily protein requirement are not related factor. The optimum dietary protein requirement is related to concentration versus activity or in other words quantity required for optimum rate of digestion and assimilation, while daily protein requirement is related to the inherent capacity of the animal to grow in synchronisation of rate of protein synthesis. Such synchronisation in speed of protein synthesis is influenced by the Eco-physiological state of the recipient.

3.7 Energy requirement/Budget of energy in fish

Food serves two purposes

(1) Maintenance and (2) Production. Maintenance sustains the organism's body utilising the main carbohydrates and lipids. Production or growth requires protein. Growth in finfish and shellfish includes : (1) tissue additon, (2) tissue repair, (3) exuvia (moulting etc.), (4) exudates (mucus) and (5) Gonadial products. However, to increase its weight the body requires a surplus of food that undergoes digestibility in organism's body mechanism to convert it into energy above that needed for maintaining itself. Therefore, digestibility, involves total quantity of nutrients consumed (ingested) and the amount of corresponding nutrients egested. Digestibility as knwon as Assimilation efficiency.

3.8 Formulae and indices considering the total nutrients of feed

1. Digestibility % (Assimilation efficiency)

$$= \frac{(C - F) \times 100}{C} = \frac{A}{C} \times 100$$

2. True digestibility %

$$= \frac{C - (F - MEN)}{C} \times 100$$

3. Food conversion ratio (FCR) = Food offered / wt. gain

4. Gross conversion (K_1) %

$$= \frac{P}{C} \times 100$$

5. Net conversion efficiency (K_2) %

$$= \frac{P}{A} \times 100 \quad or \quad A = (C - F)$$

6. Trophic coefficient

$$= \frac{C}{P}$$

7. Partial growth efficiency (%)

$$= \frac{P}{C - m} \times 100$$

m = maintenance ratio

8. Body weight gain (%)

$$= \frac{Wt - Wo}{Wo} \times 100$$

Wo and Wt are live weight at the time of starting the, experiments and at the end of the experiment for that duration of days for the size used.

9. Nutrient retention (%)

$$= \frac{Nutrient \quad gained}{Nutrient \quad consumed} \times 100$$

10. Optimum ratio of dietary energy (DE) to protein (P) (DE/P ratio)

$$= \frac{K \; cal \; / \; kg \; dry \; diet}{\% \; dietary \; protein}$$

11. Metabolisable energy (K cal/g of diet)

$$= \frac{Ae \; E - Ue \; (Ap \; . \; p - Gp)}{C}$$

Ae = apparent digestibility for energy %

E = total energy for the quantity of consumed food (k cal)

Ue = Energy loss of nitrogenous products per gram of protein deaminated (k cal/g protein)

Ap = apparent digestibility for protein (%)

P = quantity of protein in the consumed diet (g) % protein diet X feeding rate in % live wt ÷ 100

Gp = quantity of protein retained as growth (g) protein balance i.e. assimilated protein (non-faecal (exogenously excreted nitrogen) x 6.25)

C = weight of food consumed (g)

Note : Apparent digestibility

$$= \frac{Food \quad ingested \quad - food \quad egested}{food \quad ingested} \times 100$$

12. Specific growth rate (SGR) (% body wt per day)

$$= \frac{In \ wt \ - \ In \ Wo}{t} \times 100$$

`t' is the duration in days

13a. As in grown up organisms, the growth rate is constant

$\left[\frac{Log \ wt \ - \ Log \ Wo}{t}\right]$ is considered. In such case mean daily growth per day in percentage body weight (P) is calculated by following formula.

$$P' = \frac{2 \ (Wt \ - \ Wo \ )}{t \ (wt \ + \ Wo \ )} \times 100$$

13b. When growth changes in short intervals

$$P' = \left[\frac{1}{t10} (\log \ Wt \ - \ \log \ Wo \ ) - 1\right] \times 100$$

14. Average food consumption per day in percentage body wt.

$$C^{-} = \frac{2c}{(wt \ + \ Wo \ )t} \times 100$$

15. Loss due to mortality, the food consumption ratio (r) is used.

$$r = \frac{C}{(Wt \ + \ D) - (Wo \ )}$$

D = total wet weight of loss due to mortality in g.

C = total quantity of food consumed (g)

Wo = average initial weight (g)

Wt = average final weight (g)

3.8.1 Consumption (C)

Feeding rate of tropical fishes are more than temperate fishes. The tropical fishes incur more energy expenditure than temperate fishes. Consumption rate bound to increase with increase in temperature up to an optimum. Consumption is high in animals feeding on low nutritive substances. Thus larger values are met within detritus feeders and in animals which live on ooze.

3.8.2 Assimilation (A)

The portion of the nutrient not excreted as faeces out of the quantity consumed is known as assimilation. Other equivalent terms used are absorption and digestibility. Digestibility is true digestibility when metabolic nutrient excreted along with the faeces is deducted from the faecal nutrient in the calculation. It is apparent digestibility when no correction is made for release of metabolic nutrient into the faeces. Assimilation results in growth (P), plus metabolism process (R). So A = P + R. Usually animal protein assimilation range from 90-99%.

3.8.3. Metabolisible energy

It is calculated based on O_2 consumption studies. Respiratory quotient RQ and Oxycalorific quotient (Qox) are commonly understood.

$$RQ = \frac{Volume \quad of \ CO_2 \ produced}{Volume \quad of \ O_2 \ consumed}$$

$$Qox = \frac{calories \quad liberated}{unit \ O_2 \ utilised}$$

RQ for carbohydrate is 1.0, for fat 0.7, protein (in ammonotelic) is 0.9 and in (ureotelic) is 0.82. As the mean digestive values for protein, lipid and carbohydrates are 0.9 (90%), 0.85 (85%) and 0.40 (40%) respectively, the metabolisible energy for these components will be follows suggested by Brody (1945).

ME of protein (K cal/g) = (5.65 – 1.3) X 0.90 = 3.9

5.65 = calories value of each gram of protein

1.3 = protein energy lost in nitrogenous excretion as urea.

ME of lipid (K cal/g) = 9.45 X 0.85 = 8.0

ME of carbohydrates (K cal/g) = 4.10 X 0.40 = 1.6

3.8.4 Metabolism

Five major catagories of metabolism are observed in aquatic organisms like fishes. These are -

1. *Standard metabolism* (Rs) – metabolic rate at minimal maintenance or resting metabolism of an unfed fish below which physiological function would be affected.
2. *Routine metabolism* – metabolic rate at its normal activity.
3. *Maximum sustained metabolism (Rmax)* – maximum metabolic rate for sustained maximum activity.
4. *Active metabolism (Ra)* – metabolism related to swimming activity and to stress.
5. *Internal heat increment (Rf)* – Use of energy for feeding metabolism. It is also called as specific dynamic effect (SDA).

The difference of variation between maximum metabolic rate for sustained maximum activity (Rmax) and to that of standard metabolism (Rs) is knwon as metabolism scope (MS).

$$MS = (Rmax - Rs)$$

This metabolic scope varies with stages of development, environmental factors and species of fish.

Specific dynamic effect (SDA) - It is also known as heat increment and can be calculated in the following way.

$$SDA = M - (Ms + Me)$$

M = metabolic rate of just fed ones

Ms = metabolic rate of starved animals.

Me = Elevated metabolic rate due to feeding.

3.8.5 Urinary loss

Various nutritionists have indicated various per cent of urinary loss in different fishes in size, age under different environmental situations. Brafield (1985) calculated that 5.2 to 9.4% of energy intake is excreted as urinary loss. However, it is evident that urinary loss is about 7% of consumption.

3.8.6 Growth and energy budget

Based on the balance sheet of energy flow, Breet and Groves (1979) have suggested the following energy budget for young fishes.

$$\text{Carnivorous } 100C = 29\,P + 44\,R + 7\,U + 20F$$

$$\text{Herbivorous } 100C = 20P + 37R + 2\,U + 41\,F$$

Omnivorous (shrimp) 100 C =

14.32 P + 83 (R + U) + 2.2 F + 0.5 (moult)

It should be emphasied that, fish need to budget its energy. Whe metabolic demands are too high, it should meet it at the cost of growth and at times energy met even at the cost of feeding when it becomes the question of high heat increment and metabolism (Priede, 1985).

4. VARIETIES OF FISH FEEDS

Fish feeds are plants and animals origin.

4.1 Plant feeds

These are available either as grains or cakes. Their addition, could suffice to certain extent, the total nutrient content of the pelleted feed. Ricebran, Soyabean, Wheat and Maize are commonly used as grain feeds. Groundnut, mustard, til, coconut, cottonseed, linseed, rape seed, sunflower and bean cakes are often used in fish farming. Cotton seed cakes are commonly used in carp farming in USA and USSR, but not in China, though China is a good producer of cotton. However, in India, the usde of Groundnut oil cakes are commonly used in carp farming. These plant feeds provide plant protein source to the formulated fish feed. As one can not afford purified protein such as casein, egg albumin etc. becuase of high cost on commercial scale, a list of plant protein source with certain nutrient content is given in Table 6. The fatty acid composition in some of these plant oils are given in Table 7.

Table 6. Nutritional elements of various plant feeds (%)

Item	Crude protein (range)	ether extract (fat)	Crude fibre	Remarks
1. GNOC	25-30	6.3	13.0	Low in cystine. tryptophan, threonine, methionine and lysine. Rich in niacine and panthothenic acid
2. Coconut kernel meal	20	2.5-6.6	16	Protein low in lysine and histidine
3. Sunflower meal	50	3.1	12.2	Deficient in lysine

4. Cotton sea meal	44	5	12.8	Low in lysine and sulphur amino-acid viz., cystine and methionine
5. Rape seed meal	40	1.8	13.2	High in methionine, cystine and tryptophan
6. Lineseed meal	29.9	6.8	9.8	Low in methionine and Lysine, Rich in thiamine, riboflavin, nicotinamide, choline and panthothenic acid
7. Safflower	42.8	8.5	15.2	–
8. Soyabean cake	39.1	7.1	4.5	Deficient in methionine
9. Rice bran	10.8	12.0	8.2	–
10.Wheat bran	10.9	3.7	8.9	Prolamin type protein contain less lysine. Glutelin is rich in glumatic acid.
11. Barley	10.0	2.0	4.0	Deficient in methionine, histidine and tryptophan.
12. Maize	8.5	4.3	1.3	Protein is two types : (i) Zein-deficient in tryptophan and lysine. (ii) Maize glutelin-high in tryptophan and lysine, cryptoxanthine is found which is a precursor for Vit-A.

Table 7. Fatty acid composition of some plant oils.

Plant source	Saturated fatty acid	Monoun saturated	18 : 2W6	18 : 3W3
Cotton seed oil	27	22	52	0.5
Soybean oil	16	24	52	8
Safflower oil	11	26	68	1.5
Sunflower oil	10	43	44	1.0
Linseed oil	14	23	38	23
Corn oil	14	25	53	3
GNOC oil	19	40	36	3
Mustard oil	42	19	16	20
Coconut oil	83	8	4	–
Rape seed oil	56	15	16	7
Gingly	17	35	47	1

4.1.1 Green fodder

Green fodder includes aquatic plants and terrestrial grasses. These are mainly used as feed for grass carp, wuchang fish and some times for common carp, Crucian carp and tilapia. The main aquatic plants used include, Hydrilla, Vallisneria, Chara, Lemna, Spirodella, Potamogeton, *Eichhornia crassipes* (water hyacinth) and pistia. The terrestrial plants include. Elephant grass, sudangrass, rye grass, Hybrid napier, gammy grass and various leaves and vines from vegetable crops. These green fodders may be chopped into pieces and used directly as food by certain table size fishes. The other way of using aquatic weeds like water lettuce (*Pistia stratiotes),* water hyacinth (*Eichhornia crassipes*) and water peanut must be minced and mixed with rice bran and yeast as a compounded diet. As certain aquatic plants like water peanut (Alligator weed) contain Saponin, it is advisable to add table salt as an additive to eliminate the toxic effect while preparing compounded diets with such weeds. For fry and fingerlings rearing ponds, the use of aquatic plants in a form of silage or paste with appropriate size is of immense usefulness. The fry swallow the mesophyll cells in the paste, which is similar to the size of planktons. The unutilised material of the paste serves as manure after biodegradation, through benthic microorganisms. Nutritional elements especially the crude protein in some of the green fodders are given in Table 8. Aquatic weeds in their fresh state contain as much as 95% of moisture. So for compound feed preparation, the moisture content is to be reduced by sundrying followed by hot air exposure to yield a stable product. The aquatic plants usually contain 10-27% crude protein on dry matter basis. That too the rich nutritional value of some of the aquatic plant is evident from the mineral composition of weeds (Boyd, 1970), which is given in Table 9.

Table 8. Crude protein of some green fodders (%)

Plant	Crude protein % in wet basis	Plant	Crude protein % in dry basis
Wolffia	1.25	Hydrilla	17.69
Lemna	1.43	Vallisneria	16.06
vallisneria	0.61	Ottelia	14.81
Potamogeton	2.11	Najas	18.46
Pistia	1.20	Pistia	11.00

Eichhornia	1.0	Myriophyllum	13.50
Water Peanut	3.22	Salvania	11.64
Lactua tentaculata	3.02	Spirodella	14.12
Rye grass (*Lolium perenne)*	4.17	Lemna	20.81
Sudan grass (*Sorghum sudanense)*	1.78	Azolla	20.56
Romain lettuce	3.82	Ipomoea	26.56
Bunch grass	2.93		

Source IFF, China 1987 Source : Boyd; 1970

Table 9. Mineral composition of some aquatic plants (Boyd, 1970)

Plants	P	S	Ca	Mg	K
Waterhyacinth	.43	.33	1.0	1.1	4.4
Pistia	0.3	.55	2.40	1.0	3.5
Hydrilla	0.28	0.39	4.50	0.9	2.9
Ceratophyllum	.26	0.30	0.77	0.42	4.01
Typha	.14	0.15	0.76	0.15	2.65

4.2 Animal feeds

Because of the presence of methionine and lysine in animal protein which lack in plant protein, the animal feed have a superior nutritional value than plant feeds. Animal feeds are rich in protein and essential aminoacids. Common animal feeds include fish meal, trash fish, silk worm pupae, blood meal, slaughter wastes, animal byproducts, kitchen wastes, shellfish wastes, earth worm etc.

Table 10. Nutritional elements of various animal (feed (%)

Item	Crude protein	Ether extract	Ash
Fish meal	55-75	6.5-12	12-25
Crab meal	35-40	2.1	40
Shrimp meal	40-45	5-7	25-30
Blood meal	85-90	1.3-1.5	7-8
Poultry waste			
(i) Meal with viscera, feet, head	62	14	16
(ii) Feathers hydrolysed	85-90	3.0	3.5
Slaughter house meal	54	10.4	32

In Japan and China, Silk worm pupae are commonly used as animal feed to fish. But in India, the animal protein source to compounded feed of fish is solely through fish meal. However, the taste of fish freshly fed with silk worm pupae is unpleasant which is unlike to the taste of fish fed with fish meal. The nutritional elements of various animal feeds are given in Table 10. The fatty acid composition in some of the animal source is given in Table 11.

Table 11. Fatty acid composition of some animal source.

Animal source	Sardine	Codliver	Shark	Prawn head
Saturated	38.8	19	21	33
Monounsaturated	28.0	47	43	23
18 : 2W3	2	3	3	2-9
18 : 3W3	9	1	2	1
20 : 5W3	8.3	10.4	3	6
226W3	10.7	12.5	11	15

Source : Shellfish and finfish nutrition summer course CMFRI. 1987.

4.3 Antinutritional factors in feed ingradients

Although many feed ingradient sources are cited, still one of the important criteria in selecting feed ingradients for manufacturing complete and supplemental feeds relate to the presence of antinutritional factors. Because, this can significantly reduce the nutritional value of the feeds. These anti-nutritive substances are often referred to as `toxic factors' because of detrimental effect they produce when consumed by animals. However, most of these substances produces sublethal effects such as reduced growth, poor feed conversion, hormonal changes and ocassional organ damage.

Antinutritional substances are broadly of 3 groups

I. That occurred in natural feed stuffs by normal metabolism of the species from which the material originates.

II. Natural contamination like bacteria, mould, fungus etc. leading to the production of microbial toxins.

III. Artificial antagonists such as preservatives, chemical additives, pesticides, herbicides and heavy metals.

4.3.1 Effects

In group I, the effects are :

(a) depressing digestion, (b) fluctuates metabolic utilisation like minerals, (c) vitamins. Depression of digestion is due to inactivation of (i) protease enzymes, trypsin and chymotrypsin. Growth depressing effect was due to the slow release of methionine by proteolytic enzymes in the presence of trypsin inhibitors, (ii) Haemagglutinins (Lectins) inactivated by pepsin, so that haemogglutinating fraction appears to be inactivated before feed enters in to intestine for assimilation. (iii) Saponin leads to haemolysis. Alfalfa saponin inhibit succinate oxidation and digestive enzyme secretion. (iv) Polyphenolic compounds (Tannins) - Affects protein digestibility.

(b) The effects of interferring with the utilisation of mineral nutrients are due to presence of phytic acid, oxalic acid, Glucosinolates, Gossypol pigments. These interferes in utilisation of Ca^{++} and other mineral nutrients.

(c) The effect of inactivating or increasing the requirement of certain vitamins are due to the presence of Antithiamine, Pyridoxin antagonist, Antivitamin B_{12}, Antivitamin A, D, E. Antinicotinic acid, Alkaloids, Cyanogens, Algal and other marine toxins. Toxic algal forms are Gonyaulax and Gymnodinium.

In Group II, the effect are due to the contamination with *Salmonella,* Pathogenic yeast (Candida species) and fungus *Aspergillus* species. Mould growth can result in the production of mycotoxins, which can cause a wide range of pathological and physiological effects in fish. Mycotoxins are well documented with four important groups such as (a) Aflatoxins (2) Ochratoxin-A (3) Zearalenone (4) and the Tricothecenes. Aflatoxins are mainly four compounds named B_1, B_2, G_1, and G_2. Aflatoxins are potent liver toxin and carcinogen with afltoxin B_1 being the most toxic compound. Aflatoxin have cumulative effect to produce carcinogenic in the presence of gossypol and also cyclopropionic fatty acids.

Ochratoxin produced by *Aspergillus* and *Penicillum* decreased growth and fecundation. It affects the proximal kindey causing nephropathy. Zearalenone and Trichothecenes are produced by *Fusarium* species. They affect reproductive system. Because of toxins, feed refusal have been observed in fish.

In Group III, the most common contaminants are organochloride pesticides, DDT, DDE, dieldrin, endrin and industrial chemicals such are polychlorinated biphenyls (PCB); Phthalate esters and hexachlorobenzene (HCB) and heavy metals. The pesticidal toxicity results in dysplasia or sterility, loss of appetite and death being reported. Pesticides and herbicides tend to bioaccumulate or bioconcentrate (NRC, 1981).

4.3.2 Remedial measures

Identified materials for feed preparation therefore should be from non-contaminated areas. Autoclaving or heat processing, extrusion cooking and infrared cooking or micronization is effective. The effect of polyphenolic compounds of sunflower seed meal can be counter acted by methionine and choline. In oil cakes; the aflatoxin can be destroyed by ammoniation and monomethylamine treatment. Good storing of food stuffs under low humidity and temperature can control the production of mycotoxins by fungus.

Some of the commonly used feed ingredients contain antinutrients, which should be taken care of during preparation of fish feed.

Antinutrients present in fish feed ingredients

Antineutrients	Ingredient
Protease inhibitor	Soyabean, wheat, sunflower, maize
Arginase inhibitor	Sunflower
Amylase inhibitor	Wheat
Tannins	Sunflower
Gossypol	Cottonseed
Cyanogens	Cassava
Cyclopropionic acid	Cottonseed
Phytic acid	Soyabean, Wheat, Sesames, Cottonseed, Maize, R.B, GNOC
Antibiotin (Avidin) factor	Raw egg white
Antithiamine factor	Raw fish, shell fish, rice polish
Antipyrodoxin factor	Linseed
Anti-vitamin A, E, D and B12	Soyabean
Oxalic acid	Sesames
Mycotoxins (fungal toxins)	Ground nut, Maize

Heavy metals: arsenic	Poultry waste
Cadminium	Shellfish
Copper	Brewery byproducts
Zinc	Shellfish, hydrolysed feather, poultry byproducts

5. FORMULATED FEED

Rearing of spawn, fry and fingerlings until they become stockable size and their subsequent culture in grow out ponds require appropriate and nutritionally balanced diet for enhancing production. This have been one of the essential requisites in the development of Aquaculture.

5.1 Advantages

1. Proper formulated feeds are the replica or mirror image of exact nutritional requirements of fish. Therefore by understanding the nutritionally well balanced feeds which could be formulated using low cost feed stuffs available locally.
2. Ingradients of formulated feeds can complement one another and raise the food utilisation rate.
3. Proteins can supplement one another so as to satisfactorily improve most of the essential aminoacid content of the feed thereby raising the protein utilisation.
4. Large quantities of feeds can be prepared at a time with good shelflife as convenient to preserve, which can be used at the time of requirements.
5. Feed ingradient sources can be broadened with preferred and less preferred ingradients along with additive like antibiotics and drugs to control fish diseases.
6. High efficiency of feed can be achieved by Judicious manipulation of feed ingradients and can be made commercially feasible.
7. By adding a binding agent to produce pelleted feeds, the leaching of nutrients in water is diminished and wastage is reduced.
8. Dispensing over large farm areas is quite possible as formulated feeds are convenient to transport. These are

suitable for automatic feeding, for which automatic feed dispensing devices could be successfully employed.

5.2 Types of compounded feed

Commonly compounded diets are of two types

1. Purified diets for studying requirements of animals and are generally formulated using purified ingradients.
2. Practical feeds formulated by using natural ingradients available locally.

5.3 Factors considered for preparation of compounded feed

1. Selection and evaluation of raw materials which includes identification of feed ingradients available locally in large quantities for commercial production of compounded feed. That too, ingradients or materials should have consistent quality and of low cost. Feed stuffs which are primarily ment for human consumption should be avoided. The nutritional qualities are to be assessed. If the ingradient is a protein source, then the quality of protein has to be assessed by analysing aminoacid composition. The biological evaluation of the feed stuff should be carried out.
2. Physical design of the feed should be synchronising with the feeding habits of the cultured species. For example finfish generally graze the feed. So the convenient way of presenting the feed to them is in the form of wet dough. Certain fish can freely feed upon floating pellets and flaked feeds. In the case of prawns, the larvae are filter feeders and requires microparticulate or microencapsulated feeds with good suspension quality in water column. Microencapsulated diets have been tested in the laboratory for fin fishes and shell fishes.

5.4 Feed identification

Feed stuffs are classified into various catagories. Probably Hardy (1980) has classified the feed ingradients as : (1) Dry forages and roughages that have 18% crude fibre, (2) Pasture or fodders, (3) Silages, (4) Energy feed, (5) Protein feed, (6) Mineral supplements, (7) Vitamins and (8) Additives (antibiotics, hormones, medicines, coloring pigments and antioxidant). Some have simplified it as : (1) Roughages with more than 18% crude fibre, (2) Energy food, (3) Protein supplemtns and (4) Non-conventional feed such as spirulina, krill, yeast, feather meal. However, for feed preparation, the most simplified and well accepted category of feedstuffs classification is on the basis of major classes of nutrient they supply. On this basis feed stuffs are classified as : (1) Basal feed or Energy feed and (2) Protein supplements. Energy feeds are ingradients with less than 20% protein and also less than 18% crude fibre. Protein supplements are feed stuffs with more than 20% crude protein.

5.4.1 Energy feed

Usually Rice bran, wheat bran, tapioca, barley, corn, sorghum, maize, fish oil, vegetable oil and animal fats etc. are included.

5.4.2 Protein supplements

The plant protein supplements include GNOC, cotton seed cake, linseed, gingelly, coconut, rape seed cake, sunflower and soyabean cakes. The animal protein supplements include meat meal, blood meal, fish meal, slaughter house waste, prawn waste, mantis shrimp (*Oryctolagus nepa*), Silk worm pupae etc.

5.4.3. Binders

In order to avoid disintegration or dissolves away of formulated feed in the water, suitable binding agents like tapioca are added. This will not only ensure the compactness of the feed in water but also avoid pollutional problem. The common binding agents used in feed formulations are of two groups such as (1) chemical binders and (2) natural binders. The chemical binders include Agar agar, carboxy methyl cellulose, Geletin, guar gum, polyvinyl alcohol, sodium alginate starch etc. The natural binders include wheat flour, rice flour, tapioca etc.

Criteria for selection of binder

1. Binders used for feed should have good water stability for a minimum period of 3 hours.
2. The pellets when added to water, absorb water quickly and become soft. The feed remain in compact by retaining the shape at least for six hours is convenient.

3. Tapioca not only serves as a source of carbohydrate in the feed but also acts as a binder.
4. The binder to be used in feed should be inexpensive.

5.5 Formulation of feed

As no single feed material is a complete feed by itself, it is always advisable to have multi-ingradient feed formula so as to satisfy the nutritional requirement of fish to be cultured. That too, the quantitative aminoacid requirements of carps have not been worked out for which gross protein requirements has to be satisfied first. In order to initially balance the basal feed and protein supplement, the help of square method has to be taken. However, least-cost approach method for balancing the nutrients in feed is also in use. In such method linear programming includes digestibility coefficient and other nutrient composition of ingradients along with constraints which are fed to the computer in the proper form. However, square method is easier and convenient in the present state for dissemination among the rural farmers.

5.5.1 Square method

The procedure follow includes :

1. Placed in the centre of the square the percentage of crude protein needed in the feed ration.
2. Place in the upper left hand corner of the square the percentage of crude protein in the basal feed stuff or a mixture of feed stuffs.
3. Place in the lower left hand corner of the square the percentage of crude protein in the protein supplement or a mixture of protein supplements.
4. Connect the diagonal corners of the square with lines and substract diagonally across the sqaure the smaller figures from the larger. Place the answers in the opposite corner.

Example I

Suppose we have GNOC with 45% protein and wheat bran with 15% protein. We want to make a feed that contain 30% protein. Now add the figures on the right hand side of the square 15 + 15 = 30. To make the feed with 30% protein we must mix

Wheat bean $\frac{15}{30} \times 100 = 50\ \%$

GNOC $\frac{15}{30} \times 100 = 50\ \%$

Example II

To prepare a feed with 28% protein by using six ingradients such as

Fish meal	60% protein
Prawn waste	35%
Squilla (mantis shrimp)	45%
GNOC	45%
Wheat bran	15%
Tapioca	2%

Then these ingradients are grouped into basal feeds with less than 20% crude protein and protein supplements with more than 20% crude protein. The protein content in each group is averaged and plugged into the square as follows :

(1) Basal feed :	1. Tapioca	2.0%
	2. Wheat bran	15%
	Total	17%
	Average	8.5%
(2) Protein supplements	1. Fish meal	60%
	2. Prawn waste	35%
	3. Squilla	45%
	4. GNOC	45%
	Total	185%
	Average	46.25%

Basal feeds 8.5 18.25

28

Protein supplements 46.25 19.50

Total = 18.25 + 19.50 = 37.75

Now the composition will be

Basal feed = $\frac{18.25}{37.75} \times 100 = 48.3$

Protein supplement $\frac{19.50}{37.75} \times 100 = 51.7$

The final feed formula is :

(1) Wheat bran $= \frac{48.3}{2} = 24.15\ \%$

(2) Tapioca $= \frac{48.3}{2} = 24.15$

(3) Fish meal $= \frac{51.7}{4} = 12.92$

(4) Prawn waste $= \frac{51.7}{4} = 12.92\ \%$

(5) Squilla $= \frac{51.7}{4} = 12.92\ \%$

(6) GNOC $= \frac{51.7}{4} = 12.92\ \%$

Total = 100

6. DIET PROCESSING

The quality of the raw material should be checked before processing. Diet processing involves the steps of (1) Grinding (2) Mixing and homogenising (3) Steaming (4) Pelleting and (5) Drying.

6.1 Grinding

1. It is done to obtain homogenous mixture of specific particle size.
2. Improves the digestibility and pelletability.
3. Improves good water stability of the pellete.

6.2 Mixing and homogenising

1. It helps to avoid selective feeding of a particular ingradients and also achieve good pelletability.

2. If liquid ingradients are to be added, these are added in this stage.
3. If vitamins are to be added, these are added after the heat treatment step is completed as vitamins are sensitive to heat.
4. Binders like tapioca can be mixed along with other ingradients. If chemical binders are to be used, they must be dissolved/melted in cold or hot water and the solution is then added to the feed mixture.

6.3. Steaming

1. It helps in the killing of bacteria and other pathogens if any in the feed. However, cooking at higher temperature and for longer period has to be avoided to destroy many important nutrients.
2. Steaming improves digestibility.

6.4 Pelleting

1. Feed is pelletised by pressing the material through dies with different pore sizes. Pelletes with acceptable diameters by fish are suitable.
2. Pelletisation can be accomplished by compression for hard pellets, or by extrusion and adhesion for non-compact and floating pellets, With reference to the digestibility, the floating pellets are found to be higher than the hard and compact pellets. But floating pellets prepration is expensive for which compact pellets are widely used in carp culture practices.

6.5 Drying

1. The feed pellets should be dried to a moisture below 10%, otherwise the shelf life of the feed will be poor. Sometimes, hot air is blown to remove excess moisture in the feeds.

6.6 Storing

1. In order to maintain the quality of the prepared feed to a longer period, storing is necessary, In airtight containers, storing is done. Storage facility should be clean, hygenic and prevent insect infestation. In order to control the

mould, mould inhibitors are used. Mould produce mycotoxins or aflatoxins for which the common mould inhibitors used during storage are

(1) Sodium benzoate - - - - - - - - - - - - - - 0.1%
(2) Propyl-para-hydroxy-benzoate - - - 0.1%
(3) Methyl-para-hydroxy-benzoate - - - 0.1%
(4) Sorbic acid - - - - - - - - - - - - - - - - - - - No limit
(5) Calcium sarbate - - - - - - - - - - - - - - No limit
(6) Potassium sarbate - - - - - - - - - - - - - No limit
(7) Sodium sarbate - - - - - - - - - - - - - No limit
(8) Propionic acid - - - - - - - - - - - - - - - No limit
(9) Calcium propionate - - - - - - - - - - - - - No limit
(10) Sodium propionate - - - - - - - - - - - - - No limit

Irradiation by U.V. light is also done for mould control and as antifungal growth. Similarly, lipid oxidation can be controlled by antioxidants. These include

(1) Butylated hydroxyanisole (BHA) - - - -0.02%	
(2) Butylated hydroxytolune (BHT) - - - - -0.02%	
(3) Citric acid	No limit
(4) Ascorbic acid	No limit
(5) Lecithin	No limit
(6) Thio dipropionic acid	No limit
(7) Santoquin	No limit
(8) Ethoxyquin	No limit
(9) Propylgallate	No limit
(10)Tocopherol (vit-E)	—

Other antioxidants are Gallic acid, phosphoric acid etc., which are generally not used because it has some other harmful effect on the diet and the consumer.

7. MANAGEMENT OF FEEDING

Proper management of feeding in aquaculture practice is important for resulting in maxium yield, feed utilisation efficiency and reduce the waste in feed as well as the cost incured for feed to certain extent. The management of feeding involves the feeding rate as well as the freqeuncy of feeding at fixed place and fixed time. These feeding rate and feeding frequency vary with the sex, size of fish, water temperature and dietary energy level in the feed (Chiu,1989). Usually the feeding rate is adjusted either at a given precent of body weight or feeding to satiation. The former feeding rate is very common and prevalent. Feeding frequency is also positively related to the growth of fish. Fish either at short food chain at low trophic niche or at the higher feeding regime naturally grow faster although there is a maximum ingestive limit at which the increase in growth is negligible. This is defined as the optimal feeding frequency which differs from size of fish, sex, gut morphology of the species and mean size of the artificial feed.

The feeding management concept of fixed quantity and quality is to be oriented as the daily food consumption in fish is variable (De Silva *et al.,* 1986). Such daily variation in feed intake is apparent to influence on the digestibility (De Silva and Perera, 1984). Hence the management of feeding schedule should match with the diurnal variations of digestibility of the fish for proper feed utilisation and assimilation efficiency. Therefore, mixed dietary regimes of low and high protein in feeding can provide a means of reducing feed costs (De Silva, 1989) and marginal cost of fish yield.

7.1 Probiotics

Probiotics in general are described as a single or mixed cultures of selected strains of bacteria that has wide beneficial effects. The increasing trend of using probiotic feed additive in aquaculture is due to indiscriminate treatment of diseases by antibiotics, chemotherapeutants and their harmful residual effects. The term probiotic is defined by Fuller (1986) as a live beneficial microbial feed supplement which would help in their colonization as well as proliferation in the gut of the host. This greatly prevents the colonization of pathogenic organisms and produce organic and hydrogen peroxide (H_2O_2) that inhibits the growth of pathogenic bacteria in the gut of the host.

The most commonly used microbes include lactic acid bacteria (LAB) such as *Lactobacillus acidophilus, Streptococcus faecium, pediococcus* spp., selected strains of bacillus spp. and certain strains of yeast belonging to *Saccharomyces* spp. Nitrobacter, pseudomonas, Enterobacter, cellulomonas, rhodopseudomonas and photosynthetic sulphur bacteria are some of the other species that are in use (Boyd, 1990). Some latic acid bacteria such as Carnobacterium divergens and lactobacillus species are antagonistic to fish pathogen (Byun *et al.,* 1997). The production of immunoglobulins are activated by the use of probiotics. Anderson (1992) reported that their cell walls serve as excellent source for the derivation of immunostimulants (peptidoglycans, muramyl dipeptide, glucans and lypopolysaccharides.

Probiotics are classified in two categories on the mode of their application.

1. feed probiotic
2. water probiotic or bioremediations.

Feed probiotics are applied through pellet feed while water probiotics are applied directly into water.

The advantages of probiotic organisms through bioencapsulated feeding technique are:

1. To stimulate non-specific defence mechanism in the host to protect it against pathogens.
2. To produce specific compounds like bacteriocins that inhibit pathogens (Lewus *et al.,* 1991)
3. To exhibit anti cancer effects (Fernandes and Sahani, 1990)
4. Effect on growth, survival and immune status of white prawn (Uma, 1995)
5. Increase food conversion efficiency and weight gain

Uma (1999) describe about the general information on the mode of action and advantage of probiotic microbes for sustainable aquaculture. Role of probiotic in aquaculture has great potential in year to come. Babu *et al.,* 2005 reported on the efficacy of probiotic bacteria amendments over chemotherapy in shrimp culture ponds. The use of "probiotic" is gaining importance in recent years as an ecofriendly disease management tool for farming in the wake of adverse effects due to the indiscriminate use of antibiotics in the aquatic environment. Probiotics work on the principle of competitive exclusion which is a ecological process that can beneficially manipulate the microbial composition of gut of the host and environment.

In recent paper Bairagi *et al.*, 2004 reported that the addition of *Bacillus subtilis, B. circulans* in the diet of rohu that increased the growth, food conversion ratio and protein efficiency ratio. They attributed this to the extracellular cellulolytic and amylolytic enzyme production by the bacteria. In recent years feed additives and probiotics used in aquaculture sector under different trade names in India are

1. Aquamos 2. Nurtrimix 3. Nupro 4. Promarine 5. Ultimax 6. Minermate 7. Combax 8. Sporolac 9. Minerex 10. Colozin 11. Raafres AQ (Non antibiotic growth promoter). Soil and water probiotic powder namely 1. Bottom-Lact 2. Procon-PS are also in use.

The enzymes secreted by aquaculture probiotic bacteria have a very important role in the degradation of organic matter and thus act to significantly reduce sludge and slime formation. As a result, water quality is improved by reducing the bottom sediments, reducing disease incident.

7.1.1 Prebiotics

Prebiotics are basically feed for probiotics that are resistant to attack by endogenous enzymes and thus reach the site for proliferation of gut microflora. Examples of prebiotics used in animal feed is mannan oligosaccharides (MOS), fructo-oligosaccharides and mixed oligodextran. Mannan oligosaccharide (MOS) is obtained from the cell wall of yeast. Other source of MOS is the palm kernel meal. MOS interferes with the colonization of the pathogens. The concept of using prebiotics has not yet been accepted, but it has advantage to stand high pelletizing temperature in feed and also have a long shelf life.

7.1.2. Immunostimulators

Immunostimulators are natural or synthetic compound that are capable of enhancing the defence capacity of the animals. They act as barriers to infection caused by specific and non-specific pathogens. Such compounds could be used to protect the fishes and shrimps in the culture ponds from diseases. Health management in aquaculture through immunostimulants are reported (Anantharaja and Stephen, 2007).

7.1.3. Mechanism through which Immunostimulants Act

Following are some of the mechanisms of action.

1. Stimulators of T-Lymphocytes : Levamisole, Freud's complete adjuvant, glucans, muramyl dipeptide
2. Stimulators of B cells: Lypopolysaccharides
3. Inflammatory agents producing chemotaxis: silica and carbon particles.
4. Cell membrane modifiers: detergents, sodium dodecyl sulphate, quaternary ammonium compounds and saponins
5. Nutritional factor: Vitamin C and E
6. Cytokines: leuckotriene, interferon.

7.1.4. Types of immunostimulators

1. *Natural immunostimulators:* Examples - cell wall preparation, extract from tunicate and chitin
2. *Synthetic immunostimulators:* Examples - Levamisole and Quarternary Ammonium compound (QAC)

(a) Cell wall preparation

The cell wall of fungi contains

1. Muramyl peptides
2. Lipopolysaccharides
3. Lipoproteins
4. Acyloligo peptides
5. Specific bacteria peptides and
6. Glucans

Many types of glucans such as Krestin, Lentinan, Scleroglycan and schzophyllan are obtained. "Macrograd" is the commercial name of one of the glucans extracted from the fungi *Saccharomyces cerevisiae.* Injection of Macrogard improved the resistance of Atlantic salmon to the infection caused by *Vibrio spp.* Peptidoglycan and B-glucans have shown to have nonspecific immunostimulatory effects in several species including fish. Mannan oligosacharides (aqua-mos) interfere with pathogenic bacteria there by prevent initiation of infection

(b) Peptide compounds

Some of the immunoactive peptide compounds are FK 565, ISK and EF 203. FK 565 is a synthetic peptide synthesized from streptomyes *olivaceogriseus* as structural model. ISK a short chain polypeptide derived from fishery product and EF 203 prepared out of chicken egg are also in use.

(c) Levamisole is a synthetic phenylimidazothiazole has been show to have the ability to up regulate non-specific immune response in carp and in rainbow trout.

Therefore, immunomodulators have important role in disease management in aquaculture.

(d) Vitamins and other nutritional factor

It is observed that nutritional factor like vitamins and minerals help greatly in enchanting the immune status of the animals. Among vitamins Vitamin C (Ascorbic acid) and vitamin-E (Alpha-tocopherol) are identified as immunomodulators. Further vitamin-E along with selenium play a role in developing resistance against diseases.

(e) Other Immunostimulating Compound

"Eta" is a fatty acid like components extracted from tunicate. It is found to enhance resistance of American eel *Anguila rostrata* against *Aeromonas hydrophila.*

"Chitin", a polysaccharide compound obtained from crustacean shells, insect exoskeleton and cell walls of certain fungi serves as immunomodulator.

"QAC" and Levamisole are used as therapeutic agents to treat nematode in fishes.

(f) Marine Nutraceuticals

Nutraceuticals, the bioactive compounds have both nutritional and pharmaceutical effects. American Nutraceutical Association (ANA) defines nutraceuticals as the functional food with disease preventing ability and health promoting properties. These nutraceuticals impart antibacterial, antifungal and immunostimulation effects in fishes. These bioactive compounds are isolated from echinoderms, sponges, ascidians, molluscs and sea weeds.

Echinoderms such as Sea cucumber, Sea urchin contain fucoidan and fucon respectively which acts as an immunostimulant and potent antiviral agent. Beneficial effects of fucoidans are :

1. Acts as immunomodulators
2. Suppression of viral infection
3. Serum complement is activated
4. Acts as blocking agent for macrophage scavenger receptors
5. Fucoidans act as adjuvant and improve the action of immunostimulants
6. It can be applied for prophylaxis and therapy of bacterial diseases.

These bioactive compounds contain sulphated sugars such as ascophyllan and glycuronofucoglycan. Felix and Alan Brindo (2008) has given detail report on marine nutraceuticals in fish and shellfish nutrition.

Considering the positive aspects, the immunostimulants shall play a vital role not only in commercial aquaculture but also for sustainable production in the sector as whole. Smith *et al.* (2003) reported that prolonged use of immunostimulants would be detrimental to the host and safe periods for such immunostimulants be assessed in different fishes (Cited Bandyopadhyay, 2008).

7.1.5. Advantages of Immunostimulants

1. Non-antibiotic
2. Protection by triggering the host immune system
3. Less chance for development of resistivity
4. Eco-friendly and natural
5. Usually stable.

7.1.6. Nutrigenomics

It is the relationship between nutrition and the response of gene that can affect animal health. It focusses on the impact of nutrients on genome, proteome and metabolome. It explains about the application of high genomic tools in nutrition research.

8. NUTRITIONAL DISEASES

Nutritional fish diseases can be attributed to deficiency, excess or improper balance of components present in the food available. Symptoms appear gradually when one or more components in the diet drop below the critical level of the body reserves. Nutritional diseases are presented in tabular form.

Nutritional components	*Symptoms*
1. Protein	Reduced growth rate and body deformities. Tryptophan deficiency results scoliosis.
2. Carbohydrate	Depress the digestion, symptoms were similar to that of diabetes militis in warm blooded animals. Enlarged livers. Sikoki disease in carp similar to dibetic symptoms.
3. Lipids	W_3 deficiency (linolenic series) caused discoloration, hypersensitivity to shock and large liver. Fat oxidised diet caused muscular dystrophy, poor growth. Lipoid liver degeneration is characterised when liver glycogen is replaced by lipoid and ceroid produced from liver lipid through fat metabolism. Visceral granuloma, is due to auto oxidation of lipid in diet. Enteritis and Hepatoma are due to aflatoxin in diet.
4. Minerals	Thyroid hyperplasia or goiter caused by iodine deficiency. Dicalcium phosphate deficiency caused scoliosis in carps.
5. Vitamins *(a) Water soluble*	(i) Thiamine (Vit. B_1) - Deficiency resulted poor apetite, muscle atrophy, loss of equilibrium similar to that of whirling disease symptoms in trout, odema, and poor growth. (ii) Riboflavin (Vit. B_2) - Corneal vascularisation, cloudylens, haemorrhagic eye, photo phobia, dim vision, incoordination, discoloration, poor growth and anemia. (iii) Pyridoxine (Vit. B_6) - Nervous disorders, hyper irritability, anemia serous fluid, rapid gasping and breathing.

(iv) Pantothenic acid - Loss of apetite, necrosis and scarring, cellular atrophy, exudates on gills, sluggishness, cubbed gills, poor growth.

(v) Inositol - Fin necrosis anemia, distended stomach, skin lesions and poor growth.

(vi) Biotin - Blue slime patch on body, loss of apetite, muscle atrophy, fragmentation of erythrocytes, skin lesion and poor growth.

(vii) Folic acid - Poor growth, lethargy, fragility of caudalfin, dark coloration, macrocytic anemia, decreased apetite.

(viii) Choline - Anemia, haemorrhagic kidney and intestine, poor growth.

(ix) Nicotinic acid - Loss of apetite, photophobia, swollen gills, reduced coordination, lethargy.

(x) Vitamin (B_{12}) cobalamin derivative- Erratic haemoglobin level, erythrocyte counts and cell fragmentation.

(xi) Ascorbic acid - Lordosis and scoliosis. Eroded caudal fin, deformed gill operculum, impaired collagen formation.

(b) Fat soluble

Vit. A	Vit-A deficiency causes exophthalmos, ascite, odema, haemmorhagic kidney. Hypervitaminosis (A) causes necrotic caudal fin.
Vit. D	Necrotic appearance in the kidney.
Vit. K	Mild cutaneous haemorrhages due to ineffectiveness of blood clotting.
Vit. E	Exophthalmia, distended abdomen, anemia with reduced RBC numbers and haemoglobin content. Accumulation of ceroid in fish liver.

11

FISH DISEASES AND FISH HEALTH MANAGEMENT

1. INTRODUCTION

The aquatic environment possess a wide range of parameters which influences the hygenic conditions for healthy growth and reproduction of fish. As all living being in nature and environment are, as a rule, in a balance state of equilibrium like mutual interrelation, restriction and adoption, any unbalanced situation among the hosts, pathogens and environment cause diseases. In small water bodies like tanks and ponds environmental variability is very great and many man made stresses are added. Fish, therefore, need continually to adopt to these changes in regard to population density, temperature, light, dissolved gases, pH etc. Such environmental changes impose considerable stress on the **homeostatic** mechanism of fishes and cause the physiological changes, and the defence mechanism of fish gets totally lost under stress conditions. Once the environmental parameters changes, pathogenicity of pathogens become high and fish with poor health and weak resistance will be infected. Hence, the manifestation of fish disease is not an isolated phenomenon rather it is a complex iterrelation of the host, caused agent and environment. The magnitude of disease problem from preacute to chronic stages are heightened with increase in densities and intensification of culture management practices. A significant decline from 59000 tons to 8000 tons of *Penaeus monodon* production in Taiwan in 1988 was because of disease problem due to over intensification of management practices. In India, the traditional aquaculture is now being changed to scientific way involving heavy inputs and diseases of all kinds now are known to occur in increasingly large scale.

Although thousands of fish parasites are known gto occur, but only a few of them cause serious damage. Hence, a good and sound knowledge of these organisms involved is necessary to monitor them through environmental manipulation and even to control some of these virulent pathogens through chemo and immuno prophylactic measures.

2. MAJOR TYPES OF FISH DISEASES (TABLE 12)

Many workers have described various types of fish diseases in reltion with the pathogens involved primarily in diseases. If fungus is involved as pathogen, it is called fungal fish diseases and if bacteria, it is called as bacterial fish diseases. However, in certain cases when bacteria, fungi and virus are invovled as pathogens and cause pathogenecity, these are included as Infectious or Icthyomicrobial diseases. The loss of fish production from infectious diseases accounts about 60% of all diseased cases. It indicates that such a disease ranks a noticeable place among the other invasion type of fish diseases. For this reason, the study of infectious diseases is of primary significance to the development of aquaculture.

Common symptoms of diseases

(1) Unusual movements

(2) Abnormal and unhealthy look

(3) Discoloration

(4) Film like covering on the skin

(5) Eliptera swelling or spotting on the skin

(6) Pale gills, white and red spot on gills

(7) Excess slime secretion.

2.1 Infectious diseases (Fig. 49-52)

(1) This type of disease is mostly caused by the pathogens of virus, bacteria, fungi or unicellular algae.

(2) The infectious disease is characterised as preacute, acute, subacute or chronic forms depending on the magnitude and duration of infection. Preacute is recognised as mortality without gross lesions. The acute form is recognised when disease occur suddenly and remains for short period and no symptoms are seen.

The subacute form is recognised when disease remains 2-6 weeks and symptoms are seen in the latter period and the chronic form remains for long duration and visible symptoms are recognisable.

Table 12. Major types of fish diseases, causative agents and symptoms

Disease	Pathogens	Symptoms
Infectious (Bacterial)		
1. Haemorrhagic septicemia	i. *Reovirus*	Congestion, dark and slight red, Exophthalmos, congestion of operculum and fin base. It may be of red fin type, red operculum type or enteritis type.
	ii. *Pseudomonas fluorescens*	External lesions, ulcerations, exophthalmia, abdominal distension and hydrops.
	iii. *Aeromonas liquefaciens*	Gopalkrishnan (1961) described the symptoms is like of dropsy condition in Indian carps. This pathogen affects eyes, optic nerves & brain. The specimen affected shows darker pigmentation. In acute cases, haemorrhages may be pronounced in internal organs and watery fluid may ooze out if the kidney is punctured. Due to this liquification character, the causative organism, presently known as *Aeromonas hydrophila* was also known as *A. liquefaciens*. Similar conditions occurs in case of *Pseudomonas fluorescens*. Major carps are prone to infections but Silver carp and Catla are relatively more susceptible (Kumar *et al.*, 1986, Kumar and Dey, 1988). The strains isolated from diseased fish have been found to be more virulent than those isolated from pond water (Karunasagar *et al.*, 1988).
2. Furunculosis	i. *Aeromonas*	Darkening body, anoresxia (loss of apetite), gregareous near outlets.

3. Erythroderm	i. *Pseudomonas fluorescens*	Inflammation, bleeding from skin, loss of scales on the sides of the abdomen, loss of terminal tip of caudal fin, red blotches around upper and lower jaw, congestion.
4. Saddle back diseases (columnaris)	i. *Flexibacter columnaris*	Grey-brown patches, haemorrhagic spots on body.
5. Enteritis	i. *Aeromonas species*	Distended abdomen, red patches, fins congested and decayed. Anus red, swollen through which yellow mucus comes out through slight pressure
6. Gill hyperplasia syndrome	i. *Myxobacterial complex*	Cubbing gills, Aneurism in gill filaments.
7. Edwardsiellosis	i. *Edwardesiella tarda*	Emaciation, anaemia, loss of skin, peeling off and dropping of skin, necrosis, foul smelling
8. Gill rot	i. *Myxococcus*	Fish is black in appearance, inflammation on the opercular region.
9. Dropsy	i. *Pseudomonas punctata*	Body scale stretch out resembling pine cone. Hence it is called as pine cone disease or vertical scale disease, inflammation, ulceration, exophthalmos, abdominal distention.
	ii. *Aeromonas liquefaceiens*	
	iii. *A. hydrophilla*	
10. Tropical Ulcerative disease	i. *Aeromonas hydrophilla* and other pathogen in combination	Ulceration on body leading to necrotic and poss-formation.
11. Vibriosis	i. *Vibrio anguillarum, Vibrio parahaemolyticus*	Darkening of body, anorexia Ulceration, periorbital and abdominal dropsy.
Fungal disease		
12. Watermould disease (saprolegniasis)	*Aphanomyces Saprolegnia* and Achlya	Grow on dorsal musculature of tropical freshwater fishes. Mould grow on-body penetrating into muscle. Morbid muscle rot.

13. Gill mould (rot) diseases	*Icthyosporidium*	Ball shaped cyst in liver, other vital organs, gills and haemorrhages. Slime secretion.
	Branchiomyces sanguinis *B. demigrans*	Yellow brownish discoloration and disintegration of gill tissues.
Viral disease		
14. Infectious pancr-eatic necrosis virus (IPNV)	Not identified	Pancreatic cell necrosis
15. Viral haemor-rhagic septisemia (VHS)	-do-	Haemorrhages
16. Infectious haem-atopoetic necrosis (IHN)	-do-	Haemopoetic necrosis, behavioral changes
17. Carp pox	-do-	Abdominal distension.
Invasive disease and parasites		
18. Cryptobiosis	*Cryptobiosis branchialis*	Parasite generally fixes on the gill, destroy epithelial cells, excess mucus secretion, gill inflammation, Sunken eyes.
19. Myxosporidiasis and micros poridiasis	Myxobolus species. Microsporidie and Myxosporidia	Organotrophisms seen in myxobolus species. Fish skin, gills, intestines, central nervous and sensory organs are affected.
20. Ichthyophth-iriasis (white spot disease)	*Ichthyophthirius multifilis*	Skin, finrays, operculum are covered with white spores. Sick fish keep rubbing against hard substratum.
21. Trichodinasis & Trichodineliasis, Tripartiella and Glossatella	Trichodina, Trichodinella	Invade skin and Gills of Juveniles.
22. Dactylogyrosis (gill fluke) and Gy-rodactylosis (monog-enetic trematodes)	Dactylogyrus, Gyrodactylus	Gill, skin, fins are affected. Mucus secretion is more.
23. Sinergasilosis	Female sinergasilus, Ergasilus	Parasitize fish gill, Organotrophism is seen in sinergasilus.

24. Learnaeasis (Anchor worm)	Learnaea	Organotrophism is seen, poor apetite, sick, slow movement. Inflammation around the infected area.
25. Argulosis	Argulus	Crustacean parasite, seen in naked eye attached to the head and fin rays of fish.
26. Bothriocephalus	Bothriocephalus	Diseased fish is characterised by milk white and swollen capsular in the intestine. Fish lose appetite.
27. *Ichthyobodon-ecator*	Ichthyobodo	Hyperplasia of the malphigian cells and loss of goblet cells, intercellular odema or spongiosis. Gill congestion.
28. *Eimeria cyprini*	*Eimeria cyprini* *E. subepithelialis*	Found in intestinal mucosa, cause enteritis and emaciation. Yellowish nodule in the colon and rectum of affected fishes.
29. Sanguinicola (Blood fluke)	Sanguinicola	Cause thrombosis, gill haemorrhage, necrosis, nephritis, exophthalmos and loss of health
30. Posthodiplosto-mus and Diplost-omum. (digenetic trematodes)	trematode	Black nodules due to metacercarial cysts in the host body. Infext eye and cause blindness.
31. Ligula (tape worm)	*Ligua intestinalis*	Compresses visceral organs.
32. Ergasilus	Ergasilids	Damage to epithelial cells and cause local damage.
33. Albinodermasis	–	Caudal part is white like or a spot like stamp print on the body.

Fig. 49. *Haemorrhagic septicemia*

Fig. 50. *Saprolegniasis saprolegnia*

Fig. 51. Vertical scale (Pine cone) disease

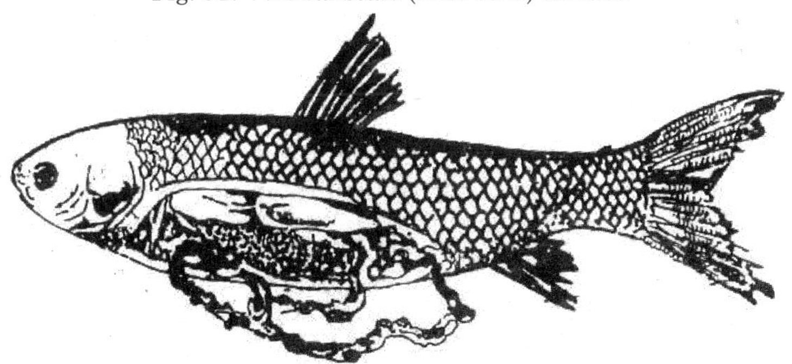

Fig. 52. Grass carp infected with enteritis.

(3) However, bacterial pathogens of infectious diseases are not strictly parasitic micro-organisms. If the conditions for parasitism is unsuitable, it leads a saprophytic life. Similarly, the pathogen for enterities may not cause any disease if the temperature of water is below 20°C, where as a temperature between 20°C to 25°C will enhance the

virulance remarkably. Accordingly the epidemic season can be identified for such diseases.

(4) Zoospores of saprolegnia often attack to the intact fish skin but do not cause disease unless the fish has injured parts, where the zoospore will rapidly grow and multiply.

(5) Most pathogens of infectious diseases show preference to certain species and to certain organs (Organotrophism).

(6) The whole course of infectious disease pass through latent period, symptomatic and attacking period.

(7) It may occur in pure form (fish infected by one kind of causative agent) and mixed infection (infection with over two kinds of pathogens on a single fish).

2.1.1. Source of infectious diseases

(i) Primary source : Sick fish serves as a primary infectious source and are the carriers of pathogens. The pathogen infects through direct contact or by discharge of morbific agents into the water.

(ii) Secondary source : Water comming from diseases ponds, contaminated silt, feeds and gears.

2.1.2. Natural resistance of fish to infectious diseases

(i) Surface texture of skin and mucous membrane of fish functions as a screen to keep the infectious micro-organisms out of it.

(ii) Lysozyme secreted from the cell can kill bacteria.

(iii) The pathogenic microbes entering in to digestive tract will be under the influence of digestive enzymes which can kill pathogens.

(iv) The phagocytotic function of white blood cells, Lymphoid cells, reticulo-endothelial cells of spleen, liver and blood vessel can eliminate foreign body as well as pathogenic micro organisms.

(v) Blood of fish contains bactericidin which can eradicate all kinds of pathogenic bacteria.

The aquatic organisms like fish, exhibit defence system which may be non-specific and specific. Non-specific defence system includes mucus, scales, epidermis and dermis. The presence of proteolytic enzymes and mucus pH levels does not provide condusive medium for pathogen multiplication and growth. The role of mucus is very great ranging from respiration, ionic and osmotic regulation, disease resistance, communication, reproduction, excretion, nest building, protection and feeding. A number of studies indicate that the mucus is constituted with

1. Glycoproteins
2. Glycosamino-glycans
3. Lysosomes
4. Immunoglobulins
5. Carbonic anhydrase
6. Lectin
7. Crinotoxin and
8. Proteolytic enzymes etc.

But it is a matter of grate controversy that these substances exist naturally in fish mucus or are secreted due to disease, injury or stress to fish. Besides goblet cells of fish integument, other secretory cell types that contribute to fish mucus include sacciform cells and acidophilic granular cells. As keratin is rarely found in fish, the most important function of mucus is defence or resistance to abrasion.

That too, the arrangement of scales in embricate manner provides protection to outer skin cover of fish from surface injury. The epidermis of fish contains goblet cells that secretes mucus, further strengthen the defence system. The dermal tissue area of fish avail the humoral defence system. All these are non-motile, permanent barrier to pathogens for infection. Non-specific defence system is followed by inflammatory system which is a dynamic alarm system. This is activated to repair the initial damage and prevent further harm. The clotting mechanism by fibrin network prevent further release of body fluid, there by preventing the entry of pathogens. The damaged cells produces defensive factor for restoration and heal the wound.

The specific defence system is due to antigen and antibody formation which is formed in response to invasion of a pathogen. The specific defence system is further strengthened due to compatibility and interlocking of receptor sites in the antigen and antibody. Antibody belong to serum protein and are immunoglobulins. Lysozyme, haemolysin or bactericidin are known to be present in blood serum that are capable of lysing foreign cells. In fish one type of Immunoglobulin, I g have been described Even I gG like activity has been described in goldfish. Antibodies exhibit specific biological activity like agglutinating, precipitating and neutralising. The functional analogues of T and B lymphocytes further add in the defence mechanis of fish.

The physiological basis of immune response involves lymphoid cells, inflammatory cells and haematopoetic cells. The complex interactions among these cells are mediated by a group of low molecular weight proteins that are collectively known as cytokines. Cytokines indirectly help in regulating the immune effector cells and some cytokines possess direct effector function of their own. Both natural and specific immunity systems of fish are in large mediated by cytokines. Cytokines are mostly produced by mononuclear phagocytes in response to antigen stimulated T cells and such molecules when activated by T- lymphocytes are called lymphokines. T-cells produce several cytokines that primarily regulate the growth and differentiation of varius lymphocyte cells, thus activate the T-cell dependent immune responses. Other T-cell derived cytokines function principally to activate and regulate inflammatory cells such as phagocytes, neutrophils and eosinophils. Cytokines function as a 1. Mediators of natural immunity 2. Regulator of lymphocyte activation, growth and differentiation 3. Regulator of immune mediated inflammation and 4. Stimualaters of immature leukocyte growth and differentiation. Hence, cytokine network plays an important role in coordinating immune effector activities in fish. Research efforts are needed on the mechanism of cytokine production and pathway of action during various systemic diseases of fish to establish the immune network in fish.

Immune System of Fishes

The specific immune system comprises two basic components such as (1) Humoral and (2) Cellular immunity. Humoral immunity is carried out by B lymphocytes where as cellular immunity is carried out by T lymphocytes. B cells produce antibodies or immunoglobulins which are protein molecules that bind the foreign substance (antigen) of (harmful bacteria or viruses) in the blood stream. These are then destroyed by phagocytosis by macrophages. T cells do not produce antibodies but recognize antigen and bound to a type of molecule on the surface of foreign cell by receptor. Hence T cells become cytotoxic or effector cells or memory cells.

2.2 Invasive diseases (Fig. 53-63)

Such diseases are caused by animal parasites, like, Trichodinasis, Icthyophthiriasis, Lernaeasis, Argulosis, Myxosporidiasis, Cryptobiosis, Dactylogyrosis or Gyrodactylosis and Sinergasilus.

Primary and secondary sources are similar to that of infectious disease :

(1) The occurrence and spreading of invasive diseases are dependant to the internal (Physiological status) and external factors (Place, climate, physico-chemical properties of water and farming skills).

(2) Pathogens show selectivity of organs and species. For example cryptobia attacks grass carp but not silver carp in the same pond, *lernaea polymorpha* parasitizes silver carp, bighead and *Learnea ctenopharyngontis* parasitizes grass carp.

For prevention of these diseases, chemoprophylaxis and chemotherapy are to be followed as described later.

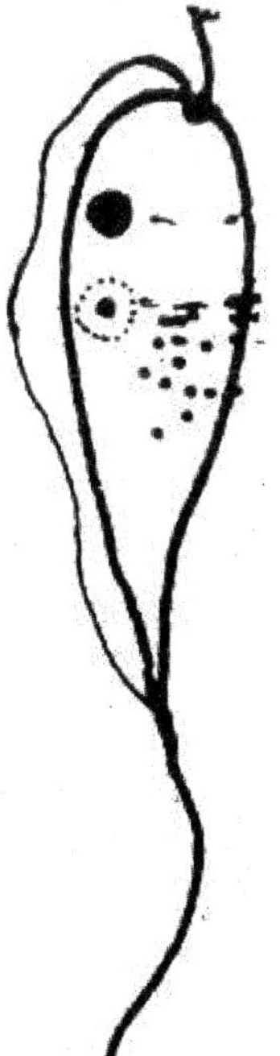

Fig. 53. *Cryptbia* sp.

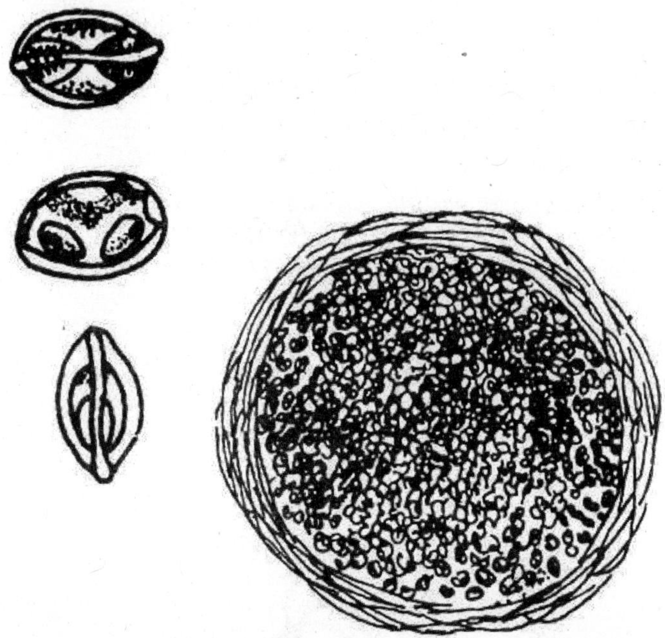

Fig. 54. Myxobolus

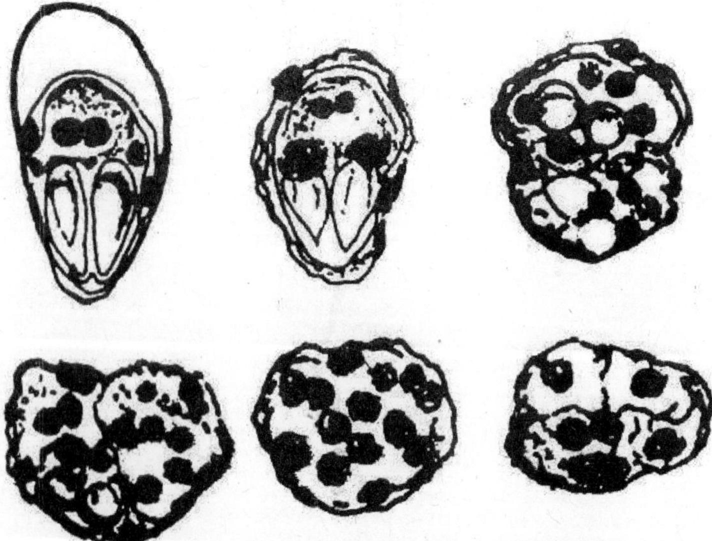

Fig. 55. Formation of spores of *Myxosporidia*

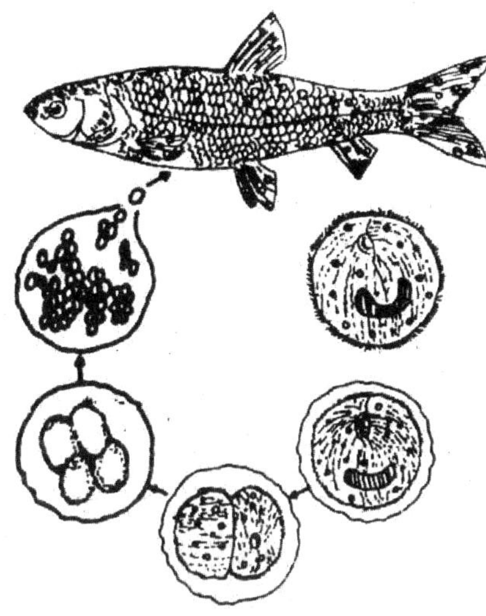

Fig. 56. Life history of *Ichthyophthirius multifiliis*

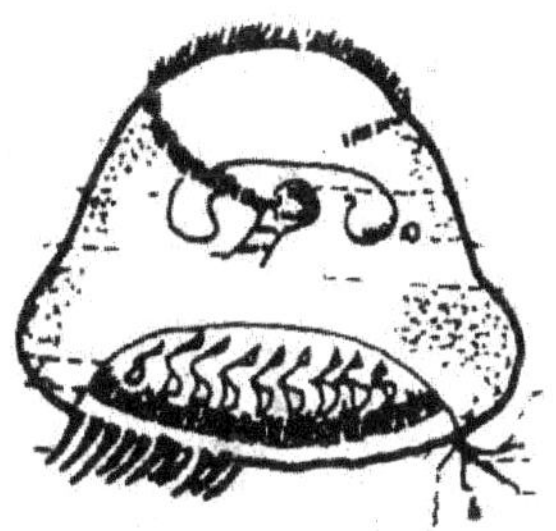

Fig. 57. The *Trichodinidae*

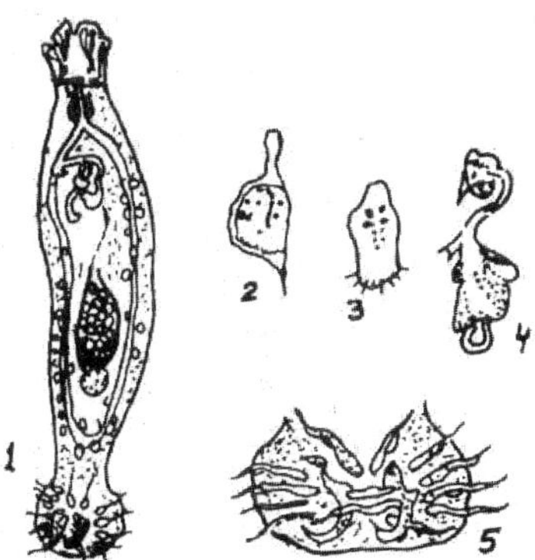

Fig. 58. *Dactylogyrus lamellatus* 1. Adult, 2. egg, 3. larva, 4. Copulatory, 5. Posterior sucking clamp.

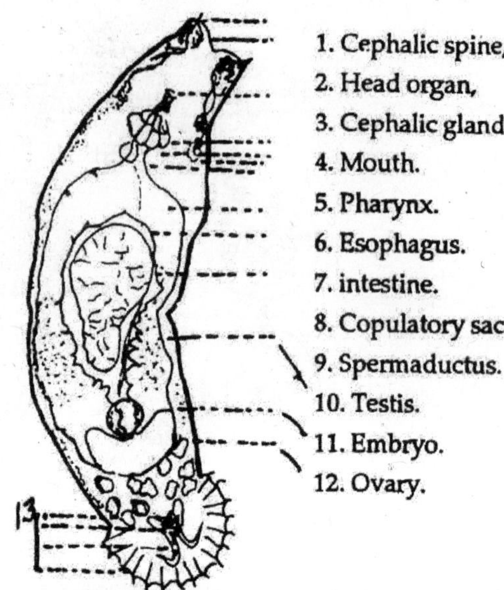

Fig. 59. *Gyrodactylus ctenopharngodontis*

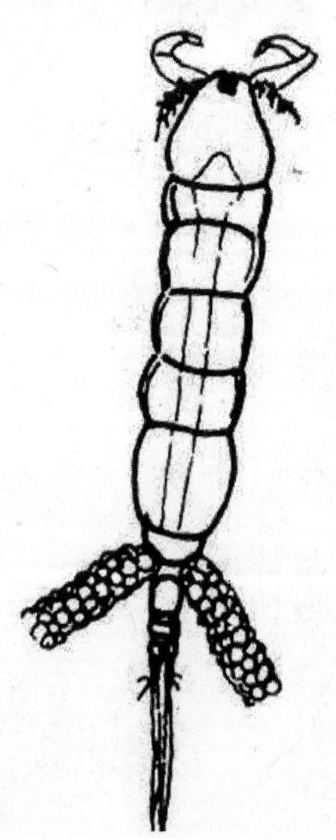

Fig. 60. *Sinergasilus*

Fig. 60a. Gill of two-year old Grass carp invaded by Sinergasilus major

Fig. 61. *Learnea*

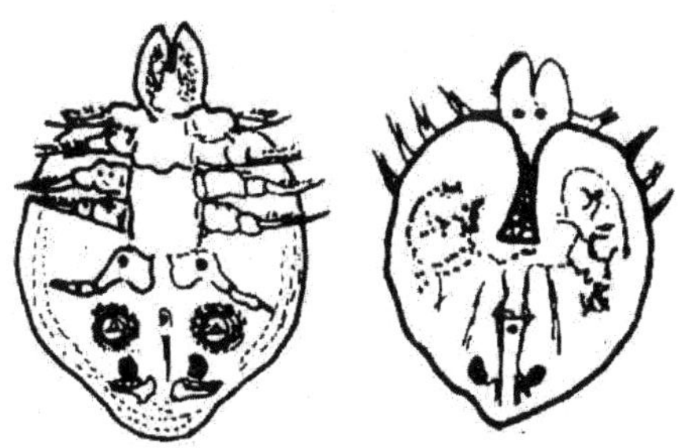

Fig. 62. *Argulus*

Fig. 63. *Bothriocephalus*

2.3 Environmental factors upon the outbreak of disease

1. *Water temperature* – Temperature fluctuations is closely related to the growth and propagation of pathogens and bring direct influence upon its pathogenicity to the host, thus causing the occurrence of disease.
2. *Water quality* - Variation of water quality relates closely to the occurrence of fish disease. Microorganisms decomposes abundantly in the pond where there is too much organic matter due to over fertilisation. As a result, toxic gases are produced. It is reported that when nitrites are more, it causes the out break of haemorrhagic septicemia in fish. If O_2 and nitrogen are supersaturated, gas bladder diseases occurs. This is because, fishes may be infected by pathogen owing to the decrease of resistance. The pathogens of gill mould disease likes to breed in the deteriorated water and trichodinasis happens usually in the pond, where large amount of green manures are fermented. Even low levels of Ammonia in water causes branchial hyperplasia in fishes.

2.4 Diagnosis of diseases

The diagnosis of diseases are accomplished by the epidemiological (epizootiological), clinical and etiological which includes pathoanatomical, microbiological and histopathological methods.

(1) Epidemiological/epizootiological investigation is made, when it is known that a problem exists. Then the problem has to be assessed in terms of location, infected population, season, environmental characteristics, clinical signs and morbid lesions.

(2) Clinical manifestation includes, behavioural changes (excitability, lethargy, erratic movement, isolated, schooling near the current of water flow, loss of apetite, excess mucus secretion, discoloration, enlarged abdomen loss of scales, ulcerations etc). which are considered to be assured indications of disease occurring. This clinical diagnosis lead to te etiological studies (causative agent).

(3) Etiological studies include pathomorphological, pathoanatomical, microbiological and histopathological studies. Pathomorphological methods include the examination of ill and freshly dead fish. The process begins with the macroscopic observations on body, gill and internal organs.

Body - Take tissue or mucous and place it on a microscope to note the parasites like Argulus, Trichodina, costia etc. The base of fins are to be checked if learnea is affected.

Gill - Gill filaments are observed for Dactylogyrus, Cryptobia and myxosporidia.

Eyes - Search for cyst of diplostomulus.

The pathoanatomical studies include the visceral organs.

Intestine - Search for myxosporidia, coccidia, tape worm and nematodes.

Cerebral cavity - Search for granule size cyst and observe under microscope if spores are present. During the process, specimens are taken for histopathological, immunological, haematological and microbiological analysis.

If visceral haemorrhages or petechial *haemorrhages* in the Serosa of the Swim bladder is noticed, this is closely associated with *Pseudomonas septicemia*. Abdominal dropsy condition with blood fringed fluid accumulation if noticed is also a characteristic feature of *Haemorrhagic septicemia.* In case of *Edwardsiella septicemia,* large cavities filled with malodrous gas and necrotic tissue is seen in fishes. Affected skin tends to be peeled off from the underlying musculature. Petechial or larger bleeding may be visible on visceral organs which is also a haemorrhagic associated sign.

(4) *Microbial method* - It consists of (1) examination of smears of skin, gill lesions, heart, liver etc. by staining technique or direct and indirect immunofluroscence and enzyme linked immunosorbant assay. This is rather a confirmatory for a presumtive diagonsis. (2) *In vitro* isolation of pathogen, (3) Antibacterial sensitivity test for chemotherapy treatment.

(5) *Histopathological lesions* -

(a) Degeneration and proliferation are the main characteristics of tissue response to injury. The common degenerative changes (regressive changes), reduced cell size or growth (Atrophy), Intracellular accumulations and cell death (Necrosis) are the processes involved during degeneration. The stages of necrosis are followed with : (i) Pycknosis (shrunken nucleus), (ii) Karyorrhexis (rupture of nuclear membrane and fragmentation of the chromatic fibres), (iii) Karyolysis (lysis of nuclear material and the cell assumes a pink colour).

(b) *Proliferation* - The proliferation in cell size and number are also due to response to injury. The increase in cell size is called as hypertrophy and the increase in the number of cells is called hyperplasia.

Degeneration and proliferation are the main cellular responses to injury. These are accompanied by excellular response to contain or remove the injurious agent and repair the damaged tissue. These extracellular processes responsible for tissue injury are called inflammation. It is a dynamic process and is of two types such as acute and chronic inflammation.

(c) *Acute inflammation* - In the process, three components such as : (i) increased blood flow (vasodilation), (ii) increased vascular permeability (exudation), (iii) neutrophils, macrophages and lymphocytes at the point of the injured tissues (emigration) are noticed. Acute inflammation involves polymorphonuclear cells (Neutrophils).

(d) *Chronic inflammation* - When acute inflammation does not restore quickly to normality and intensity of flammation goes on increasing it leads to chronic inflammation. In such inflammation lymphocytes and macrophages are developed As macrophages are the phagocytic cells, they digest cell debris and pathogens. If the pathogens are large enough to be digested, these macrophages join together to form epitheloid cells which is known as granuloma. This granuloma acts as defence mechanism in the fish to engulf the pathogens. Therefore, granuloma contain parasites, bacteria, fungus, virus and other pathogenic agents. The signs of inflammation are characterized by (1) heat increment, (2) redness, (3) swellings, (4) pain and loss of activities.

(e) *Skin* - (1) Inflammated epidermis become vascular due to migration of blood cells from dermis. (2) More mucous production and discoloration. (3) Necrosis, hyperplasia and hypertrophy are the histopathogical diagnosis.

(f) *Gills* - (1) Hyperplasia, swelling and ulceration resulting in discoloration of gill. (2) Telangiectasis or aneurism due to broken of pillar cells in lamellae. This is resulted as two sides of the lamellae are separated and blood flows with greater force to form ballon like appearance. (3) Rupture of nuclear membrane and fragmentation of chromatin fibres causing karyorrhexis. (4) Complete lysis of the nuclear material causing karyolysis. (5) spongiosis or oedema in gill lamellae. (6) Hyperplasia and fusion of lamellae and clubbing of gill tips.

(g) *Kidney* - (1) Macrophage response in haematopoetic tissue, renal necrosis accompanied with glomeruli necrosis. Necrosis is accompanied with stage of atrophy (decrease in cell size and numbers) and pycknosis (nucleus shrunks). (2) Myxosporidian cysts in the kidney with degeneration of tubules and necrosis. If Giant cells (phagocytic cells joined to form epitheloid cells) fungus infection is noticed. If haemopoetic tissue in kidney undergo variety of changes is due to viral infections.

(h) *Liver* - (1) Acute and extensive necrosis. Even focal necrosis is very common. (2) Metabolic breakdown of lipid in to ceroids and pigments in liver parenchyma enables the suspection of disease.

(i) *Pancreas* - Necrosis in pancreatic acinar cells and haemorrhages may be due to IPN virus. Scattered zymogen granules are observed in many places.

(j) *Heart* - (1) Muscle necrosis, nucleus becomes pycknotic and partially collapses, (2) Inflammation of cardiac muscle (myocarditis) due to lymphocyte congregation among the muscle fibres is indicative of parasitic infection.

(k) *Spleen* - Excessive destruction of erythrocytes lead to accumulation of haemosiderin (phenomenon is called haemosiderosis) is certainly a pathological condition.

3. SIGNIFICANCE OF FISH DISEASE CONTROL

(1) The purpose of preventing fish disease is entirely for betterment of fishery industry, improving farming production and the proliferation of fish resources.

(2) The recent out break of Epizoetic disease syndrome (UDS) has caused a severe economic losses, being in the order of US dollar 8.7 million in Thailand in 1982-83 and Dollar 3.0 million in Bangladesh in 1988. The losses due to mortality and retardation/cessation of growth in fish ponds in West Bengal is as a result of epidemic infections. Hence the significance of fish disease control is to save economic losses.

(3) Certain diseases are seen even to cross the national geographical boundary, hence necessiate the establishment of **quarantine** system in order to prevent these disease being spreading out. The occurrence of ulcerative disease in Australia in 1972, Indonesia in 1980-81, Papua New Guinea in 1972-74, 1981-82, Thailand in 1981, Lao PDR, Burma, Bangladesh 1989 and spread to India in 1989.

(4) Fish are susceptible to infections of bacteria, fungus, virus or invasions of parasites necessiates the diagnostic significance of fish disease control study.

(5) The pathological changes that are brought out by pathogens intracellularly or extracellularly necessiates the significance of such study.

4. PRINCIPLES OF FISH HEALTH MANAGEMENT

The principles of fish health management incorporates (i) minimising stress in cultivated fishes, (ii) confinement of disease outbreak to affected ponds and (iii) minimising losses from disease outbreak. This could be achieved through prophylaxis and positive treatment to the outbreak of epidemics. Because of the aquatic ambience, it is not easy to get aware of the activities of fish. It is difficult to conduct a correct diagnosis and timely treatment. This necessiates prevention of fish diseases which is more important than control of fish diseases. This signify the importance of the statement `Prevention is better than cure".

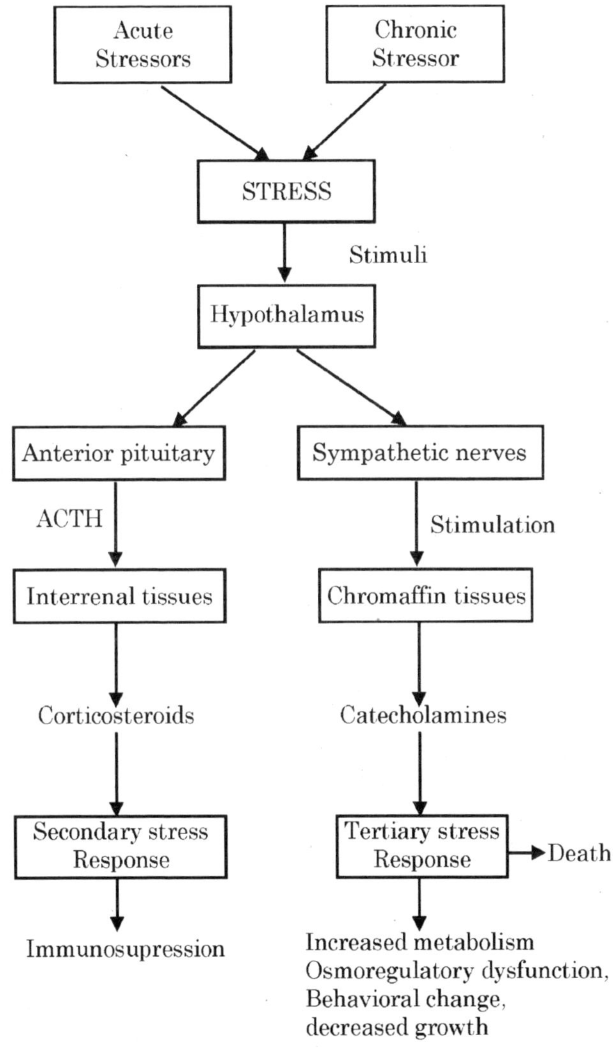

Source: Aqua International 2008: 16(3) 34
Schematic diagram of the stress response in fish
(from Maze and *et al.,* 1977)

Stress response affects adversely in fish health. Quantitative Fish Health Assessment Index (HAI) is therefore very necessary. Higher the value of HAI, lower is the health quality and vice versa. In the present context, schematic diagram of stress response in fish (from Mazeaud *et al.,* 1977) is given for better clarification.

4.1 Prevention of fish diseases

Importance - It is difficult to make out the appearance of disease in its initial stage on account of gregarious mode of fish in water which brings up difficulties in observation, diagnosis and timely treatment. Apart from this, some effective drugs and measures to cure certain fish diseases are still not known well. Therefore, perfect preventive measures must be taken since it is a key link in fish disease control.

General preventive measures

A. *Internal resistance* - Increasing the internal resistance of fish is important in the prevention of diseases. Therefore, some attentional points in fish culture are advisable.

(1) Selection of healthy fish seeds.

(2) Proper density and rational culture

(3) Careful management

(4) Qualitatively uniform ration and fresh food

(5) Good water quality

(6) Prevention of fish body from injury.

B. *Ablishing pathogens and controlling its spreading* - Existence of pathogen is one among three factors (host, caused agent and environment) in outbreak of fish disease. To abolish pathogen and control its spreading following measures can be taken.

(1) *Thorough pond cleaning and disinfection* : Bleaching powder (chlorinated lime) should be applied @ 50 ppm in the pond. It readily kill all the wild fish species, molluscs, tadpoles, crabs and disinfect pond soil and water. In nursery and rearing ponds it is desirable to use malathion @ 0.25 ppm of the active ingradient 4-5 days prior to stocking of fish seeds.

(2) *Disinfection of appliances* : Nets, gears, plastic wares hapas are to be sundried or immersed in a disinfectant solution.

(3) *Disinfection of fingerlings and feeding platform* : Disinfection with mild concentration of $KMnO_4$ solution is helpful during transferring the fingerling to stocking tanks. The feeding platform can be disinfected by hanging with bleaching powder cloth bags with mixture of copper sulphate and ferrous sulphate (ratio 5 : 2) near the feeding place. When fish come to the feeding place for feeding purpose, their skin will be automatically disinfected.

(4) *Proper feeding* : Fixed quality, quantity, time and place has to be followed for proper feeding. Any reduction in quality and quantity and variations in feed application and place may cause not only deficiency of disease but also will increase the susceptibility to many infectious diseases. With regards to different antinutritional factors in fish feeds, it is advisable to take necessary remedial measures to neutralise these antinutritional factors in feeds. For example, for application of aquatic plant feeds, bleaching powder solution at 6 ppm concentration is used or as for organic manures 120 g bleaching powder with 500 kg manure mixed together before application.

(5) *Segregation of year class fish population* : Brood and older fish may serve as carriers of disease causing organisms without exhibiting any clinical symptoms. To avoid such risk, young ones are to be segregated from the brood and older fish.

(6) *Spot removal of dead fish from the pond* : Dead and sick fish are to be removed as soon as it is located. The daily loss of fish should be recorded to provide valuable insight to the intensity of disease problem.

(7) *Chemoprophylaxis*: Effective and inexpensive prophylactic measures against wide range of parasitic and microbial diseases are advisible as chemoprophylaxis. Occasional pond treatment with $KMnO_4$ @ 2-3 ppm and dip treatments with potassium permanganate @ 500-1000 ppm for 1-2 minuts or short both in 2-3% common salt solution is safe. Some of the chemoprophylactics used in culture practices are given in Table 13. Besides, oral administration is given for preventing systemic infections.

(8) *Immunoprophylaxis*: Immunisation programme is gradually emerging as one of the most important measures for preventing infectious diseases. Vaccine to combat bacterial diseases of carps are available in developed countries. Vaccine against *Aeromonas hydrophila, Flexibacter columnaris, Edwardsiella* *tarda, E. ictaluri, Aeromonas salmonicida, Yoreinia ruckeri, Vibrio angullarum* and several viral pathogens such as IPNV (infectious pancreatic necrosis virus), CCVD (channel catfish virus diseases), VHSV (viral haemorrhagic septicemia), IHNV (infectious haemopoetic necrosis) etc. are being tried on large scale. Serodiagonistic methods that includes Flourescent antibody test (FAT), Enzyme immuno assay (EIA) and passive haemagglutination (PHA) are employed. For viral diagnosis also cell line culture techniques are employed. Study of virus, viral vaccine preparation, incubating temperature and pH are the determine factors for fish cell culture. ``Formalin inactivated vaccine" for Haemorrhagic septicemia in grass carp is adopted in China.

8.1. Vaccine : Vaccines are preparation of antigens derived from pathogenic organisms, rendered nonpathogenic by various means and aimed at stimulating the immune system in order to increase the resistance to disease from subsequent infection by pathogen. In development of vaccine, the identified protective antigens can be isolated, characterized, purified and synthesized biochemically or by using DNA recombinant technology. The two key elements of adaptive immune response which vaccination exploit are specificity and memory. The ability of fish to develop immunity to a disease by vaccination depends on (1) Size of the fish (2) Water temperature and (3) Method of vaccination. The immune response to vaccination increases with increase in size and also increasing temperature. There are three basic methods of vaccination such as (1) Immersion (2) Oral (in feed) and (3) Injection.

8.1.1. Bacterial Biofilm Vaccines: Among the various methods of fish vaccination, oral vaccination is the most preferred one by the farmers. However, oral vaccination usually elicits poor and inconsistent immune response and protection in fish, which is attributed to destruction of vaccine antigen in foregut and their failure to reach immune responsive sites of hind gut, spleen and kidney. Bacterial biofilm is a colony of cells in high density embedded in a glycocalyx matrix on a substrate which is resistant to antibiotics, chemicals and host immune system. Therefore, the resistant nature of biofilm due to the protective glycocalyx coat could be ideally exploited for developing a resistant and effective oral vaccine for fish.

8.1.2. Monoclonal Antibodies: Due to availability of limited quantity of the Conventional polyclonal antibodies, monoclonal antibodies (MAbs) have become popular. MAbs which are obtained invitro from hybridoma cells produced by the fusion of antibody producing spleen cells and myeloma cells, are epitope specific, homogenous of universal standard and available in unlimited quantity. Hybridoma technology is being extensively used in human and veterinary health management and also in aquaculture of developed countries.

AVL, the U.K based world leader in fish health has been granted a marketing Authorisation in the U.K and Northern Ireland for their product Aqua Vac™ ERM oral for rainbow trout, an orally administered vaccine against enteric red mouth disease. Further fish vaccine Aqua Vac™ vibrio oral is licensed in Greece for Sea bass and is the only licensed oral vaccine for fish in Europe.

8.1.3. DNA vaccines (Gene vaccines): Traditional vaccines are made up of protein, released from live or attenuated disease agents. These are injected in to uninfected host and trigger immune defence response. But DNA vaccine contains the gene of disease agent that codes for one or more proteins used in traditional vaccines. The DNA vaccine is injected in to muscle tissue cells of a susceptible host, which stimulates them to produce the proteins (antigens) that trigger immune system. Advantages of DNA vaccines over traditional vaccines are (1) DNA vaccines are more stable (2) Do not require cold chain as it is stable for use in warm field condition (3) Stimulus is much lower in DNA vaccine but trigger the immune response than traditional vaccines. Three fish vaccines developed in collaboration with the Institute of Aquaculture have been commercialized. These are (1) Vibriosis (2) Enteric Red Mouth and (3) Furunculosis. Research for development for Bacterial Kidney Disease (BKD) is in progress (Mukherjee, 2008).

To meet the growing challenges, the aquaculture industry is seeking various measures for enhancing its efficiency and all such options includes, Good Aquaculture Pracice (GAP), Hazard Analysis Critical Control Point (HACCP) and Application of Nanotechnology. GAP emphasizes on farming practices like pond preparation, disinfection of pond water, aeration, temperature, pH, alkalinity, salinity, feeding issues, sludge reduction, water exchange, removal of nitrogenous compounds, use of probiotics and so on. While HACCP emphasizes on the determination of critical control points and accordingly corrective measures are taken before it becomes a hazard. Routine screening of samples are necessary using PCR in managing viral diseases in culture ponds. But nano in dimensional scaling refers to 10^{-9}. Some of the upcoming application of nanotechnology in aquaculture are fish health management (through nano biosensors, nano silver, super magnetic nano particles, DNA nano vaccines), harvest and post harvest technology and water treatment in aquaculture.

Table 13. Chemoprophylactics

1. Acriflavin	3-10 ppm for pond treatment. Bath in 500 ppm for 30 minits.	Protozoan and egg disinfection.
2. Calcium hydroxide	Sprinkle in drained out pond. In perenial ponds apply @ 1-2 tons/acre.	Pond disinfection
3. Ca (OC1) Cl	25 mg-1800 mg/litre of water depending on situation.	Pond disinfectant
4. Calcium oxide	40-60 ppm or 2000 kg/ha in drained wet ponds.	Pond disinfectant
5. Malachite green	i. Dip treatment @ 66 ppm for 10-30 seconds. ii. 1-5 ppm bath for 1 hr.	Fungus prevention in eggs.
6. Malathion	0.25-3 ppm in nursery pond application	Killing of copepods.
7. Potassium permanganet ($KMnO_4$)	i. @ 5 ppm to be used in alternative days. ii. 500-1000 ppm as bath for 10-30 seconds.	Prophylaxis against external protozoa, fungi etc. Ulcertative disease.
8. Lime and $KMnO_4$	Bath @ 5 ppm for short time.	Bacteria, haemmorhagic disease, fungal infections.
9. Quarternary Ammonium compounds. Alkyldimethylbenzyl - NH_4Cl	1-2 ppm (100% concentration) product	Fungus, parasites
10. Trichlorphon and Dichlorvus	Mild dose	Ectoparasites, Argulus (Fish lice) Learnea (Anchor worm)
11.Copper sulphate	Concentration depending on the hardness of water (0.8 ppm - 1 ppm)	For ectoparasites

12. Sodium chloride	bath for 3 days	As hauling prophylactics
13. HCHO (formal dehyde)	Mild dose @ 50-100 ppm	Egg disinfectant
14. Iodine and Iodophors	Pure Iodine for 10-15 minits.	Egg disinfectant
15. Soap and oil application	18 kg soap + 56 kg diesel oil/ha.	Insect control
16. Dipterex	@ 0.3 ppm.	Argulus and ectoparasites

Chemotherapy - There is an excellent review article in fisheries chemotherapy by Alderman (1988). The term chemotherapy was introduced by Paul Ehrlich (1854-1915) cited by Smith, 1967; who was a pioneer in the development of chemotherapeutic agents. It is a procedure employed to restore normal health condition of fish. Therapy is applied in 3 ways.

(1) External treatment

(2) Systematic treatment through diet

(3) Parenteral treatment.

Antibacterial agents or antibiotics include sulfonamides, nitrofurans, furance, Tetracycline, 4-Quinolons, Erythromycine, Chloramphenicol are being used to combat fish diseases (Table 14). In 1941, the term ``antibiotic" was defined by Waksman (1946) as a chemical substance produced by microorganisms which has the capacity to inhibit the growth of bacteria and even destroy bacteria and other microorganisms on dilute solution.

4.2 Establishing Quarantine System

The quarantine work should be emphasised before the allocation and trasnsportation of fingerlings. Diseased fish could be strictly prohibited from transportation.

4.3 Prevention with Medicated Feed

During epidemic season, medicated feed is most helpful. Use 1-2 kg garlic per 100 kg of fish once a day for six consecutive days. It is advisible to blend the garlic with fish feeds and add 200 g of table salt for every 5 kg of food. Medicated feed is advisable to be given in pellete form.

Table 14. Chemotherapy

Antibiotics	Dose	Disease
1. Sulfameragine and sulfamethazine	@ feed for 2 to 3 weeks	Columnaris, Furunculosis, Gram+ve bacteria, Vibrio anguillarum; vibrio spp.
2. Nitrofurans (Feroxone, Nitrofurazone, Furazolidone, Furazone, etc.)	i. 50-75 mg/kg as feed for 20 days. ii. 1 ppm solution bath for 5-10 minits.	Bacteria diseases, sporozoan diseases, viral (vibrio), Gill rot, Bacterial septicemia, columnaris, protozoan disease.
3. Oxytetracycline	i. 50-75 mg/kg as feed for 10-15 days ii. Intraperitoneal injection @ 20-30 mg/kg of fish iii. Long duration bath in 10-20 ppm solution.	Bacterial diseases, Fungual diseases. Vibrio anguillarum
4. Streptomycine	20-25 mg/kg of fish as ntraperritoneal injection in combination with pencillin @ 20,000 IU/kg of fish.	Columnairs, Bactrerial septicemia
5. Erythromycin	25-50 mg/kg per day as feed for 4-7 days	Streptococcus
6. Chloramphenicol	i. 50-100 mg/kg fish/day for 5-10 days.	Aeromonas liquefaciens
	ii. Intraperitoneal injection at 12 mg/kg	Aeromonas
7. PVP-Iodine	i. Injection	IPN virus

Table 15.

Type of medicinal herb	Function	Dosage	Quantity of fish	Remarks
1. *Euphorbia humifusa*	Enteritis	500 g dry or 2.5 kg fresh	100 kg of fish through feed	Once a day for 3 consecutive days
2. *Acalypha australis*	Enteritis, Gill rot	125-500 g dry or 2 kg fresh	100 kg of fish	-do-

3. *Polygonum hydropiper* (water knot weed)	Enteritis, Gill rot	500 g dry or 1.5 kg fresh along with 125 g dry *Acalypha australis*	100 kg of fish	It is also effective to use 2 to 3 kinds of herbs like garlic, Euphorbia and Acalypha making a compound one for controlling such diseases.
4. *Andr-ographis paniculata*	Enteritis	2 kg dry herb or 3 kg fresh herb	100 kg of fish	Once a day for 5-7 consecutive days
5. *Portulaca oleracea*	Enteritis	1.5-3 kg fresh	100 kg of fish	–
6. *Artemisia argyi*	Enteritis. Gill rot	100 g powder of *A. argyi* and *Polygonum hydropiper* 500 gm powder or 1 kg fresh	10,000 fingerlings	–
7. Mock straw berry (*Duche-snea indica)*	Enteritis	1 kg fresh	100 kg fish and fingerlings	–
8. *Sapium sebiferum*	White head white mouth, Gill rot	250 g leaf powder	100 kg fish	Once a days for 3-6 consecutive days
9. *Pinus massoniana*	Enteritis, Gill rot, Learneasis	5 kg *Acorus calamus* + 5 kg castor oil plant, Ricinus communis. Tie then in 10 kg pine branch in 2-3 batches, grind them.	100 kg fish	Spread them in 1/15 ha, water area to control Enteritis, Gill rot. To control Lerneasis 20 kg pine per 1/15 ha water area can be applied.
10. *Cayratia japonica*	White head white mouth	1.5-3 kg leaf		Grind the leaf and spread in pond so as to make 1.5-2 ppm.
11. *Melia azedarach*	Trichodinasis and Lerneasis			1. 6 kg *Melia azedarch* + 10 kg mulberry tree leaf +

		11 kg soyabean cake + 12.5 kg dry sweet flag. Mix, grind and spread to a 1/15 ha water area. 2. Decoat (boil in water) for 1 hour about 30 kg of *Melia azedarach* and spread per 1/15 ha water area to control Trichodinasis.
12. Dry sweet flag (*Acorus calamus)*	Gill rot, Red skin, Enteritis, Saprolegnia (Water mould disease)	Application is same as per No. 9. For Saprolegniasis 2.5 kg dry sweet flag + 0.5 – 1 kg Table salt + 2.5 kg urine of cattle per 1/15 ha water area.
13. *Rheum officinale*	White head white mouth and Gill rot	1. Immerse 1 kg *Rheum officinale* in 20 lit. of 0.3 ammonical water for 6-12 hrs. and then dilute and spread in pond water per 1/15 ha. water area to make the concentration 2.5-3.7 ppm. 2. Application of *Sapium sebiferum* with *Rheum officinale* can control whitehead white mouth and bacterial gill rot.
14. *Galla chinesis*	White head, white mouth Albinodermasis, Erythrodermasis, Furanculosis, Enteritis	Grind decoated *Galla chinesis* and spray them over the pond making the pond water at a concentration of 2-3 ppm.
15. *Areca catechu*	*Bathrcephalus gowkongensis,*	One part of *Areca catechu* with 5 parts

(Pinang)	Tape warm Cestode	of feed can be applied once/day for consecutive seven days.

4.4 Application of Medicinal Herbs

The advantage of using medicinal herbs are that : (1) these are available in plenty and easily, (2) effective and cheap, (3) no adverse effect on the pond system. Because of these advantages, medicinal herbs can be used by fish farmers without any fear of killing the fishes. Rath (1989) reported that the farmers of China, have been using many kinds of medicinal herbs with good results (Table 15). At present, twenty number of different herbs are used in China for control of Gill rot, white head white mouth, learneasis, Trichodinosis, Red skin, Saprolegniasis, Albinodermasis, Erythrodermasis, Furunculosis, Cestods, Bathriocephalus, Enteritis etc. *Sabium sebiferum, Portulacea oleracea, Andrographis paniculata* are used to combat enteritis in carps especially to grass carp. *Melia azedarach* is to combat Trichodinosis and learneasis in carps.

Acorus calamus is to combat gill rot, red skin, saprolegniasis in fishes. Garlic is to combat Erythroderma in fishes.

In Indian context, some studies on herbal use in aquaculture and its mechanism for preventing fish and shrimp diseases are reported (Dash *et al.,* 1999; Parida *et al.,* 2008). Studies made by Dash *et al.* (1999) demonstrated the antibacterial activity of neem against *Aeromonas hydrophilla, Pseudomonas fluorescens*, Myxobacteria and *E. coli.* According to them all tested bacteria were highly sensitive to neem. Similarly Parida *et al.* (2008) reported about the effect of garlic as it contain antimicrobial compound allicin on *Pseudomonas* species, *Vibrio cholera, V. fluvialis, V. gazogenes, V. mimicus* that completely inhibit at 10% concentration. Neemix a product from neem tree exhibited antibacterial activity. Neem oil constitutes a broad spectrum of insecticide, miticide and fungicide derived from seeds of neem tree. Neem tree produces a compound called azadirachtin which acts as an insect growth regulator (IGR) preventing exoskeleton development and impeding the moulting process. Turmeric as it contains curcumin is found to have a wide range of therapeutic effects. It is used as an anti-inflammatory, anti oxidant, anti-helminthetic, antibacterial, anti-coagulant because of its wide range of therapeutic properties.

4.5 Epizootic ulcerative disease syndrome (EUS)

Since 1972, there have been a series of epizootic outbreak of Ulcerative disease condition among wild and cultured fishes in South East Asia and Australia. This is characterised by the appearance of large haemorrhagic or necrotic ulcerative lesions on the body surface (Sharrif *et al.,* 1988), the conditions appear to have involved a number of different species, usually of fresh or brackishwater groups and often, but certainly not always, have resulted in extremely high mortality.

Epizootic ulcertative disease syndrome (EUS) is a major thrust on the aquaculture economy in India that caused mass mortalities of pond fishes. The disease (EUS) causing ``red spot" in mullet and other estuarine species and fresh water fishes occurred in Australia in 1972 (Mackenzie and Hall, 1976), followed by Indonesia 1980-81, Papua New Guinea 1981-82, Southern rivers of PNG 1972-74, Malaysia 1980. Thailand 1981, Lao PDR 1983, Burma 1985, Bangladesh 1989 and Srilanka 1987. The disease entered into India from Bangladesh in July/August 1989. In Bangladesh EUS was first detected in the small and large water bodies at Chandipur town in the South East and spreads upstream from river Meghna, to Manikgonj and down stream to Bhola and subsequently Padma, Meghna and Yamuna.

4.5.1 EUS in India

Epizootic ulcerative syndrome (EUS) has been detected during July/August 1989 from Tripura followed by Assam, Meghalaya, West Bengal, Orissa, Andhra Pradesh etc. In Meghalaya, the epizootic has been reported from the districts of West Garo hills, East and West Khasi hills, Jaintia hills and also from Borak valley districts of Jowai. The epizootic had also been reported from three districts of Tripura such as West Tripura, South Tripura and North Tripura. Udaipur, Amarpur, Dharam Nagar, Sonamura, Khowai and Sadar Sub-divisions have been worstly affected. In Orissa, 10 districts have been severely affected. It has started in Baripada in Mayurbhanj district followed by Balasore, Cuttack, Puri, Ganjam etc. The relatively freshwater areas of Chilka lagoons and Mahanadi estuaries during monsoon near

Paradeep were also affected. The trend of infection assessed in Nov. 1990 revealed that in the districts like Cuttack, Balasore, Mayurbhanj, Puri, there is an increasing trend of this disease where as in districts like Dhenkanal, Bolangir and Kalahandi it is seen to have been checked. In districts like Sambalpur and Ganjam this disease had just started.

The EUS was initially reported from rivers, canals, lakes etc. but gradually covered the entire land locked water bodies. Most pond ecosystem which is having inlets and outlet facilities with the paddy cultivation lands are being worst affected sites.

4.5.2. Species affected

It is found that bottom dwelling species are severely affected causing large scale mortality of the fishes. Gradually it have been spread into cultivable carps, air breathing, catfishes, murrels and minnows. *Glossogobid* species. *trichogaster* sp., *Hemiramphus* sp., and mullets were also affected.

4.5.3 Symptoms

Ulceration is more pronounced in tail and head regions and sometimes abdominal regions near fins. The symptoms varies from fish to fish. In murrels and air breathing catfishes the ulcerations were more pronounced and involved mostly the head and caudal regions. It further deepens in chronic cases exposing the cranium and resulting into greyish or red necrotic areas. There is haemorrhagic spots where bloods ooz out. In chronic cases, total loss of peduncle portion and erosion extended upto posterior abdominal cavity. In Indian major carps, the haemmorphgic spots spread throughout the body followed by parasitic and fungal infection which are secondary pathogens. In *Mastocembelus* species the tail was the principal but not the exclusive site of ulcerative lesions. In *Hemirampus,* the tail portion is sometimes completely eroded in chronic cases. The swimming behaviour is abnormal in acute cases. It is found in pond that the air breathing murrels and *Puntius* species are at the surface layer to engulf the oxygen. Irratic movements are encountered. In advanced stages of infections, the internal organs show distinct pathoanatomical characters like petechial haemorrhages on the surface of liver, intestine and swim bladder. In *Heteropneustes fossilis,* the ulcer spread even 5-6 cm in length.

4.5.4. Environmental factors in the occurrence of EUS

A wide range of environmental analysis were performed on samples collected from different zones of South East Asian countries to provide detailed accounts of environment in which disease occurs. In Asia, fish are intimately associated with rice field both naturally and in some cases artificially as introduced stocks. The investigation recorded ulcerative fish over a wide geographical areas of South East Asia. The evidence from Thailand and Northern malaysia indicate that, the disease appear fairly consistency in the drier September-March period of the year when temperature are on average cooler. Some report from Burma suggest that disease out breaks tend to follow period of heavy rainfall. One interpretation for this relationship is that the rains wash harmful pesticides and fertiliser residues into the water.

4.5.5. Water condition

A wide range of analysis were made on basic water quality. Data on pH, alkalinity, CO_2, dissolved oxygen, hardness, temperature, Ammonia, Nitrate, turbidity, dissolved phosphorus were collected. Low alkalinity and hardness are normally associated with water's vulnerable to fluctuations in water quality, poor responses to fertilisation and increased solubility and toxicity to metals (Albaster and Lloyd, 1980), organic pollutants such as pesticides can also be more toxic to fish at low calcium concentration, although different compounds vary in their response to calcium hardness and alkalinity. It was found that the disease was commonly associated with water of low alkalinity and hardness that were closely related to acidic and so is in many of the area surveyed (Roberts *et al.,* 1986). Calcium is particularly important in buffering water quality. That too low alkalinity and acid waters are characterised by increased solubility and toxicity of metal ions. However, it is known that high rainfall can result in a lowering of pH, hardness and alkalinity and in these areas the disease has been associated with periods of heavy rain where such an effect might have occurred.

The concentration of various elements such as Aluminium (Al), Antimony (Sb), Cadmium (Cd), Calcium (Ca), Cromium (Cr), Cobalt (Co), Copper (Cu), Iron (Fe), Lead (Pb), Manganese (Mn), Mercury (Hg), Nikel (Ni), Potassium (K), Sodium (Na) and Zinc (Zn) analysis results indicated that the concentration of most of metals in water was very low and well within recommended levels. High peak in copper and manganese were occasionally recorded but the level of Aluminium and Iron is high, associated with high silt load of water, and acidic and aluminium rich nature of the soils of the catchment areas. The exact role of these elements in fish toxicity is uncertain but they could conceivably act as additional environmental stressors, particularly after heavy rains.

Further, the pesticidal and insecticidal residues were detectable in some samples indicated that pesticide residues were unlikely to have had a direct influence on fish health.

4.5.6 Soil

The soils of south east Asia have been studied. Most of the soil sampled in the survey were slightly acidic and relatively low in organic matter. More extreme soil type occur in South-east-Asia, such as very acidic and organically rich coastal soils of Sumatra and Malaysia. The reddish acidic clay soils of some volcanic regions of Indonesia, the high plateau of Burma and West Thailand, but these by their locations are unsuited to fresh water fisheries. A noticeable feature of the sediments collected in North eastern Thailand was their very acidic nature and low calcium content. Acidic sediments, low in calcium will produce poorly buffered acidic water in comparison with more alkaline sediments. The organic content and fertility will also influence the production of fish populations, although it seems clear that no particular characteristics of soil fertility or organic matter were related to the occurrence of the Ulcerative disease condition.

4.5.7. Histopathology

In early stages of disease infection, internal organs showed minimal changes. In the advance of severe ulceration, focal hepatic cellular degeneration, multifocal epithelial necrosis, splenic white pulp necrosis were present along with the involve-ment of bacteria and fungus in the host. The pancreas, in many cases, a hepatopancreas, also showed consistent acinar necrosis and inflammatory cell infiltration. The Glomerular occulsion and renal tubule vacuolar degeneration, haemopoetic tissue degeneration were noticed in kidney.

4.5.8 Causative organism

Among parasites myxozoan species are detected and are not significant contributor to the epizootic. Argulus and gyrodactylus were occasionally observed on diseased fish samples. A large number of Epistylis species was found causing red spots in snake heads. In acute cases, the lesions resembled fungal infestation. Apart from Epistylic species several other parasitic forms were also reported to be associated with the diseased specimen but non of them could be established as primary causative factor.

Fungal forms have been found assiociated with the ulcerated areas in limited cases. The fungus *Aspergillus* species was also found in the liver parenchymatous tissues. The fungus *Achlya* species, was reported from fish skin lesions during the severe ulcerative fish epizootic in Thailand.

A number of bacterial forms such as *Aeromonas* species, *Edwardsiella* species, *Bacillus, Arthrobacter* species, *Closteridium* have been detected from diseased fish, but none of them have been found especially associated with EUS as a primary source. Most of these bacterials forms are facultative and opportunistic invaders in nature. In Thai EUS, bacteria forms were detected in few cases. Torres (1990) suggested that *A. hydrophila* serotype I was the causative agent of EUS in fish based on virulence, numerical, taxonomy and serological studies. However, it must be noted that confusion on *Aeromonas* taxonomy hinders a great deal of progress in EUS research.

Virological studies indicate the possible role of virus in this epizootic. The cell line culture and electron microscope studies indicated the possibility of two or more groups of viruses resembling Reovirus and Picornavirus. A rhabdovirus have been isolated from grass carp (Ahne, 1975) and also from several diseased fish. However, the involvement of virus needs more research in this line to provide informations on characterisation, pathogenicity, susceptibility, fish species and environmental conditions. However, the Editorial Committee on fish health section, AFS 1990, Manila reports that, the cause of EUS remains obscure. There are two theories in the cause of EUS (1) virus (2) a new strain of bacteria/fungus. However, there is no conclusive evidence yet to any of the above theories.

Water quality have effect on fish health and monitoring of water quality can act as prophylatic measure to this epizootic diseases. Areerat (1990) described the prophylactic measures to prevent EUS which has been practiced widely in snake head culture in Thailand. This has been proven to completely check the disease. Usually the following steps should be adopted immediately if EUS is detected.

(1) Stop the flow of water into the culture pond. Until fish are harvested, avoid any inflow/exchange of water into the pond.

(2) Affected fish in the pond should be removed.

(3) Calcium hydroxide should be applied immediately at 375 kg/ha meter of water and there after at three week intervals until fresh-water supply can be introduced.

(4) Nacl is applied at 1250-1875 kg/ha a week after the water has been stopped. This treatment is repeated every 3 weeks interval until freshwater can be introduced.

(5) To avoid excessive accumulation of metabolites, water and feeding should be reduced.

Jhingran (1990) suggested that a preliminary dose of 1250 kg CaO/ha was found effective in keeping the microbial population in check. However, at present the application of lime @ 600 kg/ha in three instalments with 7 days gap has been recommended as a prophylactic measure for such disease in India. CIFAX, is also reported to be useful to prevent this disease at 0.01 ppm without harming to even primary food chain in aquatic system. Dey and Chandra (1994) reported the application of turmeric and lime at 1.0 and 10.0 mg/L respectively, per hectare meter of water in two equal instalments at 5-7 days interval could successfully controled the EUS of fish in India. Many private fish farmers and State Fisheries Departments of India have used turmeric and lime for successful control of the disease by adopting this method.

12

APPLICATION OF GENETICS IN AQUACULTURE

1. INTRODUCTION

Genetic improvement from wild stocks have considerably produced high yielding plants and animals. This had led in increasing their productivity. Similarly there is considerable scope for genetic improvement in fish, which can significantly improve the production of fish culture. Almost all farmed aquatic animals, are genetically indistinguishable from the wild population. Hence, it is important that genetic programmes should start with the domestication of new species in order to benefit from the maximum population heterogeneity. Such heterogeneity may add in improving species vigour suited for culture purpose. The genetic aspects of fishes has been favoured in India, since the success of artificial breeding of IMC in 1957 was achieved.

1.1 Principles of Genetics

Study on heredity of character from one generation to another generation is the main principle of Genetics. It has many branches. These are

1. *Microbial genetics* - It deals with the genetics of micro-organisms (bacteria, virus etc.)
2. Mycogenetics- The genetics of fungi
3. Plant genetics
4. Animal genetics
5. Human genetics

6. *Population genetics*- Genetics of different populations of plants and animals.
7. *Cytogenetics* - It provides cytological explanation of genetic principles
8. *Biochemical genetics* - Biochemical explanations of various genetical Phenomenon
9. *Molecular genetics*- It interpretes most genetical phenomena in terms of chemical molecules.
10. *Clinical genetics*- It applies genetical analysis in diagnosting various hereditary diseases.
11. *Developmental genetics* - Genetical knowledge to the developmental biology.
12. *Radiation genetics*- Deals with genetical effects of radiations on the living Organism.
13 *Quantitative genetics*- Deals with inheritance of quantitative traits such as body weight, length, milk or egg production record.
14. *Ecological genetics*- Deals with the genetics of ecological phenomena.

1.2 Application of Genetics

1. Improvement of existing race by applying certain laws of heredity to it.The Genetical branch which has its application in the improvement and betterment of mankind is called eugenics.
2. Agriculture
3. Medical science-Various human diseases like haemophilia, haemoglobin abnormalities, pathological abnormalities.
4. Legality- Helpful in solving various legal problems.

1.3 Population Genetics

All the individuals of a species constitute a population. The genetical studies for the inheritance of phenotypic trait in a given population is called population genetics. The population genetics is a quantitative science. To calculate the results of the mode of inheritance of genes in a given population various statistical and mathematical models are employed in it. If a population is of dominant genotype (AA), then the frequency of dominant alleles in the gene pool will be relatively high and the percentage of gametes bearing recessive (a) allele will be correspondingly low. In natural environment, mating of organism is a chance or random factor and hence every male gamete in the gene pool is considered to have equal opportunity of uniting with every female gamete. From such random mating, the zygotic frequencies expected in next generation can be predicted from a knowledge of the gene (allelic) frequencies in the gene pool of the parental population. If p stands for percentage of A alleles in the gene pool and q stands for percentage of a alleles, the checker board of both alleles, may predict as $(p + q)^2 = p^2 + q^2 + 2pq$. That in random mating p^2 represents the AA genotype, 2 pq the Aa and q^2 the aa genotype or in the form of equation, $p^2 + 2pq + q^2 = 1$. This equation is originally formulated by a British Mathematician Hardy and a German Physician Weinberg (1908). They formulated Hardy-Weinberg law of equilibrium. The law states that both gene frequencies and genotype frequencies will remain constant from generation to generation in an infinitely large breeding population in which mating is at random and no selection, migration or mutation occur. So a genetic equilibrium is maintained through random mating. Hardy-Weinberg law depends on

1. Infinitely large population and random mating
2. No selection is operative
3. No immigration or emigration occurs in the population
4. No mutation is operative in alleles
5. Meiosis is normal

1.4 Factors influencing on Hardy-Weinberg Equilibrium

The changes in gene frequencies that influence on Hardy-Weinberg equilibrium can be produced by

1. Reduction in population size
2. Selection
3. Mutation
4. Genetic drift
5. Meiotic drive
6. Differential migration

Population genetics has provided great support to the idea of organic evolution. Various aspects of evolution in terms of population genetics are

(1) Speciation (formation of new species),

(2) Stratification (due to mutation, selection, isolation, migration). The population is stratified in to many sub- population.

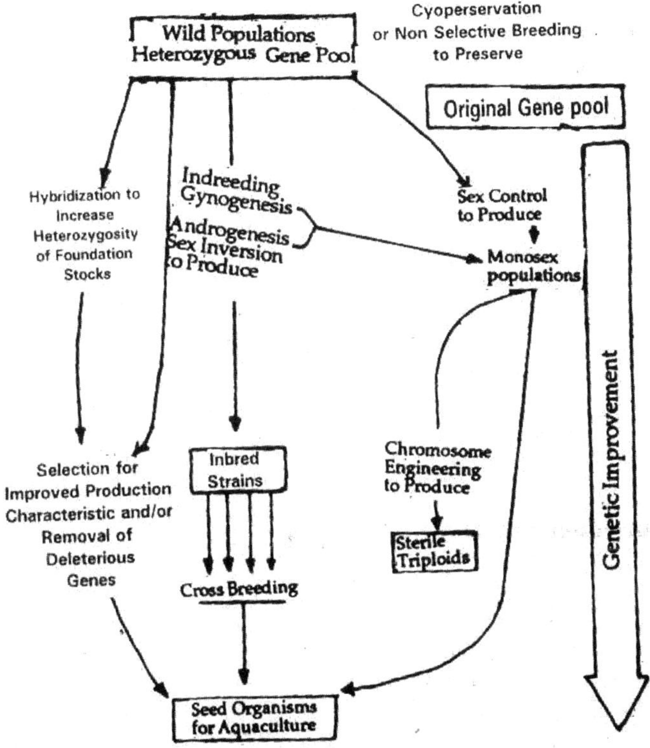

Fig. 64. Genetic improvement of aquacultural species.

2. GENETIC IMPROVEMENT OF STOCK

Genetic selection for phenotypic improvement such as body colour, fin and body shape have yielded dramatic results. However, the food producers would prefer to use genetic selection to improve performance characteristics such as growth rate, food conversion eficiency, disease resistant, fecundity, egg size and so on.

The experiments with salmonids and molluscs suggest that some selective improvement of growth may be possible. All these finding and the break through in induced breeding of cultivated carps has intitiated the hereditary breeding study, genetic basis of selection and other disciplines of genetics. These areas are given high priority in fisheries research project.

These researches in genetic improvement (Fig. 64) includes following objectives to some extent.

Fig. 65. Normally pigmented and albino channel catfish Source : Jack Turner

(1) Methods of selection and selective breeding.

(2) Qualitative and quantitative phenotypes.

(3) Hybridisation.

(4) Multiploid breeding (gynogenesis, polyploidy and mutagenesis).

(5) Nuclear engineering.

(6) Gene engineering.

(7) Genetic markers in breeding programme and genetic analysis.

2.1 Selection

The aim of a farmer is to achieve maximum profit posible from the cultivated species undertaken by himself. This can be achieved by genetic improvement of the fish stock by selection. Based on this principle, CIFA, Kausalyagang, Orissa released genetically improved Rohu named ``Jayanti Rohu" that showed 13% enhanced growth over other stocks. This achievement was made by the efforts of CIFA (ICAR), India in collaboration with AKVAFORSK, Norway by undertaking selective breeding of four reverine stocks and one farm stock of *Labeo rohita* since 1992. The improved strain was released for dissemination on 13th October, 1997 by Honourable Union Minister of Agriculture, India, Mr. Chaturanan Mishra (Source : NAGA 1998, 21 (1) : 78). In order to implement the programme on fish breeding and genetic improvement of stock, maintenance of adequate brood stock is essential. Selection can be based on single desirable trait (character) or combination of such traits.

There are prerequisites before a genetic selection programme is undertaken. These include :

(1) The minimum number of traits (characters) that determine the value of a fish species are to be defined as specifically as possible.

(2) The entire life cycle of the animal should be under control so as to identify the parents and progeny. The absence of identification of family would lead to full sib or half sib mating resulting in inbreeding depression. The individuals within a tested population should be identified by color markers or distinct biochemical genetic marker.

(3) Biological and economical character inter-relationship of the species should be fully understood. In other words, the traits that are to be selected, their relative economic value should be established.

(4) For the selected traits, the phenotypic and genotypic correlations should be known. Aquaculture organisms have very high reproductive rates and large phenotypic variations. Hence, large size brooder slection does not guarantee genetic improvements in offspring and in indirectly may result even poor performance. Even selection of a large size brooder is not always effective unless age of fish is taken into consideration. That too, the economic and genetic value of a large but aged fish is reduced on account of its greater maintenance cost and reduced number of eggs.

(5) The environmental variation influences phenotypic expression of the organisms. Hence, such variations should be well understood.

Therefore, the most common breeding goal is to develop the breeds which have some of the following performing characters.

(i) Efficient utilisation of food for growth that is high food conversion efficiency.

(ii) Short food chain, so as to utilise maximum natural fish food organisms in the system for better conversion of solar energy.

(iii) Resistance to disease and higher per cent of survival especially in larval stages.

(iv) Tolerant to fluctutated physico-chemical characterstics of water such as low oxygen levels, extreme temperature and low pH etc.

(v) Superior meat quality, high lean meat per cent and low refuges so as to improve consumer's preference.

2.2 Basis for selection

In order to have superior qualities of offspring, the genetic basis of selection for reproduction is highly essential. The basis of selection in breeding programme should include.

(1) Individual selection.

(2) Pedigree selection.

(3) Progeny testing selection.

2.2.1. Individual selection

The individual selection is made on the basis of phenotypic characteristic performance shown by the individual. If selection is identified on the basis of one trait (character) like total weight of fish at one year of age, the individual ones which show higher values than mean can be selected. No weightage is given for the performance of parent or offsprings. Historically, this selection is the main method used in fish breeding but in general, success has been limited for production characteristics. This is primarily because the fundamental requirements for selection have not been known. Particular problem have included.

(i) Poor correlation between growth rate and age. Selection of large fingerling size for stocking in ponds may not improve the overall growth rate of marketable size or selectional marketable size may not improve the growth rate at fingerling size. This is because growth at different ages is influenced by different factors and the heritability (character transmitted from one generation to other generation). Even size, often increases 2-3 times after the fingerling stage.

(ii) The selection of aggressive individuals sometime, results in insignificance of individuals of economically important in food assimilation and conversion ratio. Obviously, if a population is composed of entirely aggressive individuals its overall performance will decline.

(iii) In absence of family identify, the sib and half-sib mating results in inbreeding.

Individual selection can be more effective if unrelated populations are used for breeding, so as to produce heterogenous gene pool from which further selection can be done.

2.2.2 Pedigree selection (Family selection)

When heritability is low, the phenotypic expression is relatively low in the individual. Therefore, it is better to consider the genetic bases of family performance as a selection criteria. This is also called as pedigree selection in which on the basis of performance of the parents and grand parents, the progeny are selected to be used as parents for the next generation. This has lot of applications in farm animals where pedigree history of each individual in maintained. But in fishes it can not be practiced as no record of pedigree are available. However based on the performance of full or half sibs of the individuals, selection can be made for breeding.

2.2.3. Progeny testing selection

A prerequisite for selection by progeny testing is artificial fertilisation of eggs. The progeny are first compared under laboratory conditions and then in ponds. The raising of progeny must be repeated for several spawnings. In selection careful consideration should be taken in regard to (1) correlated changes in other characters (2) genotypic influence with environment and their interactions. Genetic variance (GV) is resulted due to cumulated addition of variations caused by additive gene (VA), dominant gene (VD) and by epistatic effect (VI). Hence $GV = V_A + V_D + V_I$. However, genetic variance (GV) affect on phenotypic variance (P_V). Such a phentypic variance (PV) is not only due to genetic variance (GV) and environmental variance (EV) but also due to interaction that exists between the genetic and environmental variance ($V_{G\text{-}E}$). Hence, $P_V = G_V + E_V + V_{G\text{-}E}$. Such a interction that exists between the genetic and environmental variance can be expressed in coefficient correlation (r).

Hence, it can be further simplified as $\sum^2 p = \sum^2 G + \sum^2 E + \sum^2 r\ 2GE$. It is seen that selection for one character may lead to correlated changes in other characters because some characters are genetically correlated. Such genetical inheritance of characters are due to pleitrophy and linkage. If genetic inheritance of character is due to pleitrophy, there is no possibility of changing the relationship between these characters for subsequent generation. However, if linkage is the cause, then the relationship will alter when appropriate cross over takes place. The genetic correlation between two characters is within the limit of +1 to -1 and can be used in heritability. Therefore, selection by progeny testing must be followed because correlated changes occur in other productive characters.

2.3 Methods of selection

It is of great interest to incorporate more possible number of traits of a fish through the method of multiple character selection. This enables the production of superior fish species by improvement in a number of traits from generation to generation. There are three main methods of selection.

(1) Tandem or individual selection

(2) Independent culling level method

(3) Selection by means of an Index

2.3.1 Tandem or individual selection

This method is the simplest one for selecting one particular trait at a time for several generation. When the desired levels of that particular trait is achieved in the population, selection for second trait is followed for a further several generation. Once that desired level of trait are achieved, the others are followed on priority. This method is rarely used, because it takes very long time and some simultaneous change of character may be required. That too characters are influenced by environment and may be inversely correlated.

2.3.2 Independent culling level method

In this method selected traits are taken up simultaneously and a culling level is fixed for each particularl trait. For selection, each individual should cross the culling levels in all the characters fixed under consideration and others are harvested (culled). The advantage of this method is that many characters can be improved at a time. The disadvantage is that if a particular individual is very good in one character but slightly less than the culling level of the other character, it can not be selected. Therefore, the individual has lost the advantage of high value shown for one particular character.

A-A' = culling level for growth rate

B-B' = dres out value (meat value)

Only those individuals falling within the hatched area will be selected.

2.3.3 Selection by means an Index

This method involves the combination of measurements of 2 or more characters in to a single index value for each individual. Such an index is alotted to all the desirable traits according to their merits and a final index score is decided for selection. This method is based on progeny testing. Progeny of different individuals in question are reared in different environments and the average index is calculated for the individual. The final score is taken into account to select or cull an individual.

The disadvantage of this method is that, calculation of a single index value is not easy and when more than two characters are involved, sophisticated computation facilities are required.

It has been found both theoritically and in practice that, individual selection (considering one character at a time) is the least effective method than indepedent culling level and index method. However, index method is relativdely more efficient than independent culling method.

2.4 Response to selection

As soon as the method of genetic improvement selection is over by any of these above methods, the subsequent question lies how much is the response of selection, that has been transferred to the next generation. In other words, the trait improvement in the progeny can be reflected when the selected individuals are bred. This is known as response to selection.

The response (R) to selection over a unit time is dependant on many parameters. These parameters are,

(1) heritability

(2) selective intensity

(3) phenotypic standard deviation

(4) generation time

(5) genetic correlation when more than one trait is involved

Responses (R) to selection can be calculated from the equation.

$$R = \frac{Sh^2}{L}$$

Where R = response : change in traits of population mean from generation to generation.

S = selection dfferential : Difference between mean of selected parents and population mean.

L = generation interval which means the time required for the individual to reach at 1st sexual maturity and breed. Selection differential (S) includes the selection intensity denoted as ``I" which means the per cent of the population chosen as parents for the next generation and phenotypic standard deviation denoted as ``Σ " of selected parents.

Hence, S = i Σ

$$s_o \quad R = \frac{Sh^2}{L} = \frac{\Sigma\, ih^2}{L}$$

Where h^2 is the heritability. Heritability is defined as the proportion of additive genetic variation in the total phenotypic variation. In other words it is the ratio of genetic variance to phenotypic variane.

Hence $$h^2 = \frac{\Sigma^2 G}{\Sigma^2 P}$$

This heritability enables the degree to which a character is heritable from one generation to the other. If heritability is high and close to 1.0 most of the variation in a trait is heritable and response to selection will be highly effective. If environmental factors have caused most of the variation, the value of heritability will be low and if heritability (h^2) is zero, no genetic gain can be obtained by selection. Therefore, in order to improve the genetic variance in the population, inbreeding should be avoided as it leads towards homozygosity in the population thus reduces the response to selection. Response to selection decreases from generation to generation and reach a limit after that further improvement may not be possible. At this position, new and better genotype has to be introduced for better response of selection.

Selective breeding is required to improve the fish stocks for further propagation. Normally fish breeding is done with the aim of producing good quantity of fish seed without considering the quality. But the major goal of a fish breeder should aim not only to produce good quantity but also good quality of fish seeds. The quality seed should have the desired values of quantitative traits like growth rate, survival rate, disease resistance etc. To improve the desired traits the population sould be studied in statistical terms like means, standard deviation, variance etc. Therefore, to have a better breeding plan for genetic improvement of fish stocks, different methods of breeding need to be well understood.

2.5 Different methods of breeding

(1) Inbreeding method and Heterosis method

(2) Out breeding method

(3) Cross breeding method

(4) Selective breeding method

(5) Random breeding method

2.5.1 (a) Inbreeding method

When two more closely related individuals than the average relationship of the populations is bred, it results in homozygosity there by causing inbreeding. Inbreeding results in uniform sequence of genetic materials in the chromosome causing homozygosity thus reduces the genetic variance. The loss of genetic variance results in certain phenotypic losses of the individual. The degree of loss of the traits is dependant to the intensity of inbreeding. Usually much of the character is lost due to intense inbreeding than mild inbreeding.

Mile inbreeding	*Intense inbreeding*
A X B	A X B
C X D	A X C
E X G	A X D
H X I	A X D
A X J	G

K

Inbreeding is calculated as inbreeding coefficient (F). This measures the degrees of relationship between the mating pair. Increase in inbreeding is denoted by ΔF.

The coefficient of inbreeding is the probability of two identical genes at one locus by descent. So inbreeding coefficient gives a comparison between the population in question and the base population. Usually base population is that population, where the coefficient of inbreeding is zero. Inbreeding coefficient in any given generation is calculated in relation to inbreeding coefficient of earlier generation.

Suppose 20 progenys are produced from 10 individuals.

So N = 1/2 N

and F_{to} = Inbreeding coefficient at t_o population = Say zero

F_{t1} = Inbreeding coefficient = $\frac{1}{2N}$

$$F_{t1} = \frac{1}{2N} + \left(1 - \frac{1}{2N}\right) F_{to}$$

In F_{to} there is no inbreeding or inbreeding is zero. So $F_{t1} = \frac{1}{2N}$

Hence inbreeding coefficient ΔF in one generation will be equal to $\frac{1}{2N}$ in large random mating population.

$$\Delta F = \frac{1}{2N}$$

So earlier equation will be

$$\Delta F_{t1} = \Delta F + (1 - \Delta F)\, F_{tO}$$

$$\Delta F = \frac{F_{t1} - F_{t0}}{1 - F_{t0}}$$ will be increment per generation.

In a large random population, the ideal one in that where the inbreeding coefficient is zero. Usually in fish farms of farmer's where no large brood fish population is maintained due to space limitations, one should consider the effective population size (Ne). Effective population size can only be decided after calculating the extent of deviations from the idealised base population by statistical computation. Ne is considered twice the harmonic mean of number of males and females in the population due to mortality and spawning success.

$$So \quad \frac{1}{Ne} = \frac{1}{4\,Nm} + \frac{1}{4\,Nf}$$

Where = Nm is male and Nf is female.

$$So \quad Ne = \frac{4\ Nm\ \ Nf}{Nm\ +\ Nf}$$

As increase in inbreeding coefficient (ΔF) is 1/2N in terms of effective population size

Hence, $$\Delta F = \frac{1}{2\,Ne}$$

Thus, $$\Delta F = \frac{1}{8\,Nm} + \frac{1}{8\,Nf}$$

This can be explained that an effective population size between 68 to 344 per generation is sufficient for raising fish farming. Considering the mortality as 50% and 60% spawning success, an effective population size of 344 would require $\frac{344}{0.5 \times 0.6} = 1147$ fish.

For effective breeding size (Ne) following calculation is made. For example if 2 males are used to fertilise 110 females the

$$Ne = \frac{4\ (2 \times 110\ )}{112} = 7.85$$

Effective breeding size is less than 8 although actual population size is 112. Inorder to prevent significant inbreeding no brood stock should be less than 30 male and 30 female chosen at random for each generation. Ne sould be 40-60.

It is thus obvious that acquisition of brood stock is the most important step because it acts the foundation population. This foundation population or base population determines the biological potential of the future population to some extent on the initial genome constitution. The best way to obtain new stocks is to have some fish from each of as many spawnings as possible or if from the wild form as wide as geographical range as possible.

The decrease or loss in the phenotypic mean value of quantitative traits due to inbreeding is known as inbreeding depression. Such inbreeding depression decreases the biological viability, fertility and the total fitness values of the individual. This is because in increase in the number of homozygous genes, particularly when harmful recessive genes become homozygous.

2.5.1 (b) Heterosis method

It is opposite to the phenomenon of inbreeding. In other words it is known as hybrid vigour. When different inbreed lines are crossed randomly, the offsprings show an increase in mean phenotypic values and can come close to the base population. So by heterosis, the lost fitness due to inbreeding is restored by heterosis. The amount of heterosis is the difference between the cross breed and inbreed means. Heterosis is expressed in percentage and can be calculated in the following way.

$$Heterosis = \frac{Average \quad reciprocal \quad of \ Fi \ hybrids \quad - Averagepar \quad ents}{Average \quad parents} \times 100$$

A simple arithmetic problem on heterosis can be calculated by the following example.

Suppose :

Sl. No.	Group	Average weight (g)
1.	Channel catfish	450
2.	Blue catfish	430
3.	Channel catfish female x blue catfish male	600
4.	Blue catfish female x male channel catfish	460

What is the heterosis in this group of experiment

Step I : Average weight of parental group

$$\frac{450\ g + 430\ g}{2} = 440\ g$$

Step II : Average weight of hybrids

$$\frac{600 + 460}{2} = 530$$

Step III : Heterosis = $\frac{(530 - 440)}{440} \times 100$

H = 20.45%

The heterosis varies in respect to differences in inbred lines. However such inbred line when crossed show general performances which is called as general combining ability of the line. Besides, certain deviation in general performances are noticed due to specific interaction. This interaction is known as specific combining ability of the two lines in combination.

2.5.2 Out breeding method

It is defined as the breeding of two individuals which are less closely related than the average relationship of population. In fish, the tracing of pedigree or ancestral parents and grand parents are not possible. This out breeding can be practiced only by mating individuals reared at distant farms.

2.5.3. Cross breeding method

Cross breeding is used in terms of hybridisation by crossing Indian major carps to produce fertile hybrids. This can be practiced in (1) Three way crossing, (2) Four way crossing and (3) Back crossing. The aim of cross breeding is to improve the general performance of the hybrid through general combining ability of inbred lines. Once the estimated value is achieved in F1 generation, further improvement can be done by introducing new lines.

(1) Three way crossing

A X B

F1 AB X P

ABP - F2

(2) Four way crossing

A X B P X Q

F1 - AB PQ - F1

ABPQ-F2

(3) Back crossing

In this method the F1'S of two lines are crossed to one parental line whose character is more desired in the offspring.

A X B

F1 (50%) AB50% X A100

A B

75% 25%

2.5.4 Selective breeding method

In selective breeding the mating individuals from a large population are selected on the basis of individuals, pedigree or progeny performance. This is the conventional method for genetic improvement of stocks.

In fishes mostly family, combined or mass selection is followed. In mass selection only a single character such as growth is used while in family or combined selection different traits such as growth, flesh quality, disease resistance, less intra muscular bones, fecundity, reduced fat content etc., are considered. The advantage in selection is that, it develops superior progeny from generation to generation through accumulation of efficient genes.

Erst while USSR is one of the pioneers to utilize genetic potential of fishes through selective breeding. The pioneering work carried out by the formerly soviets resulted in the production of superior rainbow trout with manifold increase in yield. Some of the European countries like Hungary have developed superior commom carp strain (Ukrainian) through selection, that can perform much better than ordinary one. Norway achieved several fold high production of Atlantic salmon and rainbow trout through selective breeding. Selective breeding of channel catfish (lctalurus punctatus) in USA and Nile tilapia in Phillippines are examples where potentials of genetics were taken advantage. Nile tilapia evolved strain, under the project entitled genetic improvement of farmed tilapia (GIFT) which is referred as Gift tilapia. This genetically improved strain is reported to grow 60-70% faster than the locally farmed strain. Work during 1990's on selective breeding of rohu reported on an average over 40% selection response at 4th generation level and the developed strain is know as "Jyanti Rohu". This developed rohu shows 12% faster growth rate than the normal ones. Fishes those maturing in one year provide a great opportunity to take the advantage of the genetic potential through selective breeding that would give a big boost to aquaculture and the indigenous species on demand are:

1. *L. calbasu* 2. *L. fimbriatus* 3. *L. bata* 4. *L. gonius* 5. *L. dyocheilus* 6. *Cirrhina reba* 7. *Puntius sarana* 8. *H. fossilis* 9. *C. batrachus* (Tripathi, 2008).

2.5.5 Random breeding method

This typs of breeding practices are mostly followed in fish farms. In such cases individuals are taken randomly and bred. Biasness is encountered in such breeding because individuals which show better conditions of maturity are considered.

The breeding experiments depend not only the effective population size or effective breeding size (Ne) but also on the breeding efficiency (Nb). Breeding efficiency is defined as the probability of the progeny being selected as parents in next generation for seed production. In ideal conditions, an equal number of progeny from different breeding pair should be kept randomly. Then from these mixture of progeny, some individuals are to be selected for retaining at the farm for seed production in next generation. The breeding efficiency (Nb) is the ratio of effective breeding number (Ne) and the breeding population (N).

$$So \quad Nb = \frac{Ne}{N}$$

Breeding efficiency (Nb) is affected by sex ratio as well as the effective population size. The breeding efficiency (Nb) and inbreeding coefficient (ΔF) with different sex ratio in a population of 100 individuals are given on tabular form. If a breeding programme is carried out to spawn 90 females and 10 male fishes by random mating, the breeding efficiency is only 36%. By adjusting the sex ratio of 70 females and 30 males or vice versa the breeding efficiency can be raised to 84%. So the sex ratio of 70 female : 30 male is 2.33 times more efficient than the sex ratio of 90 female : 10 male.

It is important to maintain a large breeding population in order to check inbreeding and genetic drift. Such genetic losses may be avoided by importing new brood stock from other geographical areas. If maintenance of large population or import of new fish population stock is not possible the index given in the table will help to maximise effective breeding size (Ne) within the constraints of fixed population. These thoughts and techniques are very important for brood stock management as they can be used for genetic improvement of a population. Hence, it is important to improve the genetic quality of fish stocks, because poor quality fish will perform poorly.

Changes in effective breeding size (Ne), Breeding efficiency (Nb), due to sex ratio.

No. of male	No. of female	Effective breeding size (Ne) $\frac{4\,(Nm\quad Nf)}{Nm + Nf}$	Total population (N)	Breeding efficiency in percent (Nb)	Increase in inbreeding coefficient F $= \left[\frac{1}{8\,Nm} + \frac{1}{8\,Nf}\right]$
100	0	0	100	–	–
90	10	36	100	36	0.0139
80	20	64	100	64	0.0071
70	30	84	100	84	0.0059
60	40	96	100	96	0.005
50	50	100	100	100	0.005

2.6(i) Mendel's law

Mendel's discovery could be represented by

1. Law of Segregation: Law of segregation states that in a heterozygote a dominant and a recessive allele remain together throughout the life (from zygote to the gametogenesis stage) without contaminating or mixing with each other and finally separate during gametogenesis so that each gamete receive only one allele either dominant or recessive. In monohybrid cross the genotype of F_1 hybrid remains the same from zygote stage to the gametogenesis stage of multicellular plants. These F_1 hybrids by self fertilization produce tall and dwarf plants in 3 : 1 ratio during F_2 generation.

2. *Law of Independent assortment:* It states that when the gametes are formed, the members of the different pairs of genes (factors) segregate quite independently of each other and that all possible combinations of the genes (factors) concerned will be found among the progeny. The Mendel's dihybrid cross of pea plants between yellow round seed X green wrinkled seeds explains this law of independent assortment in which F_2 hybrids are in 9:3:3:1 ratio.

2.6(ii) Genetic of qualitative phenotypes

Manipulation in gene sequence or the genetic variations in fish individuals is measured indirectly through a fish's phenotype. Inorder to exploit the biological potential of a population, it is necessary that the phenotypic variance be analysed and understood. Phenotypic variation are qualitative and quantitative types. Both these qualitative and quantitative phenotypic variations are dependant on the basic biological unit - the gene. But in the present context, the qualitative variations are discussed. Most qualitative phenotypes are controlled by one, two, or three genes. However, the qualitative phenotypes are expressed depending in the number of genes and their mode of action at each locus.

Genes can be located on either the autsomes (Somatic chromosomes) or the sex chromosomes. It is important to know the gene or genes that produce the phenotype are located on an autosome or a sex chromosome, because the inheritance of a phenotype controlled by an autosomal gene differs from that of one controlled by a sex linked gene.

In general, autosomal genes express themselves in either an (1) additive or (2) nonadditive manner. In additive gene action, each allele of gene of a locus produces an equal unidirectional phenotypic effect. In nonadditive gene action, one allele is expressed dominantly than the other. Such a dominant allele of gene has greater influence on the phenotype. In contrast, the other allele is recessive and the phenotype produced by the recessive allele is known as recessive phenotypes.

Phenotypes controlled by single Autosomal Genes with complete Dominant gene action has been studied by different authors (see Tave, 1986) in channel catfish, common carp, rainbow trout, goldfish, guppy, medaka, platy fish, sword tail etc.

Complete dominant action can be explained by illustrating the albinism in channel catfish, which is controlled by a simple autosomal recessive allele (a). But the normal channel catfish produces melanistic pigmentation which is controlled by autosomal dominant allele (+). Because (+) is completely dominant over (a), there are three possible genotype at this locus, but these genotypes can produce only two phenotypes.

Genotype	Phenotypes
+	Normally pigmented
+a	Normally pigmented
aa	albino

2.7(a) Possibilities of genotype, phenotype in F1 and F2 generations

When a normally pigmented channel catfish with (++) homozygous dominant genotype is mated to an homozygous recessive genotype albino (aa), the progeny in F1 generation will all be normaally pigmented due to inheritance of (+a) gene in F1 offspring. But situation will differ when F1 offspring in heterozygous (+a) normally pigmented are mated, they will seggregate in 3 : 1 ratio. Genotypic and phenotypic ratio can be determined by using punnett square.

(1) Normally pigmented X albino

(++) (aa)

	+	+	
a	+a	a+	–F1
a	+a	a+	

+a = Normally pigmented

(2) Normally pigmented (+a) Normally pigmented (+a)

	+	a
+	++ Normal pigmented	+a Normal pigmented
a	+a Normal pigmented	aa albino

In F.2 generation the ratio is 3 : 1

As both parents are heterozygous, they produce offspring which have all possible genotypes and all possible phenotypes due to different types of gene action. These are given in tabular form.

	Types of gene action	F2 phenotypic ratio
1.	Single Autosmal gene	
	i. Complete dominant	3 : 1
	ii. Incomplete dominant	1:2:1
2.	Two autosomal gene each producing different phenotypes	
	i. two genes with complete dominant	9 : 3 : 3 : 1
	ii. two genes : one is complete dominant	3 : 6 : 3 : 1 : 2 : 1
	iii. two genes : incomplete dominant	1:2:1:2:4:2:1:2:1
3.	Two autosomal genes produce phenotype by additive interaction	
	i. additive	1 : 4 : 6 : 4 : 1
4.	Two autosomal genes producing phenotypes through *epistatic* interaction.	
	i. dominant epistatis	12 : 3 : 1
	ii. recessive epistatis	9 : 3 : 4
	iii. duplicate genes with cumulative effects	9 : 6 : 1
	iv. duplicate dominant genes interaction	15 : 1
	v. duplicate recessive genes interaction	9 : 7
	vi. dominant and recessive interaction	13 : 3

(Autosomal genes with 2 allelas per locus. There are no lethal genotypes assumed).

2.7.1 Single autosomal gene -

(i) Complete dominant (Fig. 65) : phenotypic ratio 3 : 1

Normall pigmented female X Normally pigmented male

+a			+a
	+	a	
+	++	+a	
a	+a	aa	

Genotypic ratio = 1++

2+a

1aa

Phenotypic ratio = 3 normally pigmented
1 albino

(ii) Incomplete dominant - Phenotypic ratio 1 : 2 : 1
blue female x blue male
Siamese fighting fish

Vu		Vu
	V	u
V	VV	Vu
u	Vu	uu

genotype = 1 VV : 2 Vu : 1 uu
Phenotype = VV = 1 steel blue
Vu = 2 blue
uu = 1 green.

2.7.2 *Two autosomal genes each producing different phenotypes*

(1) Two genes with complete dominance

Phenotypic ratio in F2 = 9 : 3 : 3 : 1

	Grey and normal spined guppy female (Gg, Cucu)		grey and normal spined guppy male (Gg, Cucu)	
	G, Cu	G, cu	gCu	g cu
G, Cu	GG, CuCu Grey normal	GG, Cucu Grey normal	Gg, CuCu Grey normal	Gg, Cucu Grey normal
G, cu	GG, Cucu Grey normal	GG, cucu Grey curved	Gg, Cucu Grey normal	Gg, cucu Grey curved spine
g, Cu	Grey norml spine Gg Cucu	gG, Cucu Grey normal	gg, CuCu Gold normal spine	gg, Cucu Gold & normal spine
g, cu	gG, Cucu Grey & normal	gG, cucu Grey & curved	gg, Cucu Gold normal	gg, cucu gold & curved

Genotypic ratio 1 GG, CuCu : 2GG, Cucu : 2 Gg, CuCu :
4 Gg, Cucu : 1 GG, cucu : 2 Gg, cucu :
1 gg, CuCu : 2 gg, Cucu : 1 gg, cucu.

Phenotypic ratio 9 grey and normal spine
3 grey and curved spine

3 gold and normal spine

1 gold and curved spine

If each phenotype will be considered separately, then the mating of two heterozygous guppies in the preceding example produces 3 grey : 1 gold and 3 normal spine : 1 curved spine. This is similar to earlier case of single autosomal genes with complete dominant gene action. Similarly in the preceding punnett square 12 grey guppies along with 4 gold guppies are produced and 12 normal spined guppies along with 4 curved spines are also produced making the ratio 12 : 4 which is the same as 3 : 1 ratio.

Hence, Genotypic and phenotypic ratio can be ascertained once the mode of gene action for each phenotype is knwon.

2.7.3. Two autosomal genes produce phenotype by additive interaction - The phenotypic ratio is 1 : 4 : 6 : 4 : 1.

The additive gene interaction along with two autosomal genes can produce some phenotypes. Melanistic body coloration of Black-molly is an example of a phenotype that is controlled by two genes with additive gene interaction.

If the genotype is MM, Nn and Mm, NN, this can produce 2 different types of gametes,

Genotype	*Gametes*
MM, Nn	Mn, mN
Mm, NN	MN, mN

If the genotype is Mm Nn, can produce 4 different types of gametes,

Genotype	*Gametes*
Mm, Nn	Mn, Mn, mN, mn

Only one type of mating will produce all possible genotype and phenotypes : color class III b male X color class III b female. The punnett square will be as follows :

III-b female (Mm, Nn)		X	III-b male (Mm, Nn)	
	MN	Mn	mN	mn
MN	MMNN IVb	MNMn IVa	MNmN IVa	MmNn IIIb
Mn	MMnN IVa	MM, nn IIIb	Mm,nN IIIb	Mmnn II

mN	mN, NN IVa	mM,Nn IIIb	mm,NN IIIa	mmNn II
mn	mM,nN IIIb	mM,nn II	mm,nN II	mm,nn I

Genotypic ratio = 1 MM, NN : 2 MM, Nn : 2Mm,NN : 4Mm, Nn : 1 mMnn : 2 Mm, nn 1mm,NN : 2 mm, Nn : 1 mm, nn :

Phenotypic ratio = IV b color class - 1
IV a color class - 4
III b color class - 4
III a color class-2
II color class - 4
I color class - 1

Since, the phenotype between III a and III b are slightly distinguishable at birth, but by maturity the two are difficult to distinguish and the ratio becomes 1 : 4 : 6 : 4 : 1.

2.7.4. Two autosomal genes producing phenotypes through epistatic interaction

Epistasis is the interaction of alleles from two or more loci to produce a phenotype that neither gene produces by itself. Epistatic interaction between two loci produces variations on the general 9 : 3 : 3 : 1 F2 phenotypic ratio and F2 phenotypes are usually reduced from four to either two or three depending on the type of epistasis. Scale pattern in common carp, body color in Goldfish and Siamese fighting fish are controlled by epistatic interaction.

(i) *Dominant epistasis* - Dominant epistasis occurs when a dominant allele at one locus (epistatic locus) produces a particular phenotype, regardless of the genotype at the second locus. The dominant epistasis produces a 12 : 3 : 1 F2 phenotypic ratio.

Dark goldfish female (Mm, Ss)		X		Dark goldfish male (Mm, Ss)
	MS	Ms	mS	ms
MS	MM,SS dark	MM,Ss dark	Mm, SS dark	MmSs dark
Ms	MM,Ss dark	MM, ss dark	Mm,Ss dark	Mmnn dark

ms	mM, SS dark	mM,Ss dark	mm,NN light	mmNn light
ms	Mm,Ss dark	mM,nn dark	mm,sS light	mm,ss albino

Phenotypic ratio = 12, dark : 3, light : 1, albino.

If the epistatic locus is dominant lethal the F2 phenotypic ratio deviates from the classic 12 : 3 : 1 F2 phenotypes. The scale pattern in common carp explains such statement (Fig. 66).

The S gene controls the scaliness and the N gene modifies the pattern. The N allele is lethal in homozygous dominant state. The all possible genotypes and phenotypes for scale pattern in common carp are :

Genotypes	*Phenotype*
SSnn	Scaled
Ss, nn	Scaled
ss,nn	mirror

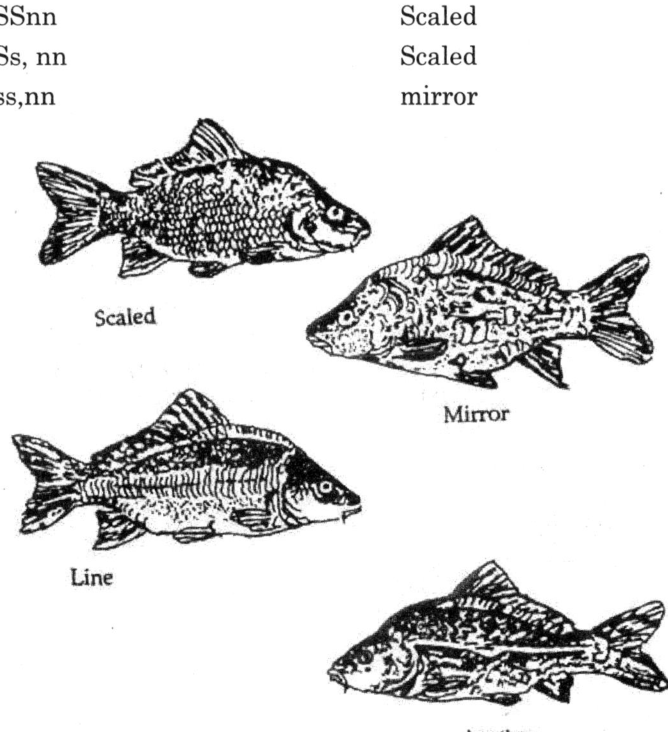

Fig. 66. Scaled, mirror, line and leather common carp.

Source : After Kirpichnikov (1981). Courtesy of V.S. Kripichnikov and nauka publishers.

SS,Nn	line
Ss,Nn	line
ss,Nn	leather
SS,NN	death
Ss,NN	death
ss, NN	death

By putting punnett square in mating two heterozygous (Ss Nn) line common carp will give the following F2 phenotypic ratio.

Line common carp female Ss, Nn		X	Line common carp, male Ss, Nn	
	SN	Sn	sN	sn
SN	SS,NN Death	SS,Nn line	Ss,NN death	Ss,Nn line
Sn	SS,Nn line	SS, nn scaled	Ss,Nn line	Ss,nn scaled
sN	Ss, NN death	Ss,Nn line	ss,NN death	ss,nN leather
sn	Ss,Nn line	Ss,nn scaled	ss,Nn leather	ss,nn mirror

Phenotypic ratio : 4 death : 6 line : 2 leather, 3 scaled : 1 mirror

(ii) *Recessive epistasis* – It occurs when the recessive genotype at one (epistatic locus) suppresses phenotypic expression by the other locus. Recessive epistasis produces a 9 : 3 : 4 F2 phenotypic ratio. Eye color in mexican cave characins is an example of a phenotype controlled by recessive epistasis. The punnett square and F2 phenotypic ratio for the mating of two heterozygous ab+, bw+ black eyed mexican cave characins are :

Black eyes female (ab+, bw+)			Black eyes male (ab+, bw+)	
	+	+bw	ab+	abbw
+	black eye	black eye	black eye	black eye
+bw	black eye	brown eye	black eye	brown eye
ab+	black eye	black eye	pink eye	pink eye
abbw	black eye	brown eye	pink eye	pink eye

Phenotypic ratio : 9 black : 3 brown : 4 pink

(iii) *Duplicate genes with cumulative effects* - Duplicate genes with cumulative effects occur when both genes produce the same phenotype and one genotype has atleast one dominant allele to produce a third phenotype. This is called a cumulative phenotype. Duplicate genes with cumulative effects produce a 9 : 6 : 1 F2 phenotypic ratio. Trunk striping in the sumatran barb is an example of a phenotype controlled by duplicate genes with cumulative effect (Fig. 67). The punnett square and F2 phenotypic ratio for the matting of two heterozygous Aa, Bb completely banded sumatran tiger barb are as follows.

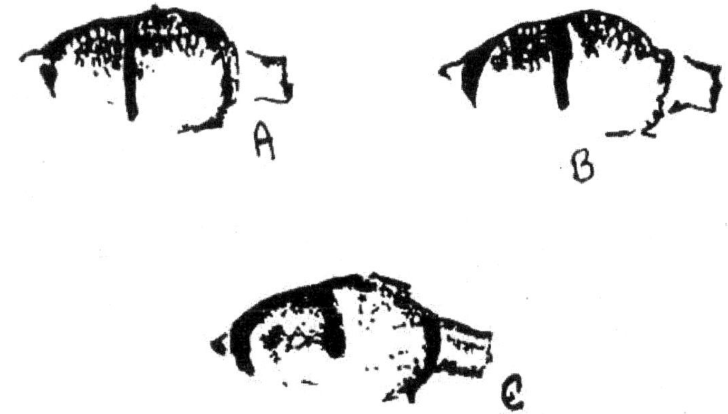

Fig. 67. Trunk striping in the Sumatran tiger barb (A) completely banded : (B) incompletely banded : (C) half-banded.

	Completely banded (Aa, Bb)		X Completely banded male (Aa, Bb)	
	AB	Ab	aB	ab
AB	Complete band	Complete band	Complete band	Complete band
Ab	Complete band	Incomplete band	Complete band	Incomplete band
aB	Complete band	Complete band	Incomplete band	Incomplete band
ab	Complete banad	Incomplete band	Incomplete band	Half band

Phenotypic ratio = 9 completely banded

6 incompletely banded

1 half banded

(iv) *Duplicate dominant gene interaction* – In duplicate dominant gene interaction, the dominant alleles at the two loci produces the same phenotype, but there is no cumulative effect. Duplicate dominant gene interaction produces 15 : 1 F2 Phenotypic ratio.

Transparent scales due to depigmentation of melanophores in goldfish is an example of a phenotype controlled by Duplicate dominant gene interaction. The punnett square and F2 phenotypic ratio for the mating of two heterozygous (Dp1 dp1 Dp2 dp2) transparent scaled gold-fish is presented below.

Transparent scaled female (Dp1 dp1 Dp2 dp2)		X	Transparent scaled male (Dp1 dp1 Dp2 dp2)	
	Dp1 Dp2	Dp1 dp2	dp1 Dp2	dp1 dp2
Dp1 Dp2	Transparent scale	Transparent scale	Transparent scale	Transparent scale
Dp1 dp2	Transparent scale	Transparent scale	Transparent scale	Transparent scale
dp1 Dp2	Transparent scale	Transparent scale	Transparent scale	Transparent scale
dp1 dp2	Transparent scale	Transparent scale	Transparent scale	Pigmented scale

Phenotypic ratio : 15 transparent scale : 1 pigmented scale

(v) *Duplicate recessive gene interaction* - In duplicate recessive gene interaction, the two recessive genotypes at each locus produce the same phenotype. Only when a dominant allele is present at both loci, another phenotype is produced. Duplicate recessive gene interaction produces 9 : 7 F2 phenotypic ratio.

(vi) *Dominant and recessive interaction* - In dominant and recessive interaction, the dominant genotype at the first locus and the recessive genotype at the other locus produces the same phenotype. Such dominant and recessive interaction produces a 13 : 3 F2 phenotypic ratio.

2.7(b) Linkage

Morgan's concept of linkage is the linear arrangement of genes in the chromosomes. The chromosome theory of linkage of Morgan and Castle states that.

1. The genes which show linkage are situated in the same pair of chromosomes.
2. The linked genes remain arranged in a linear fashion on the chromosome. Each linked gene has a definite and constant order in its arrangement
3. The closely located genes show strong linkage than the widely located genes which show weak linkage.
4. The linked genes remain in their original combination during the course of inheritance.

(i) Arrangement of Linked genes

When two pairs of genes are linked, the linkage may be

(1) two dominant genes (RR_o) in one and two recessive genes ($r\ r_o$) in other

(2) one dominant and one recessive gene ($R\ r_o$) in one and one recessive and one dominant gene ($r\ R_o$) in other chromosome.

The first arrangement with 2 dominant genes on the same chromosome is called Cis arrangement and the interrelation of linkage of genes (RR_o/rr_o) is called coupling phase. The second arrangement of one dominant and one recessive gene on the same chromosome is called trans arrangement and the interrelation of linkage of gene ($R\ r_o/\ r\ R_o$) is called repulsion phase.

(ii) Kinds of Linkage

1. *Complete Linkage:* When linked genes are inherited for two or more generations in a continuous and regular fashion then they are called completely linked genes and phenomenon of inheritance is called complete linkage.
2. *Incomplete Linkage :* When linked genes do not stay together but exchange due to crossing over, the linked genes are widely located in chromosomes and have chances of separation by crossing over are called incompletely linked genes and the phenomenon of their inheritance is called incomplete linkage.

(iii) Linkage Group

Linkage groups of a species corresponds with haploid chromosome number of that species. For example Drosophilla has 4 pairs of chromosomes and 4 linkage groups. Similarly Man has 23 pairs of chromosomes and 23 linkage groups.

(iv) Significance

Linkage reduces the possibility of variability in gametes unless crossing over occurs.

2.7(c) Crossing Over

The reciprocal exchange of genes between chromosomes of homologous pairs is performed by a process called crossing over. Crossing over produces new combinations of genes by interchanging of corresponding segments of homologous chromosomes.

(i) Kinds of Crossing over

1. *Single crossing over:* When only one chiasmata occurs only at one point of the chromosome pair it is called single crossing over.
2. *Double crossing over:* When two chiasmata occurs at two different points of a chromosome pair it is called double crossing over.

Factors affecting crossing overs are age, sex, temperature, water content, ionizing radiation, certain chemicals and chromosomal aberration such as inversion etc.

(ii) Significance

1. For constructing genetic maps of the chromosome.
2. Evidence on linear arrangement of linked genes in chromosomes
3. Increases frequency of genetic variation.

2.8 Sex linked genes

Qualitative phenotype not only influenced due to autosomal gene interaction but also by sex linked genes. The phenotypic variation due to sex linked gene are different to that of the phenotypic variation due to autosomal genes. It is because, a species of one sex is homozygous (female = XX) while the other sex is heterozygous (male = XY). Sex linked phenotypes are studied in the guppy and platyfish. Till date, the sex linked phenotypes in fishes are controlled by genes located in X and/or Y chromosomes.

Involvement of W or Z chromosomes in sex linked phenotypes have not been discovered.

2.8.1. Y-linked genes

The phenotypes controlled by Y-linked genes in guppy include-

(1) Maculatus pigmentation

(2) iridescens pigmentation

(3) armatus pigmentation

(4) sanguineus pigmentation

(5) pauper pigmentation

(6) oculatus pigmentation

(7) ferrugineus pigmantation

(8) double sword tail

(9) filigran pigmentation

2.8.2 X-linked genes - X-linked genes in guppy includes -

(1) tigrinus pigmentation

(2) coccineus pigmentation

(3) vitellinus pigmentation

(4) cinnamoneus pigmentation

(5) luteus pigmentation

(6) elongatus pigmentation, lengthening of the caudal fin

(7) nigrocaudatus pigmentation, type II

(8) caudalis pigmentation

The involvement of Y-linked genes can be explained in mating grey female guppy (XX) with maculatus male (XY ma).

Grey female	X	maculatus male
XX		XY ma
female gamete		male gamete
	X	Y ma
X	XX	Xyma

Genotypic ratio = 1 XX : 1 Xy ma

The involvement of X linked gene can be explained by mating caudalis female guppy (Xch XCp) with caudalis male guppy (XCp, Y).

		male gamete XCp	Y
Female	XCp	XCp XCp Caudalis female	XCp Y caudalis male
	Xch	Xch XCp caudalis female	Xch Y transparent tailed male

Genotypic ratio = 1 XCp XCp : 1 XCp x ch

1 XCp y : 1 Xch Y

Phenotypic ratio = 2 caudalis female

1 caudalis male

1 transparent tailed male.

2.9 Sex limited phenotypes

X-linked allele follow definite genetic pattern but the phenotypes do not follow the expected ratio. This is because some X-linked phenotypes (and autosomal phenotypes) are sex-linked. Many sex limited phenotypes need testosterone in order to be expressed. Most of the X-linked genes in the guppy produce phenotypes that are sex-limited to the males. The addition of methyltestosterone to the water or feed will allow the phenotypes to be expressed in females.

2.10 Genes with multiple alleles

In a population, the number of alleles for a given gene can vary from one to well over a dozen. Many genes have three or more alleles.

Such genes with multiple alleles can give possibility of different phenotypes and genotypes. The body color of medaka through melanophore deposition is an example of an autosomal gene with three alleles. Sex-linked genes in medaka have multiple alleles, that are responsible for carotenoids in the xanthophores for color determination. In the platyfish, a gene with numerous alleles even up to 9 is known which controls the spotting pattern (Fig. 68). The spotting pattern may be as one spot, dot type, moon type, crescent type, twin spot, comet type, moon complete,

complete crescent or even unspotted (wild type) are reported. (See Tave, 1986).

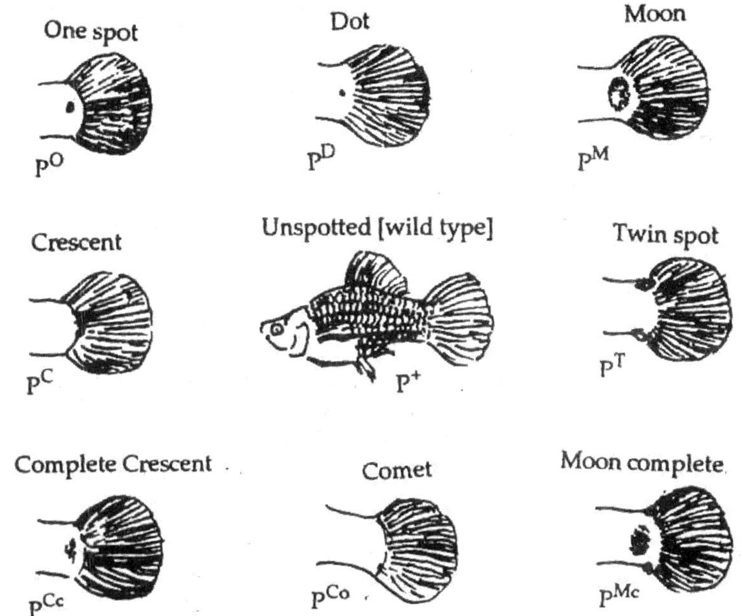

Fig. 68. Spotting pattern in the platyfish controlled by the P gene. There are nine alleles : P^O, P^D, P^M, P^C, P^T, P^{Cc}, P^{Co}, P^{Mc}, and P^+. The P^+ allele is recessive to all other alleles which are all co-dominant.

2.11 Genes exhibit pleiotropy

It is believed that a gene produces a specific phenotype in a one to one process. In other words a particular genotype usually produces a particular phenotype. But this assumption is not always correct. It is because that even a single gene can influence multiple phenotypes. This is possible, because genes affect biochemical pathways. As biochemical pathways are interconnected, genes influence phenotypes in multiple ways. Replacement of one allele for another allele can affect more than one biochemical pathway and thus phenotypes can be altered. Such a secondary phenotype is due to pleiotropic effects. Pleiotropy has been extensively studied in common carp. The pleiotropic effect on color genes in common carp is often used as genetic markers to identify a particular group. Besides this the scale patterns such as scales, mirror, leather and line common carps are in some way influenced due to pleiotropic effect. The pleiotropic effect in Tilapia causing vertebral deformities, abnormal pelvic, pectoral and anal fins and many other phenotypes are well documental by different authors (see Tave, 1986).

2.12 Quantitative phenotypes

Quantitative phenotypes are those that are measured such as length, weight, gill rakers, pharyngeal teeth, fin rays on fin etc. A sizeable carp has good demand in market than the inferior ones. Similarly numerous intramuscular bones of many fish species have a negative effect on marketing. So carps with less intramuscular bones have relatively good demand. There are ways of breeding carps for less intramuscular bones. By systematic selection it is possible to reduce the number of intramuscular bones (14) on carps. But much success in this line needs extensive research. In many trout hatcheries, through selective breeding, high fecundity, large egg size, high hatching percentage, rapid growth and early maturity have been obtained in some countries. Even production of ``super trout" with almost every character of importance to the culturist is met through selective breeding. Besides finfish, quantitative phenotypic improvement are taken up in shell fishes also. Research in selective breeding of *Macrobrachium,* Lobster (*Homarus americanus*), Oyster (*Crassostrea virginica*) are in progress in many countries. Selection for disease resistance in fish particularly infectious dropsy in common carp has been reported. Similarly river strain of *Salmo salar* against vibrio and some strains of brown trout and brook trout against furunculosis and ulcer diseases are well documented. The behaviour characteristics of sport fishes are also improved by selection. Genetic variation in the ability of common carp to escape from fishing nets is reported. The wild European carp and Chinese big belly carp are able to escape seine much better than their relatives.

The quantitative phenotype in quantitative way can be explained in the following examples.

Question - Find out the selection differential in weight of the F1 channel catfish offsprings considering the average Weight of selected female parent as 604 g and average Weight of selected male as 692 g and the average Weight of the channel catfish population as 454 g from where these above parents were selected. The heritability for channel catfish is 0.5 as reported by Dunham and Smitherman 1983 (see Tave, 1986).

Answer - **The selection differential = S**

$$s = \frac{average\ wt.\ of\ selected\ female + ave.\ wt.\ of\ selected\ male}{2} - population\ average$$

$$S = \frac{604\ g + 692\ g}{2} - 454\ g$$

S = 194 g.

The response to selection (R) = $\frac{Sh^2}{L}$

The generation is considered as 1 in this case and heritability as 0.5 in channel catfish.

$$R = \frac{194\ g \times 0.5}{1} = 97\ g$$

So the F1 generation of channel catfish on an average should be more than 97 g. than the average parent population.

Hence, F1 = P1 average weight + response to selection.

So F1 = 454 g. + 97 g.

F1 = 551 g.

In the above case only weight of the selected individual was considered. If more parameters will be taken for quantitative phenotypic estimation, the selection index has to be employed. This is explained in the following example.

Question - The following values are given with reference to phenotypes, population average and relative importance.

Phenotype	Popl. ave.	Relative importance
Weight	454 g	50%
dressing %	60%	30%
gain day	3 g/day	20%

Find out the approximate selection index for breeding programme. Compare it with following two populations with average value as :

Group I	*Group II*
Wt. 544 g.	589

dressing 63% 62%

gain/day 4g/day 3.2 g/day

Answer –

$$\text{Important factor} = \frac{relative \quad impor \tan ce}{pop \ln average}$$

$$\text{I weight (Iw)} = \frac{50\ \%}{454\ g} = 0.1101322$$

$$\text{I dressing (ID)} = \frac{30\ \%}{60\ \%} = 0.5$$

$$\text{I gain/day (IG)} = \frac{20\ \%}{3\ g\ /\ day} = 6.6666667$$

So the approximate selection index for this breeding programme is

I = Iw (wt) + ID (dressing) + IG (gain/day)

I avg. = (.11011322 X 454 g) + (0.5 X 60) + (6.66 X 3g) = 100.0

For group I

I (Group I) = (0.1101322) (544 g) + (0.5 X 63) + (6.66 X 4)
= 118.079

II (Group II) = (0.1101322 X 589) + (0.5 X 62) + (6.66 X 3.2)
= 117.210

Hence, Group I is better than Group II in terms of more quantitative phenotypic improvement. If such phenotypic changes are due to heterosis the percentage of increase in phenotypes can be expressed as illustrated earlier.

2.13 Hybridization

Hybridization is usually, termed to denote the process of crossing two different species. However, in nature hybridization occurs among fishes. Such hybridization takes place between the closely related genera. With the success of hypophysation in carps, hybridization can be operated in a planned and controlled manner inorder to improve the quantitative phenotypes of the population stock. Hence, hybridization can be defined in broad sense as a process of heterospecific insemination. Depending on the nature of the ploidy or the set of chromosome either (N, 2N) of the genetic material of the gamete, hybridization results in the production of (1) parthenogenetic, gynogenetic or androgenetic individual according to the origin of the genetic stock and (2) diploidy, triploidy or tetraploid hybrids. However, diploid hybrids are commonly produced for aquaculture on a commercial scale for increased production.

The main aim of hybridization is to obtain both qualitative and quantitative phenotypic improved fish stock than their parents with high productive potential. Hence, hybridization is usually performed to obtain the associated advantages of heterosis, which otherwise known as hybrid vigour. Hybridization produces an extensive gene pool with pronounced heterozygosity. The success of hybridization can be known by evaluating the genome status of the fish. Evaluation of genome status can be known by (1) karyotypic studies. Because production of fertile or sterile hybrids depend on the karyotypic number and their compatibility or non-compatibility. If the species having equal number of karyotypes (chromosomes), homologous and are able to conjugate in the process will result fertile hybrids. Where as, unequal chromosome number and violation of conjugation process, result the production of sterile hybrids in varying degree. Generally hybrids of Indian major carps are fertile where as the hybrids produced between Indian major carps and Chinese carps are sterile (Khan *et al.,* 1990). The sterile hybrids are very valuable for certain specific purpose in fish culture as regards to regulation of their own population. The fertile hybrids, on the other hand are valuable for obtaining new hybridigenic forms through selection in successive generations.

The process involved in artificial hybridization is the artificial insemination. The egg and milt of choosen fishes especially carps are mixed through dry or wet method. In case of common carp, the fertilised eggs are degummed by rinsing with milk or sodium sulphite + tannic acid or NaCl + urea and then wased thoroughly and repeatedly with water.

The principal ways of obtaining hybrids by artificial hybridization is the crossing of selected species. So crossing is the basis of fish improvement. It aims at overall increase in genetic variability or heterozygosity, so that the new breed have (1) improved productive quality, (2) environmental adoptability and

(3) resistance to disease. The crossing is carried out in two ways (1) commercial crossing and (2) synthetic crossing.

2.13.1 Commercial crossing

It is directed towards producing F1 hybrids for commerical purpose that exhibit high quantitative phenotypes.

2.13.2 Synthetic crossing

It is directed with the objective of developing a new breed from one generation to successive generations, still genetic stability in the new breed is established. By synthetic crossing, useful properties of the strain, varieties, species or even genera can be combined to produce a genetically stabilised new breed. The synthetic crossing can be carried out in many ways. These are (i) reproductive cross (ii) introductory cross (iii) absorbing cross and (iv) alternate cross.

(i) *Reproductive cross* - Useful traits are combined from hybrids or species. This is achieved when hybrids are completely fertile. The examples of hybrids produced through reproductive cross are (1) catla rohu (2) Hungarian carp (3) Ukraman carp breed.

A X B

F1 X F1

F2 X F2

F3 X F3

(ii) *Introductory cross* - One or several valuable traits of a species are required to be introduced into the hybrid. Hence, F1 hybrids are back crossed manytimes to improve the genetic stability.

A X B

F1 X A

F2A X A

F3 A X A

(iii) *Absorbing cross* - Here, a series of back cross is completed after the intial cross of two strains. Hence, hybrids are repeatedly crossed with donor improved strain but not of local strain.

A X B

F1 X B

F2b X B

F3b X B

(iv) *Alternate cross* - This requires intermitent crossing of hybrids with two initial breeds. After 3 or 4 generations, the alternate cross is replaced by reproductve one new hybrid breed.

A X B

F1 X A

F2 X B

F3 X A

In China, a new breed of common carp known as *Heyuan carp* is achieved upto F6 generation. This Heyuan carp is also known as Jiang carp. Jiang is a Chinese term which means established. As genome constituent of the Jiang carp is stabilised from F4 - F6 generation. This is the best breed than 19 hybrids so for produced in China. Out of 19 hybrids, following 4 hybrids, are cultured in polyculture practices in China.

(1) *Cyprinus carpio wuguanensis* X *C.C. yuenking* (Heyuan carp)

(2) *Carassus carassius gibelio* X Kuo red carp

(3) *Sarotheradon mossambicus* female X *S. niloticus* male (Fushou fish)

(4) *S. nilotica* female and *S. aurea* male = (Auni fish)

In India, interspecific hybrids such as *Labeo calbasu* X *Labeo rohita, Labeo rohita X L., calbasu L. bata X L. rohita, L. bata X L. calbasu, L. calbasu X L. gonius* were produced. Out of these interspecific hybrids, *Labeo calbasu* male X *Labeo rohita* female (Calbasu rohu), *Labeo rohita* X L. *Calbasu* (Rohu-calbasu) were extremely viable (see Choudhuri and Singh, 1984).

The intergeneric hybrids such as *Labeo rohita* X *Catla catla* and *Catla catla* X *Labeo rohita* were produced. All the hybrids show intermediate taxonomic characters. Many other intergeneric hybrids of indigenous carps have been produced by crossing 5 species belonging to 3 genera. Even outcrossing of intergeneric hybrids have been produced. These outcrossing intergeneric hybrids produced are catla mrigal calbasu, calbasu mrigal calbasu and mrigala mrigal calbasu.

Hybridization of Indian major carps with exotic carps were met with success (Khan *et al.,* 1990, Reddy *et al.,* 1990 and Gupta *et al.,* 1986). But the hybrids between IMC and Chinese and common carp are generally non-viable but most of those among IMC either interspecific or intergeneric are viable and fertile. About 35-40% mortality occurred among common carp X rohu hybrids (Khan *et al.,* 1986). That too deformed hatchlings died within 2 to 3 days. Similar observations has been reported in hybrids between common carps and Chinese carps, catla X silver carp, rohu X common carp, mrigal X common carp and common carp X mrigal (Gupta *et al.,* 1986). Common carp X catla hybrids were also produced in CIFA, K. gang under the project ``Genetic improvement of culturable carps through selection".

In regard to growth of carp hybrids, it has been reported that rohu-catla hybrid grow faster than the parent (Choudhuri, 1973) Varghese and Sukumaran (1971). In regard to flesh content, rohu X catla, calbasu catla and catla calbasu hybrids show relatively high percentage of flesh than parents. Similarly common carp X rohu and common carp X mrigal hybrids contained more flesh than common carp but lesser than rohu or mrigal (Khan *et al.,* 1986; Gupta *et al.,* 1986).

Therefore, hybridization is a rapid route to genetic improvement. Of all carp hybrids produced in India, rohu X catla and catla X-rohu are fertile and most promising. So it is suggestive that the best genotype of catla and rohu has brought significant, quantitative phenotype improvement in hybrids.

2.14(a) Sex manipulation

Certain fishes such as common carp and tilapia show prolific breeding behaviour and breed in ponds several times in a year, thereby over populate the same pond system. Such overpopulation increased the stock density resulting stunted growth of the species and other fishes stocked together. Hence, fish production is severely affected. In order to control their prolific breeding, sex manipulation for either monosex or sterile individuals are primarily required. Usually fish drain out 20% of their assimilated/food energy on reproduction which may affect on growth, survivality and flesh quality. Hence, prolific breeders drain out maximum assimilated food energy for breeding, which could otherwise be turned into flesh. Therefore, research efforts were focused since seventies on channelising this energy for somatic growth by suppressing the gonadal maturation and consequent gamete production. To achieve such objective, two techniques such as hormonal and chromosome manipulation techniques have been evolved so far. In this context, the knowledge on sexuality in fish is required.

2.14.1 Sexuality in fishes

The sexuality of fish has great significance because there are great variations in the growth rate, size, colour, breeding pattern, spawning time etc. However, within bony fishes a wide range of reproductive strategies are met. A great majority of fishes are either differentiated (e.g., platyfish, medaka) or undifferentiated (majority of fishes) gonochorists. Gonochorism means existence of either testes or ovaries in one individual. In some fishes e.g., carps, gonads first develops in to an ovary like gonad, then about one half of the individuals become males and other half females. These are called undifferentiated gonochorists. In the differentiated gonochorists fish, the gonad is directly differentiated into either ovary or testes (e.g., *Xiphophorous maculatus).* However, numerous species show hermaproditism as such fish bear recognisable male and female tissues. The hermaprodite fishes may be synchronous (balanced where both eggs and spermatozoa mature at the same time (fishes belonging to Serranidae). In some hermaprodite fishes, ovaries are first developed and function as females and there after into males. Such fishes are called protogynous hermaprodite (e.g., *Pagellus erythrinus*). However, in protandrous hermaprodite fishes (e.g., *Sparus auatus*), firstly testes are developed and serve as males and then they transform into other sex. In this process of transformation a transitory intersexual stage is encountered.

Natural sex reversal is observed in many groups of serranids, sparids and salmons. Both extrinsic and intrinsic factors play an important role in sex-reversal. Day length, water temperature, drought and crowded conditions are the main extrinsic factors responsible for natural sex-reversal in some numbers of cyprinodontids and stickle backs. The extrinsic factors influence endocrine secretion resulting in excessive release of androgens or estrogens resulting in sex reversal. Predominantly male or female populations also bring in sex reversal in labridae.

Like other vertebrates, sex in fish is genetically controlled through sex chromosomes. The sex chromosomal system in females are homogamety (XX, YY, WW) and in males are heterogamety (XY, WY, ZY). But many exceptions are on record. Tilapia males are ZZ, YY, XX (homogamety) and females are heterogametic. Similarly, medaka females are heterogametic (XY) and swordtail females are heterogametic WZ (Y) and males are ZZ (YY) homogametic. Overall, the XX : XY system for female and male respectively appears to be more dominant in fishes.

2.14.2. Hormonal sex manipulation

This studies in teleosts include three phases, such as : management, treatment and evaluation. For production of female or monosex population, the gametes from the (XX male) sex inverted homogametic individuals are mixed with the gametes of unrelated homogametic fish (XX female). The progeny would be homogametic (XX) and therefore of the same sex.

The other phase i.e. treatment includes selection of an appropriate hormone, method and dose of its application. If the dose of the hormone is high, it may lead to sterility. However, optimal sterile dosage appears to be species specificity. The treatment of the hormone should be initiated prior to the commencement of normal sex differentiation and continued until the time of morphological differentiation. Use of steroids to produce monosex are extensively studied in tilapia and common carp in the country.

2.14.3. Hormones used

Several hormones like synthetic androgen, 17 alpha methyl testosterone (MTS), 17 Beta methyl testosterone, 17-estradiol (Diethylstilbestrol-DES), Estradiol benzoate, Testosteroneacetate, etc. have been used to supress gonadal development and to induce monosex or sterile fishes (Fig. 69). Use of mibolerone in sex-reversal is also documented in fish.

2.14.4. Dosage and species

In commoncarp 17 α and 17 β methyl testosterone (MTS) was applied @ 100 mg/kg diet for a period of 36 days and produced male predominant population. In tilapia group, MTS varies from 3-30 mg/kg feed for 21-56 days and in salmon, MTS varies from 3-6 mg/kg feed for 60-90 days to produce dominant male populations.

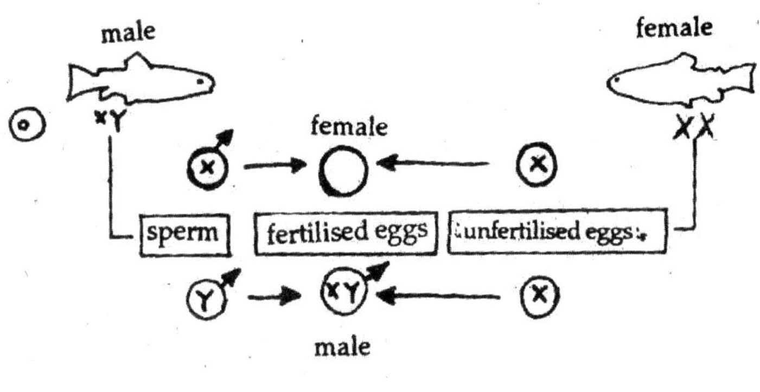

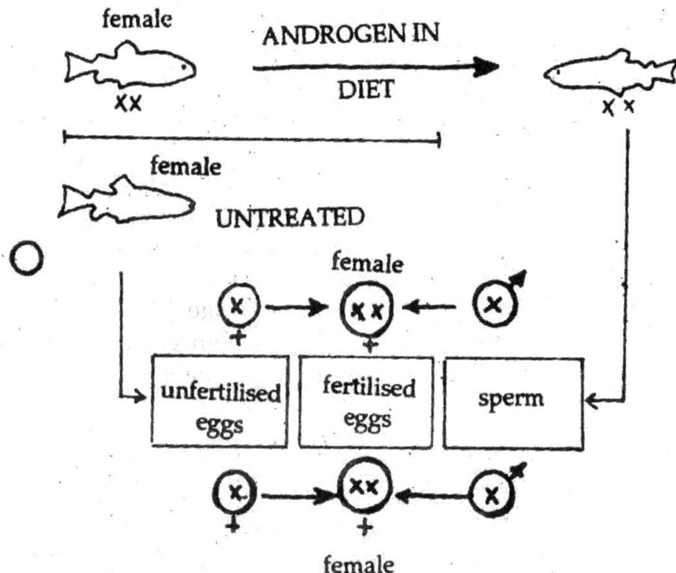

Fig. 69. Flow diagram of hormonal manipulation of sex in fish.

Also 100% sterile population of common carp can be produced by feeding 3 day old hatchlings in a diet containing 15 mg of mibolerone/kg diet for 30 days.

For predominant female population production, estradiol benzoate at 800 mg/kg of feed, and testosterone acetate at 300-400 mg/kg feed is used. The use of above hormones did not show any adverse effect in the organoleptic quality of common carp.

In Tilapia, for obtaining 100% males, artificial feed supplemented with 17 alpha MTS at the rate of 3-30 mg/kg of feed should be given to the fish fry within 10 to 20th day of hatching for 30 days.

Similarly the effective dosage of DES varied from 20-100 mg/kg diet for 25-70 days for salmon and trout and 30-50 mg/kg feed for 30-40 days for tilapia can produced 100% female population.

2.14.5 Enhanced growth through sex reversal

In fishes where growth is sex specific e.g., in Salmon, carps, trouts, mullets etc., females are known to grow faster than male, dominant female population by hormone application is advisable. The quantitative phenotypic superiority of male in population of tilapia by hormone application is advisable. Since use of steroids in enhancing growth of animals has opened up high controversy, a combination of chromosomal technique like gynogenesis (female individual), androgenesis (male individuals) and ploidy particularly triploidy sterile individuals would result better insight in sex control. Other methods to induce sterility includes surgical castration, and autoimmune gonad rejection. Surgical Castration is impractical for some organisms and not economical for farmed animal. Autoimmuno castration involves the injection of minced gonad into immature fish to encourage the production of antibodies which will destroy the fish's own gonad as it starts to develop. But this method is still unconvincing and not utilised.

2.14(b) Chromosomes

The chromosomes have been considered as the physical bases of heredity because they have a special organization, individuality, functions and are capable of self reproduction. They occur in all living beings in a specific number and usually fall in to following categories:

1. *Viral Chromosomes*- Single strand, contains DNA or RNA. Viral chromosome may be either of linear shape or circular shape.

2. *Prokaryotic Chromosomes*- Single, giant and circular chromosome. It has double stranded DNA molecules but has no protein and RNA around the DNA.

3. *Eukaryotic Chromosomes*- The chromosomes of plants and animals contain much more genetic information than virus and

prokaryotic chromosomes. Chromosome occurs as many units in eukaryotes but has always constant and characteristic number of chromosomes. For example Gold fish (*Carassius auratus*) has 2n= 94, and Man has 2n = 46 chromosomes. As per Annual report of NBFGR 2007 – 2008, the cytogenetic characteristic of few fishes are listed below: *Rita rita* 2n =54, *Aorichthys seenghala* 2n =54, *Tor tor* 2n =100, *Tor putitora* 2n=100, *Tor khudree* 2n =100, *Tor mussullah* 2n =100, *Nandus nandus* 2n=48 *Clarias gariepinus* 2n= 46, *Oreochromis niloticus* 2n =118, *Labea calbasu* 2n =50, *Puntius ticto* 2n=50, *Arius subrostratus* (Marine Fish) 2n=58, *Zanclus cansescens* 2n=48, *Thallasoma lunare* 2n=48.

Eukaryotes are diploids i.e, each somatic cell contains one set of chromosome inherited from female parent and a comparable set of chromosome from the paternal (male) parent. The number of chromosome in a dual set of a diploid somatic cell is called diploid number (2n). The sex cells of a diploid eukaryote cell contain half the number of chromosomal sets found in the cell and are known as haploid cells (n). A haploid set of chromosome is called genome. The fertilization process restores the diploid number of a diploid species.

(i) Morphology of Chromosome

The shape of eukaryotic chromosome is changeable during the process of cell growth and cell division. They are thin, coiled, thread like called chromatin threads or chromatin materials at the early stages of cell growth. These chromatin threads become highly coiled and folded to form ribbon shaped chromosome during metaphase of mitosis. These chromosomes contain a clear zone called centromere along their length. According to the number of centromere the eukaryotic chromosomes may be 1. acentric (without any centromere), 2. Monocentric (with one centromere) 3. diecentric or polycentric (with more than one centromere). The centromere has small granules and divided the chromosomes in to two or more equal or unequal chromosomal arms. According to the position of the centromere, the eukaryotic chromosomes may be rod shaped (telocentric and acrocentric), J shaped (sub-metacentric) and V shaped (metacentric). A diagramatic representation of the complete chromosome set of an individual arranged according to decreasing size and / or accepted numbering system is called Idiogram.

(ii) Chemical structure of Chromosomes

Chemically, the eukaryotic chromosomes are composed of DNA, RNA, histone and non-histone proteins and contain metallic ions. The most important enzymatic proteins of chromosomes are phosphoproteins, DNA polymerase, RNA polymerase, DPN-pyrophosphorylase and nucleoside triphosphatase. The metal ions as calcium and magnesium are in the chromosomes.

(iii) Kinds of Chromosomes

Eukaryotic chromosomes have been classified in to autosomes and sex chromosomes. Special types of chromosomes are (1) Polytene chromosomes found in the salivary gland cells of drosophila (2) Lampbrush chromosomes found in the nuclei of yolk rich oocytes of vertebrates.

2.14(c) Chromosomal aberrations

Chromosomes are the structures with definite organization. They carry genes in a definite linear order. Ordinarily chromosomes remain unchanged but under certain natural or artificial adverse circumstances certain structural changes may occur in the chromosomes, which alter the positions of gene or loss of some genes. These structural alternations affect the phenotype of the organism in various degrees and collectively called chromosomal aberrations.

(i) Types of chromosomal aberrations

The chromosomal aberrations may remain confined to a single chromosome or may extend to both of the members of the homologous pair and there fore may be of following types:

(1) Intra-chromosomal aberration: When aberration remain confined to a single chromosome of a homologous pair, they are called intra-chromosomal aberrations. These may occur due to (a) Deletion that may cause deficiency of required genes and cause lethal effect (b) Additions that may cause protection of organism and useful for evolution (c) Inversion which may be pericentric that includes the inversion of centromere or paracentric which does not include inversion of centromere that may result mutation.

(2) Inter chromosomal aberration: In such case there is interchanging of both of the homologous chromosomes. This may occur due to translocation (shifting of a part of one chromosome to another non-homologous chromosome).

Most chromosomal aberrations are caused due to the accidental, natural or induced breakage of chromosomes. The induced chromosomal breakage is caused by radiations, chemicals like Colchicines, Lysergic acid dimethylamide (LSD).

2.15 Fish Genomic

The term "genomic" is derived from the term "genome" that has been used to refer to organisms complete set of chromosomes. The genomic can be dealt as (1). **Structural genomics** that deals with the construction of high-resolution genetic, physical and transcript maps of organism. (2). **Functional genomics** that deals with functional aspects of gene

2.16 Chromosomal manipulation

Selective breeding and hybridization are the classical methos of improving fish stocks but are long term programmes. Therefore, simple genome manipulation techniques like gynogenesis, androgenesis and sex reversal would reduce the period for stock improvement to less than ten years. These techniques help in building up highly inbreed lines of desired traits in a short time. The success in chromosome manipulation depend on the operator's knowledge on gamete biology, hypophysation, irradiation of gametes and use of sex steriods. However, gamete biology, hypophysation, use of sex steriods have been discussed and in the present context irradiation of gamete is described.

2.16.1 Irradiation of gametes

For denaturing the nucleus the gametes are subjected to irradiation. UV rays offer an easier, economical and safer way of irradiation than gamma or X rays. However, UV rays have lesser penetrability power. UV irradiation takes about 15-20 minutes from a 15 watt tube at a distance of 20 metre from the sample. Chemicals like cytochalsin-B, Colchicine, dimethylsultate, ethylene, urea etc. have also been used for inactivation of the nucleus.

2.16.2 Gynogenesis

In gynogenesis, the irradiated spermatozoa enters the micropyle of the egg and activates development, but then degenerate without any paternal genetic contribution to the individual (Fig. 70). Hence, the haploid (n) embryo which contain only maternal chromosomes is produced (Fig. 71). Work on induced gynogenesis of fish started in Russia for production of gynogens of loach, sturgeon and common carp. This technology spread to U.K. and Europe for production of gynogens in flat fishes, salmon, trout, common carp, grass carp and silver carp, in USA production of gynogens in common carp grass carp sturgeon, trout and salmon are recorded. In India, gynogens in major carps and silver carps were produced.

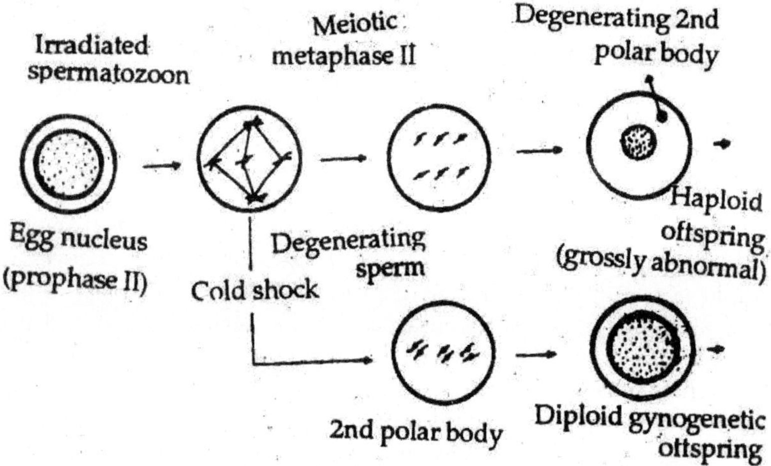

Fig. 70. Schematic representation of gynogenesis

(a) *Gynogenetic diploid production* - The gynogenetic haploid zygote is made gynogenetic diploid through administration of (i) thermal (heat/cold), (ii) hydrostatic pressure/shock, at the expected time of formation of second polar body or prior to first mitotic division of the zygote (Fig. 71). High rate of success in gynogenesis depends on (1) ripeness of egg, (2) time of shock administration as measured from the starting of activation, (3) sublethal nature of shock in terms of intensity and duration. In Indian major carps heat shocks of 30-42°C for 1-2 minit or cold shock of 12-14°C for 10-12 minit applied 4-7 minits after activation or 17-25 minits after activation of eggs resulted maximum number of gynogens. This will be 50-100% inbred (homozygous) depending on the extent of chromosomal crossing over during early meiosis. However, heat shock give better results than cold shocks. The second method to restore diploid is pressure shock. Pressure shocks are most ideal to get as high as 90% gynogens than thermal (heat/cold) shocks, through blocking of first mitotic division of the haploid zygote.

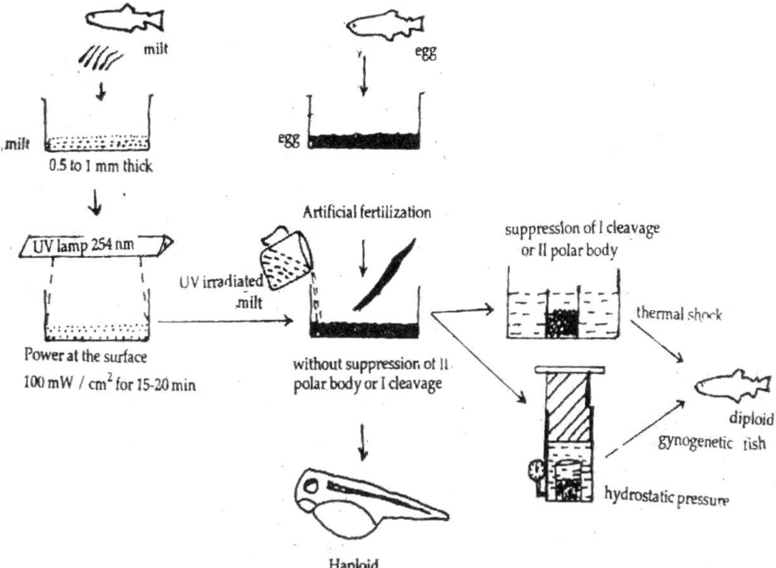

Fig. 71. Generalized scheme for inducing gynogenic, diploid and haploid in externally fertilizing fish.

(b) *Phenotypes, survivability and growth of gynogens* - Gynogenetic progeny inherit all maternal traits and is unique way of producing all female population. Gynogenes may be identified by morphological characters as they inherit all the maternal traits.

Growth of gynogenetic progeny is low due to inbreeding effects and mortality rate is higher during embryonic and larval stages. However, these can be overcomed in course of time through use of hatchery, running and race way systems.

(c) *Advantage of gynogenesis over parthenogenesis*

(1) Parthogenic individuals are haploid where as gynogenetic individuals are either haploid or restored to diploid condition.

(2) Gynogens are female individuals.

(3) Homozygosity level is higher about 45% in trout to 65% in common carp than sibmating (25%) or in self fertilisation (50%).

2.16.3. Androgenesis

It is the development of organisms entirely from the paternal chromosomes (Fig. 72). In spontaneous androgenesis, the ovum develops in to a zygote after activation by normal sperms but there is no contribution from the maternal side, as the maternal genome is not fused with sperm.

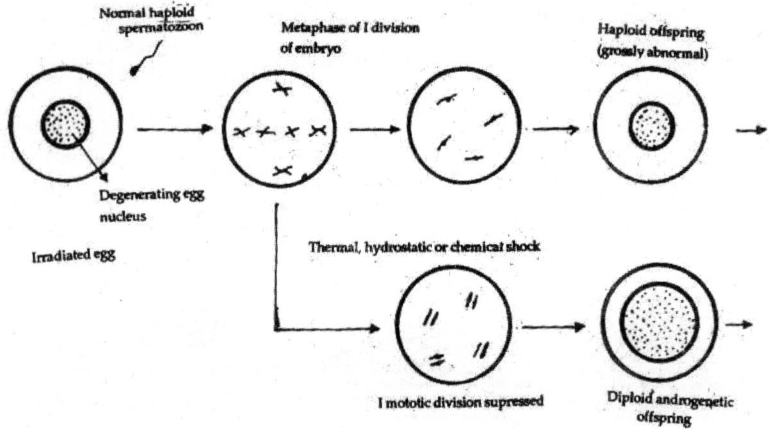

Fig. 72. Schematic representation of androgenesis

(a) Androgenetic diploid production

The androgenetic haploid zygote is made androgenetic diploid through

(1) dispermy

(2) fusion of 2nd polar body and sperm

(3) through suppression of first mitotic division.

All these processes were carried out by thermal or hydrostatic pressure shocks. The induced androgenetic progeny are produced when gamma irradiated (3.6 X 10^4 R) eggs are activated by normal sperm of desired species and diploidy is restored through application of shocks (Fig. 73). Homozygosity level in androgenetic progeny are similar to that of homozygosity gynogenes.

Induced androgenesis are known to some extent in loach, medaka, flat fishes and sturgeon, but androgenesis is extensively studied in trout, salmon and grass carp.

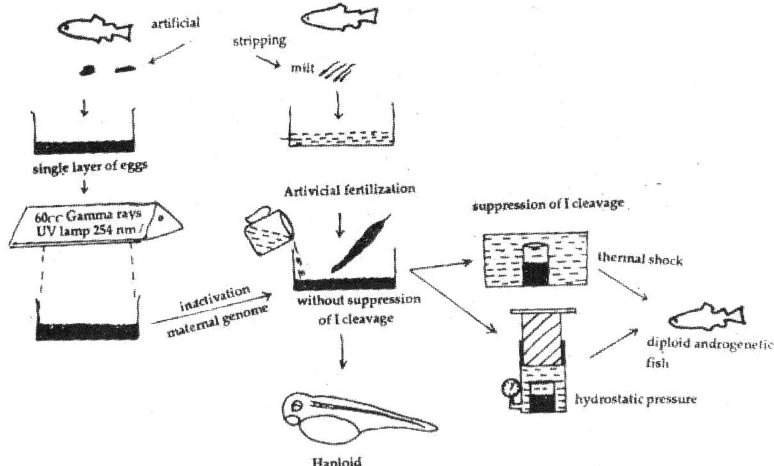

Fig. 73. Generalized scheme for inducing androgenic diploid and haploid in externally fertilizing fish

(b) Phenotype, survivality and growth of androgenic individuals

Androgenetic progeny are all males with certain paternal phenotypes, as they inherit most paternal traits. High rate of mortality have been noted during embryonic and larval stage in androgenic zygotes when subjected to thermal shocks. However, hydrostatic pressure shock (9000 pounds/sq inch for 2-4 min) as in rainbow trout showed 40% hatching rate. The growth of androgenic individuals will probably slow due to inbreeding effect leading towards homozygosity.

2.16.4 Polyploidy

In polyploidy the genetic manipulation leads to the addition of chromosomes to the normal diploid complement resulting in an increase in the genome size (Fig. 74). However, the merits of poly-ploidy is the increased heterozygosity of the resultant individual.

But allopolyploidy is a process in which there will be addition of set(s) of chromosomes to the diploid hybrid genome. Allopoly-ploidy can be achieved by preventing the extrusion of 2nd polar body.

It was believed at the beginning that polyploid individuals may have better growth due to increased number of chromosomes than the diploid ones and in most cases the triploids are expected to be sterile. Thus the energy required for the development of gonad is diverted towards somatic growth and hence sterile individuals are expected to grow faster and healthier than the normal diploids. The spontaneous triploid hybrids of grass carp X common carp, grass carp X big head and rainbow trout X brook trout are reported to be more vigour than the diploid ones. But in triploid of stickle back, no difference in size was exhibited when compared with diploids.

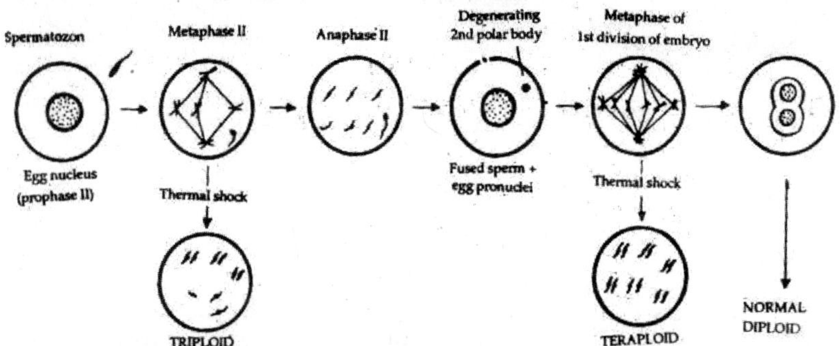

Fig. 74. Schematic representation of induced polyploidy

The growth in polyploid fish has also been studied by analysing RNA and protein synthesis per cell. In salmonids tetraploid species, the RNA and protein, level was twice than the diploid indicating greater enzyme activity. These differences in metabolism are expected to result positive effect on growth. On the contrary, in certain tetraploid cyprinids, the mean cellular RNA is same to that of diploids indicating the compensation. However, induction of triploid has a great potential utility in the control of prolific breeding tilapia and common carp.

2.16.4 (a). Method of ploidy induction

Polyploidy was reported to occur spontaneously in nature especially during hybridisation between two distantly related species. However, ploidy (tri, tetra and penta) can be induced artificially by (1) chemical treatment techniques such as cytochalasin-B, (2) thermal shocks or hydrostatic pressure.

(i) *Induced triploidy* (Fig. 75). The formation of diploid (2N) zygote is done by the fusion of haploid (n) set of chromosome from sperm with haploid (n) set of chromosome from egg. To this diploid zygote (2 n), another (n) set of chromosome is added by retaining the 2nd polar body of the ovum which contain (n) chromosome. This retention is possible by application of thermal or hydrostatic pressure shock before the extrusion of 2nd polar body. The extrusion of 2nd polar body takes place few minutes after fertilisation. Hence triploidy, (3n) is resulted.

(ii) *Induced tetraploidy* (Fig. 75) - In induced tetraploidy, the shock treatment is given before the commencement of cleavage in the diploid zygote. Hence, it suppresses the mitotic metaphase and the tetraploidy is obtained.

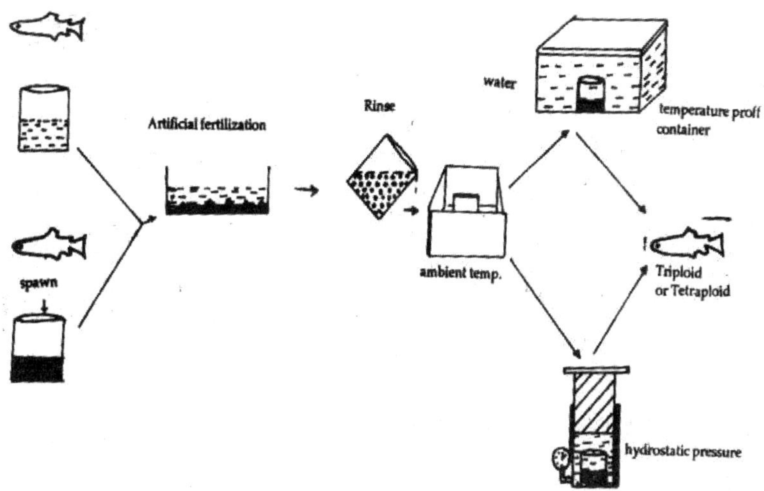

Fig. 75. Generalized scheme for inducing triploidy and tetraploidy in externally fertilizing fish.

The per cent of induced triploidy through thermal shock varies from 10-15% and 8-75% in case of producing tetraploidy. Higher percentage of polyploidy production has been reported when hydrostatic pressure shocks were applied.

(iii) *Detection of ploidy* - A number of methods have been developed to detect the nature of ploidy induced in the fish as well as the presence of paternal genomes in putative hybrids. These methods include :

(1) Direct technique of karyotypic studies like counting of chromosome numbers.

(2) Measuring DNA volume.

(3) Indirect technique of measurement of nuclear volume or area in erythrocytes, cartilage cells or retinal neurons as it was first calibrated in chromosome counts.

(4) Electrophoresis of protein and Isozyme analysis is widely used.

However, most of the direct and indirect methods are time consuming. Of late, computerised flow cytometry has become the most efficient, simple and easy means of determining the levels of ploidy but is costly. In such flowcytometry method, fluorescence of large number of stained cells is rapidly measured to estimate the DNA content.

(iv) *Viability growth and fertility of ploidy individuals* - (1) The viability of triploids appear to the varied in different species of fish. Triploid blue tilapia, stickle back show somewhat less viable than diploids. But triploid grass carp X common carp and grass carp X big head carp have better viability than diploids. Information on viability of tetraploid and pentaploids is very scanty to draw any conclusion.

(2) Pandian and Varadarajan (1988) reported the increment in somatic growth in triploid tilapia was over 20 to 30% (see Jayaraman *et al.,* 1989). However, the Indian major carps (rohu, catla, and mrigal), grass carp and silver carp are yet to be studied for ploidy induction and success in such attempts could change the scenario of commercial freshwater fish culture in the country.

(3) As per fertility, only tetraploidy rainbow trout has been reported to attain maturity.

2.16.5. Mutations

Sudden discrete or inheritable changes in gene sequence or DNA structure having no relation to the individual's ancestry causes mutation. Various classifications of mutations are indicated. These are as follows:

1. On the basis of kind of cell, mutation is classified as (a) Autosomal (b) Sex-chromosome
2. On the mode of origin, mutation is classified as (a) Spontaneous (b) Induced
3. On the basis of direction, mutation is classified as (a) Forward (b) Backward

4. On the basis of phenotypic expression, mutation is classified as (a) Dominant (b) Recessive

Some mutations mutate certain genes in such a drastic fashion that mutated gene phenotypically expressed in the death of the organism. Such mutations are called lethal mutation. All kinds of mutations may involve (i) Changes in number of chromosomes per cell (chromosomal variation) (ii) changes in structure of chromosomes (Gross mutation) (iii) changes in the DNA molecules of an individual gene (gene/point mutation). The mutations in chromosomes are called chromosomal mutations.

2.16.5(a) Gene mutation or point mutations

Gene constitutes DNA which is a most stable chemical molecule of biological world. It contains 4 nitrogen bases (Adenine, Guanine, Thymine and Cytosine). Gene mutation include very limited segment of DNA, hence are called point mutation. The point mutation may occur due to subnucleotide changes in the DNA and RNA which are as follows:

1. *Deletion mutation:* Loss of single nucleotide in a triplet codon of gene is called deletion mutation. The deletion and insertion mutation can alter the code words of gene and results an inactive or defective protein which can lead to the death of the cell.
2. *Insertion mutation:* Addition of one or more extra nucleotide during duplication, transcription or translation process to a gene are called insertion mutation. Chemicals like acridine dye and proflavin induce such insertion mutation.
3. *Substitution mutation:* When a nitrogen base of triplet codon of DNA is replaced by another nitrogen base or some derivative of nitrogen base are called substitution mutations. Substitution mutation can occur

(A) Transition: When purine base (adenine) of a codon is replaced by another purine base (guanine) or pyrimidine base (thyamine) is replaced by another pyrimidine base (cytosine) then such kinds of substitutions are called transition. The transitional substitution mutation occur due to

1. *Tautomerization* - due to tautomeric shifts in electrons and protons in the nitrogen base to convert them in their rare states.

2. *Deamination* - due to chemical substance nitrous acid that cause oxidative deamination of DNA bases.
3. *Base analogs* - due to certain chemical substances having molecular structure similar to usual DNA bases. Base analogs may be natural or artificial.

(B) Transversion: The substitution mutation when involves the replacement of a purine for pyrimidine or vice-versa, then this is called transversion mutation.

4. Spontaneous and Induced mutations

Spontaneous mutation occurs suddenly in the nature and their origin is unknown. When mutation is induced artificially by radiation, physical condition (temperature) and chemicals, such mutations are called induced mutation and the agents that cause mutations are called as mutagenic agents.

5. Forward and Reverse mutations

When mutation create a change from wild type to abnormal phenotype, then that type of mutations are called forward mutations. When the forward mutation is corrected by error correcting mechanism so that the abnormal phenotype change in to wild type phenotype then such mutations are called back or reverse mutations.

6. Significance of Mutation

The vast majority of mutations are deleterious to the organism and so are kept at low frequency in the population by the action of natural selection. Mutant types are generally unable to compete equally with wild type individuals. However, they have great evolutionary significance, because the process of speciation depends variously on it.

7. Lethality and Lethal alleles

Lethal is applied to those changes in the genome (chromosome) of an organism which produce effects severe enough to cause death. A gene whose phenotypic effect is sufficiently drastic to kill the bearer is called lethal gene. Lethal genes may be dominant, incompletely dominant or recessive. The dominant lethal allele kills the carrier individual both in homozygous and heterozygous condition. The recessive lethal allele kills the carrier individual only in homozygous condition. The completely lethal genes usually cause death of the zygote, embryo or after hatching. The lethality may be zygotic or gametic and in plants also gametophytic lethality. The lethal alleles modify the 3:1 phenotypic ratio in to 2:1 as the individual possesses lethal recessive alleles in homozygous condition.

2.17 Gene engineering

Genetic engineering is a great land mark in the field of molecular biology. This was made possible through the studies on the structure of DNA molecule which is found in the form of double helix. The two chains are long with sugars and phosphate molecules with the combination of Adenine, Thyamine, cytosine and guanin bases. Adenine and Guanine are the purine base and Thyamine and cytosine are the pyrimidine base. Adenine compliment with thyamine with triple bond where as cytosine combined with Guanine with double bond. These purine and pyrimidine bases are connected with sugar molecules and then with phosphate molecule to form a complete double helix like structure. The possibility of producing recombinant genes through gene splicing has led to the development of man made genes. These genes, which transplanted act as natural genes and reproduce successfully. This possibility of transgenic genes have tremendous application in obtaining imporved strains which are urgently needed for stepping up fish production. Gene engineering in the present context includes nucelic acid (DNA and RNA) extraction, nucleic acid hybridization, recombinant DNA, clone and gene manipulation.

Gene maps have been established in a variety of fish species such as Medaka (*Oryzia latipes*), Tilapia, Salmon, Zebra fish, Catfish etc. (Wada *et al.,* 1995; Kocher *et al.,* 1998, Hayheim *et al.,* 1998, Shimoda *et al.*, 1999). Gene mapping refers to the analysis of loci on the genome revealing the linear order of different genes on the chromosomes. Gene maps are of two types:

1. Physical map: It is based on the (a) assignment of loci to chromosomes and accomplished by somatic cell hybrid panel, In-situ hybridization etc. (b) the coordinates are chromosome region or band (c) the distance between two loci are measured in Kilobases (Kb).

2. Genetic maps: It is based on (a) studying the meiotic recombination between two or more loci through linkage analysis. This implies a reference locus is a prerequisite for genetic map. (b) genetic map do not provide exact location of loci but they reveal the genetic distance of the loci during recombination (c) genetic distance is expressed in unit of crossing over or centimorgan (cM), (d) 1 cetimorgan (cM) equal to 1% crossing over and contains approximately 1000Kb. This means, two loci which shows 1% recombination are 1cm apart on a genetic map. (e) different type of DNA markers act as reference points or land marks in mapping genes. (f) the genetic linkage map of tilapia contain 174 markers and salmon 250 markers and Zebrafish 2000 markers even. The genetic linkage map of Zebrafish (*Brachydanio rerio)* would facilitate the infrastructure for the location and positional cloning of 600 mutation genes that are crucial (Haffter *et al.,* 1996 and Driever *et al.,* 1996).

DNA marker maps are the infrastructure for the identific-ations and mapping of any gene. Therefore, development of these markers in species of interest in aquaculture would facilitate the establishment of genetic linkage map which will in turn, help identifying single gene/locus and quantitative trait locus (QTL) responsible for economic traits like growth, fecundity, age at maturity, disease resistance etc, and moved towards a marked assisted selection (MAS) in fish species. In addition to this, it has got many more advantages like mapping of disease genes, disease diagnosis, paternity testing, loss of heterozygosity in tumours, transgenic analysis etc. Thus establishment of DNA markers and linkage map could enhance cultivable fish production through reliable selection programme.

2.17.1 Nucleic acid extraction (Fig. 76)

Both DNA and RNA constitute the nucleic acid of the cell. Their extraction for studying the sequence is of immense useful for gene engineering. Nucleic acid isolation is done by homogenization of cell with acetate buffer followed by deproteinization with hot phenol and cold phenol treatment. The excess of phenol is removed by chloroform and Isoamyl alcohol. The aqueous phase is collected and mixed with sodium acetate and 95% ethanol. So DNA is precipitated. Such DNA precipitants are dissolved in TE buffer and store at 4°C for further use.

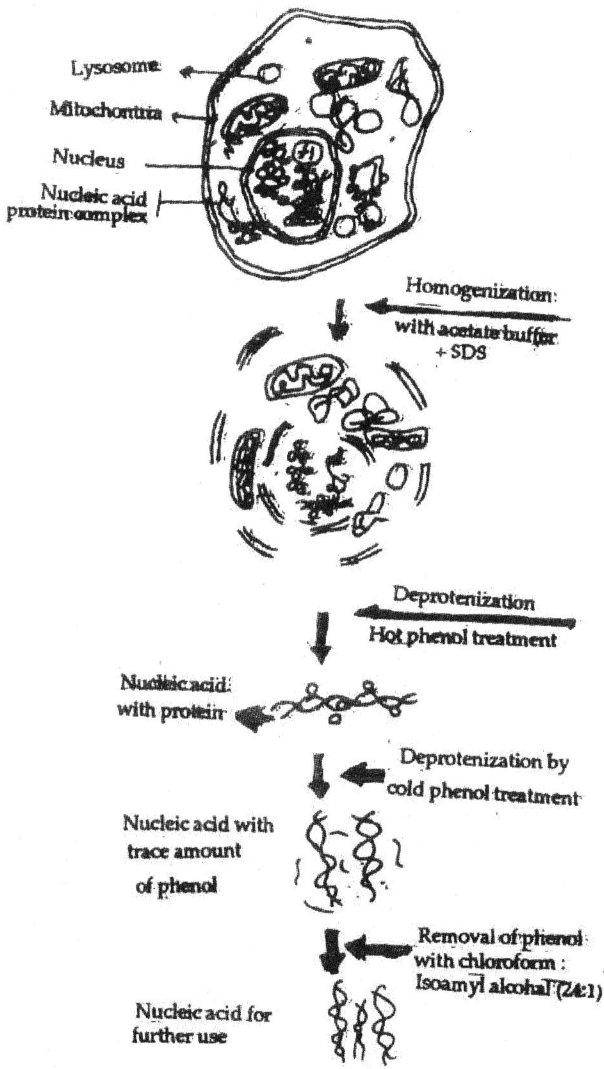

Fig. 76. Diagramatic representation of Nucleic Acid isolation.

2.17.2 RNA extraction

To isolate RNA from cells, it is necessary to free the RNA from these cellular components. The cells are first lysed with lysozymes in combination with detergents such as SDS, Sarkosyl and Triton X-100. The function of the detergent is to dissolve the cell membranes and release the RNA out of cell. Then hydrolysis of RNA is done by R-Nase enzyme. For this EDTA is generally useful to chelate Mg^{++} which is required for the activity of R-Nase. Then through successive extraction with phenol-chloroform RNA is extracted. But isolation of pure RNA is difficult due to contaminating DNA which is difficult to be fully separated.

2.17.3. Nucleic acid hybridization

It is a technique for determining the relationship between DNA sequence in the genome and the RNA products. Hybridization is carried out in Denhardt's solution or heparin sulfate (50 μg/mol) recently suggested (Singh and Jones, 1984) (see Pandian and Muthu Krishnan, 1990). Addition of such solution is to remove the non-specific back ground. Followed to this, addition of formamide in the hybridization solution reduces the reassociation temperature and make it easy to handle. This process is carried out on the surface of a nitrocellulose filter. By this process, two strands of nucleic acids are separated. But reassociation may take place between DNA-DNA, DNA-RNA and RNA-RNA depending on the concentration of DNA or RNA and size of DNA or RNA fragments.

2.17.4. Nick Translation method

The commonest means of leveling nucleic acids for hybridization is the nick translation method. The DNA polymerase I of *E. coli* having both 5' 3' exonucleolytic activity and 5' 3' polymerase activity is used for this method. The 5' 3' exonuclease activity of this enzyme releases 5' mononucleotides, Where DNA Polymerase activity takes place. If one of the 4 nucleotides is radio-labelled in the reaction mixture, the reaction progressively incorporates the level in to a duplex DNA. It is suitable for production of large quantities of probe for use in multiple hybridization reactions.

2.17.5. DNA Cloning: Tools and strategies

Genetic information is stored in DNA molecules which are located in the

1. chromosomes of nucleus

2. mitochondria of animal cells

3. chloroplast of plant cells

Usually the flow of genetic information is DNA RNA protein. The DNA sends the RNA through transcription and RNA inturn produces protein in the process of translation. It is important to know the sequence of bases in a complex pool of DNA. This could be known through (1) Molecular hybridization (2) DNA cloning. In molecular hybridization, the sequence of bases in the fragment of DNA is detected through a homologous sequence acting as a probe. (2) In DNA cloning, the desired fragment of DNA is selectively amplified to produce multiple identical copies itself to know its structure, function, regulation and expression. DNA cloning is achieved by two different ways (1) Cell based cloning (2) PCR based cloning

The steps involved in cell based cloning are

(1) identification and isolation of desired DNA fragment: Then insert preparation and purification. In insert preparation, the DNA molecules are cut to required size by restriction endonucleases (RE) and is purified. Restriction endonucleases are group of enzymes that correspond to the sissors, which recognizes specific nucleotide sequences in DNA, often 4 to 6 base pairs (bp) long and cleave both strands of DNA at /or near the recognition site. The restriction enzymes make double cleavage to the double stranded DNA creating either cohesive end or blunt end. There are three catagories of restriction endonucleases(type I, II & III). But type II restriction endonucleases vary from 4-8 bp, palindromic in nature and recogntion sequences are same in both the strands when read 5 3 direction. If the starting material is RNA, it can be double stranded through reverse transcription using reverse transcriptase and amplification using DNA polymerase, sequentially.

(2) Ligation: Ligation means joining of fragments of DNA. Ligation involves the formation of new bonds (phosphodiester) between PO4 residues located at the 5 phosphate and adjacent 3 hydroxyl moietics of double stranded DNA. This bond can be catalysed invitro by ligases, E. coli DNA ligase (cohesive) end and bacteriophase T4 DNA ligase (cohesive and blunt end). The later ligase is used for all cloning processes. Efficient ligation of vector DNA with foreign DNA depends upon the concentration of compatible DNA termini. The insert containing vectors are called recombinants. The screening of recombinants can be done by immunological method, nucleic acid hybridization, protein markers etc.

(3) Transformation: After ligation, these combined molecules are introduced in to host cells by transformation, mostly bacteria or yeast cells. The introduction of nacked DNA from the environment inside the host is called transformation. These molecules can replicate independently if put in a host cell. The replications of molecules could be provided in two ways (a) foreign fragment integrate in to host chromosome and replicate like retroviruses (b) DNA fragment is attached to extra-chromosomal replicans, which are of two types, (i) plasmids-small circular double stranded DNA molecules (ii) bacteriophages which are viruses that infect bacterial cells. Cell based cloning is mostly done by adopting extra chromosomal replicans. It is further important that vectors should have multicloning site (MCS), origin of replication (Ori), F1 origin, resistance genes, phenotypic marker selection etc.

In DNA cloning, individual gene represent only a very small part of the whole genome. A 10 Kilo base single gene is less than 0.0002% of the entire genome of a typical bony fish. DNA library, a relatively modern concept in DNA cloning can offer the possibilities of collection of DNA fragment in the form of DNA clones. Two types of DNA library methods are adopted depending on the nature of starting DNA.

1. **Genomic DNA library:** the starting DNA is the genomic DNA which is cleaved or fragmented by restriction endonucleasaes, preferably 4bp cutters such as Mbol or Sau 3 A in order to achieve fragmentation of the genome. These fragments after purification and modification can be inserted in to suitable vector (bacteriophage lambda) for cloning. Bacteriophage lambda can clone 25 Kb single gene easily.

2. **Complementary DNA library (cDNA):** The starting material is RNA which is made double stranded by enzyme reverse transcriptase and cDNAs are produced. These are then cloned in to suitable vector.

2.17.6. PCR method

Polymerase chain reaction based cloning is a rapid, versatile invivo method for amplifying a specific target DNA sequence from a pool of source DNA. PCR involves

(a) Denaturation (b) Annealing (c) Synthesis or elongation

In this technique, some prior information on the nucleotide sequence of target DNA is necessary. This is because the reaction has to be primed by two short stretches (15-30 base pairs) of oligo primers. These contain the flaking sequence information of the target DNA that can amplify discrete regions of the genome.

DNA based polymorphisms are being used for marker assisted selection strategies, parentage testing, species identification and population genetic studies. The conventional probe based DNA fingerprinting method such as restriction fragment length polymorphism (RFLP) also called arbitrarily primed PCR (AP-PCR) and micro satellites have been used for genetic analysis. An alternative new method for detecting polymorphic markers in the random amplified polymorph DNA (RAPD) assay. More recent technique called amplified fragment length polymorphism (AFLP) has been used to enhance marker density. This technique involves PCR amplification of a subset of the restriction fragment generated by digestion of genomic DNA with a combination of two restriction enzymes, usually one that cuts at 6 bp sites and one at 4 bp sites. Then this fragment is ligated to the two fragment ends by PCR amplification steps using the primers that are complementary to the adapter sequences, extends across the restriction site and end in arbitary selective nucleotide (nt). The sequence selectivity of the PCR primers at their 3' ends leads to amplification. Schematic diagram of PCR amplification is given.

Double standard DNA: **5' ——————3'**
3'——————5'

Cycle I : Denaturation **5' ——————3'**

Single strand DNA
3'——————5'
Primer annealing
5'——————3'
Introduced **3'—————— 5'**
Elongation or Polymerization
5' —————— 3'
3' —————— 5'
Cycle II : (Followed cycle 2, like 30-45 cycles)

Schematic Diagram of PCR amplifications

2.17.6(a) Application of PCR

Nucleic acid amplification by PCR has added new and revolutionary dimension to molecular biology. It has wide spread applications:

1. Amplify a target sequence, PCR significantly enhance the probability of detecting target gene sequence in a complex mixture of DNA.
2. It facilitates cloning and sequence of genes.
3. Used for direct detection of point mutations
4. Permits fusion of genes and identification of DNA sequences.
5. Used for rapid detection of pathogens
6. Used to produce large amount of probes for hybridization process
7. PCR based method namely randomly amplified polymorph DNA (RAPD) fingerprint helps to identify and characterize several bacteria species, strains and isolations.
8. Play an important role in phylogenetic /evolutionary and conservation biology.
9. Possible to monitor genetically engineered organisms by PCR.

(b) Application of restriction fragment length polymorphism (RFLP) / randomly amplified polymorph DNA (RAPD) and amplified fragment length polymorph (AFLP)

1. Useful techniques in analyzing pedigree and establishing paternity
2. Provides supporting evidences to the taxonomic relationships among fish species
3. Stock identification and characterization are possible by these techniques.
4. Provide tool to analysis interspecific gene flow and hybrid speciation. Genome analysis of fertile hybrids provides the basis of QTL (quantitative trait loci) mapping
5. Useful in MAS (marker assisted selection) to be linked with selective breeding

6. Degree of inbreeding can be measured by these techniques
7. To acertain the success of gynogenesis/ androgenesis without killing the animals using RAPD assay
8. Scx specific markers can be identified
9. Characterize the fish cells in culture
10. Diagnosis of diseases
11. RFLP has been used to study population genetic variation. This has led to give rise a term VNRTs which is the variation in the number of tandem repeats of a defined sequence modified at a define locus. Depending on the size of fragment, the tandemly repeated DNA can be grouped as satellite DNA, minisatellite DNA and micro satellite DNA. Micro Satellite DNA is simple repeats (SSR) or simple tandom repeats (STR) consisting of usually long repeat units.

2.17.7 Mitochondrial genome

Mitochondria are cytoplasmic organelles responsible for respiratory function in eukaryotic cells. Inorder to perform its cellular function, mitochondria has its own genome and accessory components. The mitochondrial genome is characterized by a single circular double stranded DNA molecule containing several genes. Unlike nuclear genome, mitochondrial DNA is haploid which is maternally inherited. In several organisms mitochondrial DNA has been used as markers for population and evolutionary genetic studies.

2.17.8 Recombinant DNA

After separation of two strands of DNA, the desired strands of DNA from chromosome are cut by treating them with a restricting enzyme. The plasmid DNA sequents from virus/bacteria isolated through ethidium-bromide-cesium chloride density gradient centrifugation or animals are mixed and pairing of homologous DNA strands taken place resulting in a recombinant DNA. However, transgenic forms are reported in Salmon, Trout and Tilapia.

2.17.9. Gene manipulation

For gene manipulation, physical and mannual techniques are employed. Such technique requires, microneedles with less than

0.1 μm tip, micropipette and holding pipette. That too the instrument is so equipped for correct micro incision and microinjection with proper slide adjustment. Micro injection of Bovine and rat growth hormone gene in Tilapia egg has been successfully carried out (Fig. 77). Tilapia egg is highly opaque and the chorion is very hard. Hence, it can not be pierced through even with very fine microneedle. Therefore, foreign DNA is injected into the germinal disc through the micropyle before the starting of cell division with a micro manipulator. Micro injection of growth hormone gene in zebra fish egg has been done by directly injecting into the cytoplasm before the first cleavage (Fig. 78). Chinese have also taken trials in nuclear transplantation in crucian carp.

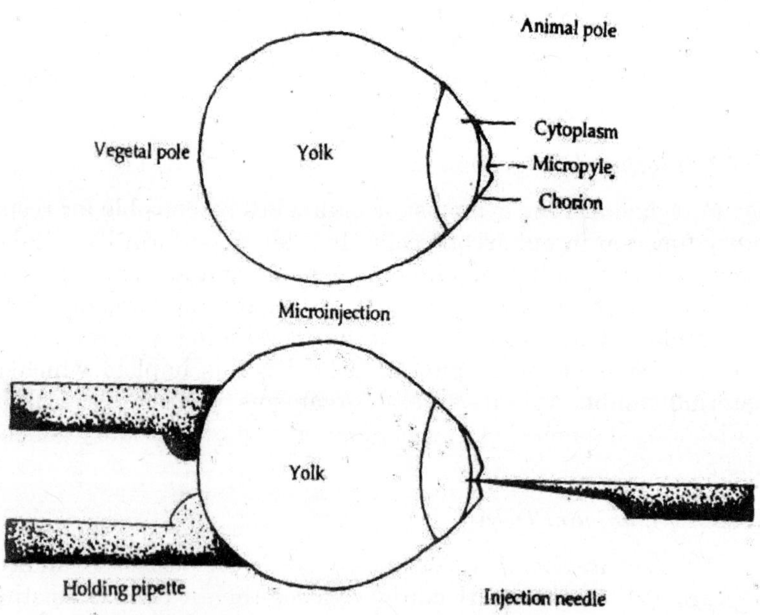

Fig. 77. Micromanipulation in Tilapia egg.

Considerable progress has been made in the field of cloning and processing of recombinant DNA in virus, bacteria, algae, insects, sea urchins, amphibians, mammals etc. In fishes only preliminary attempts are reported in case of goldfish, medaka, loach, tilzapia, trout salmon, zebrafish etc. However, gene manipulation will take up to new horizon of aquaculture in passage of time and research in gene investigations.

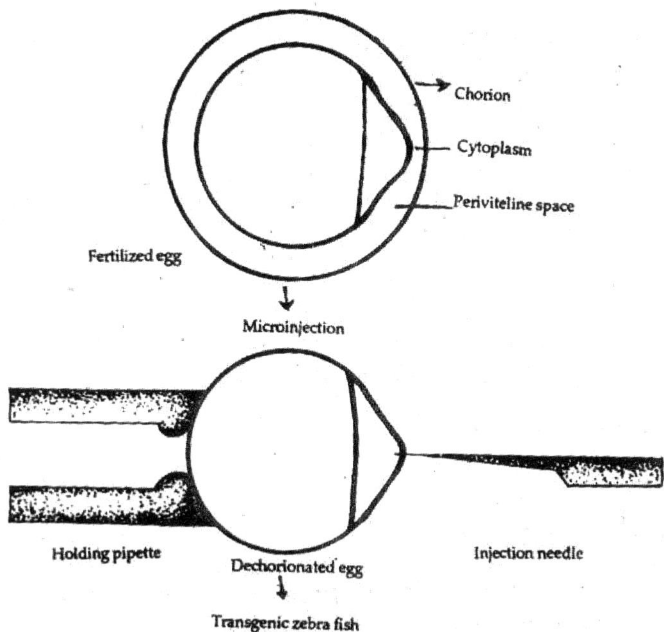

Fig. 78. Gene manipulation in Zebra fish.

2.18 Genetic markers

Genetic marker is necessary to identify individuals or group of fish in order to carryout genetic selection for breeding on a large scale. The potential usefulness of the genetic markers to the brooders may be classified in to three classes.

(1) Morphological marker

(i) fin clipping or brand marking

(ii) color dye markings

(2) Electrophoretically distinguishable protein or allozymes

(3) blood group antigens

2.18.1 Usefulness

(i) To discover the pleiotropic effect of the marker loci on productive characters.

(ii) To study on the Mendelian fashion of inheritance of marker genes on qualitative productive characters.

(iii) To estimate the degree of inbreeding in breeding operations.

(iv) To estimate the relative genetic differences between related groups.

(v) To identify inbred lines for better selective breeding.

(vi) To increase the number of genetic groups that may be tested in a limited number of ponds.

(vii) Enable the identified full and half sib offspring production.

2.18.2 Classes of genetic markers

(1) *Morphological markers* - It is done either by fin clipping, brand marking or color dying. However, fin clipping and brand markings do not serve effectively with passage of time. Because these markings do not remain for a longer period of time. The report on pelvic fin clipping of catla and common carp and painting the clipping portions with saturated solution of Sudan Black B in 70% alcohol were not effective as morphological marker (Khan *et al.,* 1985). Brand marking is done by means of a circular loop of strong wire which was heated from an accumulator such as car battery. It has a wooden handle for easy use and was operated by a switch. Depending on the application of wire loop, the fish is branded with circles and bars. A combination of these basic figures on different parts of the fish's body made it possible to identify individual fish. It is reported that these scars remain visible for several years. The natural color and scale pattern of fishes were sometimes used as morphological markers. But their disadvantages are: (1) heterozygous individuals can not be identified because of complete dominant gene action, (2) and pleiotropic effect of viability and growth rate.

The other method for morphological markers is the use of suitable colouring dye. The properties of dyes used should be -

(1) remain for longer period of time and do not fade out quickly.

(2) do not have adverse effect on growth and survivality.

(3) should not have disadvantage in marketing the color dyed fish.

Numerous reports on dye marking of fish are well documented (Jessop, 1973; Gerking 1963; Davis, 1971; Khan *et al.,* 1985).

Jessop (1973) report the use of Bismark Brown Y,m Netural red and Nile blue biological stain for marking Alewife fry. Of these Bismark Brown Y proved better than other two. In this case the fry was dipped in color dye for coloration. However, Khan *et al.,* 1985 reported the use of Indian Ink and red poster colour. The other dyes used were M-procian blue, Alcian blue 8 GS, Azocarmine G, Trypan Blue. But in these cases, the required concentration of dye was injected through size specific hypodermic needle below the skin in the lower Jaw, dorsal fin, Isthmus and along the light coloured ventral surface of Asiatic carps (Catla, Rohu and Common carp). Out of these tasted dyes, M-procian blue was found to be best marking agent, hence, used for field conditions.

Injection of dye and other agents such as Incarbon, cadmium sulphide, chrome green, National fast blue 8 G X M and mercuric sulphide were used for the production of spots and lines in fishes (Arnold, 1966). Even the use of Tannery and Textile dye were used in marking the fish.

The survivality and growth rate of M-procian blue dye marked rohu, catla and common carps in field tasts were not affected and these fishes can be used safely for human consumption after removing the flesh from the marked spot (Khan *et al.,* 1985).

2. *Electrophoretic markers* – The association of protein genotypes and fish size had not been observed until 1984, when the polymorphic haemoglobin (Hbl) was found to correlate with mean length at a given age of atlantic cod (Mark *et al.,* 1984). This success led to the possibility of using such markers as a marker. Electrophoretic markers are using as an important tool by the fish breeders and the population biologists. This provides a rapid and efficient way of obtaining much of the basic genetic information crucial for the genetic improvement of cultured species. The genetic code of DNA can be translated into proteins, which can provide basic genome informations about the individual. These proteins can be defined as electrophoretic proteins which is controlled by different alleles of a single locus. As a rule, a dominant allele control the electrophoretic protein and therefore heterozygotes can be easily distinguished. That too, such proteins are controlled by multiple enzyme molecules encoded by alternative alleles at a single locus are allozymes. Hence, it is better to say as electrophoretic allozymes. Sanchez *et al.,* 1991 reported allozyme variations in the natural populations of Atlantic salmon in Asturias in which out of 24 protein coding loci, 5 were

Aat-2 (aspartate aminotransferase), Idh-3 (isocitrate dehydrogenase), mdh-3, 4 (malate dehydrogenase), Sdh-1 and 2, (Sorbital dehydrogenase) were polymorphic. The mean heterozygoasity is 2.18% hence, amount of variability in natural population is somewhat lower than the average estimated in the Atlantic salmon population. Electrophoretically identified proteins are then staind and the pattern of stained bands is termed zymogram. Then the electrophoretic data can be analysed for the calculation of (i) allelic frequencies and (ii) heterozygosity.

However, electrophoretic markers have limitations in genetic studies because -

(i) Electrophoresis is not a perfect method for detecting genetic variations since it examines only the structural genes which comprise approximately 1% of the genome

(ii) Most amino acids are electrically neutral and only less than 25% of all amino acid substitutions can be detected by electrophoresis.

(iii) Such variations in electrophoretically detected amino acids is likely be under estimated in heterozygosity and the apparent similarity between populations be over estimated.

Despite these limitations, protein variation permits some evaluation of the amount of genetic variation in natural populations and subsequent genetic changes in offsprings. A detailed study on the biochemical genetic picture of Indian major carps is currently under way using 18 isozyme systems. Some of these isozyme systems have already indicated species specific patterns. For example : in sslmonids, the association between Tryspin like isozyme pattern and fish size was first shown by Torrissen (1987) using isoelectric focusing on agarose gel. The group of *Salmo salar* with the TRP-2 (92) allele were heavier than the group without this variant. Thus the isozyme variant TRP-2 (92) may be used with advantage as an biological marker for a selective breeding programme. Torrissen (1991) reported the growth rate of three different year classes (1984, 1985, 1987) of Atlantic Salmon (*Salmo salar)* by using trypsin like isozyme pattern in the pyloric caeca as a biological genetic marker. The fish with the homozygote variant TRP-2 (92/92) and with the heterozygote variant. TRP-2 (92/100) had higher average weight than those with the genotypes which did not possess the variant TRP-2 (92). The studies on mitochondrial DNA restriction fragment length polymorphism are initiated in National Bureau of Fish Genetic resources at Allahabad. Advanced techniques like `G' bandings are being adopted to identify individual chromosomes and also to spot intra-specific variations. Studies on Nucleolus Organiser Regions (NOR) of chromosomes of some species has indicated the usefulness of the techniques in identifying interspecific and possible intraspecific variations.

(3) *Blood group antigens* - Serological reactions make it possible to reveal that fish has polymorphic blood groups. Such polymorphic blood groups are similar to those found in man. Such groups have been reported in more than 50 species of freshwater and marine fishes. These blood groups can be used as genetic markers. Their great potential advantage (over electrophjoretic marker) is the relative ease of genotype identification. But the disadvantage of such marker is that, frequently heterozygous individuals can not be identified. That too, production of large quantities of species antisera is difficult for identification of blood groups. The genetic relation to blood group can also be difficult due to multiferious gene, environmental and biochemical interaction.

2.19 Genetic conservation of fishes

Indiscriminate fishing from large water bodies, has posed certain threat in declining of natural fish population stock. The problem is further aggravated due to human interference in discharge of petrochemical, industrial, insecticides, thermal effluents and oil pollution in the natural habitat condition of fishery resources. Introduction of certain exotic fishes has resulted the decline of natural fish fauna. In Dal lake introduction of common carp, has resulted the decline of indigenous, schizothor-acids. The decline of Mahseer (*Tor species*) in hill streams and anadromous Hilsa in the Ganga river system above the farrak barrage gives best picture that certain group of fishes are already endangered. The endangered species of fishes in India has been reported by Menon (1987). This necessiates to take immediate action to conserve endangered genetic resources of aquaculture spp. being reported (Sinha and Ibrahim, 1980). Fish germ plasm resource conservation can be done broadly by two ways.

(1) *In situ* conservation - It can be taken up within natural and man made ecosystem in which they occur. Natural park and

biosphere reserves may provide less expensive protection for the wild relatives of fishes.

(2) *Ex-situ* conservation - It can be done through the technique of liquid nitrogen cryopreservation method for gametes and embryo and gene bank of gametes. These can step up genetic conservation of fishes.

13

GLOBAL SCENARIO

1. INTRODUCTION

There are still considerable difficulties at present in obtaining accurate and reasonable estimates on aquaculture fish production data. A major effort was therefore made in 1975 by the Aquaculture unit of F.A.O. Fishery Department to bring together aquaculture production data on a world-wide basis. However, there are inconsistencies in the data submitted by various countries with respect to the data on seaweeds and molluscs. Some countries may have included seaweeds from collection from the wild and some countries may give the production statistics of molluscs including even shell weight. These inconsistency in submitted statistics have created conflicting values in the data reported. Probably because of nutritional qualities of seaweeds are so important and there is increasing trend in seaweeds production in Asiatic region, F.A.O. continue to include it in its statistics as an aquaculture commodity. Even F.A.O. has attempted to estimate the values of products based on farm gate prices. But this seems to be unrealistics, as the products generally sell at a price higher than the average price for the species. It is evident that the world catch of aquatic organisms has consistently increased from 21.2 million tons in 1948-52 to 80 million tons in 1985, 92.36 million tons in 1986, 92.9 million tons in 1987 and 96.5 million tons in 1988 (Table 16) (source FISH DAB-Sept. 1989).

Table 16. Annual world landings of Aquatic Resources (Million tons) excluding Mammals and Seaweeds.

Year	World Catch
1948-52	21.9
1953-57	29.0
1958-62	39.8
1963-67	53.9
1968-72	67.1
1973-77	66.2
1978-82	73.0
1983	77.4
1984	83.6
1985	86.0
1986	92.36
1987	92.5
1988	96.5

Source : Fish DAB, September, 1989.

Developed countries have contributed 51.55 million tons and developing countries have contributed 44.95 million tons in 1988. For the purpose of recording fish catch statistics on global basis, the world has been divided into 26 international major fishing areas. These consists of 7 Inland fishing areas and 19 major marine areas covering Atlantic, Indian and Pacific Oceans. Looking to the trend of world fish catch statistics in 1985, it is seen that Japan is the top (10.65 million tons), followed by U.S.S.R. (9.54 million tons), U.S.A. (3.76 million tons), China (3.7 million tons), Chile (3.3 million tons), Peru (2.7 million tons), Norway (2.5 million tons), India (2.4 million tons), South Korea (2.3 million tons) and Indonesia (1.8 million tons). International Development Research Centre (IDRC), 1988 reported that the world catch of fish is shared among Asia, which holds slightly less than half of the total, Europe, 19%; USSR, 13%, South America, 11%, North America and Africa about 6% each. Many developing countries rank high among fishing nations : China is 3rd, Chile 4th, Peru 6th, India 7th, South Korea 13th and Mexico 15th, (FAO, 1985 data). Since 1960's, the developing countries have been increasing their share of the world catch of fish and this trend is expected to continue under the new conventions of the law of the sea (Table 17).

Table 17. Top ten fish producing countries of the world (Million Metric Tonnes)

Country	1992-93	1993-94	1998	1999
China	17.60	20.70	17.22	17.24
Peru	8.50	11.60	4.33	8.43
Chile	6.00	7.80	3.26	5.05
Japan	8.10	7.40	5.26	5.17
U.S.A.	5.90	5.90	4.70	4.75
India	4.34	4.54	3.21	3.31
Indonesia	3.70	4.00	3.96	4.15
Russia Fed.	4.40	3.80	4.45	4.14
Thailand	3.30	3.40	2.90	3.00
South Korea	2.60	2.70	2.02	2.11

Fishing chimes 2001.21 (3): 37

During 1994-95, India's total fish production was 4.78 million metric tons of which 2.69 million tons from marine and 2.09 million tons from the Inland sectors. It was further increased in 1995-96, amounting to total of 4.94 million tons, of which 2.70 million tons from marine and 2.24 million tons from Inland sector. However, FAO has estimated by 2000 AD, the demand for aqua food will reach 110 million tons. This could be supplemented partly through improved aquaculture practices. The details of total world fish production from culture and capture sector is presented in Table 18.

Table 18. World Fish production (Million tonnes)

Sector	1994	1995	1996	1997	2000	2004	2005
Inland Aquaculture	12.109	13.860	15.607	17.130	21.2	27.2	28.9
Inland Capture	6.908	7.379	7.553	7.700	8.8	9.2	9.6
Total Inland	19.017	21.240	23.159	24.830	30.0	36.4	38.5
Marine Aquaculture	8.666	10.416	10.778	11.140	14.3	18.3	18.9
Marine Capture	85.775	85.622	87.073	86.030	86.8	85.8	84.2
Total Marine	94.441	96.038	97.851	97.10	101.1	104.1	103.1
Total Aquaculture	20.795	24.276	26.385	28.270	35.5	45.5	47.8
Total Capture	92.683	92.001	94.625	93.730	95.6	95.0	93.8
Total world production	113.458	117.277	121.010	122.000	131.1	140.5	141.6

Source : FAO. World Fishing, 1998, Nov. Page - 4.
FAO.2006

Aquaculture produces about 13.2 million tonnes in 1987 nearly equivalent to 13.0% of the total world fisheries harvest. Interestingly Asia remains the leading aquaculture region in the world contributing 11.13 million tons (84.3%) of aquaculture products in 1987. At least over 85% finfish, 76% of crustaceans and 68% of molluscs come from Asia. Over 99% of seaweed production come from Asia (Table 19 and 20). Further, the top ten aqua culture produces of fard fish and the trend in world Aquaculture production and percent contribution by different groups of aquatics (Table 21a&b).

Table 19. Summary of 1987 aquaculture production by FAO regions (in tonnes)

FAO regions	Finfish	Crusta-ceans	Molluse	Seaweeds	Other	Total	World %
Asia, Oceania & China	5,704,204	433,921	1,832,930	3,126,626	27,621	11,31,302	84.28
Africa N & NE	51,397	2	286			51,685	0.39
Africa, S of Sahara	10,461	77	229			10,817	0.08
North America	266,672	44,480	138,841			449,993	3.41
Central America	9,485	6,564	50,719			66,768	0.50
South America	21,674	79,759	2,456	9,178		133,067	0.86
Caribbean	17,725	800	1,503	210	81	20,269	0.15
Europe	399,037	3,285	645,271			1,047,593	7.93
Far-East	23,816	18				23,834	0.18
USSR	288.970		159	3.459		292.588	2.22
Total	6.793,441	574,906	2,672,394	3,139,473	27,702	13207916	

Source : FAO Circular 815, Rev. 1 (1989).

ADCP Aquaculture Minutes No. 6 (1989).

Table 20. Aquaculture production in the countries of Asia 1984-1987 (in metric tons).

Country	1984	1985	1986	1987
Australia	9,805	10,897	10.418	11.201
Bangladesh	117,025	123,811	144,722	165,100
Brunei Darus	–	–	–	–
China	3,789,826	4,456,712	4,987,025	5,600,604
Fiji	6	82	877	1,395

Hongkong	8,356	7,901	9,121	9.750
India*	746,300	746,300	746,300	746,300
Indonesia	330,764	359,997	399,285	394,090
Japan	1,207,100	1,183,604	1,291,297	1,226,190
Kampuchia*	1,618	1,618	1,618	1,618
Birbati	13	15	1,530	349
Borea D.P.R.	719,000	719,000	719,000	719,788
Korea R.	680,106	795,540	993,243	876,788
Laos	2,500	2,500	2,500	2,500
Malaysia	68,490	55,726	56,108	56,034
Myanmar	3,944	5,044	5,797	5,480
Nepal	1,997	3,265	4,162	5.435
New Calendonia	54	95	65	88
New Zealand	10.545	11,735	16,977	18,680
Pakistan*	10,000	10,000	10,000	10,000
Philipines	478,335	495,634	470,923	560.970
Singapore	1,149	1,225	1,327	1,860
Soloman IS	28	–	–	6
Sri Lanka	432	682	428	428
Taiwan	245,179	250,822	266,080	305,429
Thailand	111,933	135,841	128,417	151,658
Vietnam	237,000	250,500	267,500	260,300
Total	8,781,505	9,632,546	1,534,720	11,131,075

1. Based on FAO Fisheries Circular No. 815 Revision 1.

*FAO estimates or including repeat of earlier reports.

Table 21(a). Top Ten Aquaculture Producers of Food Fish (Million Tons)

Country	Prodn 1998	Prodn 2002	Prodn 2004
China	27.07	27.76	30.61
India	2.02	2.19	2.5
Vitham	0.54	0.70	1.2
Thailand	0.56	0.95	1.17
Indonesia	0.81	0 91	1.04
Bangladesh	0.58	0.80	0 .91
Japan	1.29	0.82	0 .77
Chile	NA	0.55	0.67
Norway	NA	0.55	0.64

USA	NA	0.50	0.60
Top ten subtotal	NA	35.73	40.11
Rest of the world	NA	4.65	5.35
Total	39.43	40.38	45.46

Source FAO 2006

Table 21(b). Trend in World Aquaculture production (Million tons)

Group	1984	1987	1989	1991	1993
Fisfish	4.70 (45.03)	6.793 (51.44)	7.784 (54.01)	8.741 (52.75)	11.20 (49.44)
Crustaceans	0.24 (2.29)	0.574 (4.34)	0.674 (4.67)	0.806 (4.86)	0.93 (4.10)
Mollusca	2.00 (19.16)	2.672 (20.23)	2.914 (20.19)	3.095 (18.67)	4.10 (18.10)
Aquatic plants	3.40 (32.57)	3.139 (23.77)	2.997 (20.79)	3.904 (23.56)	6.30 (27.81)
Others	0.097 (0.92)	0.027 (0.20)	0.048 (0.33)	0.032 (0.19)	0.12 (0.52)
Total	10.437	13.205	14.41	16.57	22.65

Figures in bracket indicate percentage.

In this context, China is the top most aquaculture producer, account 58.7% of the total, which is followed by India (6.4%), Japan (6.3%), the Republic of Korea (4.6%) and the Philippines (3.4%). The total contribution of the above five countries form 79.4% of the total world production in 1993 (FAO, 1995). Europe was the next largest aquaculture producer by region (5.3%, followed by North America (2.5%), South America (1.3%), the former USSR (0.8%), Oceania (0.4%) and Africa (0.3%). Further in world aquaculture scenario, the production is growing at a much faster rate within developing countries and regions viz, Asia, South America, Oceania and Africa than within developed regions such as North-America, Europe and former USSR. The share of aquaculture production by developing countries has been steadily increasing over the past decade, from 78% to 87% for the total farmed fin fish and from 87% to 96% for total farmed crustaceans from 1984 to 1993 respectively (FAO, 1995). The per capita aquaculture product consumption in Asia and Oceanea was 2.2 kg/ ind/ year (FAO, 1989). However, in India, the per capita consumption of fish and fishery product is 11 kg/ind/year.

1.1 Fin-fishes

In Asia about 200 species of aquacti plants and animals are being raised and the main bulk of fish production from aquaculture comprises about 32 species. Fresh water finfish particularly IMC, Chinese carps, common carp continued to be mainstay of Asian aquaculture. However, carp and tilapia dominated total finfish production. The Big head and silver carps which exceeded 1 million ton each, demonstrates that the main aquaculture outputs are contributed by species low in food chain specially species with filter feeding habits. Finfish constituted about 51% of aquaculture production of Asia in 1987. Milk fish continued to be an important farm product in the Philippines, Indonesia and Taiwan. The world tilapia production was 2,14,000 MT in 1978, the major producers being the Philippines, Taiwan, China, Indonesia and Thailand in that order of importance.

1.2 Crustacean

However, recent public interest has centred very much on shrimp and prawn farming. It is evident that there is a rising trend in world crustacean production raised from 1.64 million tons in 1970 to 2.38 million tons in 1980 of which shrimp production from both culture abnd capture was 0.96 million tons in 1970 and raised to 1.30 million tons in 1980. World shrimp Production further raised to 1.76 million tons in 1983, 1.82 million tons in 1984 and 1.9 million tons in 1985, and 2.2 million tons in 1989. Further, the world catch of shrimp was 2.24 million metric tons ins 1991, 2.22 mmt in 1992, 2.21 mmt in 1993, 2.30 mmt in 1994, 2.19 mmt in 1995 and 2.08 mmt in 1996. Major shrimp producing (culture and capture) countries in 1985 were Thailand, (1.53 lakh tons), U.S.A. (1.52 lakh tons), Indonesia (1.00 lakh tons), Norway (0.91 lakh tons), Mexico (0.745 lakh tons), Malaysia (0.69 lakh tons), Philippines (0.623 lakh tons), Brazil (0.536 lakh tons), China (2.29 lakh tons) and India (1.98 lakh tons). However, shrimp production of India was 1.37 lakh tons in 1982 and 1.73 lakh tons in 1983. Such increase in world shrimp production of cultured shrimp continued to increase and significantly influences the world shrimp market (Table 22). The record cultured shrimp harvest of 4.48 lakh tons in 1987 and 4.8 lakh tons in 1988 was surpassed in 1989 and a new record high production of 5.65 lakh tons was attained. Hence, farmed shrimp now account for about

26% of the total world shrimp production of 2.2 million tons, the balance 74% being the contribution from capture fishery. About 78% of the cultured shrimp production was from the major five producing countries, namely : China, Indonesia, Thailand, Philippines and Eucuador. India with its substantial potential for shrimp culture development had a share of only 4% in 1989 and held the seventh rank. Prasad (1996), described that India's shrimp harvest (Penaeid) from the capture fisheries reached a plateau of about 2.5 lakh tonnes and a substantial increase from this sector is not expecting in the near future. In this content the farming of shrimp gained much momentum in most of the maritime states of India and the production registered 70,000 tons in 1994 and 86,634 tons in 1998-1999.

Table 22. World shrimp production (Capture and culture)

Year	Production
1983	17.69 lakh tons
1984	18.21 lakh tons
1985	19.03 lakh tons
1989	22.00 lakh tons
1991	22.46 lakh tons
1992	22.22 lakh tons
1993	22.17 lakh tons
1994	23.07 lakh tons
1995	21.90 lakh tons
1996	20.80 lakh tons

In 1988, the leading countries in cultured shrimp production were China (1 lakh ton), Eucuador (0.7 lakh ton), Taiwan (0.5 lakh ton), Indonesia (0.5 lakh ton), Thailand (0.4 lakh ton), Philippines (0.3 lakh ton), India (0.3 lakh ton), Vietnam (0.2 lakh ton) Table 23. However, in 1989, the scenario is little changed as the cultured shrimp production (Table 24) was severely affected in Eucuador and Taiwan due to certain disease problems. The world production of cultivated shrimp had increased from 1.75 lakh tons in 1984 to 4.48 lakh tons in 1987, 4.8 lakh tons in 1988, 5.65 lakh tons in 1989, 7.33 lakh tons in 1991, 7.29 lakh tons during 1992. However, the global production of cultured shrimp decreased from 7.29 lakh metric tons to only 6.44 lakh metric tons in 1993 (Table 25). Over 80% of the total world farmed shrimp were produced in Asia.

Among new records, the international trade in shrimp now exceeds 7,50,000 tons comapred to 5,00,000 tons traded in the mid 1980's. FAO, Director General-Edouard Saouma spoke this record in Committee on. Fisheries in Rome in April, 1988. Further the top ten countries of world in shrimp production is given in Table 26. India's marine product export in 87-88 was in the order of 97179 MT of which forzen shrimps accounted to 55,736 MT and the foreign exchange earned through marine product export was in the tune of 531 crores. However export of marine products in India achieved new heights of 1,10,788 tonnes in 1989-90 valued Rs. 634.76 crores as against 99,777 tonnes valued at Rs. 597.85 crores in 1988-89. This represent a growth rate of 11.055 in terms of volume and 6.17% in terms of value. Frozen shrimp export in 1988-89 was 56829 MT valued to 470.09 crores and it was further increased in 1989-90 to 57846 MT and valued at 463.44 crores. (Seafood News Letter, Vol. 89, No. 2, 1990). The foreign exchange earned through marine product export is increasing in India. The extract of the marine product export in terms of quantity and value is presented in Table 27.

Table 23. Some of the leading countries in cultured shrimp production 1988.

Countries	Production
China	1 lakh tons
Eucuador	0.7 lakh tons
Taiwan	0.5 lakh tons
Indonesia	0.5 lakh tons
Thailand	0.4 lakh tons
Philippines	0.3 lakh tons
India	0.3 lakh tons
Vietnam	0.2 lakh tons

Table 24. World aquaculture production of shrimp 1989, sea food news letter Vol. 87 (4) : 1990.

Country estimates	% of world production	Heads on production (MT)	Hectare prodn. water source	Kilogram per hectare
China	29	1,65,000	1,45,000	1138
Indonesia	16	90,000	2,50,000	360
Thailand	16	90,000	80,000	1125

Philipines	9	50,000	2,00,000	250
Eucuador	8	45,000	70,000	643
Vietnam	5	30,000	1,60,000	187
India	4	25,000	60,000	416
Taiwan	4	20,000	4,000	5,000
Central Caribbean	2	12,000	12,000	1,000
South America (Excluding Eucuador)	1	7,000	8,000	875
Others	5	30,800	1,033,00	298
Total		5,64,800	10,928,00	517

Source : World Shrimp, Farming.

Table 25. World production of cultivated shrimp (Lakh tons)

Country	1991	1992	1993
China	1.45	1.40	0.30
Eucuador	1.00	0.95	0.78
Taiwan	0.30	0.25	0.20
Indonesia	1.40	1.30	1.00
Thailand	1.53	1.63	2.08
Philippines	0.30	0.25	0.25
India	0.35	0.45	0.58
Vietnam	0.30	0.35	0.40
Bangladesh	0.25	0.25	0.30
Others	0.45	0.46	0.54
Total	7.33	7.29	6.44

Table 26. World catch of shrimp 1991-96 (000 tons)

Country	1991	1992	1993	1994	1995	1996
China	334	369	399	420	390	370
India	271	245	231	263	245	220
Indonesia	150	163	156	168	157	150
USA	146	153	135	128	134	130
Greenland	73	82	84	86	80	85
Thailand	120	127	101	92	98	80
Mexico	68	61	71	65	60	55
Philippines	45	56	63	60	57	50
Vietnam	48	53	59	44	38	40
Others	991	913	918	981	931	90
Total	2246	222	2217	2307	2190	2080

Source : MPEDA News letter 15th Feb. 1998.

Table 27. Marine Product Export from India 1950 to 2008

Year	Quantity (Thousand tons)	Value in Rupees (Crores)
1950-51	19.7	2.46
1960-61	15.7	3.92
1970-71	35.9	35.07
1980-81	75.6	234.84
1981-82	70.1	286.00
1982-83	78.1	361.36
1983-84	92.69	373.02
1984-85	86.18	384.28
1985-86	–	402.68
1986-87	–	460.67
1987-88	97.10	531.00
1988-89	99.70	597.85
1989-90	110.70	634.76
1990-91	139.40	893.37
1991-92	171.80	1375.89
1992-93	208.60	1767.43
1993-94	244.0	2503.62
1994-95	307.3	3575.27
1995-96	296.3	3501.11
1996-97	378.1	4121.36
1997-98	385.81	4697.48
1999-2000	440.47	6443.89
2000-2001	467.29	6881.31
2002-2003	412.02	6091.95
2003-2004	412.02	6091.95
2004-2005	461.33	6646.69
2005-2006	512.16	7245.30
2006-2007	612.64	8363.53
2007-2008	541.70	7621.00

Source : 1. Hand Bood on Fisheries Statistics; 2. Sea Food Export Journal

U.S. per capita shrimp consumption rose from 0.77 kg in 1980 to 1.28 kg in 1989 registered an increase of 66% in the decade. If the rate of increase is maintained, per capita consumption by the year 2000 would touch 2 kg. Asian production of farmed shrimp (culture) is predicted to reach 8 lakh ton by 2000 (see world Aquaculture, 1989, 19 (4) : 32).

1.3 Molluscs

World's molluscan production has been raised from 3.36 million tons in 1970 to 3.92 million tons in 1980 including cephalopods. FAO, Fisheries Technical paper 254, 1984 describes about 5 lakh tons of cephalopods were caught world-wide in 1945 by some 30 nations. In 1980, this amount had increased to 15.3 lakh tons and 75 countries reported their catch (FAO, 1982). The leading nation was Japan with 7.32 lakh tons, followed by Korea with 1.27 lakh tons and Spain with 1.15 lakh tons. These three countries caught about 64% of the total world catch of cephalopods in 1980. In 1981, Japanese and Spain landings declined while Korean landing increased. The world's total cephalopod catch and leading countries in cephalopod production are given in (Table 28 and 29). However, as a whole the world's molluscan production has showed a very low rate of increase. Similarly, while most aquacultural commodities like fin fish and shrimps in the Asian region has shown steady increase, the culture of molluscs has registered a low rate of increase. From 1975-84, the average rate of annual increase was about 4.1% while from 1981-85, it was about 1.6%. In fact, the production has gone down in some countries like Thailand and Malaysia.

Table 28. World's Total Cephalopods Catch.

Year	Production
1972	11.92 lakh tons
1973	10.68 lakh tons
1974	10.74 lakh tons
1975	11.82 lakh tons
1976	12.10 lakh tons
1977	12.31 lakh tons
1978	13.30 lakh tons
1979	15.21 lakh tons
1980	15.30 lakh tons
1981	13.04 lakh tons

Table 29. Some of the leading countries in world cephalopod production.

Country	Production in	
	1980	1981
Japan	7.32 lakh tons	5.66 lakh tons
Korea (Rep.)	1.27 lakh tons	1.45 lakh tons
Spain	1.15 lakh tons	1.02 lakh tons
Thailand	0.72 lakh tons	0.65 lakh tons
USSR	0.55 lakh tons	0.45 lakh tons
Itlay	0.48 lakh tons	0.39 lakh tons
Philippines	0.31 lakh tons	0.32 lakh tons
China	0.29 lakh tons	0.28 lakh tons
USA	0.15 lakh tons	0.25 lakh tons

From FAO year book of Fishery Statistics Vol. 52 : 1982.

1.4 Seaweeds

Remarkable increase have occurred in the production of seaweeds. Scientific culture techniques and utilization have also undergone considerable improvement. Seaweed production has been raised from 1.4 million tons in 1984 to 3.13 million tons in 1987, even though commerical seaweed farming is limited to only 6 countries such as Japan, Korea (DPR), Korea (Rep), China, Taiwan and Philippines. With the recognisation of their nutritional value and industrial use, more countries are making efforts to introduce seaweed culture.

Based on the recent aquaculture production trends the projected production data through culture for the year 2000 would increase to 22 million tons. This projected figure of 22 million tons is little above twice the production figure of 1985. The production of fish by Inland Fisheries and aquaculture for the various continents 1970-87 based on nominal catch statistics is given in Table 30.

Table 30. Production of fish by Inland Fisheries and aquaculture for the various continents 1970-87 based on nominal catch statistics (catch, 000 tons)

Continents	1970-74	1975-79	1980-84	1985	1986	1987
Africa (Area 01)	1271	1425	1450	1515	1646	1669
North America (02)	138	131	188	254	259	317
South America (03)	173	243	307	326	348	330
Asia (04)	3815	4154	5434	7099	7691	690
Europe (05)	250	304	383	422	437	441
Oceanea (06)	1	2	9	10	11	7
USSR (07)	858	830	838	906	927	988
Total	6507	7089	8609	10532	11319	11442

Source : Year Book of Fishery Statistics 1987 Provisional FISH DAB.

1.5 Indian Scenario

In the context of Indian Fish production, since 1961, India's fish production increase steadily year after year reaching 1.85 million tons in 1971, 1,958 million tons in 1973, 2.4 million tons in 1976 and 3.1 million tons in 1988-89 and above 3.8 million tons (1.55 Inland + 2.25 marine), in 1990. Reviewing on Inland fish production, India ranked second in the World, China being 1st and USSR being 3rd. In 1976, China, India and USSR produced 4.56, 0.87 and 0.77 million tons of inland fish respectively. In 1987-88, India continued to maintain her 2nd position after China, in Inland Fish Production. The inland fish production having increased from 0.22 million tons in 1950-51 to over 1.3 million tons in 1987-88 and 1.55 million tons in 1990. However, India's total fish production from both Inland and Marine Sources keep India's position oscillating between 7th or 8th position in World Fish Production Scenario.

In recent year 2009-10 India ranks third in the world in total fish production of 7.43 million metric tons (3.0 Marine + 4.43 Inland). Constituting over 1% of the GDP, fisheries contributes to 5.3% of the agricultural GDP. The percent share of major top ten fish producing countries in world fish production during 2005-06 is as follows.

1. Peru--- 8.17%
2. Japan -- 4.4%
3. India---------- 4.36%

4. USA --------- 3.97%
5. Indonesia- --- 3.78%
6. Chile --------- 3.6%
7. Russian fed -- 3.1%
8. Thailand ----- 2.78%
9. Norway ------ 2.45%
10. Philippines -- 1.75%

Further India continued to stand in second position in lnland fish production mainly from culture sector. Presently farmed fishes contribute to more then 1/3rd of the total fish production. During 2006-07, India produced 1.44 lakh metric tons of shrimp and 0.3 lakh tons of freshwater prawn through culture. India's current export (2006-07) from fisheries stand at 6,12,641 mt valued at 1.85 billion US $ (Rs 8363.53 crores). More than 50 different types of fish and shellfish products are exported from India to 75 countries around the world

The technical feasibility of farming different aquatic organisms in Indian waters is highly promising and some production figures, especially for Indian carp, mixed culture and green mussel are comparable and even higher than those reported from the developed countries. Cost benefit analysis also indicates a high rate of return, which is 18.1% in the case of mussel culture and more than 100% in traditional fish farming in Goa. The techniques developed in India are low cost and practicable. The Annual Yield for different aquaculture systems developed in India were as follows :

Species/water	Location/system	Annual yield/kg/ha
IMC/Fresh	Cuttack (Orissa) Average for 132 ponds.	400 kg
Mixed/fresh	CIFRI, Barrackpore Fish Farm W.B.	8750 kg
Mixed/Brackish (Fish & Prawn)	Kakinada (AP) Fish Farm	2250 kg
Mixed/Brackish (Fish & Prawn)	Goa/traditional Fish Farm paddy fields & salt pans	1060 kg
Green mussel/ Brackish	Dona Paula (Goa) Rope culture of Floating rafts.	4,80,000

With advancement in technology, even with three species of Indian carps, the production to the tune of 8-tons/ha/year has been reported from Andhra Pradesh. Under mixed culture of IMC and

Chinese carps, production to the tune of 15 tons/ha/year has been communicated from Andhra Pradesh as well as from CIFA, Kausalyagang, Orissa through rotational culture and harvest. Brackishwater prawn farming has given a production of 4.6-9.0 tons/ha/year by Hindustan Lever (see Purushan 1989, In sea food expt. Jl. (1989) March issue) and from Sandesh kali West Bengal (see fishing chimes 1988, Dec. issue).

2. SYSTEMS OF FRESHWATER FISH CULTURE

2.1 Carp culture

The major cultivated fishes all over the world belong to the family cyprinidae particularly the carps. Although principles and techniques of carp culture is basically same, still certain deviations are noticed in regard to managment adopted in line with local conditions. However, there are 3 major systems of carp culture such as Chinese system includes the culture of Chinese carps together, the Indian system includes the culture of Indian major carps and the European system includes the culture of common carp as one of the main species.

In India and China, Polyculture is more popular whereas monoculture of common carp is more popular in European countries. However, common carp is still common in carp culture practice and prevalent throughout the world. The probable reason may be seed production of common carp is easier than the other cultivable species. In China, Chinese carp such as Silver carp, Grass carp, Big head carp, Mud carp, Black carp and Common carp are cultured in Polyculture. Other varieties of carps are also now incorporated in polyculture system.

2.1.1. Composite fish culture

In India, traditional fish culture with three endemic species such as Catla, Rohu and Mrigal were carried out and in late fifties, three exotic species of carps were introduced into India for culture alongwith Indian major carps. Studies on their compatibility with IMC were carried out at the pond culture division of the CIFRI, Cuttack. Such studies brought the concept of `aquaplosion' and laid the foundation of an era of fish revolution in India through pond fish culture. Such six combinations of three Indian Major carps with three exotic carps resulted better yield than three

Indian Major carps or three exotic carps alone. This high yielding combination of six species is termed as comosite fish culture which have streadily stepped up per hectare production. The main objective of the composite culture management technique is to take full advantage of the available three dimensional pond space, optimum utilisation of fish food resources in the body of water to produce fish, augmenting it through manuring and fertilisation as well as providing artificial feeds to the growing fish.

In 1971, the ICAR initiated All India Co-ordinated Research Project (AICRP) on composite fish culture of Indian and exotic fishes under different Agro-climatic conditions in the country.

2.1.2. Fish species

However, in a pond three remain several ecological sub-niches unexploited by the above six carp species cultured. Therefore, other fishes, such as the minor cat fish (*Ompok bimaculatus*) feeding on insects, the feather back *Notopterus chitala* taking care of weed fishes and give run to cultured carps because of their predatory behaviour are introduced. Such runs help in accelerating their metabolic activity for better food assimilation and growth rate. Hence, feather back in a composite carp culture acts as a police fish to chess behind the cultivated carps and cannot feed upon carps because of their size differences. That too, the *Pangassius pangassius* feeding on molluscs could be introduced as useful combination alongwith carp species. Even introduction of grey mullet *Mugil cephalus* had been grown in composite fish culture ponds with great success in some experimental ponds. Composite culture of Indian and exotic carps with either Gourami (*Osphronemus gorami*) or Calbasu (*Labeo calbasu*) increased the production. Even culture of freshwater prawn *Macrobrachium* alongwith IMC and exotic carps are practiced in experimental ponds in India.

2.1.3 Species ratio combination

Manipulation of the species ratio is for minimising the interspecific and intraspecific competetion for available food at various trophic levels and zones in a pond. Either single species or more than one species occupying different niches could be utilised, in a pond for exploiting the available food at various zones. In IMC, such as Catla, Rohu and Mrigal in 4 : 3 : 3 ratio combination had been observed to be guide favourable. In six species combination of IMC and exotic carps, the species ratio of silver carp 2.5, Catla 1.0, Rohu 2.5, Grass carp 1.0, Mrigal 1.0, *Cyprinus carpio* 2.0 was observed to be superior to other. However, with the availability of fingerlings the species ratio is little altered and adjusted. The selection of good quality fingerlings is one of the important links in ensuring high fish yields. The large sized fingerlings of good quality have many merits such as strong adoptability, high survival rate, fast growth, short culture period, high marketing rate and economic returns. The criteria for selecting the fingerlings for stocking are as follows : (1) Strong and healthy fingerlings without abnormal shape, (2) Fingerlings should have complete scale and fin rays and smooth skin with bright color, (3) The fingerlings of the same age should be of uniform size without much length and weight differences. That to the healthy fingerlings will jump violently in hand while poor fingerlings will not. Fingerlings are to be disinfected or treated with antibiotic or disinfectants before stocking in ponds.

The factors determining stocking density is dependant on pond conditions, seed availability and farm operation techniques. Stocking density varies correspondingly with the development of production and the state of the art. Therefore, the stocking density should be determined by local conditions in order to harvest bumper yield. Stocking density is also correlated to the water quality, particularly dissolved oxygen. Carps can sustain their normal lives when dissolved oxygen is above 3 mg/litre of water. But the optimum dissolved oxygen required for carp is 5.5 mg/litre.

2.1.4 Stocking density

In India, with improvement in management practices, the stocking density of 6 species carps had been gradually increased from 5000 nos/ha (Low input technology) to 8000 nos/ha (Intermediary input technology) to 10000 nos/ha (high input technology). A higher stocking density (15,000-25,000) of fingerlings/ha through rotational culture is being experimented upon at CIFA, Kausalyagang. The results are encouraging and promising.

2.1.5 Fish production

Generally carps are harvested after one year growout period during which it reaches 0.8-1.0 kg in weight. These marketable carps are transported to near by places as well as to distant places in insulated ice vans. The price in the domestic market is influenced by demand, supply, condition of fish and consumer's preference.

Considering the varying levels of operational inputs, the carp culture practices are categorized as 1. Low input system 2. Intermediary/ Medium input system and 3. High input system.

Low input system - In this system

1. Seeds form the principal input.
2. Natural food available in culture pond is the main source of nutrition to cultured fishes.
3. Low stocking density.
4. Application of organic and inorganic fertilizer is limited.
5. Production level of 2-3 tones/ha/yr is achieved in such system. However, biogas slurry and aquatic weed based carp culture system can yield 3-4 tones/hr/yr without any supplementary feeding

Intermediary/ Medium input system

1. Apart from fertilization, supplementary feeding is given for enhancing carp production.
2. Polyculture of three indigenous carps (IMC) and three exotic carps can yield 4-8 tones/ha/yr.
3. Proved scope for diversification of farming practices like integration of agriculture and livestock components in line with local condition.

High input system

1. Balanced diet together with aeration and water exchange are the characteristics of this system.
2. High stocking density, aeration and water exchange can result a production level over 15 tones/ha/yr

Fish production of 40 tons/ha have been obtained in Israel (Tal, 1974) through pond fish culture. In China, the national average production through pond fish culture is around 15 tons/ ha/yr and in India, the highest production through pond fish culture especially carps has gone up over 15 tons/ha/yr although the average national pond fish production is 5-6 tons/ha/yr.

Under traditional carp culture, the rate of production was as low as 600 kg/ha/yr. The production results through composite fish culture clearly indicate the potentiality which are comparable with the highest pond fish productions in still water ponds of the world.

2.1.6 Growth rate

Growth potential of different carps under composite fish culture has been studied in different agroclimate conditions in India. However, the average monthly growth rate of grass carp, and common carp is more in comparsion to silver carp, catla, mrigala and rohu. In ponds where both supplementary feeding and weed as well as fertilization were carried out, grass carp showed an average monthly growth rate of 96.6 - 112.5 gm, common carp 83.0 - 92.0 gm, silver carp 65.6-92.6 gm, catla 43.0-64.2 gm, mrigal 42.2-53.5 gm and rohu 38.0-46.6 gm (Sinha, 1979).

2.1.7. Species interrelationship

(1) *Catla and Silver carp* - Both these carp species are planktophage surface feeders, though catla is predominantly zooplanktophage and silver carp phytoplanktophage. Various species ratio combination of catla and Silver carp such as equal proportions, 1 : 2.5, or 1.5 : 2 or 1 : 3 were tried. Silver carp grew better even at times double to that of catla or more are noticed. It is interesting to note that at Pune centre, where the highest fish production rate of 10,670 kg/ha/yr had been obtained, catla and silver carp attained an average weight of 1.3 kg and 2.1 kg respectively in one year when stocked in the ratio of 1 : 2.5, the total stocking density of all the six species, being 10,000/ha. The average monthly growth rate of catla has been about 108 gm and silver carp has been about 175 gm. In certain experiments, even the average monthly growth rate of catla was 122.7 gm compared to monthly increment of 191 gm shown by silver carp. The monthly increment of silver carp showed even 264 gm at Jaunpur fish farm. Unprecedented growth of silver carp to 3.3 kg in eleven months has been obtained at Krishnanagar (W.B.) (Sinha and Sharma, 1970).

(2) *Mrigal and common carp* - The growth performance of mrigal and common carp in equal stocking ratio appears to be similar in cases where supplementary feeding was not provided. However, when supplementary feeding was provided, common

carp showed better growth performance indicating its superior capability of utilising artificial feeds than mrigal (Sinha, 1979).

The monthly average increment of weight by common carp has been 147 gm and 76-107 gm in mrigal in some experimental ponds.

(3) *Grass carp and other fishes* - Association of grass carp in composite fish culture has an indirect benefit too. The excreta consisting of semidigested aquatic vegetation, serve as the food for the bottom feeders such as mrigal and common carp. However, grass carp is also well known to utilise supplementary feeds like rice bran and oil cakes. In order to avoid competetion among grass carp, rohu etc., it is advisable to provide aquatic weeds in adequate quantity to grass carp. Grass carp has shown an unprecedented growth of 5.04 kg in one year at Kulia fish farm with an average monthly increment of 416.3 gm (Sinha and Gupta, 1975).

The concept of polyculture involves judicious exploitation of all the niches available in the pond. However, the use of extraneous fertilizers and feed increases the productivity. It modifies the natural balance in the ecosystem, which needs to be judiciously exploited. It is also essential to make composite fish culture as a commercial economical and self-sustaining system through synergetic approach in line with local condition.

In China, generally speaking, five to six even eight to nine species of fish are polycultured in one growout pond. However, among the species cultured one or two species are taken as major species which are stocked in more numbers and rest are consdered as minor species. The factors determining what kinds of fish are to be taken as major species and their stocking ratio depend on the availability of fingerlings, feeds, manures, farming techniques pond conditions and market demand.

2.1.8. Harvesting and stocking in rotation

Harvesting and stocking in rotation is a procedure where by fingerlings of different sizes are stocked into the pond at the same time. With the growth of fish, the pond becomes overcrowded. Consequently, marketable sized fish are caught in batches and are replaced by an appropriate amount of smaller fish to maintain as optimal stocking density during the whole culture period so as to increase fish yield per unit water area. By such rotational stocking and harvesting of marketable size fish, the stocking density can be raised even over to the tune of 25,000 fingerlings/ha and production over to the tune of 15 tons/ha/yr had already been achieved in CIFA, Kausalyagang, Orissa. It is possible to step up further production in Indian condition.

The advantage of rotational stocking and harvesting are : (1) to balance the carrying capacity of ponds in accordance with the growth of fish, (2) to maximize the utilisation of stocking pond for interfarmng fingerlings which can maintain high and stable fish yield, and (3) to reduce seasonal variation in fresh fish supply to the market and to speed up capital return.

However, precautions to be taken in rotary harvesting system. Harvesting is usually conducted in summer and autumn. At high water temperature, fish have a high feeding intensity and active movement, so they are not able to tolerate the long period of operation and crowd. So rotary harvesting should be done when it is cool and there is no surfacing of fish. In addition to this, fish should be less fed one day before rotary harvesting to avoid mortality of fish which will jump violently during the operation. However, operation must be done quickly and gently as possible.

During harvesting, fish consume more oxygen because of violent movement and the pond water gets turbid by turning up the silt at the bottom. It is necessary to turn on aerators and fill the pond with freshwater for preventing fish from serious surfacing.

2.1.9 Biogas slurry and weed based carp culture

Use of biogas slurry in aquaculture system is a new horizon in boosting up fish production. The production as high as 6 tons/ha/yr in polyculture ponds has been reported (see Tripathi, 1990). The slurry prepared out of water hyacinth from biogas plant has yielded over 3 tons/ha/yr in 10 months time in polyculture ponds (Tripathi *et al.,* 1989).

2.1.10 Multiple grade conveyor culture system

In Guangdong province of China, this system of polyculture farming technique is practiced. This method is different from the rotary harvesting and stocking. The rotational culture is usually carried out in one pond by stages, catching the edible sized, restocking or replenishing the smaller ones, where as in multiple-grade conveyor culture is to rear fingerlings of different sizes in separate ponds based on the growth of fish, with which they are transferred in sequence into other ponds. Therefore, ponds are usually divided into five grades, each grade is for one size of fish. When the marketable sized fishes are harvested, the larger fingerlings from each grade pond are upgraded in sequence into next grade pond for further culture to the desired size. Hence, a number of ponds are needed as fishes are upgraded.

In Indian context, fish farmers have very limited number of fish ponds for production of table sized fish. Hence, this system of multiple grade conveyor fish culture is too early to be practicable at present in Indian context. This can be taken up in time considering the availability of resource potential with the farmers in India.

3. SEWAGE FED FISH CULTURE

The use of sewage for fish culture has been studied in India, Germany, USSR, Poland, Hungary, Israel, China and Indonesia. In India (Bose, 1944; Sreenivasan, 1968; Sundaresan and Muthusamy; 1976; Choudhuri, 1976; Ganapati, 1972; Ghosh *et al.*, 1979; Kutty; 1979; Sen 1986 and Sukumaran *et al.*, 1987 have studied the growth and production of fish in domestic sewage. In many countries including India, treated sewage is used for agriculture and aquaculture purposes. These domestic wastes such as sewage is now no longer considered as a waste. It can be recycled to produce wealth. Fishes are known to convert biowaste into edible protein efficiently. Sewage fed fish culture involves meagre operational expenditure but in turn, the economic return is substantial. A fish yield even 10-15 tons/ha/yr are often achieved in well managed ponds (Singh, 1990).

Verma *et al.*, (1991) estimated that the per day discharge of sewage water is in the order of 12500 million litres in India, and if all of it is recycled, it would yield 102000 tonnes of nitrogen and 30,000 tonnes of phosphorus every year. However, Banerjee *et al.*, (1989) reported that the town refuse generation rate in India is 50000 tonnes per day while the utilisation rate is only 15000 tonnes per day. The bulk surplus being disposed indiscriminately contributes 42.07 to 48.79 tonnes of pollutional load to the environment per day. These could be utilised for integrated aqua-agriculture farming systems. However, there are three types of waste water namely sewage, sullage and sludge.

3.1 Sewage

The term sewage is used to indicate a turbid fluid arising out of domestic waste containing semi-docomposed as well as decomposed organic matter and a few essential minerals. Depending on the source, sewage may be purely domestic or it may partly, contain some industrial and agricultural waste water. In this waste water, there may be an abundance of inorganic nutrients like, Ca, Na, K, P and Nitrogen that are essential for the production of phytoplanktons. Some fishes thrive well with these primary producers and also the zooplanktons.

3.2 Sullage

It is waste water resulting from personal washing (excluding faecal matter and urine), Laundry, and cleaning of kitchen utensils.

3.3 Sludge

It is the solid substance settled at the bottom of the water body containing waste. Activated sludge-single cell protein (ASCP) derived from the bio-oxidation of curde domestic sewage has been evaluated a supplementary protein and calories in diet, which is regarded as efficient diet of trout in Germany. The ASCP contain 39-46% protein range and is incorporated in trout diet.

3.4. Quality of sewage

Generally raw sewage contains 99.9% of water and remaining constituents include fat, grease, suspended solids, organic matter, bacteria, protozoans and fungi. Organic matter present in the sewage helps in rich growth of bacteria. The quality of sewage vary from place to place and also from time to time at the same place depending on the climatic conditions, availability of water and dietary habitat of the inhabiting local population. The common gases and silts present in sewage are CO_2, H_2S, NH_3 and PO_4. Such emission of foul and abnoxious gases are due to high rate of microbial decomposition. The other heavy metals present in traces are Zn, Cu, Chromium, Manganese, Nickel and lead. The dissolved oxygen content of sewage is almost nil or insignificant. Fresh sewage is generally muddy or black clouds in color due to suspended colloidal particles.

3.5 Advantage of sewage

Sewage is beneficial for aquaculture particularly fresh water finfish culture because of (1) microparticles of sewage form direct food source to zooplanktons and benthos. (2) Macroparticles of relatively larger size of sewage are directly utilised by fish. (3) Mineral inorganic and soluble organic substances are directly used by phytoplanktons for photosynthesis and being eaten by zooplankton as a food source. (4) Waste water fertilised ponds produce high yields because of the increase in natural fish food organisms produced by microbial decomposition of these wastes. Several recognisable major processes are involved in such increase in actual fish food supply. (5) Nitrogen, phosphorous and trace elements, stimulates phytoplankton production in the same manner as inorganic fertilisers. This process is limited by autoshading. (6) The increase in natural fish foods from waste water fertilisation will be reflected in a decreasing need for supplementary feeds, due to both the quality and quantity of natural foods provided. The most common phytoplanktons are microcystis, Anabaena, Ankistrodesmus, cosmorium, scenedesmus, spirulina, closterium, Navicula, diatom, pediastrum, Oscillatoria and zooplanktons are cyclops, Diaptomus, Daphnia and moina. Even red worm, chironemids are found in sewage sediments.

3.6 Waste treatment

Waste water treatment can be carried out in three ways.

(1) Mechanical = this includes screening, filtration, skimming and sedimentation.

(2) Chemical - includes chemical precipitation, deodorisation and disinfection.

(3) Biological - treatment includes bacteria for biochemical reactions resulting oxidation of organic matter into CO_2 H_2O, N_2 sulphates etc.

However, the sewage water for use in pond culture should undergo (1) sedimentation, (2) dilution and (3) storage. It is because, sewage before treatment lacks dissolved oxygen and hence not suitable for direct use in fish culture. Hence, before it is passed in to fish ponds screen filtration followed by sedimenation has to be carried out by cascading sewage through a series of cemented ponds. Such sedimentation reduce the BOD load by about 33% and almost 90% of suspended solids and 25% albuminoid ammonia.

3.6.1 Sedimentation

The sewage is first screen filtered and stagnated in the stabilisation (oxidation) pond for few days to allow the particulate organic matter to settle down. The dissolved organic matter is decomposed by aerobic microorganisms in to inorganic nutrients. The particulate organic matter settled at the bottom also unergo decomposition resulting in the release of nutrients and also abnoxious gases. As a consequence of decomposition, the fertilization status of waste water increases and helps in propagation of fish food organisms. As a result of these changes, the sewage water in the stabilisation pond in enriched with O_2 and nutrients and thus made suitable for fish culture. Certain floating weeds like *Eichhornia crassipes* is used in oxidation pond in order to remove toxic heavy metals in the city sewage.

At Kharda (W.B.), Bhilai (Chhatisgarh), Bidanasi (Cuttack) and many other seweage stabilisation (Oxidation) tanks are concrete. These could serve as a storage for incoming sewage and should be located in open space for easy air diffusion so that sewage treatment takes short time due to its contact with atmosphere. Periodical removal of sewage at the bottom of the stabilisation (oxidation pond) could be made, so that there will be least turbulence in the sewage column, particularly the supernatant which is to be taken to the fish ponds by gravitation to make it economical. Hence, the slope of the terriane should be genetly slopping. Selection for construction of oxidation pond be gently slopping. Selection for construction of oxidation pond should be placed at elevated place than the fish ponds for easy removal of supernatant sewage by gravitational force. Various sizes with 1.5-2 m depth of oxidation ponds are seen in the country, but one has to keep in mind that the size and number of the oxidation ponds should be such that, it can accomodate as much discharge of city sewage for 10 days continuously. The free board or space above the water column of the oxidation pond should be atleast 0.5 m.

3.6.2 Dilution

Various proportions of sewage and fresh water dilutions are practiced in India. However, dilution of 1 part sewage to 5 parts of fresh water or 10 parts of fresh water with 4 parts of sewage are in use. Such proportions of sewage is also dependant to the availability of prepared sewage and incoming city sewage to the oxidation pond.

3.6.3. Storage

During storage of sewage, biochemical process in microbial community continue. These processes oxidised and liquify the organic matter which renders the fluid fit for pisciculture.

In summary it can be noted that improvement of waste water treatment can be enhanced by -

(1) Increase in phytoplankton concentration in waste water

(2) treatment of waste water removing excessive nutrients that cause eutrophication and capacity of disinfection through increased oxygen and buffering action of pH in waste water system.

3.7 Classification of waste water fish culture system

Classification of waste water fish culture system has been described by various workers considering the method of feeding fish and also using the methods of reducing the biological oxygen demand (BOD). Existing fish culture systems in waste water is by reducing the BOD through the process of treatment.

3.8 Sewage benefits to fish culture

Considerable literatures are now available to illustrate the increase in fish yield using waste water. Greater fish yield with significant reduction in artificial feed application in sewage culture are also reported. The species cultured in treated sewage water included, carps, tilapia and freshwater prawn (*Macrobrachium rosenbergii*). From time to time introduction of other species of fish are under experimentation in sewage culture in India. As treated sewage water is fertile due to rich nutrient status, considerable increase in stocking density of fish species to the order of 20,000/ha could be practiced. Air breathing and catfishes especially magur may also be cultured along with carp species. Carp species of considerable larger size can be cultured along with magur and murrels so as to avoid predation. Even in stabilising ponds, tilapia and murrels can be directly cultured because of their high tolerance to untreated sewage and less dissolved oxygen requirement. That too murrels, singhi, magur have accessory respiratory system, so that these could be very well cultured with greater return in oxidation pond.

3.9 Constraints in waste water fish culture

The main constrains in waste water fish culture is the dissolved oxygen level in ponds. Usually in a highly fertile pond, sufficient oxygen is produced during the day but there is drastic fall in oxygen level in night and become minimal prior to sunrise. When this minimum is below the critical level of oxygen required for the species, mortalities occur. Under minimal dissolved oxygen for the species cultured, the fish are also susceptible to stress from substances which they can otherwise tolerate under high oxygen levels. Therefore, aeration systems are now routinely used in commercial fish culture operations on a stand by basis to ameliorate fish stress or losses during expected period of low oxygen.

Secondly, the city sewage contain toxic substances that are deleterious to fish species. Thus there are many waste waters that can not be used directly for fish culture. Even waste water in rural and non-industrial cities and towns may cause problems to fish due to detergents that form biocide. Hence, waste water must be evaluated on its own merit for fish culture.

Thirdly, the aesthetic consideration may often limit the use of sewage cultured fish. Often fish species cultured in sewage posses off-tastes and odours arise mainly from industrial hydrocarbons, although some personal observations indicate that fish grown in well treated domestic wastes are equal or even superior in taste or odour to non-waste water cultivated fish. However, off-flavour can be overycome by keeping the harvested live fish to fresh water ponds for a couple of days when objectionable odours will be removed making the fish marketable.

The other constraints believed to be the diseases and parasites that occur in sewage fed fish culture. However, patho-morphological and pathoanatomical investigation on the skin, gut and muscle of fishes grown in sewage and fresh water showed that species of salmonella were not present in any of the fishes. Shigella was isolated from fishes reared in sewage. But neither salmonella not shigella was present in cooked fish as practiced in Indian style. In general, the literature on waste water cultured fish has not reported heavy losses from fish diseases. Some studies on waste water environment suggests that some inhibitory effect on the virulence of vibrio fish pathogen occurs in waste water ponds (Allen, Busch and Morton, 1979, see Advances in Aquaculture by TVR Pillay and Dill, Edt. 1976). Therefore, the therapeutic properties of sewage water need to be investigated. In addition, public health problems, pond effluent standard, public acceptance and fish pond management for better return need to be understood in sewage fed fish culture.

3.10 Fish production rate

Fish yield with three Indian major carp : Rohu, Catla and Mrigal species combination in the ratio of 1 : 2 : 1 resulted the production, to more than 2 tons/ha/yr. However, the production of fish through sewage fed fisheries has gone up to 9 tons/ha/yr. That too magur, fresh water prawn production in polycultured sewage system is encouraging and promising in the context of India's progress towards Industrialisation and commercialisation. The use of domestic sewage for nursery and rearing tanks for raising of fry and fingerlings in India is not uncommon, promising and economical. The growth in sewage was comparable to that achieved under fertilised field conditions.

3.11 Integrated waste water systems

Integration of aquaculture and agriculture at present is better understood and gaining momentum in India and many other parts of world. Although aquaculture, agriculture, public health and water pollution control are different disciplines of science, still interdisciplinary efforts are directed, as unitary concept of fish culture or agriculture is now changed to the concept of integrated culture system because of high economic return and reduced marginal cost of production. That too, the drained sewage could be used as fertiliser input for fodder crops, vegetable crops etc. The operational expenditure on sewage fisheries is less compared to that of fresh water fish culture. It is reported that a fish farmer can obtain an estimate of higher income if murrels are cultured in oxidised ponds and the excess sewage can be utilised for the cultivation of crops, hence, the income return could be further augmented.

The above account gives an idea about the sewage fed fish culture practices prevalent in the country and the problems associated with same. Still very little is known in many aspects, in particular, the biochemical interactions under going in nitrogen fixation and nitrogen budget in sewage fed waters affecting production of fish. However, suitable management measures and stock manipulation to optimise fish production in sewage fed waters appears to pay high dividents to fish farmers. Government has now laid priority to utilise the nutrient rich water discharged from sewage treatment plants of the country for integrated fish cum-crop culture, so as to contribute significantly the per hectare production of fish from such waters.

4. WATER LOGGED AND SWAMP AREAS FOR AIR BREATHING FISH CULTURE

Many low laying areas, where rainfall is fairly heavy and topography of the land is such that waterlogging occurs. In water logged areas the water table is very high and the soil pores in the root zone of the aquatic crops becomes saturated and water retains over the soil making it unsuitable for agriculture.

As per the report of National Commission of Agriculture (1976), the extent of water logged areas in the country is estimated as follows:

S.No.	State	Water logged area in lakh hacters	Remarks
1	Punjab	19.90	
2	Haryana	6.20	
3	Uttar Pradesh	8.10	
4	Rajasthan	3.48	
5	Madhya Pradesh	0.57	
6	Karnataka	0.07	
7	West Bengal	15.50	
8	Other States	----	Not reported
	Total	57.00	Lakh hectares

The extent of swamps in India is estimated at 6 lakh hectare which can well brought under Ar breathing fish culture.

4.1. Phenomenon of water logged area

It may be a marshy land or low land flooded with water, but are not Inland swamps. Swamps are marshy area infested with aquatic vegetation. The basic difference between a water logged area and the swamps may be as under-

	Water logged area	Swamps
1	In water logged area, it is the water which dominates the process of deterioration of land.	In swamps, it is the land which dominates the process of silting up of a water sheets.
2	Peat formation does not exists or its formation is only very rare.	The huge quantity of organic vegetation on decay forms thick deposit of organic matter which alongwith incoming silt from the catachment area forms a peaty soil for swamps.

At times, larger water areas like *Bheels* and *Jheels* become infested with weeds due to eutrophication and get converted into swamps. These water logged and swamps contain considerable quantities of fish food organisms including macrophytes, benthos and are utilised for culture of Air breathing fishes. These fishes show unique environmental adaptation for direct use of atmospheric oxygen and as such they are commonly known as air breathing fishes. The common air breathing fishes of commercial value are —

Family - Notopteridae	:	*Notopterus chitala* (Ham.)
		Notopterus notopterus (Pallas)
Family - Sacchobranchidae	:	*Heteropneutes fossilis* (Bloch)
Family - Channidae	:	*Channa marulius*
		C. striatus
		C. punctatus
		C. stewartie (Playfair)
		C. gachua (Ham.)
Family - Anabantidae	:	*Osphronemus gorami* (Lace pede)
		Anabas testudineus (Bloch)

Air breathing fishes form the bulk and main stay of tank fisheries in penninsular India. They are regarded as excellent table fish in Punjab, Uttar Pradesh, Madhya Pradesh, Andhra

Pradesh, Karnataka and Kerala. The All India Co-ordinated Research Project (AICRP) on Air breathing fishes under CIFRI, had taken up the culture practices of *Channa* species in swampy water by feeding with self-generating stock of minnows and minor barbels with success. The air breathing fishes such as magur, singhi, Koi and channa are known for their nutritive, invigorating and therapeutic qualities. These are recommended by physicians as diet during convalescence. According to marketing of fish in India, production of live fishes accounted more than 17% of country's total marketable surplus of inland fish in 1980's.

Besides, there are some other varieties of air breathing fishes also which, however, are not economically important. Generally these fisehs breed in ponds, swamps, paddy fields and low laying innundated areas under natural conditions. The availability of seeds in required quantities are now made possible under controlled condition and hypophysation technique. Even natural breeding under controlled conditions has been reported in these group of fishes and also in Wallago in CIFA (ICAR).

With the availability of fish seed production technology, pond culture management, monoculture and polyculture of Air breathing fishes are practiced in low laying areas including paddy fields. The air breathing fish culture and production under All India Co-ordinated Research Project has given encouraging results.

Various stocking densities even to the order of 50,000 fingerling/ha under monoculture of murrel had been tried without any supplementary feed and fetilizer, and the production as high as 2.6 tons/ha/yr had been achieved. The production in monoculture of *Channa marulius* has been touched above 3 tons/ha/8 months at 10,000 stocking density (see Sinha, 1985). Polyculture of Singhi, Koi and Magur gave a net production of 1200 kg/ha/7 months of rearing.

Clarias batrachus commonly known as magur inhabits fresh water rivers, swamps and ponds. It is a highly esteemed food fish and is cultured in India, Thailand and Cambodia. By virtue of their hardy nature and air breathing habits, *Clarias* and *Heteropneustes* are excellent materials not only for utilising swamps, shallow derelict waters in rural areas but also for intensive culture operation in urban areas. Their production potential by and large directly proportional to input and intensity of operational management. Clarias is cultured with high stocking density of 50,000 to 70,000 nos /ha. The fishes are fed at 3-5% of their body weight with pelleted feed in feeding basket placed in different places of the pond. In monoculture system of Clarias, the average production of 3-5 tones/ha/yr is also reported. Intensive culture of magur with waste replenishment has resulted a production equivalent to 7272 kg/ha in 6-1/2 months. The intensive culture of Singhi (*Heteropenutes fossilis*) has resulted an yield of 7200 kg from 0.2 ha area which is equivalent to 36000 kg/ha production. The culture operation of Singhi and Magur indicated a ratio of 240% of gross profit over operational cost (see Jhingran, 1985). It was experimentally proved that estimated production as high as 100 tons/ha per year can be obtained under intensive monoculture of the magur or in polyculture of magur with other air breathing fishes. Even in water logged areas, carp culture has been undertaken and a total fish production was 2900 kg/ha/yr was reported which is a reasonable production from a newly reclaimed water logged area for fish culture.

4.2. Development of Open Water Fisheries

India is gifted with vast areas of open water bodies in the form of reservoirs, lakes etc. The productivity of these open water bodies are poor. For sustainable production of aquatic organisms from these water bodies, regular stocking of large seeds of cultivable species are to be made. This phenomenon is called aqua ranching

4.2.1. Aquaranching

It is a process of harvesting rich crops of aquatic organisms *viz.* fish, shrimp, mussels etc from large open water bodies. In this process, aquatic germplasm resources could be enhanced by stocking seeds of desired species in open water, providing suitable artificial shelters to reach a size where predation and mortality could be reduced considerably.

Further the germplasm resources potential of many open water bodies are declining due to over exploitation, ecological topography, introduction of exotic carps, environmental degradation etc.,. Therefore, aquaranching practice involved both culture and capture techniques together for revival of resources through rehabilitatory measures. Further the development process involved in the technique of aquaranching includes:

1. release of desired aquatic organisms in open water
2. to maintain standard rearing practices of seed in enclosed tidal ponds, nets, canvas tanks near the open water bodies.
3. to provide artificial habitat (reef) to act hideouts for evading predatory attacks.

It is felt about the great prospects of aquaranching programme in India (Bandhyopadhyay and Thakur, 1999). The open water resources of the country such as reservoirs hold tremendous potential for improving production level through aquaranching. The hill streams and lakes could be improved for sport fisheries through this process. The sea water areas of the east and west coasts have touched maximum sustainable yield and there is hardly any scope to improve the production without introducing ranching measures. To achieve such improvement, large number of aquahatcheries for production of desired seeds for stocking and raising in open water bodies of either fresh, brackish or marine are needed to be set-up in the country.

5. FISH CULTURE IN CAGES AND PENS

Scientific cage culture of fish is a recent practice devloped in many parts of world although the culture of fish in cages and pens has been practiced since long in certain parts of Asia. Probably, this system of culture originated with a view of high retrieval of stocked fish, manufacture of synthetic materials for nets, use of plastics in aquaculture and pelleted feed preparation. The present scientific practice of cage culture originated from the temporary stocking of fish in certain containers. According to report, the cage culture was first started in Cambodia from where such practice transferred to Thailand, then to Indonesia. Further it was transferred to Vietnam, USSR and Japan. In 1964, cage culture was widely spread to many countries of world.

5.1 Significance of cage culture

The significance of cage culture and application have become more relevant for those countries where suitable water supply and land for fish culture are becoming less and less available. That too, development of open water fisheries, fish culture without feeding, polyculture in cages and intensification of culture system with forced aerator and filters to increase the carrying capacity of water signifies the importance of cage culture. The advantage and biological principles invoved in cage culture is for high stocking, high rate of survival and significant production potential. In India, since 1980, detailed studies are being made on the standardization of the techniques of cage culture of carps, murrels and tilapia from fry/fingerlings to marketable size. In Japan cage culture of marine and freshwater fishes have been undertaken. In U.S.A., cage culture of channel catfish is being carried out in commercial basis. In Spain, cage culture of bass, mullet and bream are undertaken in commercial way. It is also reported that culture of cuttle fish (Sepia) will be taken up in cages in Spain (Fish Farming International Vol. 18 No. 6, 1991). This gives an indication that the species traditionally cultured were mainly local species. However, the choice of the species cultured was mainly dependant on seed availability and market demand. Among Inland fish species, Salmon (*Oncorhynchus),* trout (*Salmo*), channel cat fish (*Ictalurus punctatus*), milk fish *(Chanos chanos*) and big head carp *(Aristichthys nobilis*) were the most common species and many more species such as murrel, tilapia, magur, singhi, common carp etc. are also now taken in cage culture. Among marine fish species, sea bass, sea bream, are the most common and other species like sepia will also be taken up in cage culture in Spain being reported.

5.2 Types of cages

The cages of various shapes such as square, rectangular, circular or boat shaped are used in many countries although use of square and rectangular sizes are more common. The smaller size of cage is 1 m^3 and even a small basket or shuttle cock shaped floating net cage (0.2 m^3) was found suitable for air breathing fish culture. The largest size of cages are about 10000m^3. But the practicable sizes of cages adopted are usually 20-60 m^3. Even 40 meter wide circumferance cage is used for culture of sea bream and sea bass in Spain.

5.3 Installation

Suitable site fulfilling the environmental and water quality requirements of cultured species to be taken under cage culture has to be selected. The methods adopted for installation of cages depends on its position in the system. If floating cages are proposed to be installed, then these have to be installed in lakes/reservoirs. If fixed cages, that are to be installed in rivers, streams or fast flowing waters. Similarly submerged cages and movable cages can be installed as practiced in Japan in offshore region of the sea.

5.4 Designing of cages

This include materials used for cage fabrication, types of cages and their installation procedure in the aquatic system. Material used for fabrication of cages varies in relation to the body of the cage, frame, floats, sinkers and attachment of feeding equipment in the cage. The body of the cage is made of nylon cloth (8 mesh/mm) Knotless or knotted webbing depending on the size of fish to be reared. Frame of cage may be collapsible, made of galvanised Iron, conduit Iron tube or High density polypropylene (HDPD) pipes are used. Other common plastic polymer products used for cage culture are mainly LDPE, HDPE, polyvinyl chloride (PVC), poly propylene (PP), Polystyrene, polymide, Polyethylene, Teraphthalate, Fiberglass reinforced plastic etc. In many cases bamboo or wooden frames are used for cages although the durability is less in comparison to that of frames fabricated out of plastic polymers or Irons. The materials for floats used in cages are either plastic, wooden or other materials available cheaply. Sinkers made of clay, concrete, bricks or stones are used in keeping the position of cage. Auto feeder or demanding feeders are installed. In Spain and Austria solar powered feeders are installed in cages.

5.5 Fish species

There are about 50 species of fresh water and marine water fish cultured in cages. However, there are well known 6 families of fresh water fish, such as cyprinidae, pangassidae, claridae, ictaluridae, channidae and cichlidae cultured in cages. Under brackish and marine fishes, the fish species belonging to salmonidae, carangidae and mugillidae are most common.

5.6 Stocking density

There is no standard stocking density of fish in cage culture. This is because of various species, cage size, mode of installation, water environment and management. However, the range of stocking densty varies from 75 to 300 individuals/m^3 of water in some Asiatic countries. However, in India, the stocking density of carp fry is about 210 individuals/m^2 and for table size fish rearing the density of carp fingerlings varies from 28-49/m^2. For tilapia cage culture, the stocking density ranges from 100 to 200 fingerlings/m^2. For murrel cage culture the stocking density adopted is 40 fingerlings/m^2. Variations in stocking density has been reported. However, Japanese believe that fish stocking at initial stage should occupy 45% of cage volume so that fish can have normal growth.

Pond cage farming of Asian sea bass (*Lates calcarifer*) was made by MPEDA in karaikal, Kerala, India during 2006-07. About 24000 fingerlings of 10g average weight were stocked in 15 cages (2 X 2 X 1.3 m size each). The biomass of each cage was maintained around 20 Kg/m^3. Accordingly during culture operation, additional cages were installed which goes up to 107 cages. Sea bass attained an average size of 600g each with 88% survival. Feeding given to cage culture showed 1 : 1.25 FCR. Sea bass seeds are produced at Thoduval village, Nagapattinam District, Tamil Nadu. Harvest of cage cultured sea bass was made on 27th September, 2007 and the harvest of 12 tons/ha of sea bass has been achieved from the brackish water earthen pond through cage culture (Source: Fishing chimes (2007)- 27 (8): 58-59).

5.7 Stocking proportions

As adopted in China, the food eating fishes occupies 85%-90% of total stocking density and natural food eaters occupy 10 to 15% in cage culture.

5.8 Fish production

By practice, Chinese have proved that the yield over 75 kg/m^2/yr is produced in reservoir with 3-20 m water depth and over 5 mg dissolved oxygen in cage sites, while in shallow lakes and channels, the production is much lower (Lisifa, 1990). In India, monoculture of carp, *Catla catla* have resulted 21 kg/m^2 of cage area in 243 days of rearing (Tripathi, 1990). It is revealed that, each 12.5 meter diameter polarcirket cage with 5 m deep net has an out put of ten tons of bass and bream every two years in Spain. In Spain 50 to 60 tons of sea bream in 4 cages of 1600 m^3

has been planned to be produced in 1991-92 (Fish Farming International 1991, Vol. 18, No. 6).

5.9 Constraints

(1) Cage culture needs frequent examination of water quality and removal of dirts attached to the net.

(2) Removal of fouling agents like snails etc. from the bottom of cage net.

5.10 Penculture

The penculture has become popular in some of the Asiatic countries. The traditional polyculture concept is practiced in penculture system. High production in penculture is related to dissolved oxygen and exchange of water. As in open water oxygen in not a problem, the production can be increased in pen culture. The materials used in pens are bamboo poles fixed in circle or rectangularly, covered with nets. The sinkers are to be put into 20 cm depth of silt and sometimes anchors are placed with a distance of 0.5 m for reinforcng the pen. The bamboo poles should be about 0.6 m below the silt soil bottom. In order to prevent heavy waves in openwater pens, some aquatic plants (safely kept inside the net making weed rope like structure) is placed atleast 10 m away from the pen. In common practice, Chinese use two layers of net enclosures of same mesh size. The inner one is used for culture of fish and the outer enclosure is used as to prevent the escape of fish. In some cases, if the pen is constructed near the transporation channel, the 3rd enclosure is constructed. Chinese has estimated that the fingerling production from 180m^2 cage area will be sufficient to rear 667 m^2 pen. It means the ratio between cage and pen is 3 : 11. The average net pen yield is 4170 kg/ha/yr (Lisifa, 1990). The production can be further increased through flow rate of water in pen and management.

6. RUNNING WATER FISH CULTURE

Running water, race ways or closed circulated system are highly productive being practiced in many parts of Asia. Three different types of running water fish culture systems such as open type, semiclosed and closed circulated system are prevalent in the Asiatic region.

6.1 Structural components in running water system

The main structural components of closed running water system consist of -

(1) running water cemented rectangular tanks

(2) earthen pond area

(3) aquatic weed cultivated water tank

(4) filtered storage water tanks

(5) pump house for pumping water. Open type of running water systems are usually constructed near the river site.

6.2 Sizes

There is no specific sizes of the running water tanks. The size of each cemented running water tank is 30 sqm with 1.5-2 m in depth. It is located at the higher elevation, so that water can flow to subsequent earthen tank by gravitational flow. A series of running water tanks, each of 30 sqm sizes are fed with a water feeding channel. The inlets and outlets of running water tanks are diagonally placed for better replenishment of water. At the bottom of the running water tank, there is an outlet for removal of uneaten food and faecal matters to the subsequent earthen pond placed adjacent to these running water tanks. The floor of the running water tank is slopy towards bottom outlet for easy drainage of fish metabolites and uneaten food materials. A schematic structural placement of different ponds are given in closed circulated system as being practiced in many parts of Asia. The ratio of running water tank, earthen pond and Aquatic weed cultivated tank for water purification is in the ratio of 1 : 10 : 3 respectively.

6.3 Species of fishes

Various species of fishes such as carps, cat fishes, tilapia, eels, sea bass, and sea bream are taken. In China, the main species taken in running water system are grass carp and wuchang fish. Polyculture, along with fresh water mollusca culture were taken in earthen ponds placed adjacent to the running water tanks. The stocking density is usually 5-10 kg/m^2 of area and the stocking size of fish is 50 gms onwards. These tanks are also used for fingerling rearing in some parts of Asiatic countries.

6.4 Fish production

In closed running water system with grass carp and wuchang fish, the maximum production so far obtained in China in 45 kg/m^2. However, with *Tilapia mossambica* the yield is 103 kg/sqm per year in Puyang city in China's He Nan province (Fish Farming International 1991, 18 (6) : 5). With such encouraging results, a new pilot project of indoor fish farm in China was erected in June 1990 and has a capacity to produce 600 tons of fish per year. In running water system, if tidal water or flooded water comes to pond in a controlled rate of 2000 cubic meter of water per day, the production can be raised more than 49 tons/ha/yr and if tidal water flow into pond is reduced to 200 m^3, the production is reduced to 26 tons/ha/yr being suggested by Pearl River Research Institute, China.

<table>
<tr><td colspan="6"></td><td rowspan="2">5</td><td rowspan="2">4</td></tr>
<tr><td></td><td></td><td>1</td><td></td><td></td><td></td></tr>
<tr><td></td><td></td><td>2</td><td></td><td></td><td></td><td colspan="2">3</td></tr>
</table>

In race ways, measuring 14.5 m long X 8 m breadth X 1.5 m deep with biofilter arrangement aeration and adjustment of temperature at 25°C, the yield of Tilapia per year is expected to be around 100 kg/m^3 (Fish Farming International 1991, Vol. 18(6): 5).

6.5 Limitation

Such closed system has certain limitations because of :

(1) More demand on energy and power for water flow.

(2) The production is dependant with the water flow.

7. FISH CULTURE IN RICE FIELDS

Rice and fish culture go together as food particularly in South East Asian countries and there is also age old systems prevalent for their combined cultivation in India. The culture system in paddy field is extensive and prevalent in parts of Africa, U.S.A. certain countries in Europe, Latin America, Madagascar, Italy, Japan, Taiwan, Malaysia, and well developed in Indonesia. However, with the advent of plant genetics, high yielding varieties of paddy and the indiscriminate use of pesticides have substantially curtailed fish culture in rice fields particularly Japan, Italy, Malaysia and to some extent in India. These reasons have declined the total area of paddy-cum fish culture to negligible area for utilisation.

7.1 Essential features

A paddy field suitable for fish-cum paddy culture should have strong dykes in order to prevent the leakage of water and retain water upto desired depth. The second essential feature is the system of channels, the location and size. The third requirement is the presence of small ditches or ponds near the outlet, which can offer shelter to the fish against heat and predators. Inlets and outlets are provided with screens.

7.2 Fish species and production

Species of fish suited for fish culture in paddy fields are those that can (i) thrive well in very shallow waters, (ii) withstand fairly high turbidity, (iii) tolerate relatively high temperature and (iv) grow to marketable size in short period. The species of fish belonging to family cyprinidae, channidae, claridae, heteropneustidae, and cichlidae are suitable for culture in paddy fields.

The traditional pokkhali fields of Kerala, Bokkhali fields of West Bengal and Khazana fields of Maharashtra are the examples of paddy cum fish culture in India. However, depending on the nature and level of operations, fish production in paddy cum fish culture ranges between 100 kg to 2250 kg/ha (see Sinha, 1985).

The constraints in paddy-cum fish culture is that, it comes in direct conflict with the use of land resources. And the prospects of paddy-cum fish culture will be only feasible when agriculturists make terms with pisciculturists to use only those pesticides which while effectively combating infestations are also tolerant to fish.

8. FRESHWATER PRAWN CULTURE

India has great remakrable advantage of having some of the largest growing species of *Macrobrachium* group such as *Macrobrachium malcomsonii* and *M. rosenbargii*. With the initial success in rearing of the giant fresh water prawn, *Macrobrachium rosenbergii* by Dr. S.W. Ling of Malaysia in the 1960's followed by workers from other countries, macrobrachium farming has assumed considerable importance in Aquaculture in last two decades. In India, the giant freshwater prawn. *M. rosenbergii* inhabits most of the tidal rivers growing up to 31 cm in size are reported. Its distribution is limited to the estuarine and fresh water zones of river mouths and back waters where the salinity ranges from 0 to 20% salinity. Because of its fast growth, attractive size and better meat quality, these two species of macrobrachium are quite suited for monoculture or polyculture with fish in fresh water pond systems. The third large size species of fresh water prawn is the Ganga river prawn, Macrobrachium gangeticum (syn.M.birmanicum choprai) collected from the Ganga river, Hooghly, Padma and Brahmaputra river. Another species of large sized fresh water prawn is the M. josephi reported from the Veli Lake and Kulathoor rivulets in Trivandrum, Kerala. This species is found to reach a maximum size of 185 mm total length.. As all the species adopt well to confined pond conditions in fresh water, the modern technologies of culture can be very well adopted to these species too. Experimental studies in India started with the use of natural collection of seeds from rivers and stocking them in ponds. There also exist a traditional practice of collecting juveniles from flooded fields adjacent to the rivers in West Bengal, Orissa, Andhra Pradesh and Tamil Nadu. However, the seed production technology of commercially important fresh water prawns are developed by the CIFRI and state governments have taken up initiation to fabricate low cost fresh water prawn hatchery. The hatchery technology in the country for seed production of *M. malcolmsonii* was achieved in 1991 and for M. gangeticum during 2000. However, raising of larvae from egg to juveniles are fairly well developed for *M. rosenbergii* and Several private prawn hatcheries are already in active operaion in Thailand, Taiwan and Malaysia. Aerated and sand filtered synthetic seawater (20 PPT) is also used for larval rearing of freshwater prawns. The composition of synthetic seawater is as follows:

1. Sodium chloride	18.3 Kg
2. Sodium sulphate	3.12 Kg
3. Magnesium chloride	3.00 Kg
4. Calcium chloride	0.88 Kg
5. Potassium bromide	0.08 Kg
6. Boric acid	0.02 Kg
7. Strontium chloride	0.01 Kg
8. clean freshwater	1000 literes

The berried freshwater prawns be maintained in 5 PPT saline water for hatching of eggs (Zoea I). The system of rearing operation involves, facilities, water supply, breeding stock and berried females, incubation, hatching of eggs, rearing of larvae, density maintenance, right foods matching with stages of development, and mortality control through management till the growout marketable product.

8.1 Production

Cultured experiments indicated the specific size stocking density relationship, the size of the prawn decreased with increased stocking density. Production is proportional to the management input. In monoculture of freshwater prawn, the average production is round 1-2 tones/ha/yr being reported. In polycultured ponds, the production of *M. rosenbergii* is 709 kg/ha (see Sourvenir 1989, CIFA K. Gang). Natarajan *et al.,* 1989 indicated that under prawn fish polyculture the growth of *M. malcolmsonii*, in 6-8 months of rearing has attained from 25 gm to 180 gm. The production of prawns under polyculture system accounted to 100 to 223 kg/ha with mean weight from 45.3 to 78 gm and survivality ranged from 38 to 70%. The per hectare fish production varied from 2000-3300 kg.

Hence, the polyculture of prawn with fish seemed to be a better method of aquaculture enterprise. Prawn farming is gaining momentum among the farmers of Orissa, West Bengal, Andhra Pradesh and Tamil Nadu which is reflected by the rapidly increasing demand for prawn seeds.

9. FRESHWATER PEARL CULTURE

Like that of the marine pearl production from pearl oyster *Pinctada fucata,* similarly the freshwater pearls are produced from freshwater mussels. Japanese and Chinese were the leading to produce fresh water pearls in the world. However, Japanese are moving a head in developing modern fresh water pearl culture which plays an important role in the fisheries economy of their country.

Pearl is a biological secretion produced due to entry of any foreign material such as sand bid, glass bid artificially or accidentally into mantle cavity between the shell and mantle covering, shining nacre is secreted resulting in a natural pearl.

The fresh water pearl are small, irregular in shape compared to marine pearl. But the Jewel qualities between fresh water and marine pearl are identical. The color of the pearl varied from silvery white to pink and even pale yellowish.

In China, two species of fresh water clams such as *Hypriopsis cumingii* and *Cristaria plicata* are used for fresh water pearl production. In India fresh water pearls are first produced from fresh water mussels such as (1) *Lamellidens marginalis* and (2) *Lamellidens çorrianus* in 1987 from CIFA, Orissa (Souvenir CIFA, 1989). Other fresh water mussel species are under investigation at CIFA, Bhubaneswar. It is further reported that, *Pareysia corrugata* can be employed for developing the freshwater pearl culture technology (Janakiram, 1995). Efforts are continued to produce multinumber of pearls from a single fresh water mussels as reported elsewhere.

Fresh water mussel belonging to Unionacea contribute the major protion of freshwater pearls. In Japan, one species of mussel (*Hyriopsis schlegeli*) and in Thailand the mussel (*Hyriopsis cumingi*) contribute to freshwater pearl production. A large number of species belongs to Quadrula, Pleuroblema, Tritonia and Megalonia of USA and species of Union of Europe are more prominent species for pearl production. The world's total production of pearl is increasing year after year. Japan ranks first in the marine pearl production and other leading marine pearl producing countries are: 1. French Polynesia 2. Indonesia 3. Australia 4. Philippines 5. Cook Island 6. Myammar 7. Thailand and 8. Malaysia. However, China produces bulk of the freshwater pearl.

These potentialities in fresh water pearl production is in process to be transferred to rural women for employment opportunity and uplifting their socio-economic conditions.

10. INTEGRATED FISH CULTURE

There are several excellent review articles in integrated farming system in India (Jhingram and Sharma, 1980; Natarajan and Sharma, 1980; Sharma and Olah, 1986; Tripathi and Mishra, 1986; Sharma, 1990). Many authors have suggested that the concept of unitary culture of either fish, crop or animal husbandry has gradually been changed to the integrated culture system with the view of producing fish, meat, egg, milk, vegetables and other allied products within a farm itself on an economic scale. The basic necessity of such integration is not only to make the farm an independent unit but also to fulfil the demands as input to other structural units (Rath, 1989a, 1989b; Sharma, Das and Das, 1988). Besides this fact, the accumulation of wastes/byproducts can generate pollution to the environment. If such great amount of excreta is not disposed of, it must pollute the environment. These great amount of wastes can be recycled for food production through integrated system of fish culture (Sharma and Das, 1988). Therefore, integrated fish farming (Figs. 79 and 80) can be called s model of recycling wastes, comprehensive utilisation of various farm products, saving energy, fully utilises the natural resources and finally maintain the ecological balance (Rath, 1990).

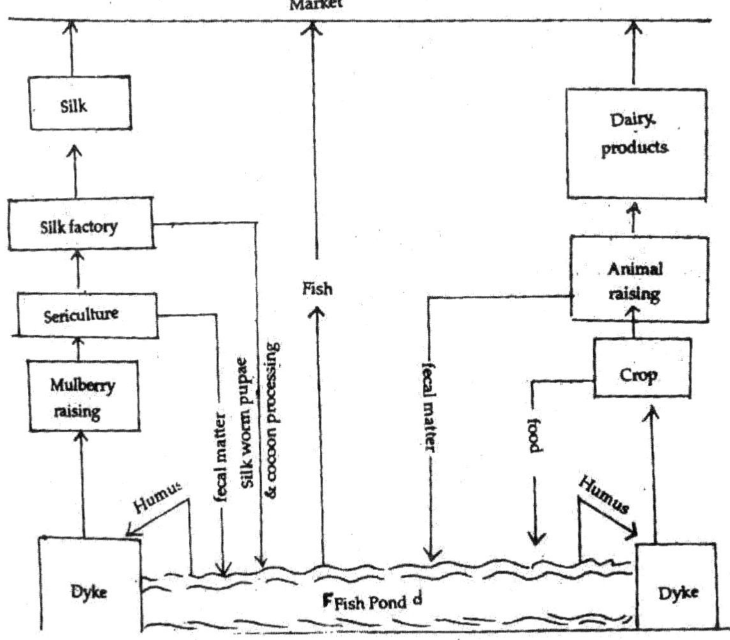

Fig. 79. Material cycling in Integrated fish farming system.

10.1 Characteristics

The characteristics of integration is the development of structural network in line with local conditions and utilisation of wastes by various ways of recycling as manure, biogas, crop production, feed etc.

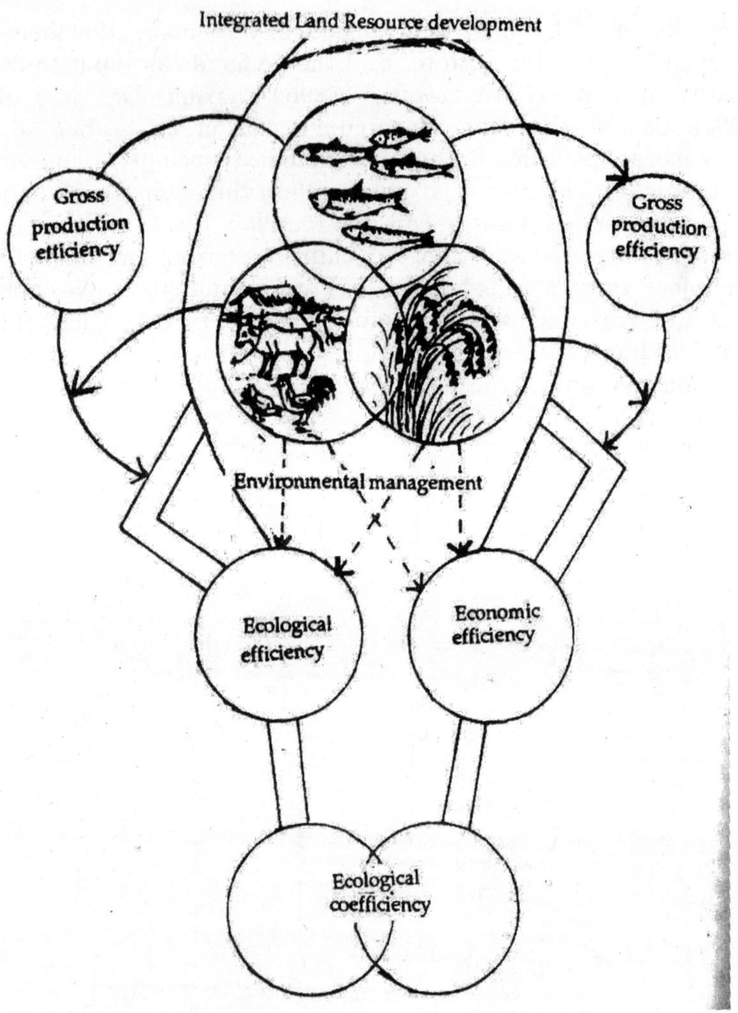

Fig. 80 Integrated farming system and efficiencies.

10.2 Type

Broadly, there are three types of integrated fish farming systems widely accepted in many countries. These are as follows :

(i) Fish-cum-crop integration

(ii) Fish-cum-livestock integration

(iii) Fish, crop and livestock integration.

10.3 Agriculture components

The agriculture components usually incorporated in an integrated system are palatable, nutritious, resistant to diseases, easy to manage, strong adoptability and well developed roots. The terrestrial agriculture components integrated are usually Alfa alfa, Rye grass, sudan grass, *Romain lettuce,* Bunch grass, Hybrid grass, Elephant grass, *Lactuca tentaculata,* Barnyard grass, gunny grass, wild grass, Soyabean, Barley, *Zea mays,* Sweet potato, Cabbage, Banana, Popaya, Cucurbita, water melon, sugar cane, Bamboo shoots, Mulberry, wheat, rice, Jute, Pineapple, coconut, orange, poi, Ipomea, Bringal, cowpea, pomplin, tomato, bean and fruit plants etc. The aquatic plant components are usually water hyacinth, water lettuce, water peanut, wolffia, lemna in China, Trapa, and Makhana in Bihar, India. Ahmad and Singh (1997, 1999) described about the cultivation of makhana (*Euryale ferox*) and aquachestnut (*Trapa bisipinasa*) with fish culture by the majority of the farmers of Bihar which gives them diversified harvest and additional income. They have described the details of fish species and models of rotation and concurrent culture with Aquachestnut. A combination of two Indian major carps such as Rohu and Mrigala and exotic carp, common carp in the ratio of 20-30, 20-25 and 40-50 is recommended for stocking at the densities of 1200 fingerligs per hectare for a production level of 1000 kg/ha. Among Indian major carps, the columnar periphyton feeder, Rohu, Omnivorous bottom feeder, mrigala and the exotic carp, common carp which is an omnivors and detritovors bottom feeder are suitable than other. The surface zooplankton feeder (Catla), surface phytoplankton feeder (Silver carp) and the macro-vegetation feeder (Grass carp) are not suitable to thrive well in the ecological conditions. In mixed culture of magur, Rata and Koi at the stocking density of 70,000 nos/ha along with Makhana yielded a production equivalent to 1200 kg/ha in 8 months in addition to yield of 320 kg of Makhana seeds (Jhingran, 1991).

The cultivation of Aquachestnut (*Trapa bispinosa*) in Bihar is an age old practice. Large scale commercial cultivation of this is done in the districts of Aurangabad, Gaya, Darbhanga and Samastipur in ponds, tanks, beels and water ways. It is cited that, 30 days after fruit setting, harvesting is done manually. Previously farmers harvest nearly 15-22 quintals/ ha but now it has gone up to 100 quintals/ha with adequate NPK fertilization and intercropping practice. The nutritive value of Trapa is as follows:

Content	Fresh	Dry
Moisture (g)	70	13.8
Protein (g)	4.7	13.4
Fat (g)	0.3	0.8
Carbohydrate (g)	23.3	68.9
Calcium (mg)	20	70
Phosphorus (mg)	150	440
Iron (mg)	0.8	2.4
Carotene (mg)	12	-
Thiamine(mg)	0.05	-
Riboflavin(mg)	0.07	-
Niacine(mg)	0.06	-
Vitamin C(mg)	9	-
Energy (Kcal)	115	336

Source: Jha.1999 (See Pradeep *et al.,* 2008).

Rohu, Mrigala, Common carp in the ratio of 20 : 25 : 40 is recommended for stocking in Fish-Aquachestnut Integration. The stocking density ranged from 3000–5000 fingerlings/ ha. Feeding comprising of Rice bran, GNOC, Soyabean meal, fish meal and mineral mixture in the ratio of 140:35:15:5:0.4 with crude protein level of 30-35% should be provided at the rate of 2-3% of fish biomass per day. The feeds should be provided in split doses 2-3 times a day. The total fish yield will be around 0.75 tons to 1 ton/ ha. The stocking of Magur and Koi in 1:1 ratio was found to give a production 1000 Kg/ha in 7 months without fertilizer and supplementary feeding (Pradeep *et al.,* 2008).

10.3.1. Fish-Mushroom

Mushrooms are fleshy fungi and its cultivation was first recorded during 1638-1715. Extensive mushroom cultivation were made particularly in different parts of Europe. Cultivation of edible mushroom in India was initiated quite recently, although the method of cultivation for some were known since long. A well organized attempt was made at Himachal Pradesh in collaboration with ICAR during 1961 and a project on development of mushroom cultivation was started at Himachal Pradesh, prior to which the entire mushroom production was from natural sources. Even today, the morel (Gucchhi), a prized mushroom is not cultivated but routinely collected from natural sources. In every country mushroom grow wild from snowy mountains to sandy deserts on all types of soils, pastures, forest land or litter, cropped or fallow land. Edible mushroom can grow in all seasons, particularly during rainy season where ever organic matter or its decomposed products are available. In literature more than 1000 species of edible mushrooms are reported and in India about 200 species are on record. In most parts of India, 3 types of Mushroom are cultivated, these are,

(1) European button (*Agaricus bisporus*)

(2) Paddy-Straw (*Volvoriella* species)

(3) Oyster Mushroom (*Pleurotus* species)

Button mushroom was introduced in mid 50's in Taiwan as a cottage industry and it expanded very quickly. Realising its importance ICAR and Government of Himachal Pradesh initiated a project in mushroom culture at Solan in 1961 and it was further strengthened by the support of FAO. Later it spread to Jammu and Kashmir, etc

Paddy straw cultivation was reported in 1943 in various parts of India. Paddy-straw mushroom is called Chinese mushroom and it has 3 edible species. These are (1) *V. volvacea*, (2) *V. diplasia* and (3) *V. esculentsa*. An average yield of 3 Kg per bed (made from 32 Kg Paddy straw with 200 g of gram powder) is normally obtained with in 45 days.

Oyster mushroom is commonly known as Dhingiri. The cultivated species are (1) *P. ostreatus*, (2) *P. flabellatus* (3) *P. sajorcaju* (4) *P. sapiduc* and (5) *P. cornucopiae*.

Being a source of most nutritious food, its alternative use in pickle preparation, medicine etc also provide good scope for job opportunities to unemployed persons in the country.

10.4 Livestock/Bird components

The livestock/Bird components usually encorporated in an Integrated system are Cattle, Goat, Sheep, Pig, Duck, Chick, Goose and Koel.

10.5. Necessity of fish-cum-crop integration

(i) It is due to the demand of fish feeds and utilisation of excess pond silt. On one hand abundant silt deteriorates the pond water and on the other hand, it is one of the high quality manure for agriculture and crop plantation. These crops inturn can be used as feeds for fish. Therefore, pond silt is a link betwen fish and crop in this integration.

(ii) Aquaculture can provide large amount of silt and fertilised water for agriculture. There still exists a potentiality of land area on fish farm. The arable area of pond dyke and slope can be used for crop cultivation. Therefore, it is necessary and feasible to integrate fish farming with crop production so as to fully utilise pond silt, arable land and water surface. As a result, of this, the need for fish feeds could be managed wholly or partially.

(iii) Aquatic plants not only serve as feed, feed ingradients, water purifier, wave checker but also used as manure in fish pond.

10.6 Necessity of fish-cum livestock/bird integration

One of the outputs in Fish cum Livestock integration is the animal dung. Animal dung is also a major component for organic farming. Hence its availability is important for farm operation. The livestock population of India, cited by De (2008) is as follows: Cows (180 million), Buffaloes (7.7 million), Goats (45 million), Pigs (11 million) and Poultry (258 million). The quantities of discharge of dung from above sources would be Cattle waste (1661 mmt), Sheep and Goat waste (20.2 mmt), Pig waste (5-6 mmt), Poultry waste (3.4 mmt) and other livestock waste (10.1 mmt). Further, agricultural residues in the form of paddy husks would be around 321 mmt. Besides these wastes, enormous amount of lignocellulose waste, sewage, sludge, slaughter refuges would add to the totality of organic wastes in the country. These wastes need to be recycled for generating wealth through production process in different productive systems.

Indian Cattle varieties are: (1) Guernsey, (2) Jersey (3) Holstein (4) Short horn (5) Friesian (6) Brown Swiss (7) Indian Zebu. Cattle breeds raised in India are (1) Gir (2) Shaiwal (3) Red Sindhi (4) Deoni (5) Tharparkar (6) Kankrej

The varieties of Goat raised in India is Surti. Similarly the Sheep variety raised in India is Kathiawari.

10.6.1 Fish cum ducks

(i) Water surface can be put into full utilisation by duck raising as ducks use both land and water as habitat.

(ii) Fish pond provides an excellent environment to ducks which prevent them from infection of any parasitic and other diseases.

(iii) Ducks feed on the predators like tadpoles, dragonfly, weed fishes thus helps the fingerlings to grow.

(iv) Duck raising in fish pond reduce the demand for protein on duck feed.

(v) Duck dropping go directly in to water bodies providing C, N, and P elements frequently, thereby increasing the biomass of natural food organisms in fish pond.

(vi) The daily waste of duck feed can be utilised as fish feed in pond or as manure resulting in relevant fish yield as well as reduces the conversion ratio.

(vii) Manuring conducted by ducks are homogenously distributed without any heaping of duck droppings.

(viii) By virtue of digging action of ducks for searching benthos, the nutritional elements of soil get diffused in water column and made available for plankton production.

(ix) Ducks are bioaerator as they swim, play and chase in pond so as to diffuse atmospheric oxygen in surface water air interface.

(x) Survivality of ducks increased when raised in fish pond than dry pen only.

Improved breeds of duck such as Indian Runners, Sylhet, Khakhi Cambell are kept in fish ponds. Experiments conducted in India has shown about 4325 Kg fish/ ha/year can be obtained with 100–150 ducks per hectare water area. Ducks are marketed when attains 2 Kg body weight. Layer ducks are kept with paddy fields as they feed on insects, weed fishes, snails etc., there by reduces feed and pesticide costs to a large extent. It is cited that an average size duck excretes about 150 g droppings per day and the droppings of 250 birds can yield 100 Kg of fish per hectare of water area.

Similarly poultry droppings and poultry litter is recycled in to fish pond with fish production of 4500 – 5000 Kg/ ha/year. Egg laying varieties of birds such as Rhode Island Red, Leghorn, Star cross shave, Kalinga Brown, BV-300 etc., are reared. However Broiler birds such as Shaver Stabro, Hybro Hebard, Vencobb etc., are much more economical. Desi breeds or local chicks in India are

Aseel, Chittagong, Ghagus. Exotic breeds of Chicks in India are White Leg horn, Red and Black Minorca, Rhode Island, Plymouth Rock, Australop, Light Sussex, New Hampshire, White Rock and White Cornis. The hybrid project for production of New Hampshire was carried out at the Central poultry breeding farms at Mumbai, Bhubaneswar, Hessarghatts and Chandigarh. It is cited that one adult bird produces about 25 Kg of poultry manure in one year and 500-600 birds are required to fertilize one hectare of water area.

10.6.2 Fish cum cow integration

(i) Cows can provide dung to be used as manure and the left over matted grass of cow shed etc. can be used for mushroom and earthworm culture.

(ii) Cow manure is nutritively rich and the levels of N and P are congenial for plankton multiplication.

(iii) The out put of cultivated carps in cow manured pond is 2-4 times more in relation to different species composition than the out put in unmanured pond.

(iv) Cow manure is very fine due to repeated digestions in stomach of cow. Therefore, it can suspend longer in the water. Such suspensibility of dung not only enables fish to get more feeds but also reduces oxygen consumption caused by manures and also avoids the formation of harmful gases.

(v) The B.O.D. of cow manure is relatively lower than other livestock manures because the cow forage has already been decomposed by micro-organisms in cow's body.

A cow weighing about 450 Kg liberates 12 tons of dung and 1095 litres of urine annually (Cited Sarkar,2002). An adult cow consumes about 10 tons of grass annually besides on other food items. About 4 numbers of cow can provide adequate amount of manure to fertilize one hectare water area . Further the manure generated from one cow can sustain 250 Kg of filter feeders.

(I) Common diseases in domestic worm blooded animals including Cattle

(A) Bacterial: Anthrax, Tuberculosis, Mastitis (Streptococcal/ Staphylococcal), Brucellosis, Salmonelesis

(B) Viral: Rhinder Pest or Cattle plague, Foot and Mouth disease, Blue tongue

(C) Fungal: Ring worm,

(D) Protozoan: Trypanosomiasis

(II) Common Poultry Diseases:

Ranikhet, Fowl pox, Fowl typhoid, Paratyphoid, Tick fever, Coccidiosis, Pullorum disease, Bronchitis, Variola mavium, New castle, Cholera, Marke's diseases, Ascariasis and deficiency diseases.

10.6.3 Fish cum pig integration

(i) The left over and residues of kitchen, aquatic plants, agriculture products and wastes are used as feed for pigs and pig excreta are inturn used as organic manure in fish pond.

A number of exotic breeds of pigs such asd White york Shine, Berkshine and Others have been introduced in India to cultivate them in close location with fish pond. Local or hybrid varieties of pigs are cultured in captivity or they are allowed to graze in open area. A run for the pigs adjacent to the pig house is essential. The washings of the run containing dung and urine are directly drained in to the pond or composted before they are used. Generally an adult pig liberates about 2 Kg of faeces per day. It is cited that about 50 Kg of pig manure results in the production of 2.5 Kg to 3 Kg of fish. The amount of dung produced by 40 pigs was found to be adequate to fertilize one hectare of water area. Fish production ranges from 6000–7000 kg/ha/year with 4200–4500 Kg of pig meat.

10.7. Fish, Livestock-cum-crop integration

(i) Utilise animal manure and pond silt for crop cultivation and utilise the crop products as feed stuff of fish.

(ii) Terrestrial and aquatic primary productivity is fully utilised.

(iii) To raise the utilisation efficiency of light energy as the photoenergy measured in kilolux is more in grass land than the photoenergy in pond depth.

11. ECONOMIC EFFICIENCY

(i) Increases the source of feed and fertiliser. Protein income is much more than unitary culture system.

(ii) Generates more employment and save energy.

(iii) Rationally utilise the labour.

(iv) Reduce the cost of expenditure.

(v) Set up reasonable ecological system.

12. MANAGEMENT

In integrated system, there is a need of integrated management. Some of the management considerations adopted in China are given below.

12.1 Allocation of crop land

The area allocation of crop field is dependant with cultivation pattern of crop and also target production of fish. It is derived by following formula.

$$S = \frac{\Sigma\ Y\ F}{P\ N}$$

S = Area of crop field

Y = Target net yield of herbivorous fish

F = Food conversion rate

P = Production of crop

N = Times of consecutive crop.

12.2 Feed demand

It can be calculated in the following manner -

M = YF

M = Feed demand

Y = Target yield of fish

F = Food conversion rate.

But in rotational plantation system, the formula can be used as follows :

$$M = Y\ Fr$$

r = Proportion between the feed intake of fish in certain period of cultivation.

12.3 Adjustment of feed demand

It is necessary, when the standard feed amount for fish is not produced within the farm itself or the feed amount is insufficient. So the adjustment of feed demand can be find out by following equation -

$$M = M' \times \frac{f}{F}$$

M = The needed feed

M' = Deficient part of the standard feed

f = The food conversion ratio of needed feed

F = The food conversion ratio of standard feed

12.4 Number of individuals/water areas -

It can be find out by following equation -

$$N = \frac{(Y_1 - nY_2)\ C}{m}$$

m = Amount of excreta per individual

C = Conversion factor of manure

Y_1 = Net yield of filter feeder and omnivorous

Y_2 = Gross yield of fish

n = Ratio of herbivorous fish fecal to filter feeder and omnivorous fish production (0.2-0.6).

12.5 Manure requirement

It can be find out by following equation -

$$M = (Y_1 - nY_2)\ C$$

12.6 Construction area of animal house

It can be find out by following equation :

$$S = \frac{N\ s}{C}$$

S = Standard area

N = Number of animals raised in whole year

s = Average construction area per animal

Cow - 7 m^2, Grazing area 15-20 m^2/one

Duck - Dry run 4-5 individuals/m^2, wet run 3-4 ind/m^2

Pig - 2 m^2/individual

Cow and Chick one cycle

Pig - Two cycles/year

Ducks-Four cycles/year.

The agriculture and animal components in an integrated fish farming system generates not only feed for fish culture but also generates high potentiality of organic fertiliser for supporting agriculture, energy and fish production. Organic manure in particular play their role in increasing food production through supply of nutrients and improvement of physical, chemical and biological properties of soil. The saving of about 29% of dung that is presently burnt as fuel in villages and its use as manure is a consideration of great practical importance in the context of the new strategy for increasing food production. As burning of cattle dung as fuel is inescapable in rural area, the alternative solution is to set up biogas plant for energy and biogas slurry for fertilisation can be practised. The policy of the farmer is to take every possible step to ensure return of all by-products and wastes to the productive system for increasing protein food production.

Most village farmers have few numbers of cattle, goats, sheep, birds, land and water areas, which can be exploited for food production. These components available with every farmer offers great scope for uplifting their socio-economic conditions. Such low input technology appears to pay high devidends for the Indian farmers by diversified activities. It is worth to mention that the pond bed provides enough humus for increasing soil productivity, crop production, consequently, helps in fish culture, animal husbandry thereby acting as an integrated web cycle in the farm making it an independent multistructural unit. Therefore, it is worth to mention that, the Integrated Fish farming a sound basis for the development of integrated rural technology which will enable to fulfil the growing need of the rural peopel.

13 ORGANIC AQUACULTURE

The concept of Organic aquaculture is being developed for 1. Maintenance of sustainable production in aquaculture system by restricting the use of harmful substances like chemicals, drugs and antibiotics which can adversely alter the ecosystem 2.

Appropriate living condition with adequate space for free movement and minimize stress by optimum stocking density as per carrying capacity of the system must be in the practices of farming. 3. The input materials such as seed, feed and fertilizers must be raised under organic management. 4. Cultivable species must be fed on organic feed to optimize good health condition. Incase, fish meal is taken as a component in organic feed; the fish meal must be from certified sustainable source. There by harvesting of products would be pollution free and also sustainable. National Organic Standard Board (NOSB, 1995), USA has declared that organic farming is an ecological production management system. It promotes and enhance biological cycles, soil microbial activity and also biodiversity. The principal guidelines of organic farming is to enhance ecological balance of natural system with the goal of optimizing health and productivity of interdependent communities of soil microbes, plants, animal and people. According to survey, IFOAM stated on 14th February 2006 that, currently more than 31 million hectares of farm land are under organic management world wide. Ausralia leads in organic farming with 12.1 million hectares followed by China (3.5 million ha) and Argentina (2.8 million ha) Most of the world's organic land is in Australia/Oceania (39%), followed by Europe (21%), Latin America (20%), Asia (13%), North America (4%) and Africa (3%). The countries with largest area under organic management are

1. Australia
2. China
3. Argentina
4. Italy
5. USA
6. Brazil
7. Germany
8. Uruguay
9. Spain
10. UK
11. Chile

The market value of organic products world wide reached 27.8 billion US Dollar, the largest share of organic products are being marketed in Europe and North America.

Scanty literatures are available on Organic fish farming some of which are Diwan and Ayyappan, 2004; Vinod and Basavaraja, 2007, De, 2008.

13.1 Principle

1. Biofertilizers can be used for increasing natural productivity of the system and plant extracts or medicinal herbs can be used to control pests and disease management. This necessitates integration of plant communities in farm management.
2. Use of hormone for seed production, genetically modified organism (GMO) and use of inorganic fertilizer are strictly regulated. No use of synthetic pesticides and herbicides.
3. Organic standards are strictly followed in the practices and materials used to produce product. Feed and fertilizer for use should be from certified organic agriculture. Although the process is organic but this does not make any claim about the end product such as quality or food safety.
4. Limitation of stocking density.
5. Preference for natural medicines.
6. Restriction of energy consumption (regarding aeration).
7. Processing according to organic farming.
8. Intensive monitoring of environmental impact.

However, the International Federation of Organic Agriculture Movement (IFOAM) has developed basic standards for organic fish farming.

Other standards developed include

1. Germany's Naturland
2. UK'S Soil Association
3. Newzealand's Biogrow
4. Switzerland's Biosuisse
5. Austria's Bioernte
6. Sweden's KRAC
7. FAO & WHO has also finalized organic crop, livestock, processing, labelling inspection and certification guidelines.

13.2 Conversion and Certification

Conversion of farming practice in the concept of Organic farming that is "the time between the start of organic management and certification of the product" requires a minimum of two years for the aquaculture production system. For the product to be considered suitable for certification, the records of practices and input materials used by the farmer must be maintained.

MPEDA has stepped in, to produce aqua products in organic practices and for this MOU has been signed with SIPPO (Swiss Import Promotion Programmer) to get their aqua-products certified and find market in European countries. Rosen Fisheries in Kerala is the first successful hatchery in India to produce organic scampi seeds as per the organic standard of Naturland, Germany (Anon, 2008).

14

POLLUTION

1. INTRODUCTION

The broad definition of the word pollution is any effect of human activity which changes the natural conditions of the environment. It is also defined as any ``alternation of the physical, chemical and biological properties of any water and atmosphere due to discharge of any liquid, gases or solid substances that likely to create detrimental or injurious to public health and aquatic life.

The word pollution is an adoptation of the latin word ``Pollutionem" meaning contamination. Different Scientists has defined pollution, some of which are quoted below.

(1) Patric, Ruth (1953) defined pollution as anything which brings about a reduction in the diversity of aquatic life and eventually destroy the balance in life in a stream.

(2) Ide (1954) defined pollution is any influence on the stream brought about by the introduction of materials to it, which adversely affect the organism living in the stream.

(3) Coulson and Forbes (1952) defined pollution as the addition of anything to water which change its natural qualities so that the organism living in it does not get the natural water to him.

(4) U.S. Public health (1962) stated that ``Pollution" means the presence of any foreign substance (organic, inorganic, radiological or biological) in water which tends to degrade the quality so as to constitute a hazard and lost its usefulness.

(5) The U.S. Conference on human and environment held at Stockholm in 1972 defined the term pollution as the

introduction of substances or energy into the environment directly or indirectly resulting deleterrious effect to living resources.

(6) Cottrell (1978) an economists defined pollution as ``the consumption of environmental quality".

(7) Natural Environmental Research Council (NERC) 1976, stated that pollution is the release of substances and energy as waste product of human activities which results in changes usually harmful within the natural environment

(8) Dasmann (1975) defined pollution as the accumulation of substances or forms of energy in the environment which exceeds the capacity of ecosystems to either neutralise or dispose them to harmless levels. He cited again that pollutants are not necessarily harmful in themselves. Hence, a critical level exists above which the substance regarded as pollutants and below it is not. Therefore, Robinson (1973) distinguished between contaminant (the material cause any deviation, local or general from the mean geochemical composition) and a pollutant (where the quantity of contaminant is sufficient to affect human or other organisms in an adverse manner).

(9) Lord Kennette has more specifically defined pollution as ``the presence at large of substances or energy patterns, which have been involuntarily produced have outlived their purpose, have escaped by accident, or have unforseen effects, in quantities which harm his (man's) health or do offend him (Quoted by Robinson, 1973).

2. TYPES

Air, water, solid and noise pollution are well cited and documented. However, all are categorised under Environmental pollution. In this, context, water pollution is more relevant which is discussed briefly. Water pollution may be occurred in freshwater, brackish and marine aquatic systems, Out of these three aquatic systems, the freshwater pollution is most relevant for discussion in the present context.

3. SOURCES

The freshwater system may be polluted through discharge of : (1) domestic sewage, (2) pesticides and insecticides from agriculture fields, (3) industrial effluents, (4) mining operations for metal ores, (5) waste heat discharged from thermal power station, (6) radioisotopes from nuclear plants, (7) solid wastes and litter wastes, (8) explosive packages and poison gas cylinder sources and (9) oil pollution. The freshwater pond fish culture system is contaminated principally from domestic sewage and agriculural wastes. However, these contaminants on the basis of chemical characteristics and ecological functions are Nutrients and Organic compounds. As nutrients are seldom a limiting factor in productivity and certainly not a pollutant, therefore it is not included. The organic compounds have profound effects on the metabolic activity and chemical constituents of organism.

Organic compounds are : (i) Hydrocarbons and their derivatives, (ii) Organic pesticides (Organochlorine and organophosphorus) and (iii) Phenolic compounds. It is now widely known that some of these groups of pesticides are most non-biodegradable and enter into food chain in any aquatic production system.

3.1 Detergents and pesticides

Detergents either natural or synthetic are toxic to fish. Nonionic detergents such as ethylene oxide do not ionize in aqueous solution. The sodium alkyl sulphate, alkyl aryl sulphonate, sodium lauryl sulphate and sodium dodecyl benzene sulphonate are proved to be decidely toxic. Sodium tetra propylene benzene sulphonate is present in washing powder, which is toxic to fish. Blast, Gamleu C.W., Santomerse D', Sterox-SE and Sterox-SK are the trade banded synthetic detergents used in certain countries for different purposes which are also toxic to fish. Detergents or soaps made from Sodium palmitate, sodium oleate. Sodium stearate have effect on aquatic systems. In the solution of the detergent containing pyridine, the respiration of the fish appeared to be affected, characterised by paralysis and loss of equilibrium. Hence, pyridine is the substance causing nervous symptoms.

The domestic wastes contain substantial quantities of adsorbed metals in addition to organochlorine pesticides and polychlorinated biphenyls (PCB). Since the success of DDT, other insecticides, pesticides, herbicides, slimecides and weedicides are produced increasingly. These pesticides production and consumption has been incresing ever since these were first produced. Consumption was 80000 tonnes during 1984-85 and 1,19,000 tonnes in 1990-91 in India. Of the pesticides that are being used. 80% constitute insecticides and of them organochlorine compound claim the major share about 40%. These organochlorine compounds are termed as ``hard" chemicals owing to their low degradablity and high persistence in the environment for period ranging from 6 months to 30 years, depending on their chemical composition, dose and environmental conditions. This very characteristics also makes them accumulate in the body fats, milk of live stock and human beings. It is now widely known that these group of pesticides are most non-biodegradable and enter into the food chain in any aquatic production system.

U.S. studies indicate that run-off from agricultural water shed is the greatest single source of pesticide. It is possible that PCB contamination of the Gulf of St-Lawrence could be a major problem in future. The direct effect of DDT and other organochl-orine compounds on salmon and trout have been well established. However, little is known about the accumulation of pesticide residues in the marine food chains of the Gulf of St- Lawerence. Due to these pesticides particularly DDT, the coastal area is suffering from the production of thin shelled eggs. Analysis of mackerel and herring from Prince Edward Island and miramichi indicate that they would contribute 0.1 to 1.0 ppm of DDT and its metabolites to predators. Limited analysis suggest that, other than mackerel, a number of commercial species of fish and shellfish from the Canadian Atlantic possess relatively low level DDT. The residue of DDT and PCB are retained in fish and fish eating birds from the Baltic and Swedist West Coast. In plains of Lebanon, Israel and Egypt, the coastal waters and even open sea receives appreciable amount of pesticides through wind action. There appears to be a general tendency to increase the use of organophosphorus and carbamate insecticides substituting the persistent chlorinated products. In Israel, Cyperus and the Nile delta, where intensive agriculture provides some probability for run off or atleast wind transport of pesticides to coastal waters, the problem will become more serious and extend over larger areas. About 40 million fish were killed when endosulfan was spilled into the river Rhine in Germany in 1969. Also in Ariake Bay, there was a large scale death of mysid shrimps in 1953 and of soft clams in 1962. The use of polychlorinated phenyl (PCP) around the area of Ariake Bay for controlling paddy insects, cause large scale death among fish in lake Biwa (Japan) at the sametime. In pesticide manufacturing industry, some time mercury (Hg) is used as an ingradient. In Finland and Sweden especially they have been used as fungicides and slimecides in the paper and pulp industry. High mercury content have been found in fish from many inland and coastal waters in Denmark, Finland and Sweden. So mercury pollution have affected inland and coastal waters. Minamate disease brokeout in 1953 at Minamata in Kumamaoto prefecture. The disease was a toxicosis of more than 100 ppm of mercury caused by eating fish and shellfish from Minamata Bay, Japan. The mercury was derived from methyl-Hg in the effluent from a chemical plant. In 1964, Minamata disease broke out again in the basin of the river Agano in Nilgata prefecture and proved lethal even to human beings after consumption of these contaminated fish materials. Even mercury content in hair from fishing communities of the state Penang, Malaysia is reported.

3.2 Chlorinated hydrocarbons or organochlorines

The common organochlorines used as agricultural pesticides are :

1. Endrin
2. DDT
3. Heptachlor
4. Dieldrin
5. Aldrin
6. Lindane
7. Methoxychlor
8. Chloridene
9. Toxaphone
10. BHC (Gammaxone)

The other chlorinated byproducts from Vinyl chloride production are -

1. 1, 2, Dichloroethane (EDC) - CH_2 Cl-CH_2 Cl
2. Trichloroethane - CHCl - CCl_2
3. 1, 1, 2-Trichloroethane - $CHCl_2$-$CHCl_2$
4. 1,2, Dichlorobutane - CH_3-CH_2- CHCl-CH_2-Cl
5. 1, 3, Dichlorobutane - CH_3-CHCl-CH_2-CH_2Cl

6. Assymetry, Tetrachloroethane - ClC_3- CH_2Cl

7. Monochlorobenzene - C_6H_5Cl

These are class of synthetic organic compounds relatively insoluble in water and highly soluble in organic solvent. Their characteristic solubility in biological lipids and other fat soluble makes them a special toxicological problem among animals. The compound accumulate in adipose tissue and absorbed by tissues with the next higher lipid content, often nervous tissues. The resultant anatomical or physiological changes which may lead to death. The toxicity of chlorinated hydrocarbon insecticides vary greatly with fish species and strains within species.

The toxicity of chlorine has brought recent attention in the aquatic environment and its impact has received ecological consideration. Residual chlorine arrives in the water body can be potentially hazardous to entrained organisms. Phytoplankton and warmwater fish have been reported killed by chlorine toxicity during pond preparation.

3.3 Organophosphate

The common organophosphates used as agricultural pesticides are -

1. Dioxathion
2. Malathion
3. Phosdrin
4. DDVP
5. Methyl parathion
6. Parathion
7. Guthion
8. EPN
9. TEPP
10. Chlorthion
11. Desyston
12. Dipterx
13. Para-oxon
14. Ronnel
15. Fention

These compounds used as insecticides are oily, relatively volatile and insoluble in water but soluble in organic solvent. The primary toxicological effect of organophosphate insecticides to fish is binding of cholinesterases, most important being acetyl-cholinestarase (AchE). AchE is essential for the destruction of excess acetyl choline (ACh) at muscle nerve synopses following muscle contractions. Hence these groups of pesticides have an indirect effect on acetyl cholinestarase in the nervous transmission of nerve impulses.

3.4 Carbamates

Carbamate insecticides are hihgly toxic to a wide variety of insects and crustaceans but have a relatively low toxicity to most vertebrates. These compounds have intermediate persistence when compared to persistent chlorinated hydrocarbon and non-persistent organophosphate insecticides. Application of carboxyl (Sevin) for example maintain insecticidal activity upto three weeks. Furthermore, its toxicity to fishes is up to 250 times less than DDT. The carbamate compound should prove to be safer insecticides for use in aquatic environment supporting fish groups than many now in use.

3.5 Herbicides

There are large number of herbicidal compound in use, each a possible pollutant to aquatic ecosystem. These may contaminate aquatic system in many ways. The common herbicides are -

1. 2,3,5 - TBA
2. 2, 3,6 - TBA
3. 2, 4 - D
4. 4, (MCPB)
5. 4-24-Des
6. Aminotrizole
7. Baron
8. C-56
9. Chlorax
10. CIPC
11. Chlorea
12. Diquot
13. Paraquot
14. Diuron
15. Dowpan
16. EDB
17. Endothal
18. F-98 (Acrolein)
19. Hyamine
20. Kuron
21. Monuron
22. Nemagon
23. Omazene
24. Phygon XL
25. Shell-D-90
26. Simazine
27. Sodium TCA
28. TCA
29. ACP-M-56 a

Out of these, several herbicides are used to remove unwanted aquatic plants from water ways. Many of aquatic herbicides are relatively persistent, for example : diquet and paraquat may be found in bottom sediments for upto 6 months after application. Some aquatic herbicides are applied to water in granular form, the formulations being designed for slow and continuous release of the herbicide to the water. Bottom feeding fish groups may be subjected to contact with herbicide compound continuously for several weeks or months. The herbicide, 2-4 dichloro phenoxy acetic acid (2, 4-D) absorbed by fish is eliminated rapidly in urine which accounts approximately 80% of the compound within 24 hours.

3.6 Phenolic compounds

The phenolic compounds which are in the effluents due to distillation of coal for the production of gases, coke and the tarry materials used for the manufacture of dyes and organic chemicals result in a watery waste known as ammonical gas liquor and the disposal of this is a cause of pollution. Ammonia-gas liquor contain free NH_3 NH_4 salts, cyanide, sulphide, thiocyanate and a variety of aromatic compounds including pyridine, phenols, cresols and xylenols. After treatment to remove NH_3, the waste is called ``spent gas liquor" and phenol or carbolic acid is the most abundant of its phenolic constituents and most dangerous to fish. The conversion of benzene to phenol by an introduction of hydroxyl group can increase its toxicity 5-10 times more and the substitution of a nitro group to form Nitrobenzene ($C_6H_5NO_2$) increases the toxicity 100 times. The combination of CH_3, an OH and two NO_2 group to form Dinitro-orthocresol (DNOC) is also an acute toxicant to fish groups. Different toxic substances present in the same solution have combining effect, which is additive they may appear to interfere with one another producing the condition known as antagonism or their separate effects, this is the condition known as synergism and occurs very rarely than antagonism.

Besides, synthetic organic compound causing pollution the natural organic pesticides used, in aquatic systems are Rotenone and pyrethrum.

3.7 Inorganic elements and compounds

These include heavy metals such as mercury, lead cadmium, copper and compounds such as fluoride, nitrate, nitrite and other dissolved solids like sodium chloride and sodium carbonate. Much of the inorganic compounds originate from industrial effluents. These compounds inhibit enzyme action in fish when accumulated in considerable quantities.

3.8 Suspended matter

The silt originating from cultivation of land, road construction and other anthropological activities may reach aquatic ecosystem and lead to high turbidity and poor light penetration. Poor primary production and clogging of gills in fishes are often reported due to high levels of suspended matter

3.9 Radioactive substances

These originates from nuclear explosives, accidents at nuclear power plants, fuel reprocessing plants, hospitals and laboratories often lead to somatic and genetic aberration in fish in aquatic ecosystem.

3.10 Thermal pollution

Often power generating plants discharge heated water to near by aquatic ecosystems. This facilitates.

1. Rate of decomposition of organic matter
2. Decrease solubility of oxygen
3. Affects on biodiversity of fauna and flora.

3.11 Pollutions and their load

The aquatic ecosystems particularly the aquaculture systems will be deemed to have been polluted and unsuitable if the following chemicals exceed the maximum recommended levels (Devraj , 1993).

Item		maximum recommended level
1.	Free ammonia	1 ppm
2.	Chlorides	8 ppm
3.	BOD	10 ppm
4.	COD	20 ppm
5.	Phenolic compounds	0.1 ppm
6.	Mercury	0.0003 ppm
7.	Lead	0.1 ppm
8.	Cadmium	0.1 ppm
9.	Copper	1.0 ppm
10.	Arsenic	0.2 ppm

4. EFFECTS

4.1 Effects on microorganisms and plants

It is reported by several workers that food and paper industries, favours the proliferation of bacteria and lower fungi. It has been demonstrated that DDT also inhibit the cellular division in phytoplanktons. The reproduction of phytoplanktons are stopped or altered by minor doses of DDT, dieldrin or endrin. The effect of hydrocarbons and detergents on the cellular membrane of algae, leading to a penetration of the pollutant in to the cell and to the extrusion of cellular fluid.

4.2 Effect on animals in relation to nutrition

Large number of pollutants which can act on different nutritional phases by (i) reducing the food resources of the environment (ii) altering certain sensorial mechanisms which change the abilities of animals to detect its prey (iii) reducing apetite of fish and other commercial animals (iv) excessive secretion of mucous, faeces and (v) nutritional behaviour.

DDT and PCB sometimes accumulated in the tissue and released in to the blood stream during starvation may give rise to serious physiological effects. Numerous pollutants probably act on the intermediate metabolism and in certain cases through the neuroendocrine system may affect the physiology.

The rate of accumulation and excretion of certain pollutants, appear to be dependant on their physico-chemical condition, concentration, environmental factors and the path way through which they are introduced into the organism.

4.3 Effect on respiration

It was shown that pollutants can act on the respiratory functions of organisms directly and indirectly through a variety of mechanisms. These include (i) Clogging of the gills, modification of the branchial cells and cause histological changes in the gill tissues.

4.4 Effect on Osmoregulation and Ionic regulation

Some pollutants such as methyl parathion, endrin and methoxychlor were shown to alter the concentration of various cations in the fish indicating an effect on the hydromineral regulation. As osmoregulatory capacity in fish and invertebrate varies with temperature and hence, thermal pollution may interfere with the osmoregulation and ionic regulation in aquatic animals.

4.5 Effect on reproduction, development and growth

Changes in chemical characteristics of water including reduced dissolved oxygen resulting from pollution can also be harmful to the reproductive functions of aquatic organisms. Pesticides were reported to affect larval development in invertebrates. Degenerative changes in testes of crabs were observed as a result of combined effect of oils and detergents.

4.6 Effect on neuro-muscular system and behaviour

Organophosphorous pesticides inhibit the acetylcholin esterase activity in the brain of freshwater fish.

4.7 Effect on cellular physiology

The application of kerosene can significantly drop the glycogen level of tissue. Methyl parathion can inhibit the esterase activity in the blood serum. Various and extensive alternations of cellular activities were reported. In this connection, reduction of the quantity of RNA in the cells and a decrease in calcium absorption were especially pointed out.

4.8 Ecological effect of pollution in the natural environment

Pollutants to some extent proliferate certain pathogenic bacteria, which might affect commercially important fish. The correlation between the release of certain wastes and fish diseases reported for some areas could be explained partly by deterioration of the physiological conditions of the fish as a result of direct action of pollutant.

4.9 Taste

Freshwater system polluted heavily by effuents from coal plants and other industries and aquatic vertebrates and invertebrates in it have a bitter taste. Because these organisms may contain aromatic compounds and it is though that the bitter taste is produced by abnormal metabolism that accompanies the accumulation of these compounds.

5. CAUSE OF PESTICIDE POLLUTION

The following are the cause of pesticide pollution.

1. Direct application to control water inhabiting pests.
2. Aerial application
3. Spray drift from routine agricultural operations.
4. Heavy rainfal immediately after pesticide application.
5. Accidental spill
6. Surface run-off and sediment transport from treated soil and disposal of waste products by pesticides processing units and other ancillary industries.

6. PREVENTIVE MEASURES AND RECOMMENDED LIMITS

1. Permissible limit use
2. Target specific pesticide should be developed.
3. Biological method an alterative to pesticide should be takenup.
4. Infact, agricultural production system and aquatic production system should supplement each other and they should not compete with each other for production.

However, these following recommended limits for toxins were extracted from water quality criteria prepared by environmental studies Board N.A. Science USA 1972 at the request of environmental protection agency.

1. Polychlorinated biphenyl (PCB) - A composite sample of fish tissue shall not contain PCB greater than 0.5 mg/kg.
2. DDT compound - The recommended limit for flesh consumed is 50 mg/kg wet weight of fish tissue.
3. Aldrin, dieldrin, Endrin and Heptachlor - The meat consumed may not have greater than 50 mg/kg wet weight of any fish tissue together (all the pesticides).

In Indian context, pollution problem is not encountered in pond fish culture except certain eutrophication. Such eutrophication is due to rich nutrients which is seldom a limiting factor for productivity and to the other biotic communities living in it. It also do not lost its usefulness. That too, the discharge of treated effluents in to water bodies are guided by pollutional control board. The research investigations in pesticide residueal pollution need to be extended from primary to secondary, and also to successive higher vertebrates. Besides, the cumulative pesticide residual effects are neither systematically studied not clearly understod from primary to tertiary level including human beings. That too, the residual toxicity of pesticides in Eco-biopopulation with regard to physiology, genetic and immune response needs much investigations. However, it is apparent that, the residual level of pesticides and insecticides in pond cultured fish tissues are at harmless level and resistable by human population for which no death hazards are encountered till date.

15

FISHERIES EXTENSION EDUCATION

1. INTRODUCTION

Aquaculture extension is a part of the technology transfer system that is primarily concerned with transmitting information and knowledge of aquaculture technology from research to farmer. Aquaculture development is closely related with the development ability of the farmer's understanding and adoption of technology. It is said that the extension umbrella in aquaculture is poor. One could agree with this observation considering the widely scattered location of ponds and tanks, religious pools in certain areas coupled with other social and economic problems and micro community of extension workers. Though extension division and units have been established at several centres, much of the extension work has been made in the form of survey of resources or constraints in adoption of technologies or socio-economic condition of fishermen etc. Lack of training and finance, nonavailability of inputs including seed, feed, fertilizers etc., are identified as some of the major constraints.

Some others talk about the lack of initiative, motivation, knowledge, skill and quality of the extension worker himself besides their small number. Still others have been all praise for voluntary agencies, youth clubs, mahila mandals rather than research institutes or the government departments. Hence there is need to ascertain the lacunae by proper and planned programmes in different areas of aquaculture extension.

2. DEFINITION

The word "extension" is derived from the latin word 'ex' meaning 'out' and 'tensio' meaning 'stretching'. Hence extension is such type of education which is stretched out to people in the rural areas far and near to provide service beyond the Institutional class room. So extension includes both education (non-formal) and services. Non-formal education is an organized, systematic educational activity carried on outside the frame work of the formal education to provide selected type of learning to particular subgroups in the population, adults according to their needs. Examples of such non-formal education are agriculture/ aquaculture extension.

2.1 Extension Education

The extension education role is generally performed by the higher learning institutions like, the Agricultural and other Universities and Colleges, ICAR Institutes, Home Science Colleges and apex level training and Extension Organisations. At the University level, extension is integrated with teaching and research, while at the research institutes, extension is integrated with research. At the other apex level organizations, extension is generally integrated with training in extension.

2.2. Extension Service

The main responsibility of extension service is with the State Government. The departments of Agriculture, Horticulture, Animal husbandry, Veterinary, Forestry, Fishery, Sericulture, Apiculture and other units like Vermicompost, Mushroom culture etc., of the State Government carry out extension work with the farmers and rural people over the entire state. The extension service has the main responsibility of educating and training the farmers, farm women, rural youth and village leaders of the state and for this purpose, they take the help of the Universities, research institutes and training and extension organization. Two trends in extension is visible in India. These are

1. Decentralization of extension through closer coordination with panchayats (Local self Govt.) and
2. Commercialisation of extension through provision of specialized services by commercial companies at a cost.

The National Commission on Agriculture (1976) refers to extension as an out- of- school education and services for the members of the farm family and others directly or indirectly engaged in farm production enable them to adopt improved practices in production management, conservation and marketing.

3. KEY ELEMENTS IN NON FORMAL EDUCATION

1. It is learner centred
2. There are variety, options and flexibility in curriculum
3. Informal human relationships are essential
4. Reliance on local resources. Both conventional and unconventional sources are used
5. Curriculum content and methodology are directly related to users (farmers) life style.
6. Local approaches to the solution of local problems, more decentralization and less bureaucratic control.

4. NEED OF AQUACULTURE EXTENSION

1. The need for extension arises out of the fact that the condition of the rural people in general and the farm people in particular has got to be improved. This is possible through extension of information from laboratory to land.
2. To bridge the extension gap between the user of technology and the technical know how possessed by the technical experts.
3. Farmers understanding and adoption of technology for aquaculture development through agency.
4. Thus there is need for an agency to interpret the findings of research to the farmers and carry the technical problems of the farmers to the research station for solution.

5 OBJECTIVE

1. To assist people to discover and analyze their problems and identify the needs felt by the farmers.
2. To develop leadership among people and help them in organizing groups to solve their problems.
3. To disseminate research information and practical importance to the understanding levels of the rural farmers and use.

4. To assist people in mobilising and utilizing the local available resources and needed materials from outside.
5. To collect and transmit feed back information for solving management problems.

6 FUNCTIONS OF EXTENSION

The functions of extension is to bring about desirable changes in human behaviour by means of education. Changes may be brought about in their

1. Knowledge
2. Skill
3. Attitude
4. Understanding
5. Goals
6. Action
7. Confidence.

However, the core functions of the aquaculture extension system includes

1. dissemination of appropriate technology (Education)
2. convincing the farming community to adopt such technologies (Motivation)
3. collect the farmer's response (Feed back)
4. refinement of technology to suit the farming situation (Assessment and Refinement)
5. act as link between the research and user system (Liaison)

7. PRINCIPLES OF EXTENSION

Principles are generalized guidelines which form the basis for decision and action in a consistent way. These include

1. Grass roots principle
2. Principles of indigenous knowledge
3. Principles of cultural differences/harmony
4. Principles of interests and needs
5. Principles of learning by doing
6. Principles of participation

7. Family principles
8. Principles of leadership
9. Principles of adaptability
10. Principles of satisfaction
11. Principles of evaluation

8. SCOPE OF EXTENSION

It includes

1. Efficiency in agriculture production
2. Efficiency in marketing distribution and utilization
3. Conservation, development and use of natural resources
4. Management on the farm and in the home
5. Family living
6. Youth development
7. Leadership development
8. Community development and rural area development
9. Public affairs

9. EXTENSION EFFORTS A REVIEW

The aquaculture extension in India has been traditionally accepted as integral part of the agricultural extension network. Several approaches for extension of aquaculture along with agriculture have been considered and implemented from time to time, but have not yielded the desired results.

Several approaches in the past such as

(i) Extensive extension (community development programme)

(ii) Intensive extension (Intensive Agriculture District programme, Intensive agricultural area programme and high yielding varieties programme)

(iii) Drought prone area programme.

(iv) Small marginal and agricultural labourers development programme.

(v) Tribal area development programme and hill area development programme.

(vi) Single purpose extension (training and visit system) have been adopted.

In addition to these some technology transfer schemes have been launched exclusively for boosting up fish production in India in the post independence period and since then aquaculture extension has indeed passed through an evolution phase. Although adequate literatures in agriculture extension are available, the literatures in aquaculture extension are scanty and sporadic. Workshop on aquaculture extension need held on 20th June 1991, organized by central institute of fresh water aquaculture, kausalyagang, along with the papers of Bhaumik *et al.,* 1993, Tripathy *et al.,* 1982, Radheshyam and Kumar 1982, Singh and Sampath 1981, 1982, 1983 and 1990, Rahiman *et al.,* 1991, Algarswami 1995, Srinath 1995, 2000, Kumaran *et al.,* 1999, 2001, 2003, Ayyappan and Biradar 2002, Sulaiman, 2003; De and saha (2005), Chakraborthy and Pal 2004, De, Saha and Srichandan (2008), Sharma and Laharia (2008) are some of the literatures solely based on the principles, methods, system statues, problems, women participation and development in aquaculture extension. Further the author's experience as Project Coordinator, KVK, Jajpur adds into the information on aquaculture extension.

10. QUALITIES OF AN IDEAL EXTENSION OFFICER/ PERSONEL

1. Sound knowledge on technical know how.
2. Leadership quality.
3. Conversant with local language and effective communication method.
4. Cooperation attitude.
5. Coordination and participation.
6. Honesty and punctuality.
7. Sociable and acceptability to cultural values of the farming community.
8. Interest and satisfaction in the job.
9. Response to feed back.
10. Responsible and should have accountability.
11. Organizing capacity.
12. Helping nature.

13. Timely action.
14. Experience in tackling the situation
15. Means of solving the specific problems.
16. Contacting attitude.
17. Knowledge on latest technology developed.
18. Liaising capacity.
19. Patience and endurance.
20. Hard working attitude.

These competencies are needed by the extension agent. Competency refers to the ability of an individual to reach a desired goal. Extension agent must be

1. Technical competency
2. Economic competency
3. Scientific competency
4. Occupational competency
5. Communication competency
6. Social competency

11. PRINCIPLES OF DIFFUSION OF TECHNOLOGY

Scientist and change agents should have basic understanding regarding principles of diffusion of agricultural technologies. The principles which are discussed below are relevant and not necessarily fixed in importance or sequence.

11.1 Learning by doing

Scientists or change agent while starting the process of transfer of technology must see that each and every stage of transfer, users of technology must be involved by users participatory approach. As Dr. Newman said "farmer, like other people, hesitate to believe themes and fact until they see with their own eye". So it is learning by doing which is the most effective technique in transfer of technology.

11.2 Ability regarding use of communication channels

In all learning situation, no single communication channel is effective. Suitable channel have to be selected keeping in view

1. The existing level of knowledge among farmer.
2. The level of understanding.
3. Comprehensive attitude of audience.
4. Availability of channel in rural area.

Communication channels include the

1. The reading materials like folders, brochures, journals and periodicals Which are the communication channels for the farmers, those can read.
2. Radio & TV channels are only for those who can view and listen.
3. Demonstrations are useful only to those who can come to see these. Farm and home visits are also valuable.

11.3 Leadership

Developing local leadership is fundamental in diffusion of technology. Develop local voluntary leadership to help the scientist in this process of diffusion of technology. Actually the involvement of leaders in technology transfer is one of the most important factors that determine its success.

11.4 Social and cultural values

The technology that is being diffused must be presented in a manner that it tallies with social and cultural values of the farming community. Farmers use to give importance to the social and cultural values towards the adoption of technology that can be easily executed in the farming community.

11.5 Cooperation, coordination and participatory approach

Team spirit of scientists are required for diffusion of technology to the user by constant persuation. In this process effective cooperation and coordination is required to be made between officials of the line department and NGO's at all the levels and effective participation by each member. Coordination and harmonious adjustment of issuing materials as input support to beneficiaries be converted into integrated effort as a whole. There must be effective coordination between various types of function such as administration, supervision, programming, action plans and programmer execution and appraisal of accomplishment. The participation of the members and farmers are of fundamental importance for success in transfer of technology. Farmers must feel that the technology that is passed to them is of benefit.

11.6 Interest and need based

The technology being diffused should be based on basic interest and need of the farming community. Interest and basic needs of the farmers must be identified in advance before transferring the technology to users. Farmer's first choice should be considered as his interest and need. Scientists must begin the interest and need as farmers see them.

11.7 Whole family

While diffusing the technology, the whole family of the farmer is considered as a unit and the scientist must assess about the benefits of technology to whole family members. The functional relationship among family members and the society due to adoption of technology should also be assessed by the scientist.

11.8 Satisfaction

Technology that is transferred to users will be fully accepted and adopted, in case they are fully satisfied with the performance. The farmers continue to act as per their satisfaction through adoption of technology that is well suited their need and resources.

11.9 Complexity and divisibility

The more complex is the technology the chances of its adoption are less. Farmers like to use that technology which is relatively less difficult to understand.

11.10 Feed back

Scientists must collect feedback during and after the transfer of technology process is over. The usefulness, drawbacks, problems faced, expected results achieved in terms of yield, cost benefit ratio, profit made should be assessed. This body of information is essential as a basis for making changes at every step of technology transfer process and at later stage modify the strategy of technology transfer. Formally feedback can be collected through survey schedules, questionnaires, tests etc. Feedback may be positive, negative or no. Positive feedback indicate about agree of farmers for adoption of technology, where as negative response indicates about disagree of farmers for adoption of technology. No response of farmers is the third category of feedback, where technology neither be made available nor modified for benefiting the farmer.

12. DIFFUSION OF TECHNOLOGY SYSTEM (DTS) (Fig. 81)

Actually, the aquacultural development depends on

1. Knowledge generating system (KGS).
2. Transfer of technology system (TOT) or knowledge disseminating system (KDS)
3. Knowledge consuming system (KCS)

12.1 Knowledge Generating System (KGS)

KGS constitute the scientist of state, central, university, affiliated college, research and production unit of private and public sectors in agriculture and its allied.

12.2 Knowledge Disseminating System (KDS)

KDS constitute the extension officers, extension supervisors, village level workers, subject matter specialists working in extension department of state, central, universities and affiliated collages. The main function of this system is to transfer the technology.

12.3 Input Supplying Agency (ISA)

Another part of this system is input supplying agencies system. Input supplying Agency (ISA) consisting of nationalized banks, fertilizer corporation, agro-industries, cooperative and corporation, etc.

12.4 Knowledge Consuming System (KCS)

KCS mainly consist of farmer's rural artisans, members of input production agencies and rural industries.

The most effective diffusion of technology system is that where all the four systems that is KGS, KDS, AIS & KCS work in close cooperation keeping in view of their common objective. The development of agricultural and allied production is the main common objective of diffusion technology system (DTS).

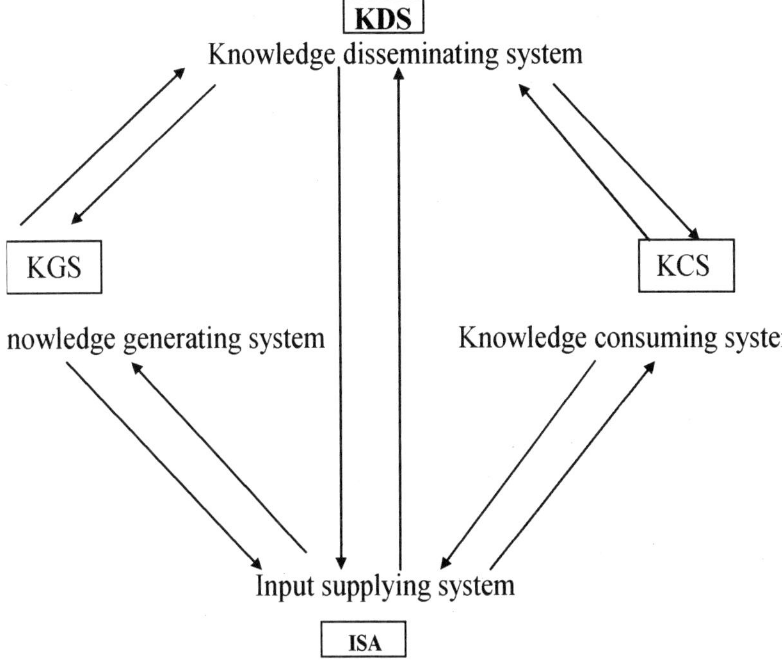

Fig. 81. Schematic diagram on diffusion technology system.

13. DIFFUSION METHOD (Fig. 82)

Technology diffusion method can be grouped as.

1. Group contact method →Demonstration, group meeting.
2. Mass contact method → Radio, television, extension leaflet, literatures and training.
3. Individual contact method → Visit, telephone. This method of technology diffusion is hardly made.

Diffusion of technology will be effective through the language used, convey ideas, understanding of facts, importance of facts and their relationship to production process and solving the problems. It is the impact on people that is the important consideration in communication.

Scientist should take strategy before going for diffusion of technology to farmers that includes

1. What is new or modification in exiting practices?
2. Who are going to be involved?
3. What is the purpose?
4. What results are expected?

Similarly scientist must take consideration on the principles of technology diffusion among farmers that include.

1. Specific objective of diffusing the technology.
2. Type of message → direct elaborate approach verses abstract approach.
3. The audience → size, interest and need.
4. Transfer technology methods → reach, effectiveness, relative costs, availability of time etc.

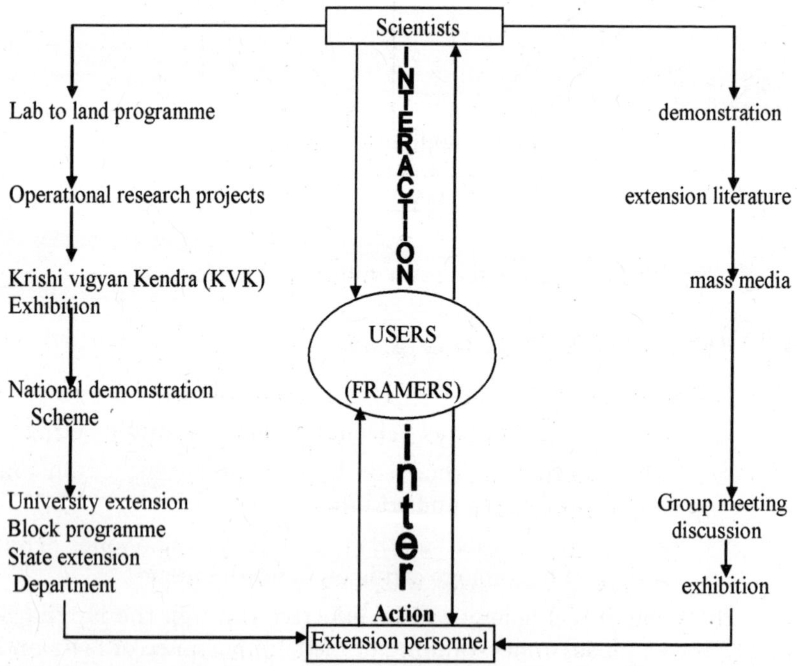

Fig. 82. Schematic diagram on technology diffusion methods.

14. AQUACULTURE EXTENSION SCHEMES

14.1 Fisheries extension units

First attempt on aquaculture extension was taken up under the "grow more food programme" of Govt. of India during 1950 to augment fish production through supply of quality fish seed to the farmers in different states. As a result of this, the first fisheries extension unit was established at CIFRI, Barrackpore in 1950. "Fish seed syndicate" Kolkata was formed by the effort of this unit to ensure supply of quality seed to the deficit states. Finding the important role of this unit in extension, nine more fisheries extension units were established during second five year plan. As extension of agriculture or aquaculture are the state concern, these total ten fisheries extension units were closed in 1976 by Govt. of India. Thereafter, State Government took up extension activity in their respective localities. Inorder to utilize the aquatic resources government took up massive seed production for culture. The fish seed production of the country during (2005-06) was 21,000 million.

14.2 Fish farmer's development agencies (FFDA)

For encouraging and popularizing pisciculture as an alternative means of employment generation and poverty alleviation the Govt. of India introduced "fish farmers development agency" during 1974 at state level. The role of FFDA was to provide

1. Training to selected person interested in fish culture.
2. Assist in utilizing suitable water resources for the purpose.
3. Arrange credit facilities from the Nationalized banks.
4. Supply fish seed and operational inputs including technical support.
5. Help in marketing their produce.

At present (2005-06), 422 numbers of FFDA are functioning. But the progress in most of the FFDA's are less than the expectation and the average fish yield is 2.2 tons/ ha.

14.3 BFDA

Brackishwater fish farmer's developments agency (BFDA) was established in 1985 with the objective of proper utilization of brackish water for fish and shrimp culture. Technological, financial and extension supports were rendered to fish farmers for increasing the aquaculture production. At precut 39 BFDA's are functioning in different states of the country.

14.4 World Bank project on carp hatcheries

During 1997, "Inland fisheries project" was launched with the assistance of World Bank for construction of modern fish seed hatcheries and granting loans to the fish farmers for improvement of fish farms and funding for fisheries extension programme. Under the project 63 fish seed hatcheries came up in the states of Uttar Pradesh, West Bengal, Bihar, Orissa and Madhya Pradesh. At present (2005-06), 1070 carp hatcheries are operating in the country to produce and supply fish seed to farmers of different states of the country.

14.5 State Fisheries/ State Agriculture Universities

During the 5th five year plan, the fisheries extension mechanism was revived in state sector. The states were encouraged for fisheries extension to help the fish farmers for augmenting fish production. Under this scheme

1. Extension services kits and
2. Audio- Visual equipments were provided by the Government of India, as a support to the state govt. to take up fisheries extension work. State fisheries departments, Extension education of States Agriculture Universities & MPEDA put attention to aquaculture extension.

14.6 Non-Government Organization (NGO)

Experience has shown that Government efforts for development activities does not always, for some reason or other, reach the deserving beneficieries. It has been felt that many government schemes fail due to lack of participation of beneficieries. These calls for some organizations of people at the grass root level. An organization not only by virtue of its structure but one which is very much concerned of the needs and aspirations of the people and of the justice and equity in the society.

Some voluntary organisations such as

1. Ram Krishna Mission
2. Lutheran world services.
3 Don Bosco society
4 Tagore rural development society.
5 Kamala Nehru Trust.
6. Nehru Yuva Kendra have been making notable contribution to the cause of transfer of technology through various schemes.

14.7 Indian Council of Agriculture Research

ICAR is basically a technology generation organization , took up the transfer of technology as one of its function through different extension projects like.

1. National demonstration project (NDP).
2. Operational research project (ORP).
3. Lab to land programmer (LLP).
4. KVK and TTC.

14.7.1 National demonstration project (NDP)

"Seeing is beleiving " is used as the guiding principle of NDP. A single convincing demonstration in the farmer's field proves to be much more effective than class room teaching. The scheme operated in 1964 and also provides validity of technology adoption in the farmer's field through demonstration.

14.7.2 Operational Research Project (ORP)

Operational Research Project is indeed an extended version of NDP in which whole village or clusters of village approach was taken for demonstration. The scheme operated during 1974. The ORP exclusively meant for composite fish culture and integrated fish farming. The production in composite carp culture has gone up from 0.45 tons to 2.5-4.0 tons/ha/yr of fish, 4000-6000 eggs and 500 kg of meat. Pig-fish farming yielded 6-7 tons/ha/yr fish and 4.2-4.5 tons/ha/yr pig meat. A total of 152 ORP centres are operating in aquaculture sector of the country.

14.7.3 Lab To Land Programme (LLP)

Indian Council of Agriculture Research launched lab to land programme in 1979 to mark the completion of 50 years of its existence. Under this project small, marginal and landless farmers were assisted to take up pisciculture by providing them technical support and critical inputs. In the field of aquaculture the impact of LLP has been quite remarkable and productivity rose leading to improvement of socio-economic status of the farmers.

14.7.4 Krishi Vigyan Kendra and Trainers Training Center (KVK/TTC)

To implement the recommendations of the education commission (1964-66), to establish Agricultural polytechnics for providing vocational education in agriculture and allied subjects, the Government of India requested the ICAR to evolve a suitable mechanism. The ICAR constituted the Mohan Singh Mehta committee to examine the issue. On the basic of the recommendations of this committee, the ICAR sponsored the project on Krishi Vigyan Kendra (Farm science center) to cater to the training needs of the farmer in 1974. The KVK is envisaged as a vocational training institution at grass root level having all the important disciplines viz., agronomy, floricultures, fisheries animal husbandry, home science etc. KVK is designed and developed to impart need based and skill oriented vocational training to the most needy sections of the rural community. The training methodology is strictly based on "Teaching by doing" and "learning by doing". The target group includes practicing farmers, farm-women, unemployed youth, school dropouts etc. First KVK was established at pondicherry as a pilot project. At present a total of 458 KVKs are operating in the country, out of which 29 KVKs are in Orissa. Under OUAT, Bhubaneswar 27 KVKs are operating in different districts of Orissa except Bolangir. One KVK at Cuttack and one at central institute of freshwater aquaculture (CIFA) are also operating. It is reported that, in fisheries sector, three important KVK's are functioning.

1. at CIFA for Inland Aquaculture
2. CMFRI (Narakkal, Kerla) for Mariculture
3. CICFRI, Kakdwip, (W.B.) for Brackishwater Aquaculture.

For smooth operation of the KVK, ICAR has categorized different States in zones. The country has been divided in to more than 130 distinct agro-climatic zones for location specific, need based research and extension relevant for specific agro-ecological situations. Orissa is under zone 7. Zonal coordinator looks after and assess the functioning of the KVKs in a zone periodically. Orissa has 10-agro climatic zones. These agro-climatic zones of Orissa are as follows:

Krishi Vigyan Kendras in Ten Agro climatic zones of Orissa

S.N.	Agro-climatic zones		District having KVYs
1	North Western Plateau	—	Sundargarh
2	North Central Plateau	—	Mayurbhanj, Keonjhar
3	North Eastern Coastal Plain	—	Balasore, Bhadrak, Jaipur
4	East and South Eastern Coastal Plain	—	Kendrapara, Nayagarh, Puri, Jagatsinghpur, Cuttack, Khurda
5	North Eastern Ghat	—	Kadhamal, Rayagada, Ganjam, Gajapati
6	Eastern Ghat High Land	—	Koraput, Naberangapur
7	South Eastern Ghat	—	Malkangiri
8	Western Undulating	—	Nuapura, Kalahandi
9	West Central Table Land	—	Bodh, Deogarh, Sonepur, Jharsuguda, Sambalpur, Baragarh
10	Mid-Central Table Land	—	Dhenkanal, Angul

KVKs of Orissa were established in different years in different districts. Thses are given as follows:

KVKs operating under OUAT

Sl.No.	*Name of the KVK*	*Location*	*Year*
1.	Keonjhar	Keonjhar	1982
2.	Koraput	Semiliguda	1983
3	Balasore	Baliapal	1983
4.	Ganjam	Bhanjanagar	1985

5.	Baragarh	Gambharipalli	1992
6.	Kandhamal	G. Udayagiri	1993
7.	Kalahandi	Bhawanipatna	1993
8.	Kendrapara	Kendrapara	1996
9.	Dhenkanal	Mahisapat	2001
10.	Angul	Angul	2001
11.	Jajpur	Barachana	2002
12.	Sundergarh	Kirei	2004
13.	Bhadrak	Ranital	2004
14.	Nabarangapur	Umerkote	2004
15.	Nayagarh	Panipoila	2004
16.	Sambalpur	Chipilima	2004
17.	Jagatsinghpur	Tirtol	2005
18.	Nuapada	Nuapada	2005
19.	Rayagada	Gunupur	2005
20.	Gajapati	R.Udayagiri	2005
21.	Boudh	Paljhar	2005
22.	Mayurbhanj	Samakhunta	2005
23.	Sonepur	Sonepur	2005
24.	Deogarh	Deogarh	2006
25.	Jharsuguda	Jharsuguda	2006
26.	Malkangiri	Mundaguda	2006
27.	Puri	Kakatpur	2006

At Bolangir, KVK is established in 2010.

The objective of KVK is to survey and identify the problem areas (2). Conduct training for target groups. The mandates of KVK is to popularize the technology through

(1) On farm trial (2) Front line demonstration (3) Learning by doing and (4) Training of in-service line department personnel. Looking to the need, many have advocated for fodder cultivation, mustard, sesames and mango orchads. Drip irrigation is also emphasized for fruit orchads. Hence KVK activities are multiferious for up-liftment of socio-economic condition of the farmers through technology adoption to utilize the natural resources available with the farmer.

14.7.5 Trainers Training Centre (TTC)

TTC's have been established in the specialized streams mainly in the ICAR institutes. The TTC impart in-service training in subject matters to the extension personnel and trainers of KVK etc. A total of 8 TTC in different disciplines are functioning in the country, out of which two are for fisheries (one at CMFRI, Kochi and one at CIFA, Bhubaneswar).

14.7.6 University Extension Block Program (UEBP)

OUAT, BBSR is operating UEBP in Puri and Khurda district of Orissa. The mandates of UBEP are:

1. Field validation of new technology.
2. Carrying out demonstration on latest proven technology.
3. Collection of field problems.
4. Conducting village level training camps for diffusion of farm innovation.
5. Farmer - Scientist interaction and organise exhibition in agriculture, farm implements and other allied sectors.
6. Creation of field laboratory for students.
7. Production, publication and distribution of farm leaflets to farmers.

In addition to these, other National Projects like

1. Rural aquaculture project (1975)
2. National agriculture extensions project (1983)
3. Women in aquaculture (1992)
4. Institute village linkage project (1996)
5. National agriculture technology project (1998)
6. Jaivigyan national science & technology mission (2000)
7. Mega seed project (2006)
8. National agriculture innovation project NAIP (2008) have substancially steered to the aquaculture development of the country. ATIC system was established under NAIP project in most of the agriculture university and other institutes. Further TACT (total aquaculture technology centres) are established for demonstration and training preferably under FFDA's and BFDA's. Although

lot of technology advancement has been made but their transfer to rural field is relatively less due to poor approach in diffusion method, system and non-participatory approach.

14.7.7. Institute Village Linkage Programme (IVLP, 1996)

This was initiated for assessment and refinement of technology in the light of biophysical and socio-economic constraints.

14.7.8 D.R.D.A

The District Rural Development Ageny has traditionally been the principal organ at the District level to oversee the implementation of different anti-poverty programmes. DRDA coordinates the officials of line department, the banks and other financial institutions, resources required for poverty reduction effort in the district. Its role is to coordinate and bring about a convergence of approach among different agencies for income alleviation through diversified activities. It takes required step to improve the awareness regarding rural development and poverty eradication particularly among the rural poor. Its role is to sensitize the different functionaries in the district to the different aspects of poverty and poverty eradication programme. It deals only with the anti-poverty programmes of the Ministry of Rural Development.

District administration use to give special emphasis on leasing out village community ponds for pisciculture to self help groups. They were also encouraged for integrated fish farming with banana and vegetable cultivation. DRDA use to make effort in training and required inputs to some extent through line departments to rural farmers.

The weakness so far noticed are

1. Many a time the lease amount is quite high for the SHGs to bid for. Good tanks slip from their hands.
2. Intensive skill training is required to the SHGs to learn the nuances of the trade.
3. Nurseries are to be promoted through SHGs for smooth and easy supply to the SHGs with a reasonable price.
4. Additional sources of income are to be identified.

5. Disseminate technical know-how in value added product and their marketing.

Many opportunities lie for increasing fish and prawn production and promote value added products of fish through SHGs. Further possibilities can be explored to finance SHGs to acquire nets and boats for fishing in large water bodies.

15. CONCEPT OF RURAL SOCIOLOGY

Human relationship and human behaviour are important factors in the rural society for carrying out extension and rural development work. Rural sociology involves the study of human relationships in rural situations.

15.1 Characteristics

The Predominant characteristics of rural society is as follows

1.	Occupation	Mostly agriculture and engaged in diverse Occupation.
2.	Work environment	Mostly related with soil, water, plants and Animal life. Close to natural environment.
3.	Weather and season	Very important to carry out their Traditional agricultural practices.
4.	Skills	Require wide range of skills for Engagement in diverse occupation also in off season
5.	Work Unit	Whole family work as a unit in Agricultural activities.
6.	Type of family	Mostly Joint family with large number of members in the family.
7.	Size of Community	Small
8.	Density of population	Low and relatively homogenous
9.	Social Interaction	Few and personal
10.	Institution	Simple and small
11.	Mobility	Low
12.	Infrastructure	Little to moderate
13.	Modern Home Appliances	Few
14.	Mass Media participation	Low
15.	Value system	Generally sacred.

15.2 Social Stratification

It is the arrangement of individuals or group of people in to hierarchically arranged strata in a community. Social stratification is usually based on the status of the rural people that includes wealth, Ancestry, Functional Utility of the individuals, religion and biological characteristics (age and Sex) etc.

15.3 Social Interaction

Two types of social interactions are noticed in rural sectors.

1. Positive interaction which includes cooperation, accommodation and assimilation
2. Negative interaction which includes competition and conflict.

16. PSYCHOLOGY

It is the science of human behaviour. Behaviour constitutes three aspects such as

1. Cognition: to become aware of or know some thing
2. Affection: to have a certain feeling about it.
3. Conation: to act in a particular way or direction after the feeling

Human behaviour is expressable. If it is a inside expression, it is called as covert. The outside expression of human behaviour is called as overt. Human behaviour can be measured through his

1. Personality
2. Attitude
3. Emotions
4. Prejudgement and
5. Stereotype (fixed images of a person)

Sharma and Laharia, 2008; have described the personal and socio-psychological characteristics of fish farmers of Haryana. Their studies revealed that change proneness, mass media exposure, experience in fish farming, extension contacts, education and caste were the most important background variables which could significantly predict the adoption level of fish farmers on different aspects of prestocking and post stocking management.

17. PRINCIPLES OF EXTENSION PROGRAMME PLANNING

There is no "Buy-Use" extension methodology, which can be recommended. However, Engle and Stone (1989) have enlisted some of the principles that should be kept in mind while developing and strengthening aquaculture extension programmes. They include

1. technology should be technically sound and adapted to local conditions
2. extension worker must be well trained and personally dedicated.
3. major emphasis of aquaculture development programmes should be on extension.
4. feed back mechanism must be established.
5. selection of extension workers should be tailored to the target group.
6. the programme should have support both from government and producers and must be stable.
7. a good reputation can be valuable to the programme or agency and
8. links between research and extension need to be strengthened and the primary goal of extension is purposeful education.

General principles on extension programme planning also includes

1. on the analysis of past experience, present situation and future needs.
2. have clear and significant objectives which could satisfy important needs of the people.
3. fix up priority on the basis of available resources and time.
4. should clearly indicate the availability and utilization of resources.
5. should have general agreement at various levels.
6. should involve people at the local levels
7. should involve relevant institutions and organizations
8. should have definite plan of work

9. evaluation of results and reconsideration of the programme
10. equitable distribution of benefits amongst the members of the community.

18. PARTICIPATORY APPROACH IN EXTENSION

All the attempts made so far in research or development process of freshwater aquaculture is unidirectional process that is the researcher played the sole role in developing technology by neglecting the user's of the society. Hence, there is a need for participatory research in aquaculture extension in which all steps of technology development can not only be solely shared by the scientists but also can be shared jointly with the user's in the problem situation in participatory research. Similar views have also been expressed time and again by many workers in India (Radheshyam and Kumar, 1982; Singh and Sampath, 1981, 1982, 1983 and 1990; Tripathy *et al.,* 1982.) Participatory research is solely in response to and for the fulfillment of the need of the people. The main out come of the participatory research is, increase of knowledge about the specific problem that will be available both with the researcher and users in the society. The establishment of location specific technical knowhow units in the farmer's village will also minimize the gap between the researcher and users at the grass root level. Better coordination and linkage need to be established among the concerned fisheries institutions so that they all could integrate their efforts in helping the farmers to increase their production through effective transfer of technology (Rahiman *et al.,* 1991; Algarswami, 1995; Srinath 1995, and 2000, Kumaran *et al.,* 1999, 2001 and 2003). Further, publication of extension materials in different languages for various categories of operatives and end users is required for creating technology awareness and for transfer of technology (Ayyappan and Biradar, 2002; Aquaculture Authority, 2001). Transfer of participatory technology will be more suitable and will generate strength among the user's of technology towards the risk taking ability and leadership development.

18.1 Trickle down system (TDS)

Trickle down system approach of aquaculture extension is a participatory farmer to farmer's extension approach. It involves an initial bottom up participatory planning and extension programme with a lateral spread of knowledge and skills of improved fish farming technology, there by ensures an active flow of technical information from the result demonstration farmer to the fellow fish farmers. In the process both categories of participatory farmers in the extension programme shall be benefitted.

19. SHIFT FROM CULTURE TO LIVELIHOOD SECURITY

With passage of time, the focus of aquaculture extension has shifted from horizontal expansion and productivity enhancement to addressing issues of food security, poverty alleviation, nutrition and livelihood system. In the light of world trade organization (WTO), it is important to adopt "farm to table" approach varying from species diversification for culture to value addition of products. Extension at present has assumed multiple roles of providing information about technologies, price, market, facilitating learning from experience, provide problem solving consultancy in-order to serve the farming community (Sulaiman, 2003).

20. EMERGING ISSUES OF EXTENSION

Main three issues of extension at present are

1. Participatory fisheries management
2. Privatized fisheries extension services
3. Gender issues in aquaculture

21. USE OF MODERN TECHNOLOGIES IN EXTENSION

1. VAST (Very small aperture Terminal) networks offer value added satellite based services capable of supporting internate, data, LAN voice/ fax communication and video conferencing.
2. GPS (Global positioning system) is a hand held satellite navigation system which helps in finding aquaculture potential in streams and locating these points on map.
3. RS (Remote sensing) is the science of collecting data by technical means on the object on or near the earth surface and interpreting the same to provide useful information.

22. SIGNIFICANCE OF ITK (Indigenous technical knowledge)

It offers low cost approaches with high benefits. Compilation of ITK will be a great contribution to conserve our traditional system.

23. NEW STRATEGIES IN EXTENSION

1. Single window delivery system by establishing ATIC (Agriculture technology Information Centre).
2. Aqua service centre, Mobile fish clinic van, Net repairing and cold chain transport system
3. One stop aqua shop
4. Farm school on the AIR
5. Initiatives in information and communication technology (ICT) applications.

Aquachoupal, a web based initiative of ICT Limited, offers farmers of Andhra Pradesh all the information, products, high quality farm inputs and services they need to enhance productivity and farm gate price realization. The M.S. Swaminathan Research Foundation has developed info-villages to provide all information on fish density in ocean to fishermen community.

24. CONCLUSION

Benefits of an effective extension system have long been recognized. The aquaculture extension has still not reached to mark as expected inspite of several advances in the technology of aquaculture. Many areas have been neglected due to non-availability of manpower with adequate knowledge on technical knowhow, poor linkage with research system, low level of job satisfaction etc.

Fish farmers being illiterate and backward require a special approach and strategy for imparting need-based skills to them. Rural women folk which form half of village population, so far been neglected should be engaged in production process during leisure time for which special attention be paid to train them in appropriate aquaculture technology. In tune with this perspective, logical thrust areas for research in aquaculture extension could be identified as under:

1. Sensitization of the target group towards innovative ideas and participatory approach.
2. Adoption behavior, yield gap, sustainability & constraint studies in aquaculture extension.
2. Studies on the impact of training and extension methods on the target group.
4. Impact assessment on recommended package of practices in aquaculture at the farmer's level.
5. Gender awareness in the aquaculture technology transfer.

Extension education has assumed greater significance in view of the emerging concerns like,

1. Sustainable aquaculture.
2. Social equality.
3. Environmental degradation.
4. Export.
5. Value addition.
6. Changing aspiration of rural mass.

Recent advancement in technology front, market led economy, shrinking investment in public sector, migration from rural to urban areas calls for reorientation of extension system. If aquaculture extension research is to be developed, their should be an interdisciplinary work and a possible networking between organizations involved in rural development, youth development, non governmental organization etc. to see the progress of aquaculture extension research and technology transfer reach the needy so that there could be an up-liftment of the rural mass above their poverty line.

16

BIOTECHNOLOGY IN AQUACULTURE

Biotechnology is defined as a technique that uses living organisms to either produce or modify in to improved plant or animal products for specific uses. Similarly aquatic biotechnology is defined as a technique that uses living aquatic organisms or parts of these organisms to either make or modify products to improve plants or animals or to develop micro- organism for certain specific uses.

1. IMPORTANCE

It has got importance to

1. Upgrade the quality of fish species for enhancing aquatic productivity.
2. Conservation and management of genetic resources of fish stocks.
3. Maintenance of genetic diversity in natural fish population.

2. DEVELOPMENT

The work in India was started with simple inter-specific and inter-generic hybridization, karyotypic studies, polyploidy and also selective breeding. The application of genetic and biotechnology principles in the 80's remain focussed on the temporate and subtropical species of fish. Modern biotechnological tools are used in India to increase production and conservation of aquatic resources for sustainable utilization.

3. STUDY ON GENETIC VARIATION

The pattern of genetic variability in natural and farmed fish population can be identified through the use of genetic markers especially by the use of molecular markers. Molecular markers namely (1) soluble proteins/ isozymes/ allozyme (2) DNA markers are employed to study on genetic variability of fish stocks, migration, nature of breeding population etc. Such information can help in designing aquatic sanctuaries, marine protected area (MPA) and endangered threatened fish species.

3.1. Isozymes

These are analogous but separable forms of enzymes encoded by one or more loci. The isozyme products of two different alleles at the same locus is know as allozymes. Allozymes are heritable and can be detected by electrophoresis by the differences in amino acid composition. For genetic variability in intraspecific population, polymorphic markers are preferred while for interspecific and inter-generic population, less polymorphic markers are preferred. From these data's, the allozyme variability can be analyzed to identify the stocks and resolving taxonomic ambiguity. Isoelectric focussing technique is also employed for allozyme variability study thereby identify the stock and taxonomic position.

3.1.1 Demerits of Isozyme / Allozyme Study

1. Differences in aminoacid composition on electrophoresis is not always adequate for detecting differences between population and individual.
2. Changes in DNA sequence of gene may not result overall changes in the corresponding protein for which many genetic variants escape from being detected by protein electrophoresis.
3. Allozymes are natural markers hence are not under selective pressure on the evolution.

3.2 DNA Markers

DNA markers are based on polymorphism detected at DNA level. Hence these are called as nuclear DNA markers or polymorphic DNA markers. This serves as landmark of anchor loci, which can identify and analyse new loci or gene in the genome (chromosome). The nuclear DNA markers are of type I or type II . Type I markers are monomorphic or slightly polymorphic. But type II are highly polymorphic.

Type II markers are

1. Minisatellites,
2. Microsatellites
3. Random amplified polymorphic DNA(RAPD)
4. Amplified fragment length polymorphism(AFLP)
5. Single nucleotide polymorphism (SNP)
6. Internal transcribed spacers (ITS)
7. Mitochondria DNA (which in maternally inherited, Circular with genes in the 16-18 kilo base pair)

3.3 Characteristics of Marker

1. Polymorphism information content (PIC) is the most important characteristics of a marker.
2. From the allelic frequencies, the variability in the population can be assessed.
3. A PIC value greater than 0.5 is considered highly informative, 0.25-0.5 is reasonably informative and less than 0.25 is slightly informative Other methods such as Restriction endonuclease (RE) banding and fluorescent in situ hybridization (FISH) techniques are used for identification of fish population stock, species and sex chromosomes in fishes.

3.4 Studies In India

1. Studies using allozyme and microsatellite DNA on *Labeo rohita* and *Catla catla* clearly established the existence of sub structuring of natural population in Ganga and Indus river system.
2. Hilsa an anadromous fish in both Ganga and Brahmaputra river system exhibit more or less genetic homogeneity, that in other words it did not exhibit significant genetic heterogeneity.
3. Using microsatellite genetic DNA (type II marker) and RADP, natural fish population stock of many fishes such

as *Etroplus, Puntius, Chitala, Labeo* and *Tor* species have been identified from different river system. Studies also indicated about genetic homogeneity in fish stock of *Mugil cephalus*, penaeid species, Pearl oyster, Indian mackrel, Oil sardine from the West Coast of India. Principal Institutes like NBFGR, Lucknow; CMFRI, Kochi; CIFA, Bhubaneshwar and NIO, Goa have made these studies by using different types of markers to assess about the genetic diversity of fish stocks.

3.5 Genotoxicity assays

Indiscriminate use of insecticides, pesticides and industrial effluents cause mortality or some genetic damages to live ones. Genotoxicity test of these pollutants can be studied by.

1. Micronucleus test.
2. Chromosomal aberration studies.
3. Sister chromatid exchanges. (SCEs)
4. Single cell gel electrophoresis.
5. Comet assay.

However, SCEs is the cytological manifestations of DNA double strand breakage and rejoining at homologous sites between the two chromatids of a chromosome. SEC s analysis using 5-bromo-2 deoxy uridine (Brd U) probe offers a new cytogenetic technique for determining the potential genetic hazards of chemical in the environment. Comet assay is a microelectrophoretic technique used for direct visualization of DNA damages in the individual cell. This assay is used to study the fate and effect of any toxic pollutants in the DNA. Such studies can provide remedial measures for conservation of aquatic biodiversity.

3.6 Sperm Cryopreservation Protocol

Cryopreservation is the ex-situ storage of fish spermatozoa, eggs and embryos without loss of viability in ultra low temperature, such as –196°C in liquid nitrogen.

3.6.1 Techniques

1. Collection of gametes from quality brood fish.

2. Testing of spermatozoa mobility. (Minimum of 70% mobile sperm)
3. Diluting the milt samples with cryoprotectant (dimethyl sulfoxide, DMSO), propylene glycol, ethylene glycol or methanol and filled in French straws (0.5 ml). Each straw containing diluted milt is sealed.
4. Then equilibrating the sample.
5. Exposing the straws over liquid nitrogen vapours for few minits in thermocol chamber.
6. Freezing rapidly and storing in liquid nitrogen.

After thawing, the milt can be activated for fertilizing the eggs.

3.6.2 Importance

1. To maintain high genetic variability
2. Production of seed even in off-season.
3. Rehabilitation of endangered fish species.
4. Improved varieties of cultured fish.

3.6.3 Progress made In India

Cryopreservation protocol has been developed for some species of fish such as (1) Catla (2) Labeo (3) Mrigala (4) Onchorhynchus (5) *Salmo trutta ferio* (6) *C. carpio* (7) *Tor putitora* (8) *Tor khudree* (9) *Ompok bimaculatus*.

Other fish species, under study are (1) Pangassius (2) Silondia (3) Etroplus (4) Schizothorax (5) Barbodes (6) Epinephelus (7) Osteocheilichthys, etc. Cryopreservation protocol provides 65-100% hatching success. But some disadvantages reported is the (1) cryoinjuries of fish spermatozoa at the temperature range of 0-40°C due to heat removal and application of cryoprotectant.

It is necessary to determine the point of equilibrium between cryoprotective efficiency of cryoprotectant and the toxicity tolerance of the cell type to be cryopreserved in order to overcome the draw back of cryoinjury.

The effects of cooling and thawing during cryopreservation may be ameliorated by optimizing the interactive factors such as gamete quality, diluents, cooling and warming protocols, freezing and storage methods.

3.7 Fish Transgenesis

Transgenesis involves the transfer of DNA construct or gene from one species to other. The animal that has a foreign or modified gene integrated in its genome is called transgenic or genetically modified organism (GMO) or living modified organism (LMO). The genetic engineering technique is used for producing transgenic fish.

3.7.1 Scope

1. Identify and construct the gene that have desirable traits.
2. To develop desired fish as per wants of aquaculturist.
3. Improved output/input ratio on basic suitable characteristics of cultivable fish.

3.7.2 Progress in India

The work on transgenic fish production in India was 1st taken up at National Institute of Immunology, New Delhi. Alok and Chelan (1995) successfully transferred human growth hormone gene in to IMC (*Labeo rohita*) using murine metallotheonein promoter (MuMT). Pandian *et al.* (1991) produced a transgenic zebra fish at Madurai Kamaraj University using rat growth hormone gene. Growth hormone gene of rohu and cat fish (*Heteropneustes fossilis*) were isolated, cloned, sequenced and confirmed in prokaryotic and eukaryotic system.

Efforts are made to produce autotransgenic catla, rohu and magur (*C. batrachus*) by using GH gene construct (both c DNA and genomic DNA).

Steps are initiated to isolate and characterize salt tolerant genes from mangrove plants, seaweed and microbes.

3.7.3 Steps for Production of Transgenic

Production of gene-modified organism is a multistage process involving:

1. Acquiring of the gene

Acqisition of an appropriate gene sequence from a gene library together with promoter and enhancer. PCR technique is also used for isolation and amplification of genes of interest.

2. Cloning of the gene

The acquired gene sequence or construct must be inserted into a plasmid or phage vector and multiplied in a suitable bacterial strain followed by harvest of the gene from the bacterial cell.The gene of interest is ligated enzymatically in to bacterial plasmid (construct). Bacteria acts as vector and enables gene to be replicated many times with in bacterial cell. The bacteria are then plated out. The amplified DNA (gene construct) is then cut enzymatically out of the plasmids. Now gene construct is ready for insertion in to eggs of the host species. Use of plasmids as vectors is a good tool in recombinant DNA

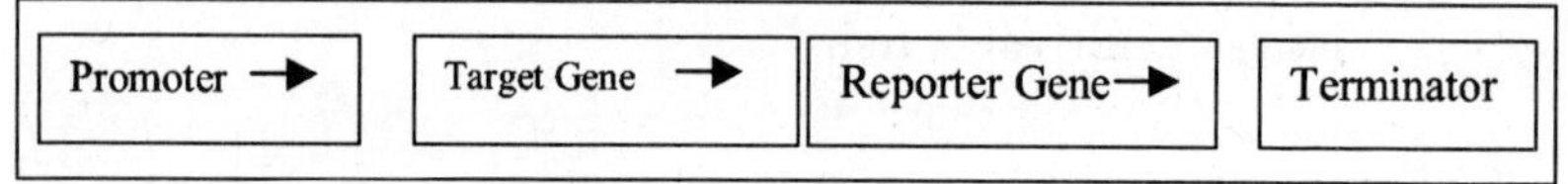

DNA sequence of typical gene for transgenic organism

3. Transfer of cloned gene construct

Now many copies of gene construct are transferred to the genome of recipient organism through several ways that may include.

(a) Microinjection with the help of a micromanipulator.

(b) Sperm mediated gene transfer

(c) Microprojectile - It is ballistic method used in artemia and seaweed species.

(d) Electroporation. In this method the cell membrane of eggs or sperm cell is subjected to high voltage electric field using an instrument gene pulsar. Such electric field shall cause temporary break down and formation of pores in the cell membrane. Through the pores, macromolecules like construct DNA be inserted to produce transgenic fish.

(e) Liposome mediated gene transfer techniques (Nucleic acid → lipid vehicle → liposome → dechorionated cells of recipient eggs)

(f) Retroviruses - use of retroviruses as vectors of construct DNA can be risky hence not advocated.

(g) Use of embryonic stem cells – This method is used for producing transgenic fish. In this method, undifferentiated blastomeres are either totipotent (to grow to a total embryo) or pleuripotent (to grow to a particular organ or tissue). So these can be manipulated invitro and reintroduced in to early embryo, where they can contribute to the germ line of the host. In this way, genes can be introduced or deleted.

4. Assaying for transgenism

The injected gene has to be detected by one of the methods, that include

1. Southern blot
2. Northern blot
3. *In-situ* hybridization

5. Gene expression

The phenotypic and genotypic expression in the off springs can be studied from generation to generation.

3.7.4 Interest Genes for Transgenic Fish

1. Growth hormone gene and growth hormone releasing factor genes.
2. Metallotheonein genes.
3. Crystalline genes- Isolated out of lens of eye of chick has been introduced in medaka fish.
4. Antifreez proteins or glycoprotein (AFGP). Successfully introduced in the genome of Atlantic salmon.
5. DNA vaccines like
    1. IHNV (infectious haemorrhagic necrovirus). Introduced in Atlantic salmon under the control of cytomegalovirus promoter (PCMV).

2. VHS (haemorrhagic septicemia virus)
3. Isozyme gene and RNA interference (RNA i)/anti sense RNA are introduced in penaeids, finfish to fight against viral diseases.
4. Green fluorescent protein (GFP). Isolated from jellyfish for creating color transgenic organism.
5. Red fluorescent protein (REP) . Isolated and cloned from sea anemone. Other fluroscent proteins BFP (blue), YFP (yellow), cyan (CFP) has been identified.

6. **Regulatory gene**- Metallotheonein gene promoter, heat stock promoters, other tissue specific promoters are some of the regulatory gene sequence that play an important role. Care be taken for matching the specific promoter and regulatory sequences.

3.7.5 Benefits of Transgenic Organism

1. Target phenotypes can be achieved through transgenesis
2. Salinity tolerance.
3. Cold tolerance (antifreeze peptide genes).
4. Sterility, feeding behaviour.
5. Predator avoidance
6. Nutritional physiology and energetic.
7. Control of sexual phenotypes.
8. Disease resistance to specifics pathogens
9. Using fish as bioaerator and production of drugs and vitamins from valuable gene.

3.7.6 Issues of genetically modified organisms

1. Gene flow- Gene flow between transgenic and wild population may pose threat to natural biodiversity.

2. Invasive species – Nuisance to environment be caused in introduction by such invasive species. Such species either dominate or could crowd out native fish population in the system. Even if invasive species is sterile but can engage in courtship and spawning behaviour there by disrupt breeding in wild population.

3. Food safety issue and health - Environmental toxins like Hg (mercury) in case absorbed by the GMO, that could pose danger to humans who eat the contaminated one. Increased allergic potential of genetic modified fish over conventional fish and shell fish may pose health problems. Studies in Tulane University has identified major allergens in shrimp.

3.7.7 Biotechnological Prospects in Aquaculture

1. For development and management of fisheries
2. Use of molecular markers to study genetic structure of wild population.
3. Improve fish by selective breeding.
4. Catalogue genetic variability.
5. Protocol development for cryopreservation of gametes.
6. Conservation of threatened and endangered species.
7. Gene bank/gene library.
8. Biosafety issue be handled carefully.
9. Physiological, nutritional and environmental performance of transgenic fish be determined.

In India context, sex reversed/sterile autotransgenic aquatic organism with secured physical containment would be a boon and appear to be an assuring approach in increasing the fish production and food security of the country.

17

ORNAMENTAL FISH PRODUCTION AND MANAGEMENT

1. INTRODUCTION

Ornamental fish keeping is emerging as one of the most popular hobbies across the world. The art of rearing and keeping fish in an aquarium is a very ancient one. It first appeared in China towards the end of 800 BC with gold fish reared in glass bowl. Their simple quality of attraction, color patterns, elegant swimming styles, hi-tech body shapes and their admirable behaviour remain as features that distinguish them from freshwater fish. Ornamental fishes are referred to as the "Living Jewels" due to their colours, shape, behaviour etc. Ornamental fish keeping and its propagation rainbow revolution has become an interesting activity for many providing not only aesthetic pleasure but also financial openings.

The name aquarium was first used by the English naturalist Henry Gosse in 1853. Aquarium is a container made of glass or with glass walls which permits easy and prolonged period of watching of aquatic animals and plants that inhabit in it as well as their care and breeding. A good aquarium is home for planned fish community where the shape, size and lay out are all-important. A rectangular glass tank either made wholly of glass or with a metal frame and glass side and a bottom of glass, slate or other rigid material, of late has replaced the glass bowl of earlier years. An ideal and favorite size of aquarium is 50 X 30 X 30 cm. Large aquarium can also be very attractive with a beautiful planting arrangement, but are expensive and difficult to manage. Tropical ornamental fishes can be housed in smaller tanks as they can survive in water with relatively low levels of oxygen. An aquarium of 50 X 30 X 30 cm. size can comfortably accommodate 6 to 8 numbers of 25-30 mm size gold fish or guppy.

1.1 Catagories of Aquaria

(1) On the manner of construction

(a) Glass aquarium (b) Frame aquarium

(2) On the basis of living habitat of the animal and plants

(a) Fresh water (b) Brackish (c) Marine

(3) On the basis of temperature

(a) Cold water (b) Warm water

1.2 Importance

(1) The aquarium should appeal to the sense of beauty of its viewers

(2) Aquarium should imitate certain natural conditions to which animals and plants are adapted to the maximum extent

(3) The aquarium should teach hobbyists and viewers to know life in nature

Life in the aquarium for the longer periods of time is made possible by adequate measures of care and healthy relationship between animals and plants through a sort of biological equilibrium. For this, proper location, installation, care and water condition for the aquarium plays a vital role. Water plays an important role for ornamental fish keeping in aquarium. Tap water is not directly used in aquarium as it is chlorinated. It is essential to keep it stagnant for a day or two before using it to fill the aquarium. This process is called conditioning and helps to get rid of chlorine. Rainwater, melted snow or distilled water is quite safe for use in aquarium. The ideal temperature for tropical fishes is 20-30°C. Cloudy and greenish ness is problem for aquarium keepers. This may lead to bacteria, fungi growth. Another effective method for correcting such green cloudiness is to place one or two large freshwater mussels in the aquarium.

2. IMPORTANT CONSIDERATIONS

2.1 Location

Aquarium setting is an art by itself. It involves scientific inputs. It needs creative imagination for setting up an attractive aquarium. It is important to choose right place for location and it should be compatible with the living environment, having a natural setting touch up. The location should receive some daylight and perfect illumination for 8-10 hrs in a day. A minimum of two hours of direct sunlight to the location of aquarium is ideal.

2.2 Sand

A layer of sand bed in the aquarium is placed to keep the rooted plants and also to provide decoration. Coarse river sand with variable grain size is the best for aquarium.

2.3 Plants

Plants not only provide decoration and natural setting but also oxygenate the water by photosynthesis. The plants should be clean and free from pathogens before putting in the aquarium. Plants can be dip treated in 2% salt solution or limewater before planting in an aquarium.

The commonly used plants are Eel grass (*Vallisenria spirals*), Sagitaria species, Ceratophyllum, Hydrilla, Verticillata, Ceratopteris, Myriophylum, Cryptocoryne, Amazon etc.

2.4 Aeration

In order to oxygenate the water, aeration is being done by the aquarist. The usual method of aeration is to release air bubbles through porous stones, kept at the bottom of the tank. Corborundum stones give the finest bubbles but they need powerful air pressure. Stones be periodically cleaned and reset to avoid any clogging. Air pumps, blowers, sprays and drips are used as effective methods of aeration.

2.5 Temperature

The favorable temperature for tropical aquarium fishes is 20-30°C. In colder places, the aquaria are heated electrically with thermostat to keep uniform temperature all the time. The thermostats are nonsubmersible and are clipped on to the sides of the aquaria. It has a glass body with the condenser inside it and a control which consists of a small adjustable screw with a non-conducting portion so that the operator can alter the setting of the bimetallic strip and hence maintain the required temperature.

Some two-in-one models of thermostat and heaters are available in the market.

2.6 Feeding

The quantity of feed should be such that it is consumed with in 5 minits and one should siphon out the left over particles. Dried uniform granulated feed particles serves better for aquarium fishes. Diets consisting of brine shrimps, white worm, blood worm, eggs, small pieces of fish, shrimp and clam, frozen foods and vegetable based food are usually provided as feed to ornamental fishes. Live feeds are also valuable. The common ones are daphnia, cyclops, tubifex, infusoria, rotifers etc.

2.7. Filtration system

For aquarium keeping, three types of filtration systems like biological, mechanical and chemical systems can be used. Among the three filtration system the biological filtration system is very important. It is suggested that, a filter with power head is essential in filtration. An external canister filter with small Ultra Violate sterilizer unit is used for small aquaria up to 400 liters. For tanks of capacity more than 400 liters, trickle filter which consist of bio-balls, ceramic media and protein skimmer with U.V unit. In case of big public aquaria exhibits, sand filters with ozonisers and calcium reactors could by critically necessary . Professional use strong water circulation cum filtration system.

3. ORNAMENTAL FISH RESOURCES

The diversified Indian aquatic environment harbours about 2118 species of fishes. Of these about 600 fish species have promising market as ornamental fish. The North East region such as Manipur, Meghalaya, Assam, Himachal Parades, Tripura are the home for around 250 native ornamental fishes. However, it has been reported that so far a total of 217 species have been identified in the Assam state, out of which 150 fishes have ornamental value (Bhattacharya *et al.,* 2003). Out of these, 66 fish species have been identified as of commercial importance as ornamental fishes in the state of Assam (Das *et al.,* 2005). The ornamental fish sector in India include two segments i.e the domestic ornamental fish industry and the export ornamental fish

Industry. During the period 2000-2001 to 2003-2004, MPEDA provided subsidies to 782 breeders in India (Kerla 295, Chennai 214, Mumbai 9 and Kolkata 191) which formed just a small percentage of breeders in India. But Ghosh 2006 cited that more than 500 culture units had come up in the southern districts of Howrah, South 24 Praganes, North 24 Pragans and Hoohly with some units in Midnapur and Maldah as well as in other parts of North Bengal. Bhaskar *et al.,* 1989 wrote on the exotic freshwater aquarium fishes and listed 261 species of egg laying and 27 species of live bearing fishes which has been introduced in India (Sekharan and Ramachandran, 2008)

Charak and Fayaz (2005) reported the native ornamental fishes in Jammu and Kashmir, Kuldip Kumar (2004) reported some of the common aquarium fishes and freshwater aquarium keeping while Christopher (2005) described on marine aquarium keeping. The reports on ornamental fishes of Kerala (Madhusood-ana, 2003), North Bihar (Singh *et al.,* 2006) West Bengal (Gupta and Banerjee, 2008) Tripura, W.B. and Meghalaya (Das and Sinha 2003) illustrates about the vast ornamental germ plasm potential of the country. Ghosh (2006) reported on captive breeding of ornamental fishes. Sunil Kumar (2005) reported on the breeding of angel fish.

Ahilan and Waikhom, 2007; Ajith Kumar *et al.,* 2007 gave detailed note on marine ornamentals in India. The island ecosystems of Gulf of Manner, Gulf of Kutch, Andaman and Nicober, Lakshwadeep and many areas on the main land act as store house for many marine ornamental resources of India. Ziauddin *et al.,* 2007 reported on ornamental fish trade and marketing in India. Further MPEDA hopes to boost ornamental fish exports in coming years.

3.1 Common Aquarium Fish

Some of the common and indigenous freshwater ornamental fishes are listed.

Freshwater

1. Black molly (*Molliensia* sp.)
2. *Molliensia latipinna* (Broad finned Molly)
3. *M. sphenops* (Short finned Molly)

4. *M. velifera* (Giant sailfin Molly)
5. Golden gouramy (Trichogaster)
6. Striped gouramy (*Colisa fasciatus*)
7. Dwarf gouramy (*C. latia*)
8. Blue gouramy (*T. trichopteris*)
9. Black tetra (*Hyphessobrycon* sp.)
10. Zebra fish – *Brachydanio rerio*
11. Guppy – *Lebistes reticulatus*
12. Fighting fish – *Betta splendens*
13. Kissing gouramy – *Helostoma temmineki*
14. Angel fish (*Pterophyllum* sp.)
15. Gambusia
16. Goldfish (Varieties: Peacock tail, Oranda, Calico Oranda)
17. Amphiprion
18. Trichogaster
19. Bass (Moon Bass, Black banded Bass, Rock Bass etc.)
20. Shubunkin (Japanese gold fish)
21. Fancy gold fish (Comets, Moors, Veil tails)
22. *X. helleri* (Sword tail)
23. *X. maculatus* (Platy)

3.2 Indigenous Ornamentals (Tripura)

1. *Chela labuca*
2. *Barilius barila*
3. *Barilius barna*
4. *Barilius bendelisis*
5. *Barilius shacra*
6. *Barilius tileo*
7. *Danio aquipinnatus*
8. *Danio dangila*
9. *Rasbora daniconius*
10. *Amblypharyngodon mola*
11. *Puntius chola*

12. *Philorhynchus balitora*
13. *Botia tario*
14. *Samileptes gongota*
15. *Xenentodon cancila*
16. *Chanda nama*
17. *Pseudambassis baculis*
18. *Badis badis*
19. *Nandus nandus*
20. *Mastacembelus oatesii*
21. *Macrognathus caudiocellatus*

3.3 Assam has several species of ornamental fishes, the most commercial ones are

1. *Lepidocephalus guntea*
2. *Macrognathus aral*
3. *Macrognathus puncalus*
4. *Pesudoambassis ranga*
5. *Puntius conchonius*
6. *Tetradon cutiutia*
7. *Botia dario*
8. *Channa barca*
9. *Hara hara*
10. *Gagata cenia*
11. *Puntius gelius*
12. *Conta conta*
13. *Brachydanio species*
14. *Danio dangila*
15. *Nandus nandus*
16. *Badis badis*
17. *Oreichthys cosuatis*
18. *Puntius sophere*
19. *Colisa sota*
20. *Colisa fasciatus*
21. *Colisa lalia*

3.4.(A) Native Ornamental fishes(Jammu)

1. *Bagarius bagarius*
2. *Barilius bendelisis*
3. *B. vagra*
4. *Botia dayi*
5. *Dani devaria*
6. *Danio rerio*
7. *Esomus dandricus*
8. *Lepidocephalichthys guntea*
9. *Mastacembelus armatus*
10. *Nemacheilus botia*
11. *Puntius ticto*
12. *Rasbora rasbora*
13. *Trichogaster fasciatus*
14. *Tryploplysa yasinensis*
15. *Xenentodon cancila*

(B) Other Freshwater Ornamental fishes In Jammu

1. *Astronotus ocellatus* (Oscar)
2. *Balantiocheilus melanopterus* (Silver shark)
3. *Carassius auratus* (Gold fish)
4. *Gymnocorymbus ternetzi* (Widow tetra)
5. *Helostoma temminck* (Kissing gouramy)
6. *Hemichromis bimaculatus* (Neon jewels)
7. *Hyphessobrycon* sp. (Tetras)
8. *Hypostomus multiradiatus* (Sucker cat fish)
9. *Labeo bicolor* (Red tailed black shark)
10. *Melanochromis auratus*
11. *Poecilia letipinna* (Mollies)
12. *P. reticulata* (Guppies)
13. *Pterophyllum scalare* (Angels)
14. *Symphysodon aequifasciatia* (Discus)
15. *Serrasamus nattereri* (Piranha)

16. *Trichogaster trichopterus* (Blue gourami)
17. *Trichogaster microlepus* (Moon light gourami)
18. *Xiphophorus helleri* (Sword tail)
19. *Xiphophorus maculatus* (Platy)
20. *Puntius tetrazona* (Tiger barb)
21. *Mylosoma plusiventre* (Silver dollar)

3.5 Indigenous Ornamental fishes (North Bihar)

1. *Colisa fasciatus*
2. *Colisa latia*
3. *Colisa chunna* (Honey gouramy)
4. *Lepidocephalus thermalis* (Stripped loach)
5. *Botia dario* (Necktie loach)
6. *Botia almorhae* (Tiger loach)
7. *Noemachilus friangularis* (Banded loach)
8. *Erethistes pushlus*
9. *Tetradon cuticutia*
10. *Alia coila* (Banspatta)
11. *Esomus danricus* (Flying barb)
12. *Chanda baculus*
13. *Chanda nama*
14. *C. ranga* (Indian glass fish)
15. *Puntius ticto*
16. *Pxentiodon cancila*

3.6 Indigenous Ornamentals (West Bengal)

1. *Chhaca chaca*
2. *Chelendon spelenodachneri*
3. *Monotrentus travancoricus*
4. *Somileptus gongota* (Jaguar loach)
5. *Lepidocephalichthys guntea* (panther loach)
6. *Acanthocobitis botia* (leopard loach)
7. *Noemachelius corcia* (spotted loach)

8. *Noemachelius savona* (banded loach)
9. *Noemachelius zonaius* (zebra loach)
10. *Nandus nandus* (Nandas)
11. *Mastacembelus pancalus* (spiny green eel)
12. *Colisa facsiatus* (Drawf gourami)
13. *Chanda ranga* (jewel glass fish)
14. *Aplocheilus panchax*
15. *Amblypharyngodon mola* (mourala)

Gupta and Banerjee (2008) reported on Ornamental fish trade in West Bengal. The Ornamental fishes are as follows:

Gold Fish (*Carassius auratus*) (Varieties-Comet, Fan tail, pearl scale, oranda, common gold fish, telescope eye, blackmoors, Ryukin, Bubble eye, Lionhead, Shubunkins)

Angel (*Pterophyllum scalare*) (Pearl scale angel, Marbel angel, Silver veil tail angel, Silver angel, Gold veil tail angel, Platinum angel, Albino angel, Black veil tail angel, Black angel, Leopard angel, Ghost angel.)

Gouramy - (Varieties-Kissing Gouramy, Blue, Pearl, Golden, Giant, White, Moonlight gouramy)

Barb - (Varieties- Rosy barb, Goldentin foil barb, Albinotin foil barb, Red tail tin foil barb, Tiger barb, Albino tiger barb, Five banded barb)

Tetra - (Varieties- Neon tetra, Widow tetra, Cardinal tetra, Congo tetra, Glow light tetra, Serpae tetra, Black neon tetra, Hockey stick tetra, Rainbow tetra)

Sharks, Cat fishes, Dollar

Oscar- (Varieties- Tiger oscar, Albino Oscar Loach, Sword tail, Platy, Molly, Guppy, Mono angel, Elephant nose

Arowana- (Varieties- Green Arowana, Silver, Golden, Black Arowana

Discus- (Varieties- Snake skin Discus, Powder blue, Red diamond, Leopard snake skin Discus, Albino, Blue, Brown Discus)

Parrot cichlid- (Varieties- Red, Green, Yellow, Blue, Bleeding heart parrot

Koi carp (*Cyprinus carpio*)

Siamese fighter fish – (Varieties- Royal blue veil tail, Green veil tail, Cherry red veil tail, Cambodian veil tail type.

3.7 Meghalaya has several species of ornamental fishes dominated by *Danio* sp., *Loaches, Puntius* and *Channa* species. Some of the food fishes like Murrels (*C. gachua* and *C. punctatus*), Climbing perch (*Anabas testudineus*), Baligirida (*Glossobius giuris*), Singhi (*H. fossilis* and *Mystus* species), chitala (N. chitala and *N. notopterus*), Kalabansi (*L. calbasu*) etc. are also reported as ornamental fishes by some authors. Moreover, Pacu (*Piaractus brachypomus*) an acid loving exotic finfish unauthorisely introduced in Tripura is more known as an ornamental fish world wide.

3.8 Indigenous ornamental fishes (Kerala)

Important indigenous ornamental fishes of Kerala include 10 species of barbs, 6 species of loaches and also cat fishes (Madhu soodana, 2003). Some are listed as below:

1. *Puntius denisonii* (Barbs)
2. *Puntius fasciatus*
3. *P. auritius*
4. *P. filamontosus*
5. *N. denisonii* (loach)
6. *Lepidocephalus thermalis*
7. *Botasio travancoria* (cat fish)
8. Mystus
9. Zebra fish
10. Killi fish
11. Rasboras
12. Glass fish
13. Leaf fish
14. Pipe fish
15. Flying barb
16. Goby
17. Malabar puffer and
18. Scats

4. BREEDING

For captive breeding of ornamental fishes, it is essential to know the sexual dimorphism and breeding behavior of fish. Ornamental fishes are sexually dimorphic as well as isomorphic. Males of many species are more colorful, larger in size and have more elaborate finnage. Vivid sexual dimorphism can be seen in live bearers such as platys, swordtail, gouramis and lake malawi cichlids.

In sexually isomorphic species, minute sexual differences are seen. Very often, the males have slim body with relatively less body depth compared to female. Females have rounded abdomen (tetras, characins, danios, rasboras etc). Some sexually isomorphic species have no identified external sexual differences. The only way to identify sex is during pairing and subsequent chasing of females by male. Examples are gold fish, barbs, some characins.

4.1 Modes of reproduction

Various modes of reproduction is seen in aquarium fishes. Some are pond spawners as they bred in natural pond environment. Other modes of reproduction are as follows.

1. Spawners oriented to hormone inducement.
2. Any time and sort period spawners.
3. Dedicated spawners

4.2 Catagories based on breeding habits

Ornamental fishes are mainly grouped in to two categories, such as livebearers and egg layers. However, egg layer category is further subgrouped as,

1. Egg scatter with adhesive eggs
2. Egg scatter having nonadhesive eggs
3. Egg depositors
4. Mouth brooders
5. Bubble nest builders
6. Egg burying ones.

4.2.1 Live bearers

The most common examples of livebearer are the guppy, molly, swordtail, platy and gambusia. There are 14 families of livebearers and there is diversity in the way the fry are born alive. Some livebearer is either ovoviviparous or viviparous.

The guppy, molly, swordtail, platy and gambusia are ovoviviparous where as the members of the family goodeidae (Goodea species) belong to viviparous category. Rath (1987) has reported the breeding behavior of red swordtail (*Xiphophorus helleri*). Felix (2006), has reported about live bearing ornamental fish production as a new avocation for women self help group. He emphasized the live bearer fish belonging to the family poecilidae are commercially important. Guppies, molly, swordtail and platy are commonly available live bearer ornamental fish. Live bearers are suitable because of (1) Small sized for breeding and handling (2) Live in shoals (gregarious, sociable and peaceful) (3) they are hardy and thrive on alkaline waters and (4) prolific breeders.

Guppy belong to genus poecilia, the important ones are as follows:

1. *Poecilia reticulata* (guppy)
2. *P. velifora* (sail fin molly)
3. *P. latipinna* (sail fin molly)
4. *P. sphenops* (short fin pointed mouth molly)
5. *P. formosa*- hybrid

Genus Xiphophorus includes varieties of ornamental live bearers. The important ones are as follows:

1. *Xiphophorus helleri* (sword tail)
2. *X. maculatus* (platy)
3. *X. variatus* (platy)

Ornamental mollies have color variants, like albino, red, chocolate, orange, marble, speckled, black, white, white and black and fin variants such as moon or round shaped caudal fins.

Ornamental swordtail have also color variants like green, red, red eye, albino, black and fin variants such as lyre tail, hi fin lyre tail, way tail.

Ornamental platy have color variants like red, blue, green, gold, black, bleeding heart, salt and pepper, sunset and rainbow and fin variants such as way tail, lyre tail.

4.2.1.1 Food and Feeding : Live bearers are omnivorous . Diet consisting of brine shrimps, white worm, blood worm, small pieces of fish, shrimp and clam, frozen foods and vegetable based food serves best for live bearers.

4.2.1.2 Water quality: The pH (7-8), temperature (27-29^{oC}) and hardness (100-150 ppm) of water is suitable for live bearers.

4.2.1.3 Maturity: Livebearers take 4-6 months to mature, however guppy and platy mature within 2 months.

4.2.1.4. Sexual dimorphism: Distinct sexual dimorphism is noticed in live bearers which is given below.

MALE	FEMALE
(1) Smaller in size.	(1) Larger in size.
(2) Brightly colored.	(2) Dull color.
(3) Dorsal and caudal fins larger.	(3) Comparatively smaller.
(4) Belly flat.	(4) Belly bulged.
(5) Anal fin is modified in to gonopodium	(5) Anal fin normal in shape.

In live bearers, fertilization is internal and gestation period is 4 to 6 weeks. After the end of gestation periods, youngs are released due to sexual play. In platys. maximum of 50 young ones can be expected. In molly maximum of 100 and in swordtail maximum of 200 young ones can be expected. After 2-3 days, the female again becomes pregnant even without the contact of male (Felix, 2006). The young fries can be fed with finely crushed dried food, brine shrimp nauplii for further rearing.

Ghosh (2007) reported parthenogenesis in golden guppy (*Poecilia reticulate*). Parthenogenesis is an unusual occurrence in live bearers. The widely known example of "pseudogamous parthenogenesis" is *Carassius auratus gibelio*. It is also reported that in golden guppy, the number of fry produced was more in higher temperature (37°C) than at lower temperature, PH varied from 7.8 to 8.2 and hardness varied from 300-350 ppm.

4.2.2 Egg layers

The most common examples of egg layers are the gold fishes, tetras, barbs, zebra fishes, koi carps, gouramis, angel fish, siamese fighter etc. Because of their different breeding habits these are catagoried as follow:

4.2.2.1 Egg scatters with adhesive eggs: Common examples are tetras, barbs and gold fishes. Rainbow fish also lays adhesive eggs.

4.2.2.2 Egg scatters with non adhesive eggs : Common examples are *danio* spp. and *brachydanio* spp.

4.2.2.3 Egg depositors : Common examples are rasboras, angel fish, other cichlids and catfish (Corridoras species). Egg depositors either show parental care or do not care the offspring. The parental care is seen by most of the cichlids and catfishes. Guarding the offspring either in open water (open spawners) or protecting them in mouth cavity or body cavity (cavity spawner) is the salient features of this group of ornamental fishes.

4.2.2.4 Bubble nest Builder : The ornamental fishes like Gouramis, fighting fish deposit their eggs by building bubble nest. Male use to pick up the released eggs of female by mouth and put it in the bubble nest. The female is taken out immediately after spawning because male drives the female away or otherwise shall chase the female till to death. The details of breeding behaviour of aquarium fighting fish (*Betta spelndens*) is reported (Rath, 1987).

4.2.2.5 Mouth brooders: It includes some cichlids in which female use to take care of the fertilized eggs by carrying them in mouth. Two categories of mouth brooders are distinguished.

(1) Egg loving mouth brooders are catagoried as ovophile, the examples are some cichlids and some of the labyrinth fishes.

(2) Larvae loving mouth brooders are catagoried as larvophile, the examples are some earth-eater.

4.2.2.6 Egg Buryers : Some of the ornamental fishes deposit their eggs in a peat moss substrate. Killi fishes are the members of this group. In this case, the eggs remain viable and fertile for more than a month even the peat moss substrate dries up during summer. In order to initiate hatching the stored peat moss substrate is immersed in soft water with vigorous aeration. For conditioning the soft water, 1-2 drops of methylene blue in added.

4.2.2.7. Common Ornamental Fish (Egg layers): **Angel fish:** The commercially important egg layer ornamental fish is the angel fish and gold fish. For production of this fish, the knowledge on breeding behaviour is essential. Breeding of angel fish has been receiving considerable attention. Angel fishes are one of the most popular of all tropical aquarium fishes. Color variants of angel fish such as black, white, blue king, semicircle, coral beauty and green are some of the common varieties. It prefers water temperature (22-30°C), feed on worms, insects, plant materials, dried food etc. They attain maximum size of 15 cm in length. It matures in one year and breeding season extends from June to September. Sexual dimorphism in angle fish is bit difficult. An 8-10 month old angle fish serves as brood fish. Parents show parental care. The eggs are deposited on slate piece. The young ones got detached from the slate piece after few hours. One-week-old hatchling swim in the tank. They thrived on the yolk deposit for a few days. These hatchling were carefully removed and reared in nursery tanks, fed on infusoria and mosquito larvae. After becoming fry, they are fed with pelleted feeds. Considering the popularity of angle fish and gold fish, rural self help women and urban women can find new employment and also generate additional income. Jha and Barat (2007) studied on the involvement of women self help group of Darjeeling in ornamental fish culture especially of angel and gold fish.

Discus

Nair *et al.,* 2008; studied on the breeding and culture of Discus (*Symphysodon aquefasciatus*). It is a egg layer ornamental fish belongs to family Cichlidae. The discus is essentially a "Round disc shaped" fish whose natural habitat is the Amazon basin of South America. Discus is the only species of fish that is born with the ability to excrete food in the form of mucus from whole body to feed their young, quite similar to our human breast feeding. In human only mother can produce milk where as in Discus both parents can produce this food for their babies. This makes Discus the only living thing on this world gifted with this unique capability.

Discus eat small shrimp, insects and insect larvae, small fish and fair amount of plant material. Dorsal and anal fin of male are pointed or some what extended on the end rear of the dorsal fin. Females tend to have round rear fins. Male Discus have less intense color but have more pattern while female tends to be more colorful but with lesser pattern. Male Discus tend to be bigger than the females. To keep Discus disease free, avoid stress in aquarium.

Oscar

Oscar (*Astronotus ocellatus*) is a tough and hardy freshwater aquarium fish. It belongs to Cichlid group. They are known under a variety of common names such as Tiger Oscar, Velvet cichlid or

Marble cichlid. Oscars are territorial, aggressive and should not be kept with other fish. Silver dollar, Pleco and Jack Dempsey fish of more or less similar size are some of the acceptable tank mates of Oscar. They can thrive in wide range of physico-chemical conditions of water. The condusive temperature of water is 22°C – 27°C with pH range of 6–8. It attains a maximum size of 13 inches and its life span is 10–13 years. It is the native of Amazon. Oscar prefer hiding places. To keep them stress free, plenty of decorative plants and gravels are put in Aquarium.

Oscar feeds on wide varieties of food including pellets, flakes and live foods. They are primarily carnivorous, even preying upon smaller fish. Blended beef heart, earthworm, shrimps, peas and lettuce constitute good foods to keep Oscar on a balanced diet.

Oscar attains sexual maturity in one year at the length of 4 inches and form life long pairs. Sexual dimorphism in Oscar is difficult. Usually female is smaller and less colorful than the male of the same age. Males have dark blotches on the base of dorsal fin. When the female is ripe and pair are choosen by each other, the ovipositor of female is distinct.

Breeding of Oscar is easier than other aquarium fishes. Singh *et al.,* 2008 reported the breeding and aquarium keeping of Oscar fish. In captivity, Oscar choose its own mate. Oscars build mounds of gravel and chase each other during spawning. The female lay eggs on a flat rock and fertilize them. Oscar lay 1000 eggs at a time. The eggs are at first opaque and then turn transparent in 24 hours. The young ones hatch out after 2 to 3 days. Like most cichlids, Oscar practice brood care, although the duration of brood care in wild is not known.

Maintenance of water quality is extremely important for breeding and preventing diseases in fish. A 25 to 30% daily water exchange is ideal. White spot (Ichthyophthirius) and Greyish cotton patches on skin (Fungus) diseases are noticed in Oscar due to poor water quality in aquarium. Hence one must treat the tank with appropriate disinfectant/ chemical at the first notice of disease outbreak.

4.2.2.8. Fancy Ornamental Fishes: A few ornamental fishes are very fancy and preferred by the aquarists over the other species even though they are costly (Venkataswamy and Athithan, 2008). This group includes (a) Arowana, (b) Flower horn cichilid, (c) Parrot cichlid and (d) Glow fish.

(a) Arowana: Arowana is an exotic freshwater aquarium fish. It is also known as "Dragon fish" and considered to be an other "Peng Shui" fish bringing luck, wealth and prosperity.

Various species of Arowanas are reported. These are

1. Asian arowana (*Scleropages formosus*)

2. Australian arowana (*S. jardinii*) and

3. South American arowana (*Osteoglossum bicirrhosum* and *O. berreirai*) as per their geographical distribution. Variety colors of ornamental fishes like Green, Silver Golden and black color arowanas are marketed by using carotenoids and lipochromes through feed. Such type of colour enhancers for ornamental fishes are reported (Alan and Felix, 2007). Spirulina can be used as red color enhancer. Cyclopeeze derived from copepod can be used as colour enhancer. Color variations are noticed in Arowana. Basically three main varieties occur in Asia viz. gold, red and green. Gold varieties are sub divided as cross back golden and red tail golden. Under the red, there are two sub varieties, blood red and chilli red. Under the green, there is only one type. There is still another natural variety called yellow tail. The details of culture and breeding of Arowanas is studied by Ahilan and Priyadarshini, 2008.

The fish attains maturity in the age of four years at 45 -60 cm length. Generally males are larger than the females. Females are smaller and have a more rounded body. The fish spawns through out the year with peak between July and December. During courtship, the male chases the female around the perimeter of tank and the pair circles each other nose to tail and female releases a clutch of orange red eggs. The suitable water temperature for spawning is 27-29°C. The male fertilizes the eggs and scoop them in to its mouth where it incubates them until the fry can swim and survive independently. Hence arowana are mouth brooders.

The arowana fish is strongly territorial and aggressive in nature. Hence, the fish should be kept in groups of 6 -10 fish to sub-due their aggressiveness. It is a good jumper and surface dweller, respiring on atmospheric oxygen besides normal respiration. They are fed daily with meat based live food such as freshwater prawn and chopped fish meat. Pellet containing 32% crude protein may be given as supplementary feed. Feeding at the rate of 2% of body weight per day can be given. Common diseases in arowana are fin rot and cloudy eye. These can be prevented by saline bath (1% salt) and 30-35% water exchange. Infection of anchor worm (Lernea) is also noticed in arowana.

(b) Flower horn Cichlid: Cichlids are the most common and widely distributed freshwater ornamental fishes present in Africa, America and Asia. Oscar and Discus are other representatives of the family cichlidae that are popular among the tropical aquarium keepers. The smallest of all cichlids is the *Apistrogramma cichlid* that grows to a maximum of 3 cm, where as the *Boulengerochromis cichlid* is the largest attaining a maximum length of one metre.

Flower horn cichlid is a fancy breed of hybrid cichlid developed in Malayasia in 1990. The presence of a hump, which is well developed in adult is the characteristic features of this species.

Flower horn cichlids are judged by the redness of the eye, length of fins and bright colors. There prevails a Feng Shui belief among the aquarists that the flower horn cichlids with right color marking will bring happiness, good luck and harmony. These are territorial and aggressive fishes. They prefer to feed on worms, crabs and small shrimps and readily accept pelleted feeds. Water exchange is made to keep these fishes healthy.

(c) Parrot Cichlid: The common colors of parrot cichlids are yellow, red, blue, green, pink, violet and multicolored. The blood parrot cichlid is one of the most beautiful cichlids that was developed in Taiwan in 1986. The blood parrot cichlid is a man made hybrid. The fish was evolved through a cross breed between South American cichlid and the midas cichlid. Blood parrots are shy fishes and can be kept with other varieties of peaceful fishes. Mid sized tetras, danios and angel fish are compatible. It feeds on variety of foods including flake, live and freeze dried foods. Sinking foods are easier for them to eat than floating foods. Parrot fish can grow to 8 inches in length and can live for several years.

(d) Glow fish: It is an ornamental fish that glows during darkness. They are transgenic, genetically modified freshwater ornamental fishes incorporated with genes of jellyfish and sea anemone. A Taiwanese company has created glowing zebra fish in 2001. The light emitting colors inducted in fishes are red, green, yellow or orange. Dr H.J. Tsai of National Taiwan University has developed such genetically modified Glow fish by incorporating a fluorescent protein from the bioluminescent jellyfish and attached it to fish embryo DNA. However, trading of glow fish is controversial as this may hamper the biotope.

5. STRATEGIES FOR DEVELOPMENT

1. State government need to prioritize ornamental fish culture sector as an avocation for employment generation especially for women inhabiting in rural and suburban areas.
2. Awareness activities in respect of aquarium setting, keeping, maintenance, species of fishes, their behaviour, life history, traits and compatibility of selected species are to be initiated.
3. Popularization of indigenous ornamental fishes in domestic and export markets.
4. Organize seminars, symposium, workshop and exhibition on ornamental fishery potential etc.
5. Status of ornamental fish breeding and culture .Species used for culture are mostly egg layers such as gold fish, angel, gouramis, and Oscar. Among live bearers, molly, guppy, platy, sword tail etc are preferred for culture.
6. Schemes on ornamental fish breeding and culture be initiated, promoted and implemented by Government, MPEDA etc. MPEDA has opened a office at Guwahati to encourage local youth and fishers to develop this sector. It provides 10% assistance on the FOB value of the ornamental fishes exported to other countries. 50% subsidy is given on loans taken to setup a unit up to Rs 40,000/-.
7. For breeding, the needed infrastructure facilities have to be set up, supported by relevant technical knowhow.
8. Rearing of commercially important ornamental species can be undertaken in recirculation and flow through water system to maintain good water quality and to stimulate natural running water condition.
9. Technologies on the production of live fish food and nutritionally balanced dry feed in pellets, powder, flakes, microcapsules etc should be developed so that they can be extended to the hobbyists and entrepreneurs.
10. MPEDA in Chennai on 16th May 1997 suggested

(a) To set up commercial ornamental fish production farms with infrastructure and technical experts.

(b) Ornamental fish culture needs to be provided with facilities and concession as given to agricultural sector.

(c) Training programme on different aspects of aquarium management be taken up.

(d) Small-scale farmers need to be assisted.

(e) Farm be equipped with fish pathology laboratory.

(f) Health centres for quarantine certification.

(g) Creating awareness on marketing of ornamental fish.

6. BRACKISH WATER ORNAMENTAL FISHES

In Brackish water, some of the fish species have great ornamental value. Some of these fishes are listed as follow:

1. Scat (Scatophagus argus)

Family- Scatophagidae

(a). Deep bodied fish with metallic yellow color having dark brown black spots.

(b). Inhabits in brackish water or salt water but can thrive in freshwater.

(c). It grows to a level of 13-25 cm in aquarium although in open water its growth is 30cm.

(d). Feed on aquatic plants, boiled egg, corn flakes and other aquarium fish food.

2. Pearl spot (Etroplus suratensis)

Family- Cichlidae

(a). Deep bodied light green colored fish with 6-8 vertical black bands and having white pearly spots all over the body.

(b). Grow to 40cm in natural water bodies .Smaller size are preferred to be kept in aquarium.

(c). Omnivorous fish and move in school.

3. Orange Chromide (Etroplus maculatus)

Family - Cichlidae

(a). Yellow body with greenish back have orange color on breast. Have 3-4 blotches on body.

(b). Attains a maximum of 10cm in length.

4. Therapan (Therapan jarbua)

Family - Theraponidac

(a). Silvery white fishes with 3-4 dark brown or black down wardly curved longitudinal stripes on the body.

(b). Dorsal fin have large black spot, caudal with dark tips and three oblique or horizontal stripes.

(c). Can remain with barbs, bluemorph etc in aquaria.

(d). Feed on prawn meat, earthworm.

5. Glass fish (Ambassis species).

Family – Ambassidae

These are elongated and laterally compressed fishes with glassy or semi-transparent body. They move in shoals and good to be kept in aquarium.

6. Needle fish (Strongylura species)

Family - Belonidac

(a). Elongated sub cylindrical or laterally compressed body

(b). Upper and lower jaw elongated to form beak which are armed with teeth.

(c). Dorsal and anal fins are posterior in position

(d). Carnivorous, can be kept in aquaria with other compatible species.

7. Puffer fish (Tetradon species)

Family - Tetradontidae

(a) These are yellow or green-bodied fishes with dark spots on them.

(b) Hardy, omnivores.

(c) Compatible with fast moving fishes like tetras, barbs etc.

Other brackish water ornamental fish includes *Monodactylus argenteus* and *M. sebal.*

7. MARINE ORNAMENTAL FISHES

Fish varieties such as butterflies, angels, other tangs, lion fish, triggers, eels, boxfish, crow fish etc together could be kept but the selection of species should be relevant to their size and their aggressive nature. As for instance, while selecting lion fish, triggers, groupers, eels etc., small fish must be avoided, as the former ones are aggressively carnivorous. Hence it is valid to keep them with other fishes of equal size or having slightly bigger size. Commercially important marine ornamental fishes are listed below.

1. *Amphiprion ocellaris*- Clown fish
2. *Zanclus canescens*- Moorish idol
3. *Holocanthus rimaculatus* – Marine
4. *Chaetodon unimaculus*- Butterfly fish
5. *Holocentrus diadema*- Squirrel fish
6. *Acanthurus glasucoparelus*- Surgeon fish
7. *Pterios volitans* – Lion fish
8. *Canthigaster margaritatus* – Puffer fish
9. *Dascyllus carneus* – Damsel fish
10. *Odonus niger* – Trigger fish
11. *Coris gaimardi* – Wrasse
12. *Nemateleotris magnifica* – Fire fish
13. *Oxycirrhiteus typus* – Hawk fish

Murugan *et al.,* 2008 reported marine ornamentals fishes in the Andaman waters which includes

(1) Butterfly Fish (*Chaetodon vagabundus*)

(2) Silver Moony (*Monodactylus argenteus*)

(3) Blue and Gold Fusilier (*Caesio caerulaurea*)

(4) Conviet surgeon fish (*Acanthurus lineatus*)

(5) Blue striped surgeon fish (*Acanthurus triostegus*).

Pillai 2008 cited that the marine ornamental fishes for which breeding and seed production technologies were developed by CMFRI are the

1 *Amphiprion sebae*

2 *Amphiprion percula*

3 *Amphiprion ocellaris*

4 *Premnas biaculeatus*

5 *Pomacentrus pavo*

6 *Pomacentrus caeruleus*

7. *Neopomacentrus filamentosus*

8 *Neopomacentrus nemurus*

9 *Dascyllus trimaculatus*

10 *Dascyllus aruanus*

11 *Chromis viridis*

8. AQUARIUM SETTING

Marine aquarium is relatively expensive compared to fresh-water aquarium. Easy steps are listed for setting up a marine aquarium. These are:

1. Place the tank at the selected location with the decorated base covering.
2. Place an under gravel filter.
3. Fill with coral gravel at least up to 3-inch height.
4. Add coral and rock decorative.
5. Place power heads and internal filter.
6. Fix external life support system.
7. Fill with clean freshwater and make it saline with artificial salt (specific gravity –1.019- 1.023).
8. Fix hood, light units.
9. Start the system and tune all the equipment at required level.
10. Add water conditioning additives and bacteria enzymes for initial maturation of aquarium water.

The best practice of getting matured system is the introduction of few small blue damsel fish as they are hardy and withstand high nitrite level due to nitrogen cycle in the aquaria. Adding small quantities of used gravel from another aquarium is also advisable for enhancing the maturation process. Gradual addition of fish and study on water quality will keep the aquarium fresh and lively year round.

8.1 Water Quality

For marine aquarium, the water quality should be as follows.

PH	– 8-8.5
Specific Gravity	– 1.019-1.023
NH_3	– 0 ppm
Nitrite	– 0 ppm
Nitrate	– <20.00 ppm
Phosphate	– 1 ppm
Copper	– 2.0 ppm
Temperature	– 24-28°C

8.2 Operational Management

(a). Feeding, observing fish behaviour and recording water temperature is the daily routine for aquarium management.

(b). With passage of time, the glass surface be cleaned from algal growth. pH, Nitrite, Nitrate, Phosphate are checked and water exchange is made if needed.

(c). The aquarium is checked for cleaning of filters, checking of electrical items, remove protein skimmer and clean internal surface, clean external canister filter in salt water and replace the carbon.

(d). Within sixth months, the U.V sterilizer tube light be changed and renew lighting bulbs, if necessary.

8.3 Feed management

Provide proteins rich feed (40-50% protein) in dry weight basis to fishes for better health condition.

Species of fish	Food items
Clown Tang Sailfin Tang Purple Tang	Algae, Lettuce and Artemia as feed.
Angel varieties	small shrimps, krill artemia, blood Worms, tubifex etc. marine feeds can can be offered.
Damsel Clown	blood worms, chopped clams.

Buller flies	selective feeding on artemia, blood worm
Lionfish and triggers	frozen peeled shrimps, small whole fish

8.4 Disease Treatment

The out side fish must be kept for two weeks in a separate treatment tank before transferring to main aquarium tank. The common marine aquarium fish disease is the white spot caused by Cryptocoryon irritans in salt water. Copper based medication is effective for control of white spot disease (Christopher, 2005). In freshwater, the common diseases in ornamental fishes are (1) Saprolegniasis (2) Branchiomycosis (3) Mycobacteriosis (4) Whirling/Ichthyophoriasis and (5) Exophila (Alan and Felix, 2008). Even Bacterial and Protozoan diseases are noticed in ornamental fishes (Mishra and Mishra, 2007). Common chemical solution bath such as Sodium chloride, Methylene blue, Copper sulphate, formalin, malachite green, potassium permanganate, Hydrogen peroxide or localized treatment of Povidone Iodine or mercurochrome is made for preventing the diseases in ornamental fishes. Antibiotics like Chloromycin and Gentomycin are found effective if the infection is characterised by ragged fins and redness at the base of fins.

8.5 Hatchery Production of Marine Ornamental

In recent years, the centre of advanced study in Marine biology, Annamalai University, Tamil Nadu and ICAR has intensified research on brood stock development, breeding, seed production, sea ranching for commercially important ornamental fishes. Technology on breeding and seed production of Clown, Damsel, Sea-horse has been developed and it is being transferred to coastal communities and other entrepreneurs. Studies on other marine ornamental are in progress.

9. ORNAMENTAL FISH TRADE

Based on the global marine data base (GMDB), the annual global trade of various groups has a range between 20 and 24 million fish, 11 to 12 million of corals, 9 and 10 million of other ornamental invertebrates (Gopakumar, 2007). Most of the world's supply of ornamental fish is from Asia. Among the Asian countries, Singapore is the largest supplier of ornamental fish. The USA is the largest single buyer of ornamental fish in the world followed by European Union and Japan. World export of ornamental fish accounts to 282 million US dollar during2005-06 (FAO, 2006). The export of ornamental fish in India accounts 1.2 million US dollar, the share being 0.4% of the world ornamental export. To increase export earning from India, there is a need for increasing capacity building (Sekharan and Ramachandran, 2008).

There are about 400 species of marine ornamental fishes belonging to 175 genera coming under 50 reef families occurring in the Indian seas (Ajith Kumar *et al.*, 2007).

The most common marine ornamental are clown, damsel, wrasse, lion, surgeon, cardinal, bat, butterfly and angel etc. It is important to develop strategies to collect these fishes in an eco-friendly and sustainable manner to avail of trade opportunities. Even though we have a very good stock of innumerable varieties with colorful designs, we are yet to perfect the technique of exploiting them in a sustainable way, with out causing damage to the productive coral ecosystem.

Utilization of tank reared fish and other ornamentals for promoting trade in live condition would be the long term solution for sustainable development. India has got the potential to emerge as one of the major source countries for a sustainable marine ornamental trade. This is possible only when appropriate policies for wild collection of species is enforced and production of ornamental through hatchery technologies for selected species are standardized and streamlined.

10. CONSERVATION STRATEGIES

1. The management of marine ornamental fisheries be given due importance in such a way that they are made biologically sustainable.
2. Care should be taken to avoid conflict with other resources and decrease post harvest mortality.
3. Establishment of marine reserves, where the collection of marine ornamental are illegal if needed and this may help in reducing this conflict.

4. Other measures like size limit, catch limit through the use of permits can reduce the pressure on marine resources.
5. The major problems in marine ornamental fish trade are
    1. Destructive collection practices.
    2. Introduction of alien species.
    3. Over exploitation.
    4. Lack of scientific information on many species collected and the threats of extinction on target species.

11. MARINE AQUARIUM COUNCIL (MAC)

There are increasing efforts to collect live specimens and to culture marine ornamental due to the growing population of aquaria world wide (Pet Industry Joint Advisory Council, 1999). There is an international effort to design a certification process, which would convince buyers. MAC is a nonprofit organization that is promoting the "certification for quality and sustainability in the collection, culture and commerce of marine ornamentals (MAC News, 2001).

12 ROLE OF MPEDA

1. MPEDA is working towards the goal by developing captive breeding technology for indigenous varieties of ornamental fish.
2. Capacity building of ornamental fish breeders
3. Standardizing methods for diagnosing viral diseases in ornamental fishes.
4. Cluster development and facilitating access to cheaper credit.

MPEDA in collaboration with the project "PIABA- conservation and management of ornamental fish resources of Rio Negro Basin" Amazon, Brazil shall take up ornamental fish breeding and export in coming years. MPEDA expects that India's ornamental fish shall get green certification.

13. CONSTRAINTS/ BOTTLE NECKS

1. Ornamental fish industry in the country is scattered and not well organized.
2. Low production volumes.
3. Inadequate transport facility, high airfreight charges and lack of packing.
4. Nonavailability of quality brood fish and lack of suitable low cost breeding technology.
5. Unavailability of funding
6. Absence of local exporting agencies.

Despite of having huge potential, the export of ornamental fish from India continues to remain negligible. Ornamental fish exports have increased from Rs. 3.2 Crore in 2001-02 to Rs. 5.6 Crore in 2006-07, but it is quite insignificant compared to the export of food fish, which is worth Rs. 8000 Crore (2006-07) per annum. Presently about 90% of the ornamental fish export was based on wild collection. Capture based export is not sustainable, hence focus be given on culture based development.

REFERENCES

Aalan Brindo, R. and N. Felix 2008. Fungal Diseases of Freshwater and Marine Aquarium Fishes and Their remedial measures. Aqua International 15(10): 22-23.

Ahilan, B and A. Nithiya Priyatharshini. 2008. Culture and Breeding of Arowanas. Aqua lnternational 15 (10) : 15-16.

Ahilan, B. and G. Waikhom 2007. Some prime Marine Ornamental fishes of India. Fishing Chimes 27(7):50-52.

Ahmed, N. 1948. J. Bomb. Nat. Hist. Soc. 47 : 586-602.

Ahmed, S.H. and A.K. Singh 1999. Integrated Aquachestnut-cum-fish culture. Fishing Chimes, 18 (11) : 35-45.

Ahmed, S.H. and A.K. Singh, 1997. Prospects of integration of "Makhana" (*Euryale ferox*) with fish culture in North Bihar. Fishing Chimes, 16 (10): 45-50.

Ahsan, S.N. 1966. Some effects on temperature and light on the cyclical changes in the spermatogenic activity of the lake chub, *Couesius plumbeus* (Agassiz). Canad J. Zool. 44 : 161-171.

Ajith Kumar, T.T., P. Nithya Jeniffer, P. Murugesan and T. Balasubramanian, 2007. Marine Ornamentals in India: Challenges and Opportunities for sustainability. Fishing Chimes, 27(6):44-48.

Akhtar, Md. Shahbaz and Ciji Alexander. 2008. Single cell proteins in Aqua feed. Aqua international 16 (7): 33-36.

Akhtar, S. Md. and W. Chimwar. 2007. Fish Transgenesis: Realities and Issues. Fishing Chimes, 27(9): 37-40.

Alabaster, J.S. and R. Llyod 1980. Water quality criterion for freshwater fish. Butterworth, London, Boston, 297 p.

Alan, Brindo, R. and N. Felix. 2007. Color enhancers for Ornamental Fishes. aqua international 15(5): 34-36.

Aldernman, David, J. 1988. Fisheries chemotherapy : A Review* in Recent Advances in Aquaculture. Vol. III Eds. James F. Muir and Ronald. J. Roberts. Croom Helm London and Sydey. Timber Press. Protland, Oregon. pp. 1-62.

Algarswami, K. 1995. Strategies for sustainable shrimp farming and transfer of technology. Proceedings of National Workshop on TOT for sustainable shrimp farming CIBA, Chennai. 29-42 p.

Alikunhi, K.. 1956. Fish culture techniques in India. In Progress of Fisheries Development in India. CIFRI and Deptt. Fish. Orissa 63-73.

Alikunhi, K.H. 1957. Fish culture in India. Fm. Bull. Indian. Coun. Agric. Res. 20 : 144.

Alikunhi, K.H. 1966. Synopsis of biological data on common carp *Cyprinus carpio* Linn. 1758 (Asia and the Far East), FAO Fish Synops 31 (1) : Pag. var.

Alikunhi, K.H. Datta, A., Singh, V.D., Singh, A. and Dubey, G.P. 1964. Observations on the breeding of carps in bundhs, near Nowgong, Madhya Pradesh, during Jly-Aug 1964. Bull. cent. Inst. Fish. education. Bombay 1 : 22.

Alikunhi, K.H. Sukumaran, K.K. and Parmeswaran, S. 1963. Induced spawning of the Chinese grass carp, *Ctenopharyngodon idellus* (C and V.) and the silver carp, *Hypophthalmichthys molitrix* (C and V.) in Ponds at Cuttack, India. Proc. Indo. Pacif. Fish. Coun. 10(2):181-204.

Allen, K.R. 1951. The Horokiwi stream : a study of trout population. Fish. Bull. N.Z. 10 : 1-238.

Amoriq, L. 1964. L activite et le phenomene Social chez. *Lebistes reticulatus* (Poecilidae, Cyprinidontiformes). Compt. Rend. 259 : 2701-2702.

Anantharaja, K. and J.S. Sampath Kumar. 2007. Health Management in aquaculture through Immunostimulants. Aqua International 15(3): 21-22

Anderson, D.P. 1992. Immunostimulants adjuvants and vaccine carriers in fish: application to aquaculture. Annu. Rev. Fish. Dis. 2 : 218-307.

Anon, 2008. Organic Aquaculture Project Gets Under way. Sea food Export Jl. XXXVIII (9): 15.

Anon. 2008. MPEDA hopes to boost Ornamental Fish exports to $ 50 m. Aqua International 16(3)8-9.

Aquaculture Authority 2001. "Shrimp aquaculture and the Environment-An Environment Impact assessment report" submitted to the Supreme Court of India. Govt. of India. Chennai, 92 pp.

Areerat, Sopa, 1990. Control of EUS. Fish health section, Asian Fisheries Society. Vol. I. No. 1, pp. 5.

Arnold, Dean, E. 1966. Marking fish with dyes and other chemicals. Bureau of sport fisheries and wild life. Technical paper No. 10-44 p.

Assem, H. and Hanke, W. 1981. ``Cortisol and Osmotic adjustment of the Euryhaline Teleost, *Sarotherodon mossambicus*" Gen. Comp. Endocrinol. 43 : 370.

Atz, J.W. 1973. Comments on ``Reproductive endocrinology of Fishes" by E.M. Donaldson Amer. Zool 13 : 929-931.

Ayyappan, S, J. K. Jena, A. Gopalkrishnan and A. K. Pandey. 2006. Hand book of fisheries and aquaculture. ICAR Publication. New Delhi-12. pp1-711.

Ayyappan, S and R. S. Biradar. 2002. Fisheries Education in India. Fishing Chimes, 22(1):45-49.

Babu, Trinatha, B., Y. Prabhakar Rao and R. Madhavi, 2005. Efficacy of Probiotic Bacterial Amendments over Chemotherapy for a healthier shrimp production. Fishing Chimes 24(10):100-105

Bagenal, T.B. 1963. Variations in plaice fecundity in the clyde area. J. mar. biol. Ass. U.K. 43 : 391-399.

Bairagi, A; Sarkar Ghosh, K; Sen, S. K; Ray, A. K. 2004. Evaluation of the nutritive value of *Leucaena leucocephala* leaf meal, inoculated with fish intestinal bacteria, *Bacillus subtilis* and *Bacillus circulans* in formulated diets for rohu, *Labeo rohita* (Ham.) fingerlings. Aquaculture Research, 35, 436 – 446.

Ball, J.N. 1960. Reproduction in female bony fishes. Symp. Zool. Soc. London 1 : 105-135.

Ball, J.N. and B.I. Baker 1969. The pituitary gland : anatomy and histo physiology pages 1-110 in W.S. Hoar and D.J. Randall eds, Fish physiology, Vol. 2, Academic press, New York.

Ball, J.N. and M. Olivereau 1966. Indentification of ACTH cells in the pituitary of two teleosts *Poecilia latipinna* and *Anguilla anguilla* : Correlated change in the interrenal and in the pars distalis resulting from administration of metopirone (SU 4885) Gen. Comp. Endocrinol. 6 : 5-18.

Bandyopadhyay, Partha. 2008. Principal and action mechanism of Probiotics for sustainable aquaculture. Aqua International, 16 (5): 25 – 26.

Banerjee, R.K. Singit, G.S. and Ray, P. 1969. Some observations on the use of poultry manure as fertiliser in rearing major carp fry. Indian. J. Fish, 16 : 29-34.

Banerjee, R.K., S.K. Saha, P.R. Sen and K.V. Srinivasan 1989. Ecological studies on town refuse in the context of environmental pollution abatement and fish production. J. Inland. Fish. Soc. India 21 (2) : 25-30.

Banerji, P.P., S. Bhattacharya and P. Nath 1989. Purification and properties of pituitary gonadotropic hormone from Indian teleosts : Freshwater murrel *Channa punctatus* and carp *Catla catla.* Gen. Comp. Endocr. 73 : 118-128.

Bara, G. 1965. Histochemical localisation of δ S-3 B-hydroxyosteroid dehydrogenase in the ovaries of a teleost fish, *Scomber scomber* (L) Gen. Comp. Endocrinol. 5 : 284-296.

Bara, G. 1968. Histochemical study of 3 B-3 α - 11B and 17 B-hydroxysteroid dehydrogenase in the Adrenocortical tissue and the corpuscles of stannius of *Fundulus heteroclitus.* Gen. and comp. Endocrinol 10 : 126-137.

Bara, G. 1972. Histochemistry of hydroxysteroid dehydrogenaes in the testis, adrenocortical tissue and corpuscles of stannius of *Pseudopleuronectes americanus.* Acta. Histochem. Bd. 44 : S. 333-347.

Barrackpore : Annual Rept. CIFRI, Barrackpore, India 1968, 39 pp.

Barrackpore : Report, CIFRI, Barrackpore, India 1973.

Barrington, E.J.W. 1964. Hormones and evolution. The English Universities Press Ltd. London.

Belova, L. 1981. Ecological physiological peculiarities of semen of pond carps. II. change in the physiological parameters of spermatozoids of some carps under the influence of environmental factors J. Ichtyol, 21 : 70-81.

Belova, L.V., 1982. Ecological physiological properties of semen of some cyprinids fishes. III. physiological biochemical parameters of the semen of some cyprinid species. J. Ichtyol 22 : 63-79.

Belova, L.V., 1983. Ecological physiological properties of semen of some cyprinids fishes IV. physiological biochemical properties of testes. J. Ichthyol : 23 : 75-84.

Benegal, T.B. 1971. The interrelation of the size of fish eggs, the date of spawning and the production cycle. J. Fish Biol. 3 : 207-219.

Benegal, T.B. 1978. Aspects of Fish Fecundity. Chapter 4. In Ecology of Fresh water fish production Ed. Shelby. D. Gerking. Black well scientific publication. pp. 75-101.

Benegal, T.B. 1966. The ecological and geographical aspects of the fecundity of the place. J. mar. biol. Ass. U.K. 46 : 161-186.

Bern, H.A. and Chieffi, G. 1968. Bibliography on the steroid hormones of fishes, Publ. Stn. Zool. Napoli 36 : 287-320.

Bhaumik, U., P.K. Pandit and J.G. Chatterjee 1993. Involvement of Women in the development of inland fisheries. Environ. Ecol. 11(3): 641-644.

Bhowmick, R.M., G.V. Kowtal, R.K. Jana and S.D. Gupta, 1986. Observations on the use of various hormones and clomiphene citrate in hypophysation of Indian major carps. Veterinarski Archiv. 56(6) : 285-292.

Bhowmick, R.M., G.V. Kowtal, R.K. Jana and S.D. Gupta. 1976. Induced fish breeding. In Annual Rept. CIFRI; Barrackpore, India.

Billard, R. *et al.,* 1986. Biology of gametes, eggs and embroys. R. Billard et. J. marcel, Ed. Aquaculture of cyprinidis, INRA, Paris.

Billard, R. Fostier, A., Weil, C. and BVreton, B. 1982. Endocrine control of spermatogenesis in teleost Fish. Can. J. Fish Aquat. Sci. 39 : 65-79.

Billard, R., Choisis, J.P. and Reinaud, P. 1983. Stimulation of spermiation in carp in response to LHRH and DAL^{6}a LHRH ethylamide. Aquaculture, 35 : 173-176.

Blackburn, R.D. 1968. Weed control in fish ponds in the united States FAO. Fish. Rep. 44 (5) : 1-17.

Bose, P.C. 1944. Calcutta Sewage and fish cultrure, Proc. Nat. Inst. Sci. India 10 : 443-450.

Botte, V. *et al.,* 1964. Osservasioni sulla istofisiologio dell interrenale dei corpuscoli di stannius di alcuni Teleostei Boll. Zool 31 : 461-470.

Boyd, C.E. 1970. Vascular aquatic plants for mineral nutrients removed from polluted water. Econ. Bot. 24(1) : 95-103.

Boyd, C.E. 1979. Trans. Ame. Fish. Soc. pp. 108.

Boyd, C. E. 1990. Shrimp pond bottom soil and sediment management. In Wyban, J.A. (Ed) Proceeding of the special session on shrimp farming. World Aquaculture Society, Baton Rouge, Lausiana. pp 161-181.

Brafield, A.E. 1985. Laboratory studies of energy budget In: Fish energetics : New perspectives Tytler, P. and P. Calow (Eds), Croom Helm. Bockenham. pp. 257-281.

Brandt, F. De. W. and H.J. Schoonbee, 1980. Futher observations on the induced spawning of the phytophagous Chinese carp species (*Ctenopharyngodon idella* (Val) and *Hypophthalmichthys molitrix.* Water SA 6 : 27-30.

Breton, B., and C. Weil 1973. Effects du EH/FSH-RH synthetigue et d'extraits hypothalamiques de carps surla secretion d' hormone gonadotrope in vivo chezia capre (*Cyprinus carpio*) C.R. Acad. Sci. (Paris) 277 : 2061-2064.

Breton, B., G. Kann, E. Burzawa-Gerard and R. Billard 1971. Dosage ratio immunologiue d'une hormone gonadotrope de carpe. C.R. Acad Sci. (Paris), 272 : 1515-1517.

Breton, B., R., Billard, B. Jalabert and G. Kann; 1972 a. Dosage radio-immunologique des gonadotropines chez. *Carrassius auratus* aucours du nycthemere et pendant l' ovulation. Gen. Comp. Endocrinol. 18: 463-468.

Breton, B.C. Weil, B. Jalabert and R. Billard 1972b. Activite reciproque des facteurs hypothalamiques de belier (Ovis aries) et de poissons teleosteen Suro la secretion in vitro des hormones gonadotropes G-HG et LH respectivement par les hypophyses de carpe et de beler. C.R. Acad. Sci. Ser. D., 274 : 2530-2533.

Bretschneider, L.H. and J.J. Duyvene de wit 1947. Sexual endocrinology of non-mammlian vertebrates Elsevier pub. Co. Inc. Amsterdam.

Brett, J.R. and T.D.D. Groves 1979. Physiological energetic. In : Fish physiology Vol. 8, Hoar, W.S. D.J. Randall and J.R. Brett (Eds.) Academic press, New York, pp. 279-352.

Brock, J. 1878. Beitrage zur Anatomic und Histo logic der Geschlecht sorgane der knochenfische morphol. Jahrb, 4 : 505-72.

Brody, S. 1945. Bioenergetics and Growth. Reinhold, New York.

Bry, C. 1982. Daily variations in plasma cortisol levels of individual female Rainbow trout, *Salmo gairdneri* : Evidence for a post-feeding peak in well adapted fish. Gen. Comp. Endocr. 48 : 462.

Buchmann, H. 1940. Hypophyse and Thyroidea im Individualzyklus des Herrings. Zool. Jahrb. Abt. Auat. Ontog. Tiere 66 : 192-261.

Buhler, C.J.D. and Halver, J.E. 1961. Nutrition of slamonid fishes. IX carbohydrate requirement of chinook salmon. J. Nutri. 74 : 307-318.

Burzawa, Gerard, E. 1971. Purification d'une hormone gonadotrope hypophysaire de poisson teleosteen, la capre (*Cyprinus carpio* L.) Biochimie 53 : 545-552.

Burzawa-Gerard, E. and Y.A. Fontaine 1972. The Gonadotropins of lower vertebrates. Gen. Comp. Endocrinol. Supll. 3 : 715-728.

Buss, D. and Larsen, L.O. 1975. Absence of known corticosteroids in blood of River Lampreys (*Lampetra fluviatilis*) after treatment with mammalian corticotropin. Gen. Comp. Endocrinol. 26 : 96.

Byun, J.W., Park, S.C., Benno, Y. and Oh, T.K. 1997. Probiotic effect of *Lactobacillus* sp. Ds-12 in flounder (*Paralichthys olivaceus*) J. Gen. Appl. Microbiol, 43:305-308.

Calderwood, W.L. 1891. A contribution to our knowledge of the ovary and intraovarian egg in telosteans. J. marine biol. Ass. U.K. 2 : 298-313.

Castell, J.D. 1979. Review of lipid requirements of fin fish. In J.E. Halver and K. Tiews (eds.) Fin fish Nutrition and Fish feed technology Vol. 1. pp. 59-84. Proceeding of world symposium. Hamburg. Germany.

Cavolini : Uber die Eraeugung der Fishe und Krebse. In stromstem, F.A. (1930-31). The Development of ovary an oviduct of the Gold Fish. Proc. Acad. Sci. 27 (1792) : 397-398. Cited by D.N. Saksena 1976. Zool. Jb. Anat. Bd. 95 (8) : 127-146.

Chakraborty, I. and Devonita Pal 2004. Participation of Women of Fisher families in Nadia District (WB) in the Decesion making process: Fishing Chimes, 24(6) 33&44.

Chandramohan, K and S. Benakappa 2008. Larvivorous fishes in freshwater realms- A check list. Aqua International 16(3): 35- 36.

Chang, J.P. and Peter, R.E. 1983 (a). Effect of Dopamine on gonadotropin release in female gold fish, *Carassius auratus.* Neuroendocrinology, 35 : 351-357.

Chang, J.P. and Peter, R.E. 1983 (b). Effects of Pimozide and Des Gly 10. D. Ala Luteinizing hormone-releasing hormone Ethylamide on serum gonadotropin concentration, germinal vesicle migration and ovulation in female goldfish *Carassius auratus.* Gen. Comp. Endo. 52 : 30-37.

Charak, K. S and F. A. Fayoz. 2005. Ornamental fishes of Jammu. Fishing Chimes, 25 (6): 24 – 25

Chaudhuri, H. and K.H. Alikunhi, 1957. Observations on the spawning of Indian carps by hormone injection. Curr. Sci. 26 : 281-282.

Chaudhuri, H. 1973. Fertility of hybrids of Indian carps and perliminary studies on the F_2 generation of carp hybrids. J. Inland. Fish. Soc. India 5 : 195-200.

Chaudhuri, H. 1976. Ecological balanced waste water disposal systems management for developing countries. Seminar on Environment and Development, Madurai University, Madurai (India), March 3-5, 1976.

Chaudhuri, H. 1976. Use of Hormones in induced breeding of carps. J. Fish Res. Bd. Can 33 : 940-947.

Chaudhuri, H. and S.B. Singh 1984. Induced breeding of carps. Indian Council of Agricultural Research, New Delhi pp. 82.

Chaudhuri, H. and S.D. Tripathi 1976. Problems of warm water fish seed production. Advances in Aquaculture 127-134. Paper presented at FAO Technical conference on Aquaculture, Kyoto, Japan, 26 May - 2 June 1976.

Chaudhuri, H. and K. Janaki Ram 1975. Nature Lond, 253 : 40-41.

Chaudhuri, H., R.M. Bhowmick, G.V. Kowtal, R.K. Jaa and S.D. Gupta, 1977. Efficiency of partially purified salmon gonadotropin (SG-G100) in induced spawning in Indian major carps. J. Inland Fish. Soc. 9 : 161-164.

Chester, Jones I. *et al.,* 1959. Comparative physiology of the adrenal cortex. In comparative Endocrinology (A. Gorbaman ed.), pp. 582-621 wiley, N.Y.

Chester-Jones, I. 1957. The adrenal cortex. Combridge Univ. Press. London and New York.

Chieffi, G., and Botte; V., 1963. Histochemical reaction for steroid 3 β dehydrogenase in the internal and corpuscles of stannius of *Anguilla anguilla* and *Conger conger.* Nature 200 : 793-794.

Chiu, Y.N. 1989. Considerations for feeding experiments to quantity dietary requirements of essential nutrients in Fish. P. 46-57. In S.S. De Silva (ed). Fish Nutrition Research in Asia. Proceedings of the Third Asian Fish Nutrition Network meeting. Asian Fish Soc. Spe. Publ. 4 : 166 p. Asian Fisheries Society. Manila, Philippines.

Cho, C.Y., C.B. Cowey and T. Watanabe, 1985. Finfish nutrition in Asia. Methodological approaches to Research and Development Ottowa, Ont., IDRC., 154 p.

Chondar, S.L. 1985. HCG a better substitute for PG for induced breeding of Silver carp on commercial scale P. 521-534 In : Proceedings of the second international conference on warm water Aquaculture. Finfish. Hawaii, U.S.A.

Chondar, S.L. 1990. Mass scale breeding of silver carp in ``Bangla Bundh' through Human chorionic gonadotropin and its combination with pituitary p. 17-20. In : P. Keshavanath & K.V. Radha Krishnan (eds) Carp Seed Production Technology, Proceeding of the workshop on carp seed production technology, 2-4 Sept. 1988. Special publication 2, 102 p. Asian Fisheries Society, Indian Branch, Mangalore, India.

Chowdhury, D.K. 1989. Breeding with homeopathic treatment. Fishing Chimes, March 89. Page 12.

Christopher, I. M. M. 2005. Marine aquarium keeping. An emerging hobby trend in India. Fishing Chimes 25(7):34-36.

Chua, Thia Eng Elsie Tech. 1990. Aquaculture in Asia : Quo Vadis? In Aquaculture in Asia (Ed). M. Mohan Joseph. 1990. Asian Fisheries Society, Indian Branch p. 13-30.

CIFA, Kausalyagang, Epizootic Ulcerative Diseased Syndrome out break in India. Summer Institute on fish Disease programme and Health management in freshwater Aquaculture system 5-24th June 1989.

CIFA; 1989. Souvenir, National seminar on forty years of fresh water Aquaculture in India. 7-9 November 1989. Association of Aquaculturists (CIFA) ICAR, Kausalyagang 55 pp.

Clemens, H.P. 1967. A reveiw of selection and breeding in the culture of warmwater food fishes in north America FAO. Fish Rep 44 (4) :67-80.

Clemens, H.P. and Grant, F.B. 1964. Gonadal hydration of carp (*Cyprinus carpio*) and goldfish *Carassius auratus* after injection of pituitary extracts. Zoologica, N.Y. 49 : 193-210.

Colombo, L., Belvedere, P.C., Marconato, A. and Bentivegna, F. 1982. ``Pheromones in teleost Fish : In C.J.J. Richter and H.J. Th. Goss (comp) proceedings of the international symposium on reproductive physiology of fish. Pudoc, Wageningen. p. 84.

Conference on Application of Hormones to Economic Fish. 1975. Experiments on inducement of spawning in domesticated fish by injection of synthesized hypothalmic luteinizing hormone-releasing hormone (LH-RH), Keue Tongbao, 20 : 43-48.

Cooperative Team for Hormonal Application in Pisciculture 1977. A new highly effective ovulating agent for fish reproduction. Sci. Sin. 20: 469-474.

Cowey, C.B. and A.G.J. Tacon, 1983. Fish nutrition-relevance to marine invertebrates. In G.D. Pruder, C.J. Langdon and D.E. Conklin (eds). proceedings of the second international conference on Aquaculture Nutrition : Biochemical and physiological approaches to shellfish Nutrition pp 13-30. Baton Rouge, Lousiana state Univ.

Cowey, C.B. and J.R. Sargent, 1979. Nutrition. In : Fish physiology. Vol. VIII Hoar, W.S.; D.J. Randall and J.R. Butt (Ed). pp. 1-69. Academic Press Publication.

Dahlgren, U. 1914. Science, 40 : 862-863.

Dalton, H. and L.E. Mortenson. 1972. Dinitrogen (N_2) Fixation (with a Biochemical Emphasis). Bacteriological Reviews. American Society for Microbiology 36 (2) : 231-260.

Das R. C and A. Sinha. 2003. Ornamental fish trade in India. Fishing Chimes 23(2):16-18.

Das, Isita, and Ray, A.K. 1989. Growth performance of the Indian Major carp, *Labeo rohita* (Ham.) on duckweed (*Lemna polyrhiza Lin*) incorporated pelleted feed : a preliminary study. J. Inland. Fish. Soc. India, 21 (1) : 1989 : 1-6.

Das, K.M. and B.N. Dasgupta, 1945. Breeding of the principal carps in Bengal. Proc. Nat. Inst. Sci. India (B), 11 : 324-327.

Das, K.M. and Tripathi, S.D. 1979. Feed protein utilisation and growth of *Ctenopharyngodon idella* (Val.). Ind. Sci. Cong. 66 (3) : 85.

Das, Manoj, Sanjay Sarma and A.K. Das 2005. Status of Ornamental Fisheries of Assam, Fishing Chimes, 25(3):13-15.

Dash, G. 1991. Herbal use in Aquaculture. Green Technology. 2(1&2): 44-46.

Datta, M. K and M. C. Nandeesha 2006. Pacu (Piaractus brachypomus) an Acid water loving Exotic fin fish enters Indian Aquaculture Scene. Fishing Chimes 26(6): 10-12.

David A., N.G.S. Rao and M.F. Rahman 1970. J. Inland Fish. Soc. India 2 : 159-160.

Davis, Clarencew, 1971. Marketing Small Atlantic herring with biological stains. Prog. Fish. Culturists. Vol. 33 No. 3 p. 160-162.

De Leeuw, R., H.R. Haibi, C.S. Nahorniak and R.E. Peter 1989. Dopaminergic regulation of pituitary Gonadotropin releasing hormone receptor activity in the gold fish, *Carassius auratus* J. of Endocrinology 121 : 239-247.

De Silva, S.S. 1989. Digestibility evaluations of natural and artificial diets, p. 36-45. In S.S. De Silva (Ed). Fish Nutrition Research in Asia. Proceedings of the Third Asian Fish Nutrition Network Meeting. Asian Fish Soc. Spec. Publ. 4, 166 p. Asian Fisheries Society, Manila, Philippines.

De Silva, S.S. and M.K. Perera 1984. Digestibility in *Sarotherodon niloticus* fry : effect of dietary protein level and salinity with further observations on daily variability in digestibility. Aquaculture 38 : 293-306.

De Silva, S.S., C. Keembiyahetty and R.M. Gunasekara 1986. Optimum ration and feeding frequency in *Oreochromis niloticus* young. In J.L. Maclean, L.B. Dizon and L.V. Hosillos (eds.). The 1st Asian fisheries forum pp. 559-564. Asian Fieheries Society, Philippines.

De, H.K. and G.S. Saha 2005. Aquaculture extension in India-few emerging issues in Development initiatives for farming community (eds) Baldeo singh and R.N. Parida Indian society of Extension Education, New Delhi. pp 418-423.

De, H. K., G. S. Saha and R. Srichandan. 2008: Technology transfer in Aquaculture. Fishing Chimes 16(3):24-26

De, Souresh Prasad. 2008. Organic Fish Farming: An Apporach, Fishing Chimes 28 (5):21-26

Dean, J.L. 1969. Bur. Sprot. Fish. Wild Tech. Pap. 24, Washington DC : 15 pp.

Deb. Diprangshu and T.J. Varghese 1987. The effect of 17 α methyltestosterone on the growth and food conversion of major carps. The first Indian Fisheries forum, Mangalore Dec. - 4-8 (1987) pp. 64 (Abstract.

Dehadrai, P.V. 1984. Carp seed production in India p. 33. In : R.C. May, R.S.V. Pullin and V.G. Jhingran (Eds). Summary report of the Asia regional workshop on carp hatchery and nursery technology 1-3 February, Manilla, Philippines. Asian Development Bank and ICLARM. Manila, Philippines.

Delgado, C. L., N. Wada, M. W. Rosegrant, S. Meijer and M. Ahmed, 2003. Fish to 2020: Supply and demand in changing global markets. International Food Policy Research Institute and World Fish Centre. Technical Report 62, 226 p.

Delmendo, N.M. 1980. A review of integrated livestock food fish farming systems. Proc. ICLARM-SFARCH Confl. integ. Agri. Aquaculture Fing. Systems : 59-71 (Manila-Philippines).

Devlaming, V.L. 1974. Environmental and endocrine control of teleost Reproduction. In : Control of sex in fishes (Ed. Schreck, C.B.) Virginia Rept. Ext. Div. Virginia polytechnic Inst., 1974.

Devraj, M. 1993. Pollution in Freshwater ecosystem and its impact on Aquaculture. Proceedings of the Regional Workshop on Aquatic Pollution held in 5-8 march 1992. Ed. by Dr G. Santhane Kumar and Dr A. M. Selvaraj., J. J. Publication 1993.

Dey, R.K. and S. Chandra 1994. New trends in fish disease management through application of herbal materials. Fishing chimes, 14 : 13-14.

Diwan, A. D and S. Ayyappan 2004. Organic fish farming. Fishing Chimes 24(6):11-13.

Dodd, J.M. 1963. The pituitary complex, pages 161-185 in p. Eckstein and F. Knowles, ed. Techniques in endocrine research. Academic press, New York.

Dodd, J.M. 1972. Ovarian control in cyclostomes and elasmobranchs Amer. Zool 12 : 325-339.

Dodd, J.M. 1975. The hormones of sex and reproduction and their effects in Fish and Lower Chordates Twenty years on, Amer. Zool. 15 (suppl. 1) 137-171.

Dodd, J.M. and Matty, A.J. 1964. Comparative Aspects of Thyroid Function in R. Pitt. Rivers and W.R. Trotter (Eds). The Thyroid Gland. Butterworths, London. 1 : 303.

Donaldson E.M.; Yamazaki, H.M. Dye and W.W. Phileo, 1972. Preparation of gonadotropin from salmon (*O. tshawytscha*) pituitary glands. Gen. Comp. Endocrinol 18 : 469-481.

Donaldson, E.M. 1973. Reproductive endocrinology of fishes. Am. Zool. 13 : 909-927.

Donaldson, E.M. and G.A. Hunter 1983. Induced final maturation, ovulation and spermiation in cultured fish. p. 351-404. In : Fish physiology (Eds. W.S. Hoar, D.J. Randall and E.M. Donaldson) Vol. IXB. Academic press, New York.

Dubey, G.P. 1983. Recent trend in seed production through dry bundhs. Workshop on Aquaculture and Hatchery engineering. 5 to 18 May 1983. CIFE, Bombay p. 81-90.

Dubey, G.P. and R.P. Tuli 1961. Observations on the breeding of major carps in Madhya Pradesh. J. Bomb. Nat. Hist. Soc. 58 (1) 81-91.

Dubey, Kiran and G.P. Dubey 1990. Mandi Dry Bundh - A case study p38-40. In P. Keshavanath and K.V. Radha Krishanan (Eds) carp seed production Technology. Proceedings of the workshop on carp seed production Technology, 2-4 Sept. 1988. Special publication 2. 102 p. Asian Fisheries Society. Indian Branch Mangalore, India.

Dutta O.K. and P.K. Goswami 1988. Investigations on cow and pig manure recycling in Fish ponds. J. Inland. Fish. Soc. India 20 (1) 37-41.

Dwivedi, S.N. 1980. Fish hatchery and fish breeding unit for raising fish seed in controlled environment. India Today and Tomorrow 8 (4) : 167.

Dwivedi, S.N. and Reddy A.K. 1986. Fish breeding in a controlled environment carp hatchery CIFE-D 81. Aquaculture, 51 (1 and 2) : 27-36.

Dwivedi, S.N. and V. Ravindranathan 1982. Carp hatchery model CIFE D. 81 a new system to breed fish even when rain fails. CIFE. Bull 3-4.

Dwivedi, S.N., A.K. Reddy and V.K. Tiwari 1988. A carp hatchery for India with controlled environment. J. Inland Fish. Soc. India 20 (1) : 6-12.

Edwards, P. 1980. Food potential of aquatic macrophytes. ICLARM. Studies and reviews. ICLAR, Manila 5 : 51 pp.

Engle, C. R and N. M. Stone. 1989. A Review of Extension Methodologies in Aquaculture, Aquaculture Development and Coordination Programme, FAO Rome 65p.

Epler, P. 1978. Effect of gonadotropin and Steroid hormones on ovarian maturity in carp. Zesz., Nauk. Akad. Roin. Kvakow, 146 : 7-51.

Epler, P. Marosa, E., Bientarz, K. 1980. Joint action of various steroid hormones on maturation of carp oocytes invitro. Zes Nauk. Akad. Roin. Kvakow. 59 : 71-82.

Epler, P., Bieniarz, K. and Marosz. E. 1985. Effect of temperature. 17 α hydroxy 20 β dihydro progesterone and prostaglandin F 2 a on carp oocyte maturation and ovulation invitro. Gen. Comp. Endocrinol. 53 : 192-201.

Evennett, P.J. 1963. The endocrine control of reproduction in the river lamprey (*Lampreta fluviatilis* L. Ph.D. Diss. University of St. Andrews.

FAO, 1995. Fishery information, Data and Statistics service. Aquaculture production statistics. FAO fisheries Curcular No. 815 (Rev. 7) : 186 p.

FAO, 2006. The State of World Fisheries and Aquaculture, 2006. 162 p

Farran, G.P 1938. On the sex and number of ova of Irish. Herrings. J. Cons. Perm. Int. Explor. Mer 13 : 91-100.

Felix, N and R. Alan Brindo. 2008. Marine Nutraceuticals. New Era in Fish and Shrimp nutrition. Fishing Chimes 28 (5): 34- 39.

Felix, N. 2006. Live Bearing Ornamental Fish Production: An Apt Venture for Women Self Help Group. Fishing Chimes, 26(8):45-46.

Fernandes, C. F. and Shahani, K. 1990. Anticarcinogenic and Immunological properties of Dietary Lactobacillus., J. Food Protection, 53, 704-710.

Fernholm, B., 1972. Neurohypophysial-adenohypophysial relations in hagfish (Myxinoidea, cyclostomata), Gen. Comp. Endocrinol. Suppl. 3 : 1-10.

Fontaine, M. 1976. Hormones nd the control of reproduction in aquaculture J. Fish. Res. Bd. Can 33 (1976) : 922-939.

Fontaine, M. and Chauvel, M. 1961. C.R. Acad. Sci. 252 : 822-824.

Fontaine, M. and Leloup-Hatey 1959. Mise en evience de cdorticosteroides dans l'interrenal d'un Teleosteen : le *Salmon salmon salar* L. J. Physiol (Paris) 51 : 468-469.

Fontaine, Y.A., and E. Gerard 1963. Purification d'un facteur gonadotrope de I'hypophyse d'un teleosteen la carpe (*C. carpio* L.) C.R. Acad. Sci. (Paris) 256 : 5634-5637.

Fostier, A., B. Jalabert and M. Terqui 1973. Action predominate d'un derive hydroxyle de la progesterone Sur la maturation in Vivo des ovocytes de la truite are-en-ciel *Salmo gairdneri,* C.R. Acad Sci 277 (1973) : 421-424.

Fostier, A.B., Jalabert, R., Billard. B. Breton, and Y. Zohar 1983. the Gonadal Steroids. In Fish physiology Edt. by W.S. Hoar, D.J. Randall and E.M. Donaldson. Vol. IX. Reproduction part. A. pp. 277-372.

Fox, D.L. 1979. Pigment transactions between animals and plants. Biol. Rev. 54 : 237-268.

Fuller, R. 1986. Probiotics., J. Appl. Bacteriol suppl., 15-75.

Ganapathi, S.V. 1972. Reuse of waste water. J. Sci. and Ind. Res. 31 : 390 : 393.

Gaur, A.C., S. Neelakanta and K.S. Dargan 1984. Organic manure ICAR publication, New Delhi.

Gerking, S.D. 1962. Production and food utilisation in a population of blue gill sun fish. Ecol. monogr. 32 : 31-78.

Gerking, S.D. 1963. Non-mutilation marks for Fish. International Committee for North-West Atlantic Fisherie, species publication 4, p. 248-254.

Gerking, S.D. 1967. Biological basis of Fresh water fish production (Ed). Black well scientific publications, Oxford.

Germking, D. Shelby. 1978 (Edt); Ecology of Fresh water fish production. Black well scientific publication. Ch. 17. The growing science of Aquaculture John E, Bardach pp. 424-446.

Ghosh, A., Banerjee, M.K. and Hanumantha Rao, L. 1973. Some observations on the cultral prospects of Silver carp in sewage fed ponds. J. Ind. Fish. Soc. India 5 : 131-133.

Ghosh, Indrani. 2006. Captive breeding of Ornamental fishes. Fishing Chimes 26(3): 23-28.

Ghosh, Indrani. 2007. Parthogenesis: An unusual occurrance in Golden Guppy, Poecilia reticulata. Fishing Chimes. 26(11): 26-28.

Ghosh, R.I., 1985. Chemical composition of the cavity fluid and ova of the grass carp and the carp. Hydrobiol. J., 19 : 84-87.

Goetz, F.W. 1983. Hormonal control of oocyte final maturation and ovulation in fishes. pp 117-170. In : Fish physiology (Eds. W.S. Hoar D.J. Randall and E.M. Donaldson. Vol. I X B. Academic Press, New York).

Goetz, F.W. and H.L. Bergman. 1978. The invitro effects of mammalian and piscine gonadotropin and pituitary preparations on final maturation in yellow perch (*Perca flavescane*) and Walleye (*Stizostedion vitreum*). Can. J. Zool. 56 : 348-350.

Gopakumar, G. 2007. Development of a sustainable trade of Marine Ornamental species from India. INDAQUA-2007 1-14.

Gopalkrishnan, V. 1961. Observations on infectious dropsy of India Major carps and its experimental induction. J. Scient. Ind. Res. 20 : 357-358.

Goswami, S.V. and Sundararaj, B.I. 1971. In vitro maturation and ovulation of occytes of the cat fish *H. fossilis* (Bloch) J. Exp. Zool. 178 : 467-478.

Grodzinski, W., R.Z. Klekowski, A. Duncan 1975. Methods of Ecological bioenergetics. IBP Hand book No. 24 Blackwell scientific publication, Oxford.

Gulidov, M.V., Popova, K.S. 1979. The hatching dynamics and morphological features of larvae of roach, *Rutilus rutilus* in relation to incubation temperature. J. Ichthyol., 19 : 87-92.

Gupta Sandipan and Samir Banerjee. 2008. Ornamental Fish Trade in West Bengal. Fishing Chimes 28 (8): 25 – 27.

Gupta, O.P. 1979. Aquatic weeds : Their menace and control Today and Tomorrows Printers and publishers, New Delhi 227 p.

Gupta, S.D., H.A. Khan, M.S. Tantia, and R.C. Das 1990. Effects of Winstrol and progesterone on invivo oocyte maturation and spawning in *Labeo rohita.* p. 24-28. In P. Keshavanath and K.V. Radha Krishnan (eds) carp seed production technology. Proceedings of the workshop on carp seed production Technology, 2-4 Sept. 1988. Special publication 2. 102 p. Asian Fisheries Society, Ind. Branch, Mangalore India.

Gupta, S.D., H.A. Khan, P.V.G.K. Reddy and G.V. Kowtal. 1986. Studies on the intergeneric hybrid common carp *Cyprinus carpio communis* LX mrigal *Cirrhinus mrigala* (Ham.). Paper presented in the symposium on conservation and management of Fish Genetic Resources in India. Allahabad.

Hajra, A. 1987. Biochemical investigations on the protein-calorie availability in grass carp (*Ctenopharyngodon idella* val.) from an aquatic weed (*Ceratophyllum demersum* Linn.) in the tropics. Aquaculture, 61 : 113-120.

Hajra, A. and S.D. Tripathi 1985. Nutritive value of aquatic weed, (*Spirodela polyrhiza* Linn.) in grass carp. Indian J. Anim. Sci. 55 (8) : 702-705.

Halder, S., S. Sen, S. Bhattacharya, A.K. Ray, A., Ghosh and A.G. Jhingran 1989. Effect of purified piscine hypothalmic gonadotropin releasing hormone (GnRH) on the maturation and breeding of Indian teleosts. Paper presented at the workshop on Brackishwater fin fish Breeding and seed production. CIBA, Aquaculture 6-7 Dece. 1989.

Halver, J.E. 1972. Fish nutrition. Acad. Press. New York and London 713 pp.

Halver, J.E. 1976. The nutritional requirements of cultivated warm water and coldwater species. FAO Technical conference on Aquaculture Kyoto. Japan 26 May to 2 June, 1976. FIR/Conf. 76/R. 131.

Hamano, S., 1957. Physico-chemical studies on the activation and fertilization of fish eggs. Memoirs Fac. Fish Hokkaido Univ. 3 : 91-143.

Hane, S. and Robertson, O.H. 1959. Changes in plasma 17-hydroxycorticosteroids accompanying sexual maturation and spawning of the pacific salmon (*O. tschawytscha*) and rainbow trout (*S. gairdneri*) : Proc. Natl. Acad. Sci. U.S. 45 : 886-893.

Hane, S., O.H. Robertson, B.C. Wexler and M.A. Krupp 1966. Adrenocortical response to stress and ACTH in *O. tschawytscha* and *Salmo gairdneri* at successive stages in the sexual cycle. Endocrinology 78 : 791-800.

Hanke W., and Chester Jones, I. 1966. Histological and histochemical studies on the adrenal cortex and the corpuscles of stannius of the European eel. *Anguilla anguilla* L. Gen. Comp. Endcr., 7 : 166-178.

Hanke, W. *et al.,* 1967. Gen. Comp. Endocrinol 9 : 64-75.

Hardy, R. 1980. Fish feed formulation. In Fish feed Technology, Aquaculture Development and Co-ordination Programme, FAO/ACDP/RCP/ 80/11p, 233-239.

Harvey, B. 1982. Cryobiology and the storage of teleost gametes 123-127, in C.J.J. Richter and H.J. Th. Goos (Comp), Reprod. Physiol. Fish. PUDOC Wageningen.

Herrick, C.L. 1891. Studies on the brains of some American freshwater fishes A. Topography comp. Neurol 1 : 228-252.

Hickling, C.F. 1962. Fish culture, Faber and Faber. London.

Hirose, K. 1973. Hormonal control of fish ovulation. Bull. Tokai Reg. Fish. Res. Lab. 74 : 67-81.

Hirose, K., 1976. Endocrine control of ovulation in medaka, *Oryzias latipes* and Ayu, *Plecoglossus altivelis* J. Fish. Res. Board of Canada. 33 (4) : 989-994.

Hoar, W.S. 1955. Reproduction in teleost fish. Mem. Soc. Endocrinol. 4 : 5-24.

Hoar, W.S. 1957. The gonads and Reproduction. In ``The physiology of Fishes" (M.E. Brown ed.), Vol. 1, pp. 287-321. Academic press, New York.

Hoar, W.S. 1969. Reproduction. In ``Fish physiology" (W.S. Hoar and D.J. Randall eds). Vol. 3 pp. 1-72.

Hora, S.L. 1943. Indian Farming, 4 : 508-550.

Hora, S.L. and T.V.R. Pillay, 1962. Hand book on fish culture in the Indo-pacific fisheries Region. FAO, Biol. Techn. Pap. (14) : 203 p.

Hora, S.L., 1951. Knowledge of the ancient Hindus concerning fish and fisheries of India. Matsyavinoda or a chapter on angling in the mana-soltara by king someswara (1127 AD). J. Asiatic. Soc. 17(2): 14569.

Hornell, J. 1920. Administrative report of the Department of fisheries, Madras for the year 1918-19. Madras Fish. Bull. 12 : 1-42.

Hornell, J. 1922. Administrative report for the year 1920-21. Madras Fish. Bull; 15 : 1-43.

Horvath, L. 1978 (b). Relation between ovulation and water temperature by farmed cyprinids. Aquaculture Hungarica (Szarvas) 1 : 58-65.

Horvath, L. 1986. Carp oogenesis and environment, R. Billard et. J. Marcel Ed. Aquaculture of cyprinids, INRA. Paris pp. 109-117.

Horvath, L. and Lukawioz, V.M. 1982. Tables with data of hatchery procedures and rearing progress of some bred wearm water fishes. Aquaculture ungarica (Szarvas), 3 : 212-219.

Horvath, L., 1978 (a). Experience in propagation of the common carp. *Cyprinus carpio* L. Out of spawning season. Aquaculture Hungarica : (Szarvas) : 1 : 66-72.

Horvath, L., and Peteri, A. 1980. The effect of oxygen content of water on the gold fish, *Carassius auratus*. Can. J. Zool. 55 : 1908-1913.

Houssay, B.A. 1931. Accion Sexual de la hipofissi en los percesy reptilies, Rev. Soc. argent Biol. 6 : 686-688.

Huet, Marcel 1971. Text book on Fish Culture Breeding and cultivation of Fish. Fishing News Books. Ltd. Parnham, Survey, England, 437 pp.

Hussain, Md. S. and K. Jamil 1989. In Sym. Dev. without destruction, NATCON pp. 73-74.

Idler, D.R and Truscott. B. 1972. ``Corticosteroids in Fish''. In D.R. Idler (ed), steroids in Nonmammalian vertebrates. Academic press. London p. 127.

Ishida, R., K. Hirose and E.M. Donaldson 1972. Induction of ovulation in ayu, *Plecoglossus altivelis* with salmon pituitary gonadotropin. Bull. Jap. Soc. Sci. Fish. 38 1007-1012.

Iwamatsu, T. 1974. Studies on oocytes maturation of the medaka *Oryzias latipes*. II effect of several steroids and calcium ions and the role of follicle cells on invitro maturation. Anat. Zool. Jpn. 47 : 30-42.

Jalabert, B. 1976. Invitro oocyte maturation and ovulation in rainbow trout (*Salmo gairdneri*) northern pike (*Esox lucius*) and goldfish (*Carassius auratus*) J. Fish. Res. Bd. Can. 33 : 974-988.

Jalabert, B. and D. Szollosi. 1975. Invitro ovulation of trout oocytes; effect of prostaglandins on smooth muscle like cells of the theca. prostaglandins 9 : 765-778.

Jalabert, B. Breton, B. Brzuska, E. Fostier, A. and Wieniawski, J. 1977. A new tool for induced spawning; the use of 17α hydroxy 20 β dhydroprogesterone to spawn carp at low temperature. Aquaculture. 10 : 353-364.

Jalabert, B., and B. Breton 1973. Invitro maturation of Pike *Esox lucius* ooxytes. Gen. Comp. Endcr. 22 : 391.

Jalabert, B., C. Bry, B. Breton and C. Campbell 1976. Action de la 17 α-hydroxy - 20 β dihydroprogesterone et de la progesterone sur la maturation et l' ovulation in vivo et Sur le riveau d' hormone gonadotrope palsmatqu t-GtH chez la Truite Arc-en-ciel *Salmo gairdneri*. C.R. Acad. Sci. Paris Serie. D. 283 : 1205-1208.

Jalabert, B., B. Breton and A. fostier, 1978. Precocious inudction of occyte maturation and ovulation in rainbow trout (*Salmo gairdneri*) problems when using 17 α- hydroxy 20β dehydroprogesterone. Ann. Biol. Anim. Bio. Chem. Biophys. 18 : 977-984.

Janakiram, K. 1995. Freshwater pearl culture. Fisheries fortune Vol. 1, No. 3 and 4, page. 26-27.

Jaoul, A., Roubaud, P., 1982. Resistance de l' oeuf de carpe commune (*Cyprinus carpio* L. cyprinidae) a des chocs thermiqyues chauds oufroids. J. Can Zool. 60 : 3409-3419.

Jayaraman, K. Saivaraj and A. Abdul Kareem 1989. Chromosome manipulation is fish. Fishing chimes vol. 9 No. 6. pp. 37-39.

Jenkins, J.T. 1902. Alterbestimmung durch otolothen bei den clupeiden. Wiss. Meeresunters, Abt. Kid. 6 : 81-121.

Jessop, B.M. 1973. Marking Alewife Fry with Biological stains. Prog. Fish. Culturist. Vol. 35, No. 2, 90-98.

Jha, P. and S. Barat, 2007. Ornamental fish (P. scalare and C. auratus) culture. Involvement of women self help group in Darjeeling District of W. B. Fishing Chimes 27(1):72-75.

Jhingran, A.G. 1990. Fish health Section Vol. 1 No. 1. 1990.

Jhingran, A.G. and Apruba Ghosh 1988. Aquaculture as a potential system of sewage disposal. A case study J. Inland Fish Soc. India 20 (2) : 1-8.

Jhingran, V.G. 1966. Synopsis of biological data on catla (*Catla catla*) Hamilton, 1822) FAO. Fish Synop. 32 Rev. 1. 78 page.

Jhingran, V.G. 1991. Fish and Fisheries of India. 3rd Edition. Hindustan Publication comp.

Jhingran, V.G. and B.K. Sharma, 1980. Integrated livestock fish farming in India. Proc. ICLARM SEARCH Conf. Intg. Agri. Aqua. Fmg. systems : 135-142. (Manila-Philippines).

Jhingran, V.G. and K.L. Sehgal 1978. Cold water fishereis of India ``care of brood Stock" Inland Fisheries Society of India. Page 1090-111.

Jhngran, V.G. 1983. Hind Pub. Corp. Delhi pp 417-418.

Jhnigran, V.G. and R.S.V. Pullin 1985. A Hatchery manual for the common, Chinese and Indian Major carps. Asian Development Bank and ICLARM. Manila. Philippines. pp. 20-36.

Jose, T.M., M.V. Mohan and M.J. Sebastian, 1989. Inudced breeding of Indian carps using Des Gly^{10} ($D\text{-}Ala^{6}$) LHRH ethlamide. Fish Technol, 26 : 67-69.

Kandler, R. and Pirwitz, W. 1957. Uber die Fruchtbarkeit der platifische im Nordsee-obtsee-Raum, Kleler meeres forsch. 13 : 11-34.

Kapur, K. 1978 (a). The effects of pesticides and heavy metals on the steroidogenesis in the ovary and testes of *Cyprinus carpio*. All India Symp. Exptl. Zoology, Jaipur, Dec. 1978.

Kapur, K. 1978 (b). Effect of prostaglandin, Synthetic LHRH, Mammalian and piscine pituitary extracts on steroidogenesis in the ovary of intact and gonadotropin deprived *Cyprinus carpio*, Satellite symposium on Hormonal steroids in fish, New Delhi.

Kapur, K. 1979. The effect of endomethacin on ovulation and spawning in exotic fish. *Cyprinus carpio*, Ind. J. of Expt. Biol. 17 : 517-519.

Kapur, K. 1981. Spawning of major Indian carps- a Review Actahydrochim. hydrobiol. 9 (2) : 163-182.

Kapur, K. and H.S. Toor 1979. The effect of clomiphene citrate on ovulation and spawning in indomethacin treated carp. *Cyprinus carpio*. J. Fish Biol. 14 : 59-66.

Karunasagar, I., K. Sagar, I. Karunasagar, P.K.M.M. Ali and G. Joy Sa. Karan, 1988. Virulence of *Aeromonas hydrophila* strains isolated from fish ponds and infected fish. In Recent Advance in Biotechnology and applied Biology. Shuting -chang, Kwong. Yu chan and Norman Y.S. Woo. (Eds) Chinese Univ. HongKong, 205-211.

Kaul, M. and K.K. Rishi, 1986. Induced spawning of the Indian. Major carp *Cirrhina mrigala* (Ham) with LH-RH analogue or pimozide. Aquaculture, 54 : 45-48.

Kerr, T. 1942. A comparative study of some teleost pituitaries. Proc. Zool. Soc. London. 118 : 973-988.

Kestemont, P. 1988. Effect of hormonal treatment on induced ovulation in gudgeon *Gobio gobio*. Aquaculture 68 : 373-385.

Khan, H. 1938. Ovulation in fish (effect of adminstration of anterior pituitary gland) Curr. Sci. 7 (5) : 233-234.

Khan, H. 1942. Spawning of carps and their spawning grounds. J. Bombay Nat Hist. Soc. 43 (3) : 416-427.

Khan, H. 1945. Observations on the spawning behaviour of carp in the Punjab. Proc. Nat. Inst. Sci. Ind. 11 (3) : 315-320.

Khan, H. and V.G. Jhingran 1975. Synopsis of biological data on rohu, Laboe rohita (Ham). FAO Fish Synop. 111.100 p.

Khan, H.A., P.V.G.K. Reddy and S.D. Gupta 1986. Studies on some aspects of biology and morphometry of Integeneric hybrid *Cyprinus carpio* communis LX *Labeo rohita* (Ham.) Proc. Nat. Acad. Sci.

Khan, H.A., S.D. Gupta and P.V.G.K. Reddy 1988. Suitable dye for marking Asiatic carps. Ind. J. of Animal Sci. 58 (8) : 1003=-1006.

Khan, H.A., S.D. Gupta, P.V.G.K. Reddy, M.S. Tantia and G.V. Kowtal 1990. Production of sterile intergeneric hybrids and their utility in aquaculture and reservoir stocking p 41-48. In : P. Keshavanath and K.V. Radhakrishnan (eds) carp seed production Technology. Proceeding of the workshop on carp seed production technology. 2-4 Sept. 1988. Special publication 2, 102 p. Asian Fisheries Society. Indian Branch, Mangalore. India.

Khanna, S.S. 1988. An Introduction to fishes. Central Book Depot. Allahabad. pp. 180-183.

Kirpichnikov, V.S. 1981. Genetic bases of Fish Selection (Ch. V : Biochemical genetic of fishes) page 143-199. Springer verlag. Berling-Heidelberg, New York.

Klekoswski, R.Z. Prus., T. and Zyromska-Rudzka, H. 1967. Elements of energy budget of *Trilobium castaneum* (Hbst) in its developmental cycle In. Petrusewicz, K. (Edt.) Secondary productivity of Terrestrial Ecosystems. PWN Wars Zawa Krakow. 2, 859-879.

Knowles, F.G.W. and L. Vollrath 1966. Changes in the pituitary of the migrating European eel during its Journery from rivers to the sea Z. Zellforsch Mikroskop Anot 75 : 317-327.

Krivobok, M.N. and Sterozhuk A. ye. 1970. The effect of egg size and age of volga sturgeon on the weight and chemical composition of the mature eggs. J. Ichthyol 10, 761-771.

Kruger, De, J.C.W., Smit, G.L., Van Vuren, J.H.J., Ferreiro, J.T. 1984. Some chemical and physical characteristics of the semen of *Cyprinus carpio* (L) and *Oreochromis mossambicus* (Peters). J. Fish Biol. 24 : 263-272.

Kuldip Kumar, 2004. Maintenance and Management of Fish Aquarium. Fishing Chimes 24(4): 46-48

Kumar, D. and R.K. Dey 1988. Bacterial septicemia in Silver carp (*Hypophthalmichthys molitrix* val). Vet. Arhiv. 58 (6) : 277-283.

Kumar, D., J. Farkas and V.R.P. Sinha 1986. Bacteria from diseased fres-hwater fish from India. Aquaculture Hungarica (Szarvas) V : 113-118.

Kumaran, M., Ponnusamy, K and Kalaimani, N. 2001. "Diffusion and Adoption of shrimp farming Technology" Paper presented in the International Symposium on Fish for Nutritional security in the 21st century at CIFE, Mumbai, 4-6th December 2001.

Kumaran, M., Ponnusamy, K., Harbir Singh., M. Krishnan and P.S. Birthal 1999. Extent of information source utilisation among the shrimp farmers of "Tamil Nadu". Paper presented in the inter insti-tutional work-shop on "Aquaculture Development in India: Prospects and Retrospect" held at NCAP, New Delhi, 6-7th August 1999.

Kumaran, M., V. S. Chandrasekaran and N. Kalaimani. 2003. Aquacu-lture extension: the neglected aspect in aquaculture development. Fishing Chimes, 23(6): 29-34

Kuo, C.M. *et al.* 1973. Aquaculture 1, 429-432.

Kutty, M.N., 1979. Utilisation of sewage for fish culture. V. International Symposium on Tropical Ecology. Kualalaamopur, Malaysia p. 106.

Kutty, M.N., Natarajan, V. and Venkataramanjuam, K. 1979. Recycling sewage for fish culture. Proc. Symp. on Inland aquaculture, CIFRI, Barrackpore, Feb. 12-17, 99-100.

Kyle, A.L., Stacey, N.E., Peter R.E. and Billard, R. 1985. Elevations in gonadotropin concentrations and milt volumes as a result of spaw-ning behaviour on the gold fish. Gen. Comp. Endocrinol., 57 : 10-22.

Lagler, F. Karl. 1956. Fresh water Fishery Biology, W.M.C. Brown Company, Dubuque Iowa.

Lake, J.S. 1967. Rearing experiments with five species of Australia Fresh water fishes. I. Inducement to spawning. Aust. J. Mar. Fresh water Res. 18 : 137-153.

Lam, T.J. 1982. Application endocrinology to fish culture. Can. J. Fish. Aquat. Sci. 39 : 111-137.

Lam, T.J., S. Pandey and W.S. Hoar 1975. Induction of ovulation in gold fish by synthetic Luteinizing hormone releasing hormone (LHRH). Can. J. Zool. 53 : 1139-1192.

Larsen, L.O. and B. Rothwell 1972. Adenohypophysis. p. 1-67 kn M.W. Hardisty and I.C. Potter eds. The biology of lampreys. Vol. 2. Academic press, New York.

Lawrence, J.M. 1968. Aquatic weed control in fish ponds. FAO Fish Rep. 44 (5) : 76-91.

Leatherland, J.F. 1982. Environmental physiology of the teleostean thyroid Gland : a Review, Env. Biol. Fish. (7) 83.

Lederis, K. 1977. Chemical properties and the pharmacological Actions of Urophysial peptides. Am. Zool. 17, 823.

Lewis, M. and J.M. Dodd. 1974. Thyroid functions and the ovary in the spotted dog fish *Scyliorhinus canicula* J. Endocrinol 63 : 63.

Lewus, C. B., Kaiser, A. and Montville, J. J. 1991. Inhibition of Food borne Bacterial pathogens by Bacteriocins from Lactic acid bacteria Isolated from meat, Appl. Microbial, 57, 1683 – 1688.

Li Sifa, 1990. Recent Advances in Freshwater Aquaculture in China. In Aquaculture in Asia (Ed.) M. Mohan Joseph. 1990. Asian Fisheries Society Indian Branch, p. 141-161.

Liley, N.R. 1966. Ethological isolating mechanisms in four sympatric species of poecillid fishes. Behav. Suppl. 13, 1-197.

Little, J.M. and J.A. Dawson 1989. Advances in finfish induced breeding from Labroatory to field. Paper presented at the marine Biotechnology conference, August 1989. Tokyo, Japan.

Lone, Khalid, p. 1988. Inland Fisheries and Aquaculture in Pakistan. A review PARC, Directorate of Scientific Information, Pakistan, ARC P.O. Box 1031, Islamabad.

Lopinot, A.L. 1991. Leaky ponds. In Farm Pond Harvest. published by professional sports man's publishing company vol. 25 No. 1 pp. 16.

Ludwig, G.M. and Lange E.L. 1975. The relationship of length, age and age-length interaction to the fecundity of the northern mottled sculpin cottus b. bairdi. Trans. Am. Fish. Soc. 104, 64-67.

Lyamin, K.A. 1956. Investigation into the life cycle of summer spawning herring of Iceland. Spec. Scient. Rep. U.S. Fish Wildl. Serv. Fisheries. 1959. No. 327, 166-202 (Translation from Trudy polyar mauchno. issled. Inst. Morsx ryb. Khoz. Koeanigr. No. 9).

Macer, C.T. 1974. The reproductive biology of the horse mackerel (*Trachurus trachurus* (L)). in the North sea and English channel. J. Fish. Biol. 6, 415-438.

Mackenzie, R.A. and Hall, W.T.K. 1976. Dermal ulceration of mullet (*Mugli cephalus*) Aust. Vet. J. 52 : 230-231.

Madhusoodana kurup, B. 2003. Popularization of aquarium keeping and promotion of Ornamental fish culture and Marketing: Kerala model of development, Fishing Chimes, 23(9): 31-34.

Mammen, T.A. 1962. Live fish transport using modified splashless carrier with compressed in aeration. In . Training course on Live Fish Transport, No. 19-21, Hyderabad Fisheries Extension Unit. Govt. of India. 4 pp.

Marr, A., Mortimer, M.A.E. and Lingen-I-Vander 1966. Fish culture in central East Africa : 160; Rome : FAO.

Matty, A.J. 1960. ``Thyroid cycles in Fish". Symp. Zool. Soc. London. 2 p.1.

Matty, A.J. 1985. Fish Endocrinology. Timber press, Oregon pp. 181-197.

Matty, A.J. 1987. Growth promoters used in Aquaculture. Proceeding in 1st Indian Fisheries Forum. Dec. 6-10, Mangalore, India.

Mc Geachin R.L. and J.W. Debnam Jr. 1960. Amylase in freshwater fish. Proc. Soc. Expt. Biol. Med. 103 : 814-815.

McBride, J.R. and A.P. Van Overbeeke 1969. Hypertrophy of the interrenal tissue in sexually maturing sockeye Salmon *O. nerka* and the effect of gonadectomy J. Fish. Res. Board. Can. 26 : 2975-2985.

Menezo, Y., plouidy, M.G., Billard R. 1983. Sur la composition en acides amines libres et peptidiques du liquide seminal de carpe (*Cyprinus carpio*). C.R. Acad. Sci. Paris. Serie 1, 296, 67-70.

Menon, A.G.K. 1987. Rare and endangered Indian Freshwater fishes symposium on the impact of current land use pattern and water resources development on riverine fishereis. CICFRI, Barrackpore, April 25-27, 1987.

Millennium Task Force on Hunger, 2004. Halving hunger by 2015. A Frame work for action. Interim Report. Millennium Project, New York.

Millikin, M.R. 1982. Qualitative and quantitative nutrient requirements of fishes. A review. Fish. Bull. 80 : 655-686.

Mishra, S. K and Salini Mishra. 2007. Ornamental fish trade and Art of aquarium keeping. Aqua International 15(5): 23-26.

Mohanty, S.N., Swamy, D.N. and Tripathi, S.D. 1990. Growth, nutritional indices and carcass composition of the Indian major carp fry. *Catla*

catla, Labeo rohita and *Cirrhinus mrigala* fed four dietary protein levels. Aquaculture Hungarica, 6 : 211-17.

Moitra, S.K. and S.K. Sarkar, 1975. A new method of induced breeding y hypophysation of some major carps in dry bundhs of Bankura, West Bengal Proc. Zool. Soc. Calcutta, 28 : 41-50.

Mookerjee, H.K., S.R. Mazumdar and B. Dasgupta 1944. Observation on the breeding grounds and spawning habits of certain Indian carps in the midnapore District, Bengal with suggestions for their breeding, collection of egg and rearing of fry. J. De. Sci. Calcutta Univ. 11 : 383-384.

Morizawa, K., Suzuki K., Shimizu, H., Morisawa, S., Yasuda, K. 1983. Effects of osmolality and potassium on motility of spermatozoa from freshwater cyprinid fishes. J. Exp. Biol., 107 : 95-103.

Mork, J., Giskeodegard, R. and Sundnes, G. 1984. The haemogrlobin polymorphism in Atlantic cod (*Gadus morhua* L.) genotype differences in Somatic growth and in maturing age in natural populations. Flodevigen Rapporterer, 1 : 721-732.

Moses in Baraj, R and S. Haider 1988. Invitro Effectiveness of Estradiol 17b, Androgens, Corticosteroids, Progesterone and other pregnene Derivatives on Germinal vesicle Breakdown in Oocytes of the Exotic Common Carp *Cyprinus carpio* (L.) Ind. Jl. of Experimental Biology. Vol. 26 : 583-585.

Moses, T.L., S. Samuel and P. Clement Rassel. 2008. Fish – mushroom Integrated Aquaculture system in India. Aqua international 16 (7): 37 -38.

Mukherjee, S.C. 2008. Health Management in Aquaculture through application of Biotechnology (Paper presented at the International seminar on Aquaculture at ILDEX (International livestock and Diary Expo), India, 2008 held at Pragati Maidan, New Delhi on 23. 8. 08), Fishing Chimes 28 (6): 34 – 36.

Murugan, M., Maloy Kumar Sahu, M. Srinivasan, Kamala Vevi, L. Kannan. 2008. Marine Ornamental Fishes of the little Andaman waters. Fishing Chimes 28 (8):44-45

Nagahama, Yoshitaka, 1983. The function morphology of teleost gonads. In fish Physiology Edt. By Hoar, W.S., D.J. Randall and E.M. Donaldson Vol. IX A. Academic press pp. 223-275.

Nagai, M. and S. Ikeda. 1971. Carbohydrate metabolism in fish. I. Effect of starvation and dietary composition on the blood glucose level and the hepatopancreatic glycogen and lipid content in carp. Nippon suican Gakkaishi, 37, 404-409.

Nair, Shiva, Supal, B. R and Satam, S. B. 2008. Discus the King of Aquarium Fish. aqua international 16(1&2): 26-28.

Nandeesha, M.C., K.G. Rao, R. Jayanna, N.C. Parker, T.J. Varghese, P. Keshavanath and H.P.C. Shetty 1990b. Induced spawning of Indian major carps through single application of ovaprim C. p. 581-585. In : R Hirano and I. Hanyu (Eds). The second Asian Fisheries Forum proceedings. Asian Fisheries Society, Manila, Philippines.

Nandeesha, M.C., P. Keshavanath, T.J. Varghese, H.P.C. Shetty and K. Gopal Rao 1990a. Alternate inducing agents for carp breeding : progress in research, p. 12-16. In P. Keshsavanath and K.V. Radha Krishnan (Eds). carp seed production Technology. Proceedings of the work shop on carp seed production Technology, 2-4 September 1988. Special publication 2, 102p. Asian Fisheries Society Indian Branch, Mangalore, India.

Nandi, J. and Bern, H.A. 1960. Corticosteroid production by interrenal tissues of teleost fish. Endocrinology 66, 295-303.

Natarajan, A.V. and B.K. Sharma, 1980. Integrated composite fish culture livestock farming, workshop All India Co-ordinated Research Project. National Demonstration Pondicherry.

Natarajan, A.V. and V. Pathak. 1983. Pattern of Energy flow in fresh water Tropical and Subtropical impoundments. CIFRI, Barrackpore, Bull. No. 36.

Natarajan, M.V., G. Chandrasekaran and S. Durairaj. 1989. Introduction of Indian River prawn. *Macrobrachium malcolmsonii* in to fish Culturc ponds. Fishing chimes Vol. 9. No. 3. pp. 47-48.

Natarajan, V., Venkataramanujam, K., Kutty, M.N. and Rahjesekharan, P., 1977. Fish production through recycling waste waters. Proc. Seminar on Environmental Impact on Developmental activities and Adoption of Suitable Guidelines. Page 16.1-16.10. Hyderabad, Dec. 19-20, 1979 (India).

National Commission on Agriculture 1976. Report, part I, Review and Progress, and part XI, Research, Education and Extension. Ministry of Agriculture and Irrigation, Government of India New Delhi.

National Research Council 1981. Nutrient requirements of cold water-fishes Nat. Acad. Press. Washington, D.C.

National Research Council 1982. Nutrient requirements of warm water-fishes and shellfishes. Nat. Acad. Press, Washington, D.C.

Neess, J.C. 1949. Development and status of pond fertilization in central Europe. Trans. Amer. Fish. Soc. 76 : 335-358.

Nikitina, M.M. and Godovich, P.L. 1983. Dynamics of hormones in the blood of male and female carps, *Cyprinus carpio* (Cyrpinidae) during the spawning period. J. Ichthyol. 23 : 118-123.

Nikolskii, G.V. 1963. The Ecology of fishes : 352 p. London, New York. Academic Press.

Nikolskii, G.V. 1969. Theory of Fish population Dynamics as the Biological Back ground for Rational Exploitation and Management of fishery Resources. (Translated by Bradley, J.E.S.) Oliver and Boyd. London.

Norris, H.W. 1941. The plagiostome hypophysis general morphological and type of structure Grinnel, Iowa.

Oguri, M. 1960. Studies of the Adrenal Glands of Teleosts-3. On the distribution of chromaffin cells and Interrenal cells in the HEad Kidney of fishes. Bull. Jap. Soc. Science. Fish. 26 :443.

Olivereau, M. 1949. L-activite thyroidienne de *Scyllium canicula* aucorrs du cycles sexual C.R. Soc. Biol. 143 : 247-250.

Olsson, R. 1959. The neurosecretory hypothalamus system and the adenohypophysis of Myxie I. Zellforsch. 51 : 97-107.

Ozton, N. and Gorbman A. 1960. J. Morphol. 106. 243-261.

Palay S.L. 1945. Neurosecretion VII. The preoptico-hypophysial pathway in fishes J. Comp. Neurol 82 : 129-143.

Pandey, S. and W.S. Hoar. 1972. Induction of ovulation in gold fish by clomiphene citrate. Can J. Zool. 50 : 1679-1680.

Pandian T.J. 1989. Protein Requirements of Fish and Prawns cultured in Asia P. 11-22. In S.S. Desilva (ed). Fish Nutrition Research in Asia. Proceedings of the Third Asian Fish Nutrition Net work Meeting. Asian Fish Soc. Spec. Publ. 4 : 166 p. Asian Fisheries Society, Manila, Phillippines.

Pandian, T.J. 1987. Fish energetics. P. 357-465. In T.J. Pandian and F.J. Vemberg (eds). Animal energetics Vol. 2. Academic press. New York.

Pandian, T.J. and J. Muthukrishanan 1990. Research methods for Gene and Chromosome manipulation in Fish. A training programme sponsered by DBT, Govt. of India. Madurai Kamraj Univ. Madurai, pp. 1-58.

Parameswaran, S.P., Kumaraiah and G.S. Singit 1988. Use of LHRH analogue, buserelin acetate for induced spawning in fishes, P. 119-121. In M. Mohan Joseph (ed). The first Indian Fisheries Forum Proceedings. Asian Fisheries Society, Indian Branch Mangalore.

Parida, S. K., G. Dash and S. N. Sahoo. 2008. Herbal mechanism for combating fish and shrimp diseases. aqua-international. 16 (1&2): 19-21.

Paromita Bandhyopadhay and N. K. Thakur (Late), 1999. Aquaranching, Fishing Chimes, 19(1): 97-98.

Parrish, B.B. and Saville, A. 1965. The biology of the North-East Atlantic herring populations (Oceanogr. mar. biol.) A. Rev. 3 : 323-373.

Partridge, B.L., N.R. Liley and N.E. Stacey 1976. The role of pheromones in the sexual behaviour of the gold fish. Anim. Behav. 24 : 291-299.

Patnaik, S. 1972. Observations on the control of *Pistia stratiotes* in fish ponds by ``Gamoxone : Proc. Indian. Acad. Sci. (B) 75 (1) : 15-22.

Patnaik, S., K.M. Das and K.C. Pani 1988. Large scale clearance of water lily (*Nymphaea nauchali* Burm). By 2, 4-D and subsequent management of the ponds through fish culture. J. Inland. Fish Soc. India 20 (1) 1988 : 57-62.

Patra, B.C. and A.K. Ray 1988 (a). Observations on the influence of three organic manures on the growth and yield of Indian major carps under pond culture. J. Inland. Fish. Soc. India 20 (2) 1988 : 61-63.

Patra, B.C. and A.K. Ray 1988 (b). A preliminary study on the utilisation of the aquatic weed *Hydrilla verticillate* (L.f.) Royle as feed by the carp, *Labeo rohita*, growth and certain biochemical composition of flesh. Indian Biologist. 20 (1) : 44 - 50.

Patro, J.C. 1968. A note on a new fish carrier, double barrel type (mimeo); 6 p.

Pet Industry Joint Advisory Council, 1999. US Ornamental Aquarium Industry "Pet information Bureau, Washington, DC. 2p.

Peter R.E. 1970. Hypothalmic control of thyroid gland activity and gonadal activity in the gold fish. *Carassius auratus* Gen. Comp. Endocr. 14 : 334-356.

Peter, R.E. 1980. Serum gonadotropin levels in mature male goldfish in response to Luteinizing hormone-releasing hormone (LH-RH) and des-Gly 10-D-Ala-LHRH ethylamide. Can. J. Zool. 58 : 1100-1104.

Peter, R.E. and Paulencu, C.R. 1980. Involvement of the preoptic region in gondaotropin release inhibition in goldfish. *Carassius auratus* Neuroendocrinology 31 : 133-141.

Peter, R.E., H.R. Lin and G. Vander Kraak. 1988. Induced ovulation and spawning of cultured fresh waterfish in China, Advances in application to Gn RH analogues and dopamine antagonists, Aquaculture, 74 : 1-10.

Peter, R.E., M. Sokolowska, C.S. Nahorniak, J.E. Rivier, W.W. Vale. 1987. Comparison of (D-Ala^6, Trp^7, Leu^8, pro^9 NEt)- Luteinizing hormone releasing hormone (SGn-RH-A) and D-Ala^6, Pro^9, Net) luteinizing hormone releasing hormone (LHRH - A) in Combination with pimozide in stimulting gonadotropin release and ovulation in gold fish, *Carassius auratus*, Can. J. Zool. 65 : 987-991.

Philipose, M.T. 1968. Present trends in the control of weeds in fish culture waters of Asia and the Far East. FAO. Fish Rep. 44 (5) : 26-52.

Philipose, M.T. 1972. Fifty years of aquatic weed control in India. In C.K. Varshney and J. Rzoska (eds.) Aquatic weeds in South East Asia. Dr. W. Junk, B.V. Publishers. The Hague : 215-223.

Phillips, A.M., Jr. A.V. Tunison and D.R. Brockway 1948. The utilisation of carbohydrate by trout. Fisheries Res. Bull. 11. State of New York conservation Department. Albany.

Pickering, A.D. and Pottinger. 1983. Seasonal and Diel changes in plasma cortisol levels of the Brown trout *Salmo trutta.* Gen. Comp. Endocrinol. 49 : 232.

Pickford, G.E. and J.W. Atz. 1957. The physiology of the pituitary gland of Fishes. N.Y. Zoo Soc. N.Y.

Pillai, N. G. K. 2008. Mariculture Research in India – Status, Constraints and Strategies for future Development. "Brain Storming Meet on Aquaculture 2025" 7-8 June 2008, CIFA, Bhubaneswar pp 33- 46.

Pillay, T.V.R. 1990. Asian Aquaculture : An overview. In Aquaculture in Asia. (Ed). M. Mohan Joseph 1990. Asian Fishereis Society. Indian Branch. p. 31-42.

Plouidy, M.G., 1982. Composition chimique des liquides d'accompagnement des gametes de carpe. Memoire find'Etudes I.S.A. Beauvis, 79 pp.

Pradeep Kumar, Raju Baitha and M. Venkatasamy, 2008. Scope for Integrating Aquachestnut cultivation with fish culture in Bihar. Aqua international 16 (8):36-38.

Prasad, G. 1996. The World Aquaculture production : The present and future role of India. Seafood Export Jl. Vol. XXVII, No. 7 : 13-19.

Priede, I.G. 1985. Metabolic scope in fishes. In : Fish Energetics New perspective. Tytler, p. and p. calow (Eds). Croom Helm. Beckenham, pp. 33-64.

Pruginin, Y. 1968. Weed control in Fish ponds in the Far East. FAO. Fish Rep. (44) 5 : 18-25.

Radheshyam and D. Kumar 1982. Fisheries extension for Rural Aquaculture. In Souvenir. Workshop on the Development of inland Fisheries in Orissa through institutional financing FFDA Balasore 6-8 March 99-105 pp.

Rahiman, O. A., A. M. Tampi and C. Bhaskaran 1991. Transfer of technology system: financial linkages between and among four subsystems Journal of Extension Systems. 7(1): 91-96.

Raitt, D.F.S. and Hall, W.B. 1967. On fecundity of the red fish *Sebastes marinus* (L.) J. Cons. Perm. int. Explor. Mer. 31 : 237-245.

Raj. B.S. 1916. Notes on the freshwater Fish of Madras, Rec. Indian Mus. 12 pp.

Ramachandran, V. and Ramaprabhu, T. 1968. Investigations on aquatic weed control with special reference to the use of chemicals. FAO. Fish Rep. 44 (5) : 92-108.

Ramaswami, L.S. 1962. Endocrinology of reproduction in Fish and Frog. Gen. Comp. Endocrinol. Suppl. 1 : 286-299.

Ramaswamy, L.S. 1978. Vertebrate neurosecretion - Review. Proc. Indi. Natn. Science B. 46 Supplement pp. 1-78.

Ranganathan, V., 1961. A review of Chanos fry transport. In Training course on Fish Fry transport, June 5th to 10th. 1961. Mandapam Camp. Fisheries extension Unit. Govt. of India 2 p.

Ranganathan, V.A. Sreenivasan, K. Ummar and N.K. Vasudevan 1967. Two interesting cases of spwaning of major carps in confined cement tanks. Cur Sci. 36 (5) : 129-131.

Rath R. 1989 (b). Integrated fish farming system in China, an analytical review part II. Ibid. 9 (6) : 20-27.

Rath, R. 1980. Biology, reproduction and effect of hormones on the female gonad of freshwater oesteichthys, *Channa punctata* (Bloch). Ph.D. Thesis. Utkl University. India.

Rath, R. 1987 (b). An approach to the problems of induced breeding of carps in few fish farms of Orissa. National seminar in Freshwater Aquaculture. Nagarjunsagar. Dec. 27-30. Andhra Pradesh (Abstract).

Rath, R. 1987(a). Effect of hormones in the reproduction of Fish. Fishing Chimes. 7 (5) : 33.

Rath, R. 1987. Breeding behavior of aquarium fighting fish, Betta splendens and Red sword tail *Xiphophorus helleri*, Fishing Chimes, 7(8):10-11.

Rath, R. 1988 (a) Induced fish breeding with H.C.G. Fishing chimes 7 (12) : 28-29.

Rath, R. 1988 (b). Problems encountered during hatching of carp eggs in Chinese hatchery at Digapahandi Fish Farm (Berhampur). Sci. and cult. 54 (9) : 299-301.

Rath, R. 1989 (a). Integrated fish farming system in China, an analytical review part I. Fishing chimes 9 (4) : 31-39.

Rath, R. 1989 (c). Organic Fertilizer augments Fish production. Indian Farmers Digest. Vol. XXII, No. 8 : 31-33.

Rath, R. 1989 (d). Prevention and control of fish diseases by herbal medicines, a review. Fishing chimes Vol. 9, No. 11 pp. 38-42.

Rath, R. 1990. Integrated Fish Farming a sound basis for the development of Integrated rural technology. National seminar on Aquaculture Development in India. Problem and Prospects. 27-29 Nov. 1990. Thiruvananthapuram (Abstract) No. TS 1 : 12.

Rath, R. 1991 (b). LRH-A, an inducing agent for the breeding of Black carp *Mylopharyngodon piseus* in Yixing Fish Farm. China. National semi-nar in Developmental Biology. Utkal Univ 23-25 Feb. 1991 (Abstract).

Rath, R. and P. Mohanty Hejmadi (1991a). Cytochemical studies on the development of protein yolk in oocytes of fresh water murrel. *Channa punctata* (Bloch). In Proc. Nat. Symp. Fresh water Aqua. CIFA, Kausalaygang, Orissa. 1991 pp. 51-52.

Rath, R. and P. Mohanty-Hejmadi 1976. Fecundity of *Ophiocephalus punctatus* (Bloch). Jl. of Zool. Soc. Ind. 26 (1 and 2) : 1-6.

Rath, R. and P. Mohanty-Hejmadi 1984. Morphological staging of ovary in freshwater snake head murrel *Channa punctata.* (Bloch), Pranikee, 5 : 22-30.

Reddy, A.K. and N.K. Thakur 1998. Ovatide - a new Hormonal preparation for induced fish breeding. Fishing chimes. 18 (4) : 28-30.

Reddy, P.V.G.K., H.A., Khan, S.D. Gupta, M.S. Tantia and G.V. Kowtal 1991. On the ploidy of three inter generic hybrids between common carp (*Cyprinus carpio communis* L.) and Indian Major carps. Aquaculture Hungarica (Szarvas) 1990 Vol. VI pp. 5-11.

Redfield, A.C., Kethum, B.H. Richards, F.A. 1963. The influence of organisms on the compositions of sea water. In : Hill, M.N., ed., The sea John Wiley and Sons Ltd., New York. NY. USA pp. 26-77.

Renukaradhya, K.M. and Varghese, T.J. 1986. Protein requirement of the carps. *Catla catla* (Ham) and *Labeo rohita* (Ham). Proc. Indian Acad. Sci. (Animal Sci.), 95 (1) : 103-107.

Report of the Expert consultant ulcerative Fish Diseases in the Asia Pacific Region (TCP/RAS/4508 Project) Bankok, Thailand 5-9 Aug 1986.

Rivier, W.W. Vale., J.A. King and R.P. Miller 1985. Structure- activity relationship of mammalian, Chicken and Salmon gonadotropin releasing hormones invivo in gold fish. Gen. Comp. Endocrinol. 58 : 231-242.

Roberts, R.J., Donald, J. Macintosh Kamon Porn Tonyuthai, Sidhi Boonyaratpalin, Nuansri Tayaputch, Michael J. Phillips and stuart D. Millar. 1986. Field and Laboratory Investigations in to ulcerative fish diseases in the Asia pacific region. FAO of the united Nation Bankok, 1986. Technical report of FAO project TCP/RAS/4508.

Robertson, O.H. and A.P. Rinferet, 1957. Endocrinol. 60, 559-561.

Robertson, O.H. and Wexler, B.C. 1957. Pituitary degeneration and adrenal tissues hyperplasia in spawning pacific salmon. Science 125, 1295-1296.

Rodhe, W. 1958. Primarproduction Und. seetypen, Verh. Int. Verein Limnol. (13) : 121-140.

Ross, E. 1971. Hawaii Farm Sci. 20 : 11-12.

Roth, W.D. 1957. The pars distalis of the adenohypophysis of the sea lamprey *Petromyzon marinus* Anat. Rec. 130-366.

Ryther, J.H. 1956. Photosynthesis in the ocean as a Function of light intensity, Limnol and Oceanogr, 1, 61-70.

Ryther, J.H. 1957. Fecundity of the North American Salmonidae. U.S. Fish and Wildlife Serv. Fish. Bull. 57 : 451-468.

Saha, K.C., D.P. Sen., A.K. Roy Chowdhury and S.K. Chakravarty 1957. Studies on the factors influecing spawning of Indian major carps in bundhs fisheries. Indian J. Fish 4 (2) : 284-294.

Sahu, A.K. and N.C. Biswal 1988. Use of pimozide in combination with LHRH-A on the induced spawning of Indian major carps *Catla catla, Labeo rohita* and *Cirrhinus mrigala.* p. 113-114. Proceedings of the National Symposium on current status of General and comparative Endocrinology, 25-27 November, Delhi.

Sanchez, J.A., Blance, G., Vazquez, E., Garcia, E. and Rubio, J. 1991. Allozyme variation in natural population of Atlantic Salmon in Asturias (Northern spain). Aquaculture, 93 : 291-298.

Sankaran, T., 1976. Aquatic weeds in South ėast Asia. Dr. W. Junk pub : pp. 323-327.

Sar, U.K. 1991. A new tool *Simyra conspersa* (Lepidoptera noctuidae) For biological control of *Nelumbo nucifera* in Fish ponds. Proc. Nat. Sym. Fresh Water Aqua. 117-119 pp.

Sarkar, S. K. 2002. Freshwater Fish Culture Vol. I Daya Publishing House Delhi. pp 1-557.

Satapathy, B.B. 1990. Nutrient dynamics versus fertilizer application in aquaculture management. Techniques of Aquaculture management in large water bodies KVK/TTC., CIFA, K. Gang. pp. 35-43.

Sathyanesan, A.G. 1969. Stereometric structure of the hypothalamo neurohypophysial complex in some Indian catfishes. Neuroendocrinology 5 : 10-23.

Scahrrel, E. 1930. Uber sekretorisch Tatige Zellen im Thalamyus Von *Fundulus heteroclitus* L. (Unter-Schungen uber das zeischenhirn der Fische. (II) z. vergl. physiol. 11 : 767-773

Schaperclaus, W. 1933. Text book of pond culture : rearing and keeping of carp, trout and allied fishes (Tr.F. Hund), Fish. Leaflet wash. (311) : 260 p.

Schaperclaus, W. 1959. Die karp fenteichswirtschaft in der Deutsche Demokratischen Republik. Stiz. Ber. VII (7). Berlin Deutsche Akad. Landwirtschaftwiss.

Schmidt, P.J., and D.R. Idler. 1962. Steroid hormones in the plasma of salmon at various stages of maturation, Gen. Comp. Endocrinol., 2 : 204-214.

Seaman, D.E., and W.A. Porterfield. 1951. Weeds, 12 : 87-92.

Sekharan. N. Mini and A. Ramachandran. 2008. Capacity building in Ornamental Fish sector in India. Sea Food Export Jl. XXXVIII (4): 19-21.

Sen, P.R. 1986. Fishing wealth from waste water. Aqua. World, No. 6 pp. 219-221.

Sen, P.R. and Chatterjee, D.K. 1976. Enhancing the production of Indian major carp fry and fngerlings by use of growth promoting substance. In: Advances in Aquaculture Edt. by T.V.R. Pillay and Dill. A. (1976). ``FAO and Fishing News Books Ltd., New York.

Sen, P.R. Rao., N.G.S. Ghosh, S.R. and Rout, M. 1978. Observations on the protein and carbohydrate requirements of carps. Aquaculture 13 : 121-129.

Shah, K.L. and B.C. Tyagi. 1985. Hydrobiologia, 122 : 243-245.

Shariff, M. J.L. Torres, L.A. Theem and M.N. Shamsudin. 1988. Research News. UPM 2 pp.

Sharma, B.K. and M.K. Das 1988. Integrated fish livestock crop farming systems in M. Mohan Joseph (Ed.). The First Indian Fishereis forum, Proceedings. Asian Fisheries Society, Indian Branch, Mangalore pp. 27-30.

Sharma, B.K. 1990. Integrated Fish farming-various possibilities and economic consideration. Techniques of Aquaculture Management in large water bodies. CIFA (ICAR) Training Report No. 3 pp. 66-76.

Sharma, B.K. and J. Olah. 1986. Integrated Pig farming in India and Hungary. (Presented in the symp.) on Aquaculture on carps and allied species held in Paris on September, 1985). Aquaculture 54 : 135-139.

Sharma, B.K. Dilip Kumar, Das, M.K., Das, S.R. and Chakraborty, D.P. 1979. Observations on swine-dung recycling in composite fish culture. In : Symposium on Inland Aquaculture (Abstract) Feb. 12-14 1979. CIFRI, Barrackpore.

Sharma, B.K., M.K. Das and S.R. Das 1988. Carp farming integrated with pig raising in India. Environ Ecol. 6 : 159-168.

Sharma, Vikram and S. N. Laharia 2008. Fish farmer: Fish Production Technology Adoption level. Fishing Chimes, 28(2): 49-56.

Sherwood, N.L., Eiden, M. Bownstein J. Spiess, J.E. Rivier and W.W. Vale, 1983. Characterization of a teleost gonadotropin releasing hormone. Proc. Nat. Acad. Sci., U.S.A. 80 : 2794-2798.

Shiraishi, Y. and Fukuda, Y. 1966. The relation between the day length and the maturation of four species of salmonid fish. Bull, Freshw. Fish. Res. Lab. Tokyo 16 (2) : 103-11.

Shirgur, G.A. 1973. Observations on efficiency of large capacity vertical flow bins for hatching major carp eggs. J. Indian Fish. Assoc. Vol. 3 : 80-85.

Shirgur, G.A. 1978. Chronological account of devclopment of carp egg hatcheris, from protable glass jar models of 3 and 6 compartment linear aluminium hatcheries with multipurpose egg vessel (1974-78). J. Biol. Sci. 21 : 20-41.

Shirgur, G.A. and S.T. Indulkar, 1999. Comparative growth rate of *M. rosenbergii* with different growth promoters. Fishing World, 6 (5) : 8-12.

Shirgur, Girish, A. 1972. Observations of efficiency of vertical flow jars, (glass and transparent polythene) for hatching of major carp eggs. J. Indian Fish. Asso. 2 (1 and 2) : 90-101.

Shyama, S. and P. Keshavanath. 1987. Growth response of mahseer (*Tor khudree*) and Silver carp (*Hypophthalmichthys molitrix*) to steroid incorporated diets. The first Indian Forum. Mangalore Dec 4-8 (1987) pp. 66 (Abstract).

Singh, A.K. and T.P. Singh 1976. Effect of clomid, Sexovid and prostaglanding on induction of ovulation and gonadotropin secretion in freshwater catfish *Heteropneustes fossilis* (Bloch). Endokrinologie, 68 : 130-136.

Singh, B.N. 1989. Protein requirements of young Silver carp, *Hypophthalmichthys molitrix* (Val.). In. Sym. modern trends in fish biology research, Bhagalpur Univ. Bhagalpur 17-20 June. Abstract.

Singh, B.N. and Bhanot, K.K. 1988. Observations on the protein requirement of *Catla catla* (Ham). fry. p. 77-78 (Ed. M. Mohan Joseph). The First Indian Fisheries Forum, proceedings Asian Fisheries Society, Indian Branch, Mangalore.

Singh, B.N. Sinha, V.R.P., Kumar, K. and Swamy, D.N. 1986. Observations on the feed formulations and fortification of conventional fishfeed for rohu, *Labeo rohita* and Mrigal, *Cirrhina mrigala* fry and fingerlings, Intl. J. Acad. Ichthyol, 7 (1) : 31-34.

Singh, H. 1987. Bundh Breeding of Indian Major carps. Fishing Chimes. October 1987 pp. 57-59.

Singh, S.K., C.A. Singh and Y.J. Singh. 2008. Breeding and Aquarium keeping of Oscar fish. Aqua International, 16 (5): 27- 30.

Singh, S.B. 1962. Preliminary experiments on the relative manurial values of some aquatic weeds as composts. Proc. Ind. Pacific Fish. Coul. 10 (2) : 146-53.

Singh, S.B. 1990. Status of Aquaculture in the socialist Republic of Vietnam. Aquaculture in Asia (Ed.). M. Mohan Joseph, 1990. Asian-Fisheries Society, Indian Branch p 371-384.

Singh, Tun,Tun, Durga Prasad., J. P. L. Das and Shambu Prasad, 2006. Wetlands of North Bihar: A Potential Resource for wild collection of Ornamental fish. Fishing Chimes, 26(9): 53-54.

Singh,V. D. and Sampath, V. 1990. A well organized fisheries extension service system- a pre-requisite for development of aquaculture for higher production. Punjab. Fish. Bull. XIV (2):1-9.

Singh,V.D. and Sampath, V. 1981. Introduction of fisheries extension service for development of inland fish farming in India. Pub. Fish. Bull. 29-34 pp

Singh,V.D. and Sampath, V. 1982. Role of fisheries extension in Rural areas with particular reference to Uttar Pradesh. Indian Farmers Digest, XV (1): 41-43

Singh,V.D. and Sampath, V. 1983. Total Aquaculture Technology Center (TACT). A field guide. Fishing Chimes, 3(1)37-39.

Sinha, V.R.P. 1971. Induced spawning in carp with fractionated fish-pituitary extract. J. Fish. Biol., 3 : 263-272.

Sinha, V.R.P. 1972. Compiste Fish culture in India. Indian Farming. 22 (6).

Sinha, V.R.P. 1984. Principles of composite Fish culture. Seminar on Aquaculture, Cuttack 11-12 Feb. Sourvenir, Orissa Fishereis Deptt. pp. 11-20.

Sinha, V.R.P. 1989. Fisheries Development in India, Fishing chimes Vol. 9. No. 3 pp. 23-29.

Sinha, V.R.P. and B.K. Sharma 1976. Composite fish culture in large sheet of water, Indian Farming.

Sinha, V.R.P. and K.H. Ibrahim 1980. Genetic resources of fish in India. Tiger paper Vol. VII No. 1 (FAO) Regional office of far Asia and the Far East) pp. 21-22.

Sinha, V.R.P. and M.V. Gupta, 1975. On the growth of grass carp (*Ctenopharyngodon idella* Val.) in composite fish culture at Kalyani, West Bengal (India), Aquaculture, 24 (4) : 283-290.

Sinha, V.R.P. and V. Ramachandran 1985. Fresh water fish culture. ICAR publication 80 pp.

Sinha, V.R.P., 1979. Contribution of supplementary feed in increasing fish production through composite fish culture in India. Proc. World Symp. on Fin Fish Nutrition and Fish feed Tech; Homburg 20-23 June 1981.

Sinha, V.R.P., V.G. Jhingran and S.V. Ganapati 1974. A review on spawning of the Indian major carps. Arch. Hydrobiol. 73 (4) : 518-556.

Sjafei, D.S., 1985. Conservation et vieillissement in vitro des ovocytes, ovules de la carpe commune (*Cyprinus carpio* L.) these 3e cycle. Univ. Paris VI. 189 pp.

Smirnov, N.N. 1960. Hydrobiologia : 208-224.

Smith, Alice Lorraine, 1968. Microbiology and pathology. 9th Edition. The C.V. Mosby company, USA 147.

Sokolowska, M., Popek. W. Bieniabz, K. 1978. Synthetic releasing hormones LH/FSH-RH and LHRH : effect of intracerebral and intramuscular injections on female carp (*Cyprinus carpio* L.) maturation. Ann. Biol. Anim. Biochem Biophys., 18 ": 963-967.

Sreenivasan, A. 1967. Fish production in some rural demonstration ponds in Madras (India) with an account of the Chemistry of water and soil. FAO Fish. Rep. (44) 3 : 179-197.

Srinivasan, R. and Chako, P.I. 1952. The control of aquatic vegetation with 2, 4-D. J. Bomb-nat, Hist. Soc. 51 (1) : 164-169.

Stacey, .E. and N.R. Liley : 1974. The regulation of spawning behaviour in the female gold fish (*Carassius auratus*). Nature, 247, 71-72.

Stacey, N.E. and F.W. Goetz. 1982. Role of prostaglandins in fish reproduction. Can. J. Fish Aquat. Sci. 39 : 92-98.

Stacey, N.E. and N.R.S. Pandey 1975. Effects of endomethacin and prostaglandins on ovulation of goldfish. Prostaglandina : 597-608.

Stacey, N.E. and R.E. Peter. 1979. Central action of prostagladins in spwaning behaviour of ovulation in gold fish *Carassius auratus*. Gen. Comp. Endocrinol. 37 : 246-249.

Stahl, A. 1953. La neurosecretion chezles poissons teleosteens. Contribution al'eudefde la neurohypophyse chezles mugilidaes. C.R. Soc. Biol.-paris 147 : 841-844.

Sukumaran, K.K. 1969. Growth, maturation and fecundity of cultivated fishes. FAC/UNDP Regional seminar on induced breeding of cultivated fishes. Calcutta FRI/IBCF/5-53 p.

Sukumaran, N., K. Venkataramanujam and M.N. Kutty 1987. Growth of Common carp hatchlings in Domestic sewage. Fishing chimes pp. 25-26.

Sulaiman, V. Rasheed. 2003. Innovatives in agricultural extension in India. S. D. Dimensions FAO June 2003.

Summer Institute in Recent Advances in Fin Fish and shell Fish Nutrition. CMFRI, Cochin, 11-30 May 1987.

Sundararaj, B.I. and S.V. Goswami 1977. Hormonal regulations of in vivo and in vitro oocyte maturation in the catfish. *Heteropneustes fossilis* (Bloch). Gen. Comp. Endocr. 32 : 17-28.

Sundarasan, B.B. and Muthusamy, S. 1976. Environmental pollution and Ecologically balanced system. Water conservation technology Training seminar. T.N.A.U. Nov. 15-17 (India).

Sunil Kumar, P. 2005. Breeding of Angel fish (*Pterophyllum scalare*). Fishing Chimes, 25(6):24-25.

Swain, P., S. K. Nayak and P. K. Nanda. 2008. Potential applications of nanotechnology in aquaculture. Aqua International 16 (8): 24 – 27.

Swamy, D.N. Mohanty, S.N. and Tripathi, S.D. 1989. Dietary protein requirement of catla, *Catla catla* (Ham) fingerlings. Proc. Nat. Sem. Forty years of freshwater aquaculture in India. 7-9 Nov. 1989. Abstract No. SVN/3.

Swingle, H.S. 1957. Proc. 5th weed control. 10 : 11-17.

Swingle, H.S. 1967. Standardisation of chemical analysis for waters and ponds muds. FAO. Fish Rep. (44) 4 : 397-421.

Swingle, H.S. 1970. History of warm water pond culture in the united states. In A century of fisheries in North America, Spec. Publ. No. 7 I (N.G. Benson ed.) pp 95-105. Amer. Fish. Soc. Bethesda, Maryland.

Swingle, H.S. and E.V. Smith 1938. Fertilisers for increasing the natural food for fish ponds. Trans. Am. Fish. Soc. 68 : 126-135.

Swingle, S.W. 1956. A repressive factor controlling reproduction in fishes. Proc. Pacif. Sci. Congr. 3 : 865-71.

Symposium on hormonal steriods in Fish. 1978. V. Internation congree held in Delhi Indian National Science Academy.

Tacon, A.C.J. and C.B. Cowey. 1985. Protein and aminoacid requirements. In. P. Tytler and P. Calow (eds) Fish energetics, new perspectives, pp 155-184 Croom Helm Press, London.

Tal, S. 1974. Preliminary observations on new methods of fish culture in Israel. Bamidgeh. Vol. 26 : No. 3. 51-56.

Tan, Y.T. 1970. Composition and nutritive value of some grasses, Plants and aquatic weeds tested as diets. J. Fish. Biol., 2 : 253-267.

Tave, Dauglas. 1986. Genetics for fish hatchery managers. AVI Publishing Co. West Port, connecticut, USA 209 page.

Tavolga, W.N. 1956. Visual, Chemical and Sound stimuli as cues in the sex discriminatory behaviour of the gobiid fish, *Battygobius sopartor.* Zoologicae, 41 49-64.

Thakur, N.K. 1984. Recnet advances in Aquaculture practices in Thailand, In Seminar on Aquaculture, Cuttack 11-12 Feb. Souvenir, Govt. of Orissa pp. 45-49.

Thomas, K.J., 1975. Biol. J. Linn. Soc. 7 : 243-247.

Timms, A.M., and H. Kleerekoper. 1972. The locomotor responses of males *Ictalurus punctatus,* the channel catfish, to a pheromone released by the ripe female of the species. Trans. Am. Fish. Soc. 101, 302-310.

Todd, J.H.J. Atema and .e. Bardach. 1967. Chemical communication in Social behaviour of a fish, the yellow bull head (*Ictalurus natalis*). Science 158, 672-673.

Toor, H.S. 1983. Water soluble phosphate : observed effects on embryonic development, hatching time and survival of common carp. Prog. Fish. Cult. 45 : 134-135.

Torres, J.L. 1990. Studies on motile *Aeromonas* species associated with healthy and epizootic ulcerative syndrome positive fish. Ph.D. Thesis, Faculty of fisheries and marine science, Univ. Pertanian, Malaysia, Serdang. Fish health Section News letter, AFS Manila Vol. 1 No. 2. pp 3.

Torrissen, K.R. 1987. Genetic variation of trypsin like isozymes correlated to fish size of Atlantic Salmon (*Salmo salar*), Aquaculture, 562 : 1-10.

Torrissen, K.R. 1991. Genetic variation in growth rate of Atlantic Salmon with different trypsin enzyme patterns. Aquaculture, 93 (4) : 229-312.

Training manual on Integrated Fish farming RLCC, Wuxi, China 1987.

Tripathi, S. D. 2008. Brain storming meets on "Aquaculture-2025, Challenges and Opportunities. Wither aquaculture 2025? Organized by CIFA, K. Gang and Association of Aquaculturist (June 6-8). Pp 15-20.

Tripathi, S.D. 1990. Fresh water Aquaculture in India. In Aquaculture in Asia, (Ed) M. Mohan Joseph. 1990. Asian Fisheries Society, Indian Branch. P. 191-222.

Tripathi, S.D. and H.A. Khan 1990. Carp seed production Technology - A review, P 1-11. In : P. Keshavanath and K.V. Radhakrishnan (eds) carp seed production Technology. Proceedings of the workshop on carp seed production Technology 2-4 September, 1988. Special publication 2, 102 p. Asian Fisheries Society, Indian Branch, Mangalore, India.

Tripathi, S.D. and Mishra, D.N. 1986. Synergistic approach in carp polyculture with grass carp as a major component. Aquaculture, 54:157-60.

Tripathi, S.D., Aravindakshan, P.K., Ayyappan, S., Singh. R. Mohanty, A.N. Muduli, H.K. 1989. Carp polyculture based on cowdung and water hyacinth based biogas slurry as manures. Proc. Nat. Semi. On forty years of Freshwater Aquaculture in India, 7-9 Nov. 1989. Abst. No. SIV/1.

Tripathy, N. K., Kumar, D. and Kumar. K. 1982. Concept of Community Fish Farming. In Souvenir. Workshop on the development of inland Fisheries in Orissa through institutional financing, Balasore 23 pp.

Uma, A. 1999. Feed probiotics for sustainable aquaculture. Sea food Export. JI., 30(3): 9-11.

Uma, A., Abraham T.J., Jeyaseelan, M.J.P. and Sundararaj, N. 1995. Influence of a commercial probiotic feed supplement on the growth, survival and immunity of Indian white shrimp, *Penaeus indicus.*

Varghese, P.U. 1967. Prolongation of spawning season in the carp, *Cirrhina reba* (Ham). by artificial treatment. Cur. Sci. 36 (17) : 465-67.

Varghese, T.J. and G.P.S. Rao 1976. On the use of marine catfish pituitary glands for induced spawning of Silver carp and Catla. Curr. Sci. 43 : 302.

Varghese, T.J. and Sukumaran, K.K. 1971. Notes on hypophysation of catla X rohu hybrid : Annual Day Sourvenir, CIFE, Bombay pp 16-17.

Varghese, T.J., Deveraj, K.V., Santharam, B. and Shetty, H.P.C. 1976. Growth response of the common carp *Cyprinus carpio* var. *communis* to protein rich pelleted feed. In Symposium on the development and utilisation of inland fishery resources. Colombo, Srilanka, IPFC 1976/ Sym/40 p.

Varghese, T.J., G.P. Satyanarayan Rao, K.V. Devaraj and B. Chandrasekhar. 1975. Preliminary observation on the use of marine catfish pituitary glands for induced spawning of the Indian major carps. Curr. Sci. 44 : 75-78.

Venkataswamy, M and S. Athithan, 2008. Fancy Fresh-Water Ornamental Fishes. Aqua international 15(11): 35-36.

Venugopal, G. 1990. Field observations the efficacy of ``portable circular hatchery. A rural model" p 57-60. In. P. Keshavanath and K.V. Radhakrishnan (eds). Carp seed production Technology. Proceedings of the workshop on carp seed production Technology, 2-4 September. 1988. Special publication 2, 102 p. Asian Fisheries Society, Indian Branch, Mangalore, India.

Verma, R.S., A.K. Singh and N.C. Tripathi 1991. Irrigating trees with raw sewage water. Indian Farming, Vol. 41 (1) : 27-28.

Villadolid, D. 1936. Bull Phillip. Agric. Com. 7 : 1-18.

Vinod, B. H and N. Basavaraja 2007. Organic Aquaculture. Fishing Chimes 22(4):37-40

Von Ihering, R. 1937. A method of inducing fish to spawn. Prog. Fish. Cult. 24 : 15-16.

Waksman, S.A. 1946. Antibiotic Substances- contribution of the microbiologists Ann. N.Y. Acad. Sci. 48 : 35 : 1946.

Wallace, C.R. 1970. The use of pheromones and sound during reproduction by the black bull head. *Ictalurus malas*. State Nebr. Game Parks. Comm. Job. Proc. Rep. March 1, 1969. February 28, 33-38.

Watanabe, T. 1987. Lipid nutrition in Fish. Comparative Biochemistry and Physiology 73 B : 3-15.

Weil, C. *et al.* 1986. Induced spawning (ovulation and spermiation) in carp and related species, R. Billard et. J. Marcel Ed. Aquaculture of cyprinids, INRA, Paris 1986. pp. 119-137.

Welcomme, R.L. 1967. The relationship between fecundity and fertility in the mouth breeding cichlid fish. *Tilapia leucosticta*. J. Zool. Lond. 151, 453-468.

Welcomme, R.L. 1969. The biology and Ecology of the fishes of a small tropical stream. J. Zool.Lond. 158, 485-529.

Wilson, H.C. 1909. Artificial propagation of Hilsa in the coleroon, Madras Govt. Revenue Deptt. order No. 1219, Madras.

Withler, F.C. 1982. Cryopreservation of spermatozoa of some freshwater fishes cultured in South and Southeast Asia. Aquaculture, 26, 395-398.

Witschi, E. 1955. Mem. Soc Endocrinol. 4 : 149-165.

Wood, H. 1930. Scottish herring shoals. Prespawning and spawning movements. Rep. Fish. Res. Bd. Scot. Sci., Invest. 1 : 1-71.

Workshop on Aquaculture Extension Research Needs: 1991. Concept papers. Organized by Central Institute of Freshwater Aquaculture, K. Gang, BBSR. pp-1-12.

Woynarovich, E. and Horvath, L. 1981. La reproduction artificiella des poissons en eau chaude : mannuel de vulgrisation. FAO Doc. Tech,. Pendhes, 201 : 191 pp.

Woynarovich, E. and L. Horvath 1980. The artificial propagation of warm water fin fishes - a mannual for extension, FAO, Fish, Tech. pap. 20 183 pp.

Yamamoto, K. and Yamazaki F. 1967. Hormonal control of spermiation and ovulation in gold fish. Gunma Symp. Endocrinol. 4 : 131-145.

Yamamoto, K., Nagahama, Y. and Yamazaki, F. 1966. A method to induce artificial spawning of gold fish all through the year. Bull Jap. Soc. Sci. Fish. 33 : 977-983.

Yamamoto, T. 1961. Physiology of fertilization on fish eggs. Int. Rev. Physiol. 12 : 361-405.

Yamamoto, T., 1954. Physiological studies on fertilisation and activation of fish eggs. V. The role of calcium ions in activation of Oryzias egg. Exp. Cell. Research, P. 56-68.

Yamazaki, F. 1969. The gonadotropin of fishes. Bull. Jap. Soc. Sci. Fish. 35 : 695-709.

Yamazaki, F., 1965. Endocrinological studies of the reproduction of the female gold fish, *Carassus auratus* L. with special reference to the function of the Pituitary gland. Mem. Fac. Fish. OKKAIDO Univ. 13 : 1-64.

Yaron, Z. 1971. Gen. Comp. Endocrinol. 17 : 247-252.

Zawisza, J. and Backiel, T. 1970. Gonad development, Fecundity and egg survival in *Coregonus albul* L. Biology of Coregonid fishes. (C.C.

Lindsey and C.S. woods Eds) pp. 363-397. University of manitoba press, Winnipeg, Canada.

Zhukinskiy, V.N. and Bil'ko V.P. 1984. Effect of semen pH on embryo viability in some cyprinid fishes. J. Ichthyol, 24 : 64-76.

Ziauddin, G; C. Dutta and A. Goswami, 2007. Ornamental fish trade and marketing in India, Fishing Chimes, 27(9):44-46.

SUBJECT INDEX

C

D

E

F

G

H

Q

R

S